# Policy Instruments for Environmental and Natural Resource Management

## THOMAS STERNER

RESOURCES FOR THE FUTURE
  *Washington, DC, USA*

THE WORLD BANK
  *Washington, DC, USA*

SWEDISH INTERNATIONAL DEVELOPMENT
COOPERATION AGENCY
  *Stockholm, Sweden*

Printed in the United States of America

An RFF Press book
Published by Resources for the Future
1616 P Street, NW, Washington, DC 20036–1400

A copublication of Resources for the Future (www.rff.org), the World Bank (www.worldbank.org), and the Swedish International Development Cooperation Agency (www.sida.se).


Library of Congress Cataloging-in-Publication Data

Sterner, Thomas, 1952–
    Policy instruments for environmental and natural resource management / Thomas Sterner
      p.   cm.
    Includes bibliographical references and index.
    ISBN 1-891853-13-9 (hardcover: alk. paper) — ISBN 1-891853-12-0 (pbk. : alk. paper)
    1. Environmental policy.  2. Conservation of natural resources—Government policy.  I. Title.
    GE170 .S75 2003
    333.7′2—dc21                                                                    2002151330


f e d c b a

The paper in this book meets the guidelines for permanence and durability of the Committee on Production Guidelines for Book Longevity of the Council on Library Resources.

The geographical boundaries and titles depicted in this publication, whether in maps, other illustrations, or text, do not imply any judgement or opinion about the legal status of a territory on the part of Resources for the Future or of any other organization that has participated in the preparation of this publication.

The text of this book was designed and typeset by Betsy Kulamer in Bembo and Gill Sans. It was copyedited by Pamela Angulo. The cover was designed by Rosenbohm Graphic Design. Cover photograph of fishing by Curt Carnemark, from the World Bank Photo Library.

Available from RFF: ISBN 1–891853–13–9 (hardcover) and ISBN 1–891853–12–0 (paper)
Available from the World Bank: ISBN 0–8213–5381–0 (paper)

# About
# Resources for the Future
# and RFF Press

RESOURCES FOR THE FUTURE (RFF) improves environmental and natural resource policymaking worldwide through independent social science research of the highest caliber.

Founded in 1952, RFF pioneered the application of economics as a tool to develop more effective policy about the use and conservation of natural resources. Its scholars continue to employ social science methods to analyze critical issues concerning pollution control, energy policy, land and water use, hazardous waste, climate change, biodiversity, and the environmental challenges of developing countries.

RFF PRESS supports the mission of RFF by publishing book-length works that present a broad range of approaches to the study of natural resources and the environment. Its authors and editors include RFF staff, researchers from the larger academic and policy communities, and journalists. Audiences for RFF publications include all of the participants in the policymaking process—scholars, the media, advocacy groups, nongovernmental organizations, professionals in business and government, and the general public.

# Resources for the Future

# Brief Contents

## PART III
### Selection of Policy Instruments

## PART IV
### Policy Instruments for Road Transportation

## PART V
### Policy Instruments for Industrial Pollution

## PART VI
### Policy Instruments for the Management
### of Natural Resources and Ecosystems

## PART VII
### Conclusion

# Contents

## PART II
### Review of Policy Instruments

## PART III
### Selection of Policy Instruments

## PART IV
### Policy Instruments for Road Transportation

## PART VII
### *Conclusion*

# *Foreword*

$M$ANY COUNTRIES—both industrialized and developing—face serious problems of natural resource protection and environmental management. The design and implementation of policies that respond to important challenges to economic, ecological, and social sustainability has been debated for many years by two related groups: by governments and other organizations and by academic economists and other policy analysts.

The design and implementation of environmental and natural resource policies has been the focus of growing intensity throughout the world. What began at the 1972 United Nations Environment Summit in Stockholm and was encouraged at the 1992 Earth Summit in Rio de Janeiro assumed center stage at the Rio+10 Summit in Johannesburg in 2002. Governments (particularly in the developing world), national and multilateral development institutions, the private sector, and nongovernmental organizations have been taking action, and the resulting policies often have a regulatory or command-and-control flavor. The effects have been mixed: some successes and a number of disappointments.

On a parallel track, since the 1960s, the design and implementation of environmental and natural resource policies has been the focus of increasingly fruitful research. Much of this work has focused on what can be described as incentive-based policies, which attempt in varying ways to rely more on economic motivations and to provide more flexibility than do traditional regulatory approaches.

Progress in understanding both types of policies has been made, and promising applications can be found today in many developed countries, as well as in the developing world. However, experience in developing countries is more limited, and significant skepticism remains about the applicability of incentive-based policies for the developing world.

In this book, Thomas Sterner successfully advances both conceptual and practical understanding of what needs to be done if good environmental and natural resource policies are to be devised and implemented in different countries, both developed and developing. Based on solid economic theory and consideration of other social and political aspects, the book begins with a thorough presentation and analysis of the menu of available policies. It then proceeds to a broad discussion of actual and potential applications (pollution, natural resources, and transportation) in different types of economies.

Sterner's survey of the policy landscape develops several important lessons for analysts, policy practitioners, and students. These include the following:

- Properly designed incentive-based policies can and do work, both in protecting the environment and natural resources and in lowering the cost of achieving that goal. Blanket resistance to the use of such policies in some parts of the world, whether due to philosophical stance or lack of information, needs to be reconsidered.
- Badly designed incentive-based policies can be just as ineffective as the alternatives their advocates seek to replace.
- No policy regime, incentive-based or otherwise, can accomplish much without the necessary underlying economic, legal, and technical institutional capacities and an appropriate social milieu. Often these capacities are quite limited in the developing world.
- To succeed, therefore, incentive-based and other policies must be tailored to the existing social context and institutions, and their application needs to be accompanied by capacity building. Analysis and experience teach much, but simple cookbook answers are unlikely to be very successful.

We hope that this fine book will serve as a valuable resource for those practitioners considering and evaluating concrete policy options and for those analysts who seek to provide the intellectual base for such efforts. Given the importance of both protecting natural resources and the environment and doing so cost-effectively, we see no higher priority on the environmental and natural resource policy agenda today and in the future.

John A. Dixon
*The World Bank*

Michael A. Toman
*Resources for the Future*

# *Preface*

$T$HIS BOOK IS INTENDED to be used by individuals who are interested in the selection and design of policy instruments for the environment: university professors, undergraduate or graduate students, analysts who advise policymakers, and, particularly, people in countries that have not yet made extensive use of market-based policy instruments. Its purpose is to pull together the distinct experiences of policymaking that have evolved in the United States, Europe, and other countries of the Organisation for Economic Co-operation and Development (OECD) and also in non–OECD countries, including some formerly planned economies. A wide range of environmental and natural resources issues illustrate points that are ecologically important or good examples of the principles of policy design.

To be sure, this book is not an encyclopedia of resource and environmental problems, a pure textbook in environmental economics, or a mere description of policies. Many important issues are not covered at all, or at least not in proportion to their importance. Theoretical issues are presented as refreshers for readers who have studied some economics rather than as rigorous training for future environmental economists. If it were possible to write a "cookbook" with "recipes" for environmental policymaking, then it really ought to be done. However, the ecological, technical, social, and economic realities of environmental policymaking are so complex that there are no simple guidelines. Instead, in-depth understanding of both the economics and the environmental science is necessary to successfully design good policies.

Only a small amount of mathematics is presented in this book. Although mathematical proofs can greatly aid some readers, they can just as well frustrate others. The text was written to be intelligible even to readers who skim the formulas, so detailed mathematical explanations are presented as supplemental information.

The topics of this book include the key theoretical issues, worldwide applications, and various "brown" and "green" issues. My own personal experience and prior work have necessarily affected my choice of emphases. Although I have a keen interest in other social and natural sciences, I am first an economist. And even though I tried to cover as many countries as possible to illustrate various economic systems and income levels, because of my experience, cases from Sweden and the United States are overrepresented among the industrialized countries.

Scientific analysis of policy instruments is not new. However, such analysis tends to concentrate on one issue or aspect at a time and tends to be written for specialists. Literature that systematically covers the whole menu of resource and environmental policy issues in different countries is more sparse, and this book is intended to fit in that niche.

# Acknowledgements

DURING THE PAST DECADE, I have been engaged as a teacher in Sweden and in capacity-building efforts worldwide. The material presented in this book has been extensively tested on students, in courses on natural resources and environmental policy instruments at the University of Gothenburg (Gothenburg, Sweden), at teaching workshops at the World Bank (Washington, DC), and at the African Economic Research Consortium elective course on environmental economics (Nairobi, Kenya). The manuscript or parts thereof were presented at several seminars in economics at the University of Gothenburg, the department of political science at the University of Gothenburg, Resources for the Future (RFF), the Harvard Institute for International Development, and the World Bank. I thank all the participants at these events, who are too numerous to mention. Part of the manuscript also was presented at the Institute of Economic Growth (Delhi, India); thanks to Kanchan Chopra, B.N. Goldar, Shubash Gulati, Srikant Gupta, and M.N. Murthy for valuable comments.

The Environmental Economics Unit at the University of Gothenburg has an extensive collaboration with the Swedish International Development Cooperation Agency (Sida) that includes the Beijer Institute at the Royal Academy in Stockholm, Sweden, as well as several networks for capacity building on various continents. This collaboration has enabled me to work with colleagues and students from countries in Africa, Asia, and Latin America. I acknowledge these coauthors as well as friends and colleagues who have read the chapters or the whole manuscript carefully and provided valuable comments: Hala Abou-Ali, Shakeb Afsah, Wisdom Akpalu, Tekye Alemu, Francisco Alpizar, Greg Amacher, Jessica Andersson, Christian Azar, Razack Bakari, Allen Blackman, Gardner Brown, Dallas Burtraw, Fredrik Carlsson, Nasima Chowdhury, Frank Convery, Håkan Eggert, Anders Ekbom, Jorge Garcia, Gautam Gupta, Henrik Hammar, Winston Harrington, Lena Höglund, Bill Hyde, Olof Johansson-Stenman, Per Kågeson, Beatrice Kalinda, Vinnish Kathuria, Michael Kohlhaas, Gunnar Köhlin, Sandra Lerda, Martin Linde Rahr, Åsa Löfgren, Susanna Lundström, Minhaj

Mahmud, Peter Martinsson, Alemu Mekonnen, Katrin Millock, Adolf Mkenda, Edwin Muchapondwa, Samuel Mulenga, Astrid Nuñez, Wilfred Nyangena, Peter Parks, Daniela Roughsedge, Ulaganathan Sankar, Daniel Slunge, Churai Tapvong, Tom Tietenberg, Martine Visser, and Mahmud Yesuf. Thanks!

I also benefited from discussions with many other colleagues, students, and friends from many countries and disciplines: Milford Aguilar, Sara Aniyar, Jaap Arntzen, Collins Ayoo, Thinus Basson, Mohamud Belhaj, Gunnar Bengtsson, Anders Biel, Alan Carlin, Partha Dasgupta, Henk Folmer, Don Fullerton, Lena Gipperth, Haripriya Gundimeda, Joachim Häggström, Michael Hanemann, Tomas Kåberger, Eseza Kateregga, Lennart Lundqvist, Nelson Magbagbeola, Karl-Göran Mäler, Marco Martinez Negrete, Vivekananda Mukherjee, Elinor Ostrom, Remy Paris, Jack Pezzey, Jorge Rogat, Maximo Rossi, Bo Rothstein, Mikael Söderbäck, Dan Strömberg, Jeroen van den Bergh, Mike Warren, and Wang Zhongcheng. I hope that the many others will forgive my not mentioning their names explicitly. Thank you all for interesting discussions, policy examples, comments, and other assistance related to this book.

I started to write this book during a sabbatical spent in Washington, DC (1998–1999). I spent it partly at RFF, which was kind enough to offer me a Gilbert White Fellowship, and partly at the Environment Department of the World Bank. In different ways, both experiences were stimulating. Several colleagues at RFF (in addition to those already mentioned) helped me to understand U.S. policymaking: Ruth Bell, Jim Boyd, Joel Darmstadter, Carolyn Fischer, Suzi Kerr, Ray Kopp, Alan Krupnick, Molly Macauley, Virginia McConnell, Dick Morgenstern, Richard Newell, Wally Oates, Karen Palmer, Ian Parry, Paul Portney, Roger Sedjo, and David Simpson.

Many people who either worked at or passed through the World Bank contributed to my understanding of developing-world policies through countless lunchtime and coffee-break discussions: Adriana Bianchi, Hans Binswanger, Thomas Black, Jan Bojö, John Briscoe, Ken Chomitz, Paul Collier, Maureen Cropper, Susmita Dasgupta, Gunnar Eskeland, Asif Faiz, Per Fredriksson, Annika Haksar, Kirk Hamilton, David Hanrehan, Philip Hazelton, Narpat Jodha, Ian Johnson, Magda Lovei, Kseniya Lvovsky, Carl-Heinz Mumme, Stefano Pagiola, Julia Peck, Klas Ringskog, Ina Marlene Ruthenberg, Claudia Sadoff, Lisa Segnestam, Ronaldo Seroa da Motta, Priya Shyamsundar, Joe Stiglitz, Vinod Thomas, Lee Travers, Alberto Urribe, Hua Wang, Yan Wang, Bob Watson, David Wheeler, and Jian Xie.

At RFF, Porchiung Benjamin Chou and Sara Gardner provided excellent research assistance on principal–agent models and fisheries, respectively. John McClanahan at the NYU School of Law did some excellent research for me concerning property rights. Ramón López and I collaborated closely on many issues. At Harvard, several friends invited me to give a seminar and commented on chapters of the manuscript or contributed examples: Randy Bluffstone, Theo Panayotou, Bob Stavins, Jeff Vincent, and Cliff Zinnes. A special thanks to John Dixon and Michael Toman, who invited me to Washington and not only helped me find my way around but also participated in many key discussions about this book.

Finally, many people have helped me with references, secretarial tasks, and research assistance, and I am particularly grateful for their good help: Mlima Aziz, Elisabet Földi, Gabriela Gitli, Katarina Renström, Pauline Wiggins, and Fredrik Zeybrandt. I also thank RFF, the World Bank, and Sida for financial support for this book. Additional support from Handelsbankens Forskningsstiftelser for the research projects and assistance closely intertwined with this book is gratefully acknowledged. At Sida, Mats Segnestam has been not only a source of constant encouragement but also an important contributor to discussions concerning poverty and the environment. Don Reisman and the staff of RFF Press have been a constant source of understanding and help. My thanks to the copyeditor, Pamela Angulo, and the anonymous referees that RFF employed on my behalf.

*To Greta for a good past*

*To Gustav, Erik, and Kalle for a sustainable future*

*To all of you and to Lena for sharing the present*

# Abbreviations

| | |
|---|---|
| CAAA | Clean Air Act Amendments |
| CARB | California Air Resources Board |
| CDM | Clean Development Mechanism (of the Kyoto Protocol) |
| CEO | chief executive officer |
| CERCLA | Comprehensive Environmental Response, Compensation, and Liability Act |
| CFC | chlorofluorocarbon |
| CO | carbon monoxide |
| $CO_2$ | carbon dioxide |
| COP | Conference of the Parties (to the UNFCCC) |
| CPR | common property resource |
| DDT | (an effective but dangerous pesticide) |
| DNA | deoxyribonucleic acid (essential molecule for genetic information) |
| EEC | European Economic Commission |
| EPA | U.S. Environmental Protection Agency |
| EPCRA | Environmental Protection and Community Right to Know Act (United States) |
| ESMAP | World Bank's Energy Sector Management Assistance Program |
| ETBE | ethyl *tert*-butyl ether |
| E.U. | European Union |
| GATT | General Agreement on Tariffs and Trade |
| GDP | gross domestic product |
| GEF | Global Environmental Facility |
| GEMI | Global Environmental Management Initiative |
| GPS | Global Positioning System |
| gWh | gigawatt-hour |
| HC | hydrocarbon |
| HCFC | hydrochlorofluorocarbon ("soft" CFC—partially chlorinated, and less damaging to the ozone) |

| | |
|---|---|
| HCl | hydrochloric acid |
| $H_2O$ | water |
| IDA | International Development Agency |
| IPCC | Intergovernmental Panel on Climate Change |
| ITQ | individual transferable (fishing) quota |
| kWh | kilowatt-hour |
| LEV | low-emission vehicle |
| MBI | market-based instrument |
| MEY | maximum economic yield |
| MSY | maximum sustainable yield |
| MTBE | methyl *tert*-butyl ether |
| mWh | megawatt-hour |
| NAFTA | North American Free Trade Agreement |
| NGO | nongovernmental organization |
| $NO_x$ | nitrogen oxides |
| NPSP | nonpoint-source pollution |
| $O_2$ | oxygen |
| ODS | ozone-depleting substance |
| OECD | Organisation for Economic Co-operation and Development |
| OTC | Ozone Transport Committee |
| PCB | polychlororinated biphenyl (a persistent organic chemical) |
| PM | particulate matter |
| ppm | parts per million |
| REP | refunded emissions payment |
| SEPA | Swedish Environmental Protection Agency |
| Sida | Swedish International Development Cooperation Agency |
| SIP | state implementation plan |
| SKr | krona (Swedish currency; 10 SKr $\approx$ US\$1) |
| $SO_x$ | sulfur oxides |
| TAC | total allowable catch |
| TCE | trichloroethylene |
| TEP | tradable emissions permit |
| TRI | Toxics Release Inventory (United States) |
| TSPs | total suspended particles |
| UAE | United Arab Emirates |
| UNDP | United Nations Development Programme |
| UNEP | United Nations Environment Programme |
| UNFCCC | United Nations Framework Convention on Climate Change |
| VA | voluntary agreement |
| VOC | volatile organic compound |
| WHO | World Health Organization |
| WTO | World Trade Organization |

# Policy Instruments for Environmental and Natural Resource Management

# CHAPTER I

# Background and Overview

$I$N AUTUMN 1999, the United Nations announced that the human population had reached 6 billion individuals. It is not clear whether this announcement was a cause for celebration or alarm.

Global population is growing fast—almost 80 million people per year—and has doubled since 1960. Most of that growth is in poor countries. India's population has passed 1 billion, and India may become the world's most populous nation within a few decades. Recent projections indicate that the rate of growth is slowing somewhat, but world population is still projected to reach 9 billion within a few decades. This population growth poses considerable challenges for resource and environmental management.

## Definitions, Concepts, and Challenges for Policymaking

The links among population, poverty, growth, resources, and environment are complex, and the mechanisms that determine human fertility and mortality (and thereby population dynamics) are an interesting topic of study.[1] The harsh-but-effective Chinese policy has shown the world that policy mechanisms can affect human fertility and mortality, but can population growth be affected by policies that do not infringe so heavily on personal liberties?

Interestingly enough, population growth appears to be decreasing quickly in most countries. The global average number of children per woman has fallen from about 6 in 1950 to 2.9 in the 1990s. In the richer countries, fertility is typically around 2 children per woman, which means that population will stabilize or in fact slowly decline. Income and education are particularly important determinants of fertility, and thus "development" automatically brings some decrease. The speed of transition depends on many cultural and institutional factors that may lock countries into a form of "demographic trap" in which poverty is both cause and effect of fast population growth. Results of studies exploring the links among institutions such as property, marriage, and inheritance law as well as the more subtle cultural determinants of fertility indicate that policies can and do have a large effect on household decisions, such as whether to marry and how many children to have (Dasgupta 1993). This finding indicates that policymakers may be able to successfully affect the fertility issue, but the sociocultural and personal aspects of fertility and mortality make it a difficult area for policy application.

Besides population, other major determinants of human impact on ecosystems are level of consumption and choice of technology. This concept is neatly summarized by the $I = PAT$ equation, whereby impact depends on population, affluence, and technology (Ehrlich and Holdren 1971).

## Market Failures

One frustration of many environmentalists is that seemingly simple solutions to serious environmental problems exist but are never implemented. In this book, I write about policy instruments that are designed to ensure implementation. To begin, policymakers must understand why environmental policy is needed. The reasons include market and policy failures that are interlinked with the evolution of property rights.

*Market failure* is a technical term that roughly refers to conditions under which the free market does not produce optimal welfare. It is thus a "failure" compared with the abstract model economists make of a perfect market economy. Important examples of such failure include external effects (externalities), public goods, common pool resources, poorly defined or defended property rights, noncompetitive markets, and imperfect (or asymmetric) information. *Policy failure* may appear to be a simpler concept, but a seemingly neutral concept of welfare underlies it. Policies reflect economic interests, and in some cases, there may not be a single policy that is "optimal" for every group in society. One can sometimes distinguish between corrupt policy and bad policy. The corrupt policy is one that claims to be in the interest of the whole country but actually serves the interest of one group (and may actually do that very successfully). A bad policy is one that intends to enhance welfare in a reasonable way but fails due to ineptitude. Property rights are institutions that can be affected by policy, although the process is typically very slow.

*Externalities* are nonmarket side effects of production or consumption, such as soil erosion caused by unsuitable agricultural practices (particularly on hillside slopes). The silting of dams and the destruction of coral reefs are real costs, but these costs are not borne by the individuals or corporations that cause the damage. Such situations can be seen as consequences of incomplete property rights: if waterways had owners with a right to clean water, then those owners could sue those who caused the soil erosion and thus internalize the effects.

*Public goods* are products or services that are enjoyed in common, such as defense and air (clean or dirty). The market tends to undersupply these goods because it is hard to exclude those who do not pay. Instead, political processes are needed, such as the election of a government that collects taxes and finances public goods. *Common pool resources* also have costly exclusion, but the goods produced with these resources are consumed individually (as *private goods*). Examples include firewood and fodder, and the resources are often managed as *common property*. Free riding and other mechanisms that lead to the undersupply of public goods may also lead to the overuse of common pool resources unless institutions are strong enough to limit access by the users. *Noncompetitive markets,* monopolies, and oligopolies usually result in nonoptimal supply (e.g., too little may be sold at too high a price).

Of all the market failures, *asymmetric information* is perhaps the most pervasive. Economists typically point out that there are no "free lunches" yet commonly assume that information is freely available to everyone. Information is costly, and lack of information stops the market from operating perfectly. Understanding information asymmetries not only helps us design policy instruments to address monitoring difficulties; it also goes to the heart of the most essential dilemma: how to promote social goals such as equity without destroying incentives for work and efficiency. Because policymakers do not have reliable data on pollution damages and abatement costs, for instance, they cannot design policies that are both efficient (with respect to resource allocation) and fair (in sharing the burdens of all the costs involved). If policymakers need the cooperation of individuals who have "inside" information, then they must accept that those individuals may be able to earn something in return for disclosing information.

## Social Rights and Norms Concerning Nature

The concept of *environmental problems* sounds simple enough, and depending on one's background, it may bring to mind issues such as factory smoke, soil erosion, and dam siltation. However, at a deeper level, the concept is difficult to comprehend because it touches on the relationship not only between human beings but also between humans and nature.

To determine what an environmental problem is and what needs to be remedied, policymakers must understand not only technology and ecology but also the sociology, economics, and politics of property rights. Rights, policy instruments, and politics are interlinked in ways that vary between economies, and information also plays a pervasive role. One everyday illustration of rights is cigarette smoking.

A few decades ago, individuals had the right to smoke almost wherever they pleased. People who suffered from the effects of secondhand smoke had no alternative but to try to avoid smokers. Over time, increased information and other factors have changed this situation so much that today, in some countries, the rights have been reversed: individuals have the right to enjoy a smoke-free environment. This sea change has permeated even the private sphere, so smokers visiting private homes kindly ask permission to smoke, or they simply go outside before lighting up. The use of instruments such as no-smoking zones, tobacco taxes, prohibition of tobacco advertising, and legal suits against the tobacco companies has strongly affected the general perception of rights regarding cigarette smoking. Whereas some policy instruments are only possible thanks to changes in individual rights, instruments also can help to change the structure of rights by changing moral and ethical perceptions.

*Current Problems and Warning Signals*

A few examples illustrate the kinds of problems that face humanity:

- Earth's protective stratospheric ozone layer has been degraded by the emission of toxic synthetic chemicals into the air.
- Synthetic chemicals and toxic metals have spread to the supposedly most inaccessible corners of the planet, including the Antarctic; some have accumulated in the food chain and have penetrated the genetic makeup of the human population.
- Already in the 1980s, human activities used about 40% of the primary energy transformation through photosynthesis, which is the basis of all life on Earth. This consumption level does not leave much for natural ecosystems and biodiversity (Vitousek et al. 1986, 1997). Energy consumption, especially of fossil fuels, poses threats at local and global levels. Its potential effects on climate are a topic of international concern.
- Water scarcity is a threat to agriculture and consumers in many countries. The level of some of the world's major waterways (e.g., Nile, Indus, Ganges, Colorado, and Yellow Rivers and the Aral Sea) has fallen visibly as a result of industrial, agricultural, and residential use, and water tables in many regions of the United States, India, China, and other countries are being drawn down rapidly.
- Soil degradation, loss of forest cover, and threats to the marine and coastal ecosystems (e.g., mangroves and coral reefs) have created considerable risk to biodiversity as well as to the sustainability of the food chain.
- Yields of many of the world's fisheries are decreasing. To keep up catches, earnings, and employment, fishermen have stepped up efforts by using larger boats, nets with smaller mesh, and sophisticated technologies such as sonar and satellite navigation. Instead of encouraging restraint, many policies "help" the fishermen by subsidizing the purchase of boats and technology, thus lowering costs to fishermen and increasing the overall fishing effort—thus exacerbating the problem rather than resolving it.
- The energy crisis of the 1970s spurred research into technologies for saving energy (e.g., fluorescent lighting, heat pumps, "hypercars," and thyristors) and for alternative methods of producing energy (e.g., wind power, solar power, and biofuels); good technologies have been developed for efficient energy use in transportation, lighting, heating, and industrial processes. However, sometimes the consumer price of energy is too low to make the alternative technology commercially viable. External costs related to local and global environmental problems (e.g., health and productivity costs of getting asthma and bronchitis in urban areas) usually are not included as part of the cost of electricity or gasoline. If consumers were required to pay the real total cost of energy, they would be more motivated to adopt energy-efficient techniques.
- People whose livelihood depends on natural resources (e.g., grazing lands) typically know their resources well and would have the knowledge to manage those resources rationally, even optimally, if given the opportunity and the means. However, absolute poverty makes the risk of variations in yield unacceptable and can result in unsustainable behavior. Instead of investing in new

productive and sustainable technology, for example, poor individuals might continue to use methods that damage the ecosystem. These methods may be individually rational adaptations that fill the place of missing markets or institutions for savings and insurance, thus showing the detrimental effect of this market failure.

- The income and equity aspects of environmental issues and policy instrument design are often crucial. Imposing taxes to reduce herd size, overfishing, or vehicular traffic can solve congestion and overuse problems but may still be resisted because they leave the users with less welfare if the taxes collected are siphoned off for purposes that are perceived as unproductive for the local users. Policy instruments must give local users a price signal that internalizes externalities without transferring the money out of the local community. There are numerous ways of doing this—for example, through permits that are allocated freely to local users, or by levying charges rather than taxes and then using the charges for local environmental or resource funds, which then can be allocated locally. Many environmental fees in developing countries operate in this way (see Chapter 24).

- In many instances where environmental policy is warranted, polluters have more information and typically greater resources at their disposal than the policymakers do; informational instruments may be an important first step toward successful policy. By collecting and disseminating information, an agency can create a baseline for future action; encourage transparency in implementation, so that individual inspectors cannot "make deals" with polluters outside the law; and clear the way to inform and empower customers, workers, investors, neighbors, and other concerned groups (see Chapter 24).

### Applying Theory to Nature

*Environmental economics* (or *ecological economics*[2]) addresses the interface between economics and the life support system of Earth. *Natural resources economics* addresses both geological resources such as oil and minerals and, increasingly, biological resources such as forests and fisheries. It can be considered an integral part of environmental economics, even though it often is treated as a separate discipline. To take advantage of the lessons that these two areas can provide for each other, I discuss them jointly as far as possible. Environmental policy is interdisciplinary; although economic theory can make a fundamental contribution to the understanding of policy instruments, it can do so only in conjunction with natural science, technology, and other social sciences.

Some people doubt that the conventional paradigm is forceful enough to manage the many serious environmental problems that now face global society, but an increasing number of powerful policy instruments are now available in the conventional tool kit: taxes, charges, permits, deposit–refund systems, labeling schemes, and other information provision systems. The main problem is that these instruments are rarely used properly. Historical examples of serious attempts at environmental policymaking are quite rare. Rather than worry about whether the available policy instruments will ultimately be sufficient, policymakers should make larger scale use of them.

In attempts to avert ecological disasters, policymakers must remember that "disaster" is already an apt description of everyday life for many people in developing countries. Many of the problems that low-income people face are deeply intertwined with the degradation of natural resources and, in some cases, the spread of pollution. Policymakers must focus on the interaction between poverty and ecosystem resources and take particular care to study the distributional characteristics of environmental and resource issues, especially of proposed policy instruments.

Many developing countries lack the resources needed to implement ideal market-based instruments; for the same reason, they also lack the ability to manage other policy instruments. (Regulations need monitoring, enforcement, and occasionally sanctions, which are not necessarily easier to implement than taxes.) At the same time, the welfare effects of environmental degradation can be the worst and the urgency of economic efficiency the greatest in developing countries. The selection and design of policy instruments is more complicated and more important in developing than in developed countries.

A body of scientific analysis on policy instruments already exists. The many references in this book are only a sample of the available literature. Much of this work concentrates on one issue or aspect at a time and tends to be written for specialists; the seminal article by Weitzman (1974) is a fine example. Other work includes the popular textbook by Baumol and Oates (1988). Tietenberg wrote a series of empirical and conceptual analyses (e.g., Tietenberg 1990), and Xepapadeas (1997) wrote a recent theoretical book of great clarity. Other central works on instrument selection include that by Bohm and Russell (1997) (see also Supplemental Reading).

There is also a specialized literature on environmental policymaking in developing countries. Several important contributions stem from the Harvard Institute for International Development (e.g., Panayotou 1998; Vincent et al. 1997). The World Bank (2000) provides an exciting summary and discussion of many new policy initiatives that its research department has been following and, in some cases, fostering. The authors of such books commonly are either proponents or strong skeptics of "economic" policy instruments in developing countries. To some extent, this debate may center on whether the glass is half-full or half-empty. However, one should not be too quick to reach general conclusions about which type of instrument is best suited. Choices should be made carefully, on a case-by-case basis.

## Overview of the Book

Parts One to Three comprise the theoretical portion of the book, defining the need for policy instruments, reviewing the policy instruments available, and discussing the selection of policy instruments under various conditions, respectively. Parts Four to Six illustrate the theoretical concepts by looking at instrument choice and design for road transportation, industrial pollution, and natural resources management, respectively.

## Part One

Chapter 2 presents the classical issues of growth, welfare reform, market failure, and externalities. Chapter 3 is a discussion of public goods, congestion, and asymmetric information and uncertainty. Some of this material is traditional public economics and may be familiar to economists, who may want to skim this part as an introduction. An understanding of public economics is essential to seeing environmental policymaking as one kind of public policy reform. Chapter 4 addresses intertemporal, spatial, and ecological complexities that are sometimes underestimated in applying economic models to environmental policymaking. Chapter 5 treats the evolution of rights, which is fundamental to the functioning of markets, the existence of market failures, and the design of market reform.

## Part Two

The main role of Part Two is to illustrate the range of available policy instruments and how they operate. Its starting point is the *policy matrix* that organizes information about various policy instruments and their applications in different areas. Direct regulation is presented in Chapter 6. Other instruments discussed include permits (Chapter 7), taxes (Chapter 8), and subsidies and other instruments (Chapter 9). Details in instrument design are important, so I differentiate kinds of permits depending on how they are allocated. Charges that are refunded to the polluters are treated separately, because they result in a different distribution of the cost burden, and thus the politics of implementation changes. Chapter 10 interprets the notion of "instrument" in a broad sense, including common property resource management and the creation of property rights in general, and Chapter 11 shows how legal, informational, and political instruments are affected by local factors in developing national policy and the building of appropriate institutions.

## Part Three

Part Three concerns the selection and design of instruments. Chapter 12 focuses on the efficiency of policy instruments under different conditions concerning abatement and cost curves, the character of technical progress, and so on. Chapter 13 examines the role of uncertainty and information asymmetry. The next few chapters present economy-wide (general equilibrium) effects (Chapter 14), effects that are related to income distribution (Chapter 15), and effects of property rights, politics, culture, and psychology on instrument selection (Chapter 16). Chapter 17 is a discussion of international aspects and interaction between policies, and Chapter 18 synthesizes the information presented in Part Three for application to policy design.

## Part Four

Part Four concerns the road transportation sector. The environmental damage caused by transportation is presented in Chapter 19, including a discussion of the damage function related to each mile of driving. In Chapter 20, I describe environmentally differentiated road pricing as the corresponding "first-best" policy instrument. Chapter 21 turns to the "second-best" policy instruments that are used, which range from regulations and fuel taxes to some fairly advanced road-pricing schemes. Chapter 22 addresses the issues of fuel quality, including the phaseout of lead from gasoline, vehicle inspection and maintenance programs, and urban planning in developing-country cities. Chapter 23 is a collection of lessons learned from policy experience in road transportation.

## Part Five

The focus of Part Five is the design of policy instruments for industrial pollution. Chapter 24 recounts the experience of developed countries (mainly those in the Organisation for Economic Co-operation and Development [OECD]), comparing taxes and permits for acidifying emissions and comparing regulation, prohibition, taxation, and information provision for hazardous chemicals. Global issues related to CFCs (chlorofluorocarbons) and climate change are discussed briefly. Chapter 25 focuses on the experience of developing and transitional countries. Taxes and differentiated tariffs are important, but the focus is on the use of the tax proceeds and on distributional effects. Voluntary agreements and information provision are prominent instruments, whereas monitoring, funding, and the building of institutional capacity in the environmental protection agencies are fundamental concerns or constraints for policymakers.

## Part Six

The overarching theme of Part Six is the management of natural resources and ecosystems: water (Chapter 26), waste (Chapter 27), fisheries (Chapter 28), agriculture (Chapter 29), forests (Chapter 30), and ecosystem services (Chapter 31). These issues are of the greatest significance to people and countries with low incomes, because natural resources can be the main source of livelihood and future prospects for people in countries that have little industry. However, the underlying technology and science is complex and often poorly understood, and many categories of users have fairly insecure or unclear rights highlighting the importance for welfare of distributional issues. In addition, the political and cultural setting can be complex and conflictive.

The text ends with Chapter 32, which attempts to summarize some of the main issues of environmental policymaking and their potential solutions.

## Additional Materials

Even though abbreviations are defined in the text, a list of common and technical abbreviations appears toward the front of the book. Because readers of this book come from different academic backgrounds, Supplemental Reading lists are provided at the end of most chapters. The comprehensive References list toward the end of the book includes full citations of the bibliographical references cited in text as well as the Supplemental Reading listings.

## Supplemental Reading

**Environmental Economics and Policy**
Dasgupta and Mäler 2000
Freeman 1993
Hanley, Shogren, and White 1997
Kolstad 2000b

**Environmental Policymaking in Industrializing Countries**
Aaltonen 1998
Anderson 1990
Blackman and Harrington 2000
Bluffstone and Larson 1997
Ekins 1999
Eskeland and Jimenez 1992
Huber, Ruitenbeek, and Seroa da Motta 1997
Lvovsky 1996
Seroa da Motta, Huber, and Ruitenbeek 1999

**Global Climate Change**
Climate Strategies 2002
Toman 2001
UNFCCC 2002

**Relationship between Population and Resources**
Dasgupta 2000
Jodha 1988, 1998

**Selection of Policy Instruments**
Dijkstra 1999
Nordic Council of Ministers 1999
OECD 1989
Russell and Powell 1996
Stavins 2001
Sterner 1994
U.S. EPA 2001

## Notes

1. The most important link is perhaps between the environment and the number of rich people, who consume more and thus exert more pressure on the ecosystem. This effect is sometimes referred to as an *ecological footprint*. It has been estimated that if 6 billion people were to enjoy a North American standard of life, then the equivalent of another two planets would be needed to meet the demand for resources (Rees and Wackernagel 1994). This eye-opening observation can be misleading because it assumes constant technology, whereas the ecological impact of economic activity depends crucially on technical progress. In other words, many extra planets would already be needed to meet current consumption if the technologies prevalent 50 years ago were still in use.

2. Ecologists and natural scientists tend to call themselves "ecological" economists, whereas economists appear to prefer the term "environmental." For some researchers, there is an ideological difference between the terms, which are not identical but do overlap strongly; the distinction is not emphasized here.

# PART ONE

# *The Need for Environmental and Natural Resource Policy*

THE FIRST PART OF THIS BOOK IS DEDICATED TO explaining why there are environmental and natural resources problems. This task is not as trivial as it may sound, because there are many potential answers. In fact, a term like "environmental problem" is really a misnomer. It is not the environment that creates problems for society but society that creates problems for itself by not understanding how to interact with the environment. To address these issues, a chemist or physicist might concentrate on the spread of certain compounds, whereas an ecotoxicologist might analyze the resilience of an ecosystem to certain disturbances, a social scientist might study laws and norms, and an architect might be concerned with town planning. Each of these viewpoints contributes a vital element to our collective understanding.

In the next four chapters, I concentrate on conveying how economists address environmental and natural resources problems, starting in Chapter 2 with the most important concept, that of market failure. Economists generally believe that markets can be very efficient at allocating resources, but under many conditions, their attractive efficiency actually breaks down. In the area of the environment, this scenario is unfortunately common; therefore, it is important to consider very carefully what role markets can play and how their efficiency in allocation and their fairness in distribution can be enhanced by policymaking. Chapter 3 reviews the economics of public policy and information. Chapter 4 focuses on adapting economic models to the complexity of ecosystems, and Chapter 5 discusses the evolution of property rights to ecosystem resources.

## Consequences of Economic Growth

The debate on the consequences of economic growth for humanity dates back at least to Thomas Malthus (1766–1834), who prophesied that population would grow exponentially and resources would be constant or grow more slowly (linearly), so that people would be doomed to live in poverty (Malthus 1803). More recently, the Club of Rome researchers (often referred to as "modern Malthusians") have warned that even the current population and economic activity on Earth is unsustainable.

The first Club of Rome report discussed shortages for such vital metals as lead and mercury that have not materialized (Meadows et al. 1972). Today, there is widespread concern about the toxicity of these chemicals and their abundance in the ecosystems, not their scarcity. This example illustrates an important economic mechanism: with market prices, tendencies toward a shortage—or increased demand that could lead to a shortage—tend to raise the price, which leads to substitution away from that particular commodity and increased resources spent on discovery and developing alternative supplies. To date, this mechanism appears to have been sufficient to avoid shortages of lead and mercury. However, the mere existence of this mechanism does not negate the importance of a country's carefully considering how to manage its natural resources to give maximum benefit, which includes formulating contracts with contractors, designing regulations, and determining resource rents.

The fact that the Club of Rome researchers were wrong about the supply of certain minerals does not mean that they will be wrong about ecosystems in general. In the realm of renewable resources (e.g., fisheries, forestry, water, and agriculture), overuse is a real possibility because of the combination of complicated ecology and inappropriate property management systems. Poor countries must develop policies that both bring about economic growth and address environmental concerns (World Commission on Environment and Development 1987). Global environmental threats include climate change, ozone layer depletion, acidification (of lakes, forests, and more), and the spread of both synthetic (human-made) chemicals and toxic elements from the geosphere into the biosphere. Local environmental threats, which often have health effects, include noise, air pollution, unsanitary working conditions, and infectious disease related to poor water and waste management.

Additional environmental issues include natural resource degradation, overfishing, the destruction of forests, damage to marine ecosystems, soil degradation, and overgrazing of commons. Although local in focus, these issues create a global problem in that the resources are essential to the livelihoods of billions of people in communities around the world. These issues also are inextricably related to each other and to other environmental problems in numerous ways: for instance, soil erosion can exacerbate air pollution and water pollution, which affects fishing, recreation, and human health. Intensive agriculture requires pesticides, which may cause ecological damage and human health problems while leading to decreased overall biodiversity (yet another important global issue). Deforestation dramatically increases the damage from hurricanes and other natural disasters.

The above list is by no means exhaustive. Unfortunately, many other environmental and resource problems exist—many of which are still poorly understood, and some of which have not yet been identified.

## Institutional and Policy Failure

Much of Part One is dedicated to the discussion of market failure. However, market failures are not always the most serious threats to ecosystems. Institutions are also imperfect, and one of the most important institutions is government. The monumental failure of state ownership has already been mentioned. The notion that the state is a neutral and perfect agency to enforce the general well-being of society is very naïve.

One example of imperfect government policy is the formerly planned economies of eastern Europe, where the banishment of "short-sighted profit interests" was hailed as an opportunity to implement policies truly geared to maximizing welfare; yet, the policies really achieved the exact opposite goal, partly because of a simplistic application of the Marxist theory that value is created only by labor. By treating natural resources as free goods of no value, the intrinsic value of those resources was, in many cases, effectively destroyed. The Aral Sea is a sad symbol of such policy. More than half of its area is gone. This giant lake has been turned into a poisonous dust bowl as a result of irrigation, poor management, and excessive cotton production. Ships now lie in the sand, many miles from the current shoreline.

Another example of imperfect government policy is fisheries, because the absence of property rights to the sea leads to the risk of a tragedy of open access. It is a case of considerable market failure in which political policies are badly needed, but the wrong policies make matters worse. Subsidies intended to "help" fishermen actually exacerbate the market failure that they are supposed to address.

In the future, the public sector must not be analyzed as if it were a monolith. In reality, *government* is a series of public-sector bodies with distinct structures, motivations, and modes of operation at different levels. Furthermore, governments are not the only institutions that can fail. The family is also an institution at the micro level, and some form of rational division of labor within it (as well as some fair division of proceeds among its members) is often assumed. However, it is not always the case. In many poor communities, men hold a large share of formal rights and power while women and children are effectively dispossessed and exploited within the contexts of their own families. It is not uncommon for the government to exacerbate such tendencies by automatically granting titles to a male "head of household" when traditional rights may have been much more equal, even where agriculture was managed primarily by women.

Some faulty government policies are a real threat to the sustainable use of natural resources and ecosystems, for example, subsidies for goods, services, or practices that cause severe environmental degradation. Such subsidies are so common that many economists (particularly those working with developing countries) rank "subsidy removal" as one of the main instruments of environmental policy. Given this sadly ironic state of affairs, one might expect that this section ought to be one of the longest in this book. However, undoing bad policies is not a new category of policy; rather, it is part of a general process of policy optimization or adjustment.

One possible, and indeed plausible, explanation for the prevalence of bad policy is lack of information or understanding about the ecological, technical, and economic relationships that are used to choose and design policy instruments. Other explanations stem from the fact that policies are not designed only by altruistic welfare-maximizing policymakers who are free from personal economic or political interests. The truth is, policies are formed by the interplay of conflicting political and economic interests, and a thorough understanding of the political economy of policymaking is required to analyze any set of instruments used in a specific context. In countries where state institutions are weak, the risk of policy capture by various groups is particularly strong. Presumably, the abundance of damaging subsidies for agriculture and industry in developing and developed countries should be seen in this context.

# CHAPTER 2

# Classical Causes of Environmental Degradation

$H$OW ARE ENVIRONMENTAL THREATS related to economic development and growth?[1] At one extreme, some researchers claim that growth has already exceeded the sustainable level of activity on Earth (Meadows et al. 1972). At the other extreme, some researchers believe that technical progress will make it possible to meet the demand that will result from increased population and per capita income (Kahn et al. 1976). In some overly simplified analyses, growth is the main culprit, whereas in others, it is the principal panacea for environmental issues. Neither of these simple positions is tenable.

## Growth and the Environment

The essential determinant of environmental stress is not the average rate of growth but the technology used and the composition of growth or of the economy itself. Whereas increased consumption of polluting cars, pesticides, and chemical-intensive products could become a problem for sustainability, increased consumption of music, Internet information, ecotourism, and organically grown food probably would not. However, in a free market economy, the consumers—not the "social planners" or the ecologists—decide the composition of output (and, indirectly, production).

Policymakers can influence the path of the economy by using policy instruments. To do this, they must first understand the fundamental determinants of economic development. A great deal of research has focused on the composition of the economy and its development over time, which is often associated with the environmental Kuznets curves (EKCs). The idea behind EKCs is that with economic growth, emissions typically follow the inverted "U" curve (as illustrated for emissions in Figure 2-1). According to this hypothesis, the early phases of economic growth inevitably imply increased pollution, but as incomes increase, emissions peak and then decline. The curve for the quality of ecosystem resources

15

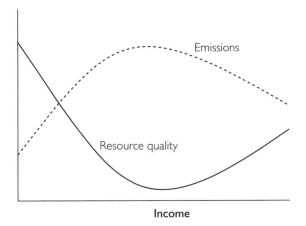

**Figure 2-1.** *An Environmental Kuznets Curve*

would be the inverse of that for emissions—that is, an upright "U"—signifying deterioration followed by gradual improvement. However, the stock character of these resources complicates the issue.

Explanations for the EKCs concentrate on several general factors, including the development of technology, relative prices, the income elasticity (which describes how demand for goods and services tends to vary with income) for a clean environment, and income distribution. Income distribution also varies with economic growth, as illustrated in the original Kuznets curves (Kuznets 1930), adding another layer of complexity to the relationship between environment and growth.

In fact, the relationship between growth and environment need not take any particular shape. Several functional forms are possible. For some pollutants (in the relevant income range), the curve appears to be a constant increase (e.g., carbon dioxide emissions) or a constant decrease (e.g., bacteria in drinking water). In principle, this difference might result from variations in the turning points of the curves. Thus, it is possible that there is an inverted "U" curve for carbon dioxide emissions, but a constant increase is observed because no countries have yet reached the peak. In such cases, for all practical purposes, EKCs are irrelevant.

EKCs are not inevitable or fixed development paths. The effect of policies may often dominate over the variations described by EKCs, which is encouraging because a strict belief in EKCs would lead policymakers to accept as true that pollution increases were inevitable in the short run. In fact, fairly inexpensive measures often can limit pollution considerably, even in the short run. It would be dangerous if policymakers were to believe that pollution and destruction of natural resources did not matter because the damage would automatically be reversed later, farther along the curve. Experience shows that "repairing" ecosystems and "replacing" natural resources is much more expensive than prevention, and in some cases, the damage is irreversible.

The central recommendations for successful and sustainable development in *World Development Report 1990* (World Bank 1991) include formulating the correct macroeconomic policies, creating a market-friendly orientation, being open to trade, and investing in people through health and education. During the

1990s, many countries were quick to implement a market-friendly orientation and macroeconomic policies (López, Thomas, and Wang 1999). Barriers to trade and finance were broken down; price controls and deficits were reduced. Some countries significantly increased education and health expenditures, and many countries experienced economic growth as well as declining poverty. The 1990s also showed how easily advances in some areas could evaporate into economic crisis and the enormous price (as environmental damage) of economic progress in some countries. It also became strikingly clear that corruption not only is an issue of morals but also entails enormous economic costs (López 2000).

In East Asia and Latin America, openness to trade and knowledge have been important factors in economic growth (Thomas and Wang 1998). An even distribution of human capital, as reflected in the high-quality public education systems in East Asia, is a primary factor behind rapid growth, particularly in conjunction with openness to trade. Global financial integration has implied great benefits for some countries but may imply risks if other economic policies are not appropriate. Despite setbacks in recent years, this rapid growth has been accompanied by rapid decreases in poverty, with a fairly even distribution of income. However, development in East Asia has not been positive for the environment; 9 of the 15 world cities with the worst air pollution are located in East Asia. Some 20% of the vegetated land suffers from soil degradation caused by water logging, erosion, and excessive grazing. Deforestation rates are high, and 50–75% of the coastal areas and protected marine environments are classified as highly threatened (Worldwatch Institute 1996).

Economic growth and environmental sustainability are complex aggregates, determined by the interplay of numerous factors; it would be foolhardy to believe that there is a deterministic relationship between the two. Technology and output composition are important, but these parameters cannot be determined directly;[2] they are determined endogenously in the economy. The composition of output tends to develop in certain ways that reflect factor endowment, tastes, and comparative advantages during certain periods. However, this development results from not an iron law of physics but social behavior, which can be considerably influenced by suitable policies. Similarly, technology choices are made by economic agents and can be highly influenced by suitable policies.

The importance of a good environment for business may be greater than the availability of finance, because the latter will simply come if the conditions are appropriate. A good environment for business is definitely not the same as a good natural environment, but the two parameters are far from contradictory. Of foremost importance for a good business environment appears to be a transparent, predictable, and reasonable legal and political structure. It must be free from corruption and exaggerated bureaucracy but also structured enough to avoid the costly uncertainty of contract enforcement. A good business environment also requires a reasonable natural environment; employees can hardly thrive or be healthy in a deteriorated environment. The distribution of environmental quality is crucial. If the living environment of the poor is so degraded (e.g., through disease or malnutrition) as to inhibit their productive development, then the economy experiences not only a direct decrease in human welfare but also a loss of productive potential.

To reconcile economic freedom, growth and ecological constraints may require a careful blend of policy instruments to influence the composition and the technology of consumption and production (e.g., Carlsson and Lundström 2000). Research and policy experience show that certain policy instruments work better under some circumstances than in others. The choice and design of policy instruments is an important and promising area for future work.

## Welfare and Policy Reform

Economists often assume that the well-being of an individual ($i$) can be expressed as a utility ($U$) function that depends on income, consumption, leisure, working conditions, environment, and other factors. Analogously, it is also often assumed that social welfare ($W$) depends on all these individual utilities, as in the social welfare function $W(U_1, ..., U_i, ..., U_n)$. Economists typically do not know much about the shape of these functions, but for simplicity, they are sometimes assumed to be linear in the sense that net monetary income would reflect utilities and welfare. Another common assumption is convexity of the functions, which implies that an increase in income for the poor is more important than an increase in income for the rich. Even with general welfare functions, the assumption that economists often make (and that I basically agree with) of the desirability of maximizing welfare still depends on one of several possible value judgments. Other people might prioritize an egalitarian society or a society with some other goals. For most purposes, welfare maximization is still a general goal. The welfare of future generations may be included as well as distributional concerns and concerns about long-run sustainability.

One of the main lessons of economics is that the market mechanism is efficient at allocating resources. Economics attempts to formally illustrate this efficiency by building mathematical models of the economy. Models show that under "perfect" conditions (i.e., a market with free competition and without noncompetitive markets, public goods, or external effects), a market will automatically achieve a (Pareto) optimal outcome. This hypothesis is often referred to as the First Theorem of Welfare Economics.[3] *Pareto optimality* is an efficiency concept that implies that the economic situation of one individual can be improved only if the economic situation of another individual is worsened. Intuitively, this concept can be understood by considering the opposite: that one individual's economic situation could be improved at no expense to anyone else. Most people would agree that this is an unnecessary deprivation. However, even this seemingly technical and neutral efficiency criterion hinges on value judgements. Making one rich person richer (with constant incomes for everyone else) may not be considered desirable. It is thus conceivable that a welfare function would decrease in some individuals' incomes for certain values.

In general, a given economy has many possible optimal outcomes, and different starting conditions (notably, of income distribution) will give different Pareto optima. Choosing between them necessarily requires some value judgement. Criteria are needed to judge the desirability of different states of the economy that imply gains for some groups but losses for others (e.g., taxing the wealthy to help the poor). Such criteria are an expression of the social welfare function.

The Second Theorem of Welfare Economics states (under fairly restrictive conditions) that any desirable and feasible outcome of the economy that one chooses with the help of a social welfare function can be achieved as the result of a competitive economy. It implies that any outcome can be "decentralized," that is, achieved by the market agents themselves, if the state arranges appropriate conditions (e.g., by a lump-sum redistribution of the initial endowment). It means reallocating money—taking money from some individuals and giving it to others—but otherwise leaving the economy and its mechanisms intact. Such redistributions do not always work in practice (partly because taxes and subsidies influence people's behavior), but in some cases, policy instruments can decentralize the outcome.[4]

A real economy could not be a "pure market" in the absolute sense of having no state interference. The mechanisms that make the market work (e.g., the definition and enforcement of property rights and civil laws that govern contract enforcement) are public goods that have to be provided for by a public body. These and other necessary mechanisms, such as the maintenance of social order and defense, also are costly activities that make at least some level of taxation an inevitable feature of the economy. Taxation requires resources, and its implementation tends to distort the price signals of the market, which modifies the optimal properties that can be derived in a simple and abstract market with no outside interference. As soon as one aspect or area (such as taxation) deviates from the simple textbook model of the "perfect market economy," the conclusions and recommendations from that model may no longer apply. Policies or outcomes that would be best, given that some imperfections already exist, are called "second-best" by economists. Economists' ability to analyze the optimality of second-best situations is limited; however, it is commonplace (and perhaps, on the whole, reasonable) to believe that at least some of the "first-best" efficiency properties of market solutions remain valid; an instrument is "first best" when it would be optimal under some set of ideal (often unrealistic) market conditions, which often implies that the instrument is not optimal in the real (imperfect) world.

The virtual collapse of many formerly planned economies that attempted to rush the transition to the free market without due attention to building the necessary institutions is a good, although unfortunate, illustration of two important facts:

- Economies with an excessive degree of state intervention fail severely in attaining efficiency.
- Economies with an excessively free and unregulated market may fail abysmally on both efficiency and social issues.

Ironically, some economies (e.g., the Russian Federation and other countries of the former Soviet Union) rapidly moved from what could be characterized as "excessive state" to insufficient or perhaps inadequate state institutions. They learned the hard way that the market is a social institution and one that requires considerable enforcement from a state strong enough to defend property rights and uphold a necessary degree of trust and impartiality in civil law if entrepreneurs are to feel comfortable investing in the economy.

Although neither absolute anarchy nor totalitarian planning has many serious proponents, policymaking is carried out against a backdrop of intense academic and ideological conflict over the optimal extent of state intervention in the economy. The proponents of free markets focus on efficiency as the engine of economic welfare, whereas the advocates of state intervention emphasize that the markets are imperfect without adequate policies to regulate them and to maximize welfare. Preaching the virtue of environmental policymaking is a challenge when the main message among many development economists and macroeconomists is "deregulation and reduced state influence." Consider the Russian experience: economic policymaking need not contradict environmental stewardship. To promote development, it is necessary to eliminate the regulations that stifle growth, not all regulations. In the economic area, a general absence of rules would lead to stagnation; in the ecological area, it could lead to expensive abuse of originally productive resources.

The relative importance of institutions may vary depending on such factors as cultural norms and the characteristics of technology. Less political enforcement may be needed in extremely structured cultures with a tradition of high work ethics and an emphasis on honesty; feedback from economic outcomes and political institutions gradually changes social norms. The institutions needed in an economy where technological progress is slow and labor-intensive will be distinct from those needed in an economy where technological progress is rapid and capital-intensive. Other important factors are the market structure, the size and openness of the economy, and issues related to risk and information asymmetries.

Over the past few decades, the market mechanism has shown its many strengths in real-world economies. The exceptions are nevertheless important, especially as applied to the management of natural resources and ecosystems. This area is characterized by externalities, public goods, common pool resources, imperfect foresight, and other types of market failure. The concern for social welfare implies a special focus on the poor, which in turn tends to imply that risk and uncertainty are given greater weight than maximizing expected return. Variation in income is not an acceptable risk for people who live in danger of starvation; it is far from the "first best" world in which all marginal costs and utilities are equalized.

One particular problem for analysts is the glaring lack of good studies on the efficiency or even the cost-effectiveness of environmental policies.[5] It is important to ensure that public money is used efficiently, but unfortunately, this task is far from simple. Other areas of public spending face the same problem; few good studies report whether a marginal increase in funding of police, military, or intelligence services will be efficient in the sense of optimally promoting public security. Nonetheless, some comparisons are possible, and international comparisons may be the most promising method in many cases.

Considerable literature is available in the area of policy reform. One common problem is that the process of policy reform (e.g., redesigning taxes) changes the relative prices and incomes in a way that makes comparisons with the original state difficult. A successful process of policy reform may well entail taking temporary steps that are not efficient (Guesnerie 1977). These issues are particularly well illustrated in developing countries, which have large disparities in income (and thus in the marginal utility of money). One empirical study that attempted to design optimal tax reforms for India used a technique referred to as the *inverse optimum,* which entails determining which set of welfare weights would have made the observed state of the economy an optimum (Ahmad and Stern 1984). If the calculated weights do not match the welfare weights that can be assumed for the decisionmaker (especially if they contain weights that are blatantly impossible—such as negative weights for some groups), then tax reform that increases welfare must be possible.

## Market Failure

The conditions under which the welfare theorems hold (i.e., a "perfect market") are convenient analytical abstractions that provide a starting point for economic analysis. A situation in which those theorems fail to hold is called a *market failure* and is very common. In this section, I focus on three kinds of failure: noncompetitive markets, external effects, and public goods.

The idea of a market is that people engage in mutually beneficial trade, and to do so, they must have clear ownership rights and information. In a perfect market, every good and resource has an owner and a price, and the agents have full information of the options available to them. Production and consumption technologies are characterized by the absence of indivisibilities and increasing returns to scale (i.e., the rate at which output changes as the quantities of all inputs are varied)—or, more formally, by the absence of nonconvexities in production and consumption sets. This seemingly technical explanation is illustrated in Figure 2-2.

The intuitive meaning of *convexity* is that you can combine goods (or production inputs) at will within the set of possibilities. This assumption is often taken for granted, and understanding how its absence can render optimality unachievable is simple: suppose that you want to buy bread and cheese, but the smallest package of each is so large that you would have to spend your entire budget on only one item. Although this example sounds trivial, indivisibilities are sufficient to prevent the attainment of optimality. Similarly, the high minimum-efficient scales of refineries or steel mills cause problems in small countries; small plants are not cost-efficient, big plants may have insufficient demand, and no plant means that the country will be 100% dependent on imports. The same applies to infrastructure investments such as bridges, health programs, and railways. A country may desperately need infrastructure but find the available systems excessively large and prohibitively expensive. The absence of intermediate-scale solutions implies constrained choice, and a decentralized market will not lead to an optimum. Such a nonconvexity may be created by increasing returns to scale. It may mean that a project of size $A$ is feasible, but a project one-half the size of $A$ is too expensive.

*Noncompetitive markets* such as monopolies and oligopolies are a kind of market failure that usually leads to excessively low production volume being sold at too high a price. This situation diverges from the optimum because an increase in

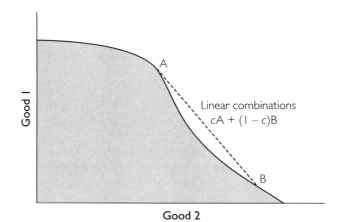

**Figure 2-2.** *Convex Combinations in Consumption Space*

*Note:* Input or consumption bundles *A* and *B* are both feasible, but combinations such as one-half of *A* plus one-half of *B* (or more general linear combinations) are not attainable because of the nonconvexity shown.

production would be possible and would be valued more highly than the additional cost. The existence of monopolies in the economy is partly related to underlying cost structures, such as increasing returns to scale. They are generally accepted as something that policymakers should regulate.

Typically, *external effects* also create this kind of nonconvexity. Consider two goods, *A* and *B,* that have strong negative externalities between them. You can have *A* or *B*, but obtaining both is difficult. Similarly, by definition, *public goods* imply that everyone has to consume the same quantity: there can be one state of the economy in which everyone gets none of the public good and there can be a state in which everyone gets a quantity $z$ of the same good, but there cannot be a state in which some people get none and some get $z$. This restriction, too, means that the decentralized market will not lead to an optimum. Other causes of nonconvexities include common pool resources, congestion, and joint production.

The simplest economic models are deterministic and atemporal, but the real-world economy takes place in real time, and its outcomes are stochastic; the best one can hope for is to know their probability distribution. In general models of the economy, a "space" of goods—each of which is labeled by a probability and a date—can be defined. In such a model, there can be many more sources of nonconvexity, including varying degrees of myopia (nearsightedness), uncertainty, risk aversion among the agents of the economy, and transaction costs.

For a market to be perfect, all property rights must be fully allocated. No externalities, public goods, or other nonconvexities should exist. There also must be markets for all goods and resources, including, most importantly, future goods (i.e., future markets) and full information about all these markets. The mere existence of a market failure does not automatically warrant the implementation of a given policy because the costs of market failures must be weighed against the potential for "policy failures." This comparison must be carried out within the specific context of general policymaking in the economy to be studied (see Chapter 5).

## Externalities

An *externality* can be defined in different ways, but it typically is an unintended and uncompensated side effect of one person's or firm's activities on another. Good examples are the health effects of smoke emissions from vehicles, factories, and cigarettes. These side effects occur because of a technical interdependence in consumption or production (see Box 2-1, Definition of Externality). They are not intentional damage per se, and they typically are difficult to avoid. Note that this interdependence must also be a nonmarket dependence to qualify as an externality. If many people are lined up to buy a good (e.g., medicine or water), the price of which consequently increases, then the effect (commonly referred to as a pecuniary externality) is not an external effect, because it is perpetrated through the market mechanism. If most people in an area take antimalaria drugs and this action ultimately decreases the number of malaria mosquitoes (and thus the malaria risk to individuals who do not take tablets), then people who do not take tablets are the unintended beneficiaries of an external effect.

Another approach to defining *externality* is to suppose that ownership rights or markets are missing for the particular resource in question. For example, if there were private ownership of the air, then people would have to buy the right to pollute it with smoke, and passive smoking would be internalized through the market. Practical barriers to the establishment of such rights and markets, however, are likely.

*Externality* is perhaps the most basic concept in environmental economics. It has long been recognized as a problem but originally was seen as a minor one. The classical economists wrote of the soot from factories in Manchester and Liverpool, England, that dirtied the laundry hanging on the line and of the bees who pollinated neighboring farms' orchards. Sometimes the beauty of a rose garden, enjoyed by not only the owner but also passers-by, was used as an example. These examples

---

## Box 2-1. Definition of Externality

A general definition of an *externality* is the existence of some variable that enters into the utility or production function of an agent (an individual or a firm *i*) in the economy, although it is controlled by another agent (*j*) who does not take effects on *i* into account and does not pay compensation. For the case of utilities, the utility ($U$) of individual *i* depends not only on his own consumption but also on the consumption (or some other variable) of another individual *j*:

$$U_i = U_i(x_i, x_j) \tag{2-1}$$

One example of this function is second-hand cigarette smoke.

When many agents affect each other, the function can become quite complicated. However, any information about the nature of these functions can be included. For example, if a number of agents ($j = 1, \ldots, m$) emit smoke ($s_j$) and this emitted smoke is perfectly mixed in the atmosphere, then it is the sum of all the smoke ($S = \Sigma s_j$) that affects utility, allowing a simplification of the equation:

$$U_i = U_i(x_i, s_1, s_2, \ldots, s_m) = U_i(x_i, S) \tag{2-2}$$

clarify the issue but also border on triviality. However, the environmental issues that confront us today—contaminated drinking water, smog in developing-world cities, destruction of the ozone layer, acid rain, and global warming—are far from trivial. Unfortunately, the dependencies might be long-ranging, which makes defining property rights or negotiating difficult. Examples include cases in which the polluter and victim are separated by long distances (e.g., the effects of soil erosion on coral reefs) or in time (e.g., the risks posed to future generations by nuclear waste).

The environment and the various services provided by ecosystems enter into ordinary production and consumption in ways similar to

## Supplemental Reading

### Environmental Kuznets Curves (EKCs)
Dinda et al. 2000
*Environment and Development Economics* 1997
Grossman and Krueger 1995
López 1994

### Welfare Economics/Resource Economics
Arrow 1951
Debreu 1951
Hartwick and Olewiler 1998
Mäler and Vincent 2001

### Policy Reform
Atkinson and Stiglitz 1972, 1980
Diamond and Mirrlees 1971a, 1971b
Feldstein 1976
Hanemann 1995
Murty and Ray 1989

those of other inputs. Thus, they should be included in economic accounts at corporate and national levels. They often are not, because there is no "owner" or because the environment has the characteristics of common property or a public good. The absence of property rights is related to scarcity. Classical writers such as Marx noted that water might have great "use value" but little "exchange value" when plentiful. Without water, there would be no production and no life, and thus, in a sense, water is "infinitely valuable." In countries where it is abundant, it is practically free. Similar arguments apply to many natural, environmental, and ecosystem resources. Oxygen, phosphorus, DNA, chlorophyll, iron, and biodiversity—to name only a few—are individually infinitely valuable.

Trying to estimate the total value of global ecosystems is pointless because the value is infinite. Only marginal changes can be studied. History shows that property rights and market values appear only when use value is coupled with scarcity. Thus, the existence of external effects is intimately tied to the absence of markets, and this absence, in turn, is the result of a certain social and historic condition. In fact, the absence of property rights or of markets is an alternative way of defining externalities. At one time, there were no rights to land anywhere; today, most land is claimed, and agents are staking out rights to radio waves, geostationary parking slots, genetic codes, and even property lots on the moon. (Property rights are discussed further in Chapter 5.)

Externalities are commonly distinguished as depletable or nondepletable. The manure from horses is a depletable externality because if one person takes it, another cannot. However, the odor of horse manure is a nondepletable externality because one person's exposure does not reduce the exposure to others. (This concept also applies to congestion and is essentially the same as the nonrivalry of public goods.)

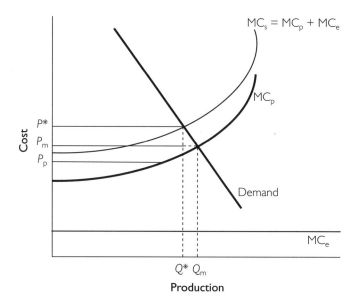

**Figure 2-3.** *Externalities and Their Effect on Markets*

*Note:* $P$ = price; $Q$ = quantity; MC = marginal cost. Asterisk indicates optimum value. Subscripts m, p, s, and e represent market, private, social, and emissions, respectively.

The effect of externalities on resource allocation is illustrated in Figure 2-3 for production externalities. It shows the usual market analysis for a certain product, with supply determined by ordinary private production costs ($MC_p$). Market equilibrium is determined by the intersection of demand and supply curves ($Q_m$, $P_m$). If each unit of production gives rise to a certain (for the sake of simplicity, constant) external effect, then there is an extra cost to society ($MC_e$) that is not borne by the producer. This damage would be measured as the sum of the decreases in utility due to the external effect for all individuals or firms affected. For individuals, it would be $MC_e = \Sigma_j U_{ji}'$ (i.e., the sum of marginal disutilities for all $j$ of acts carried out by $i$). If internalized, it would be a social marginal cost of production ($MC_s$), $MC_p + MC_e$. The intersection of this curve with the demand curve gives the social optimum ($Q*$, $P*$). The analysis is often more complicated, because several production methods give rise to different quantities of externalities. Also, the same physical emissions might cause different amounts of damage, for instance, depending on the location of the pollution source.

## Notes

1. The relationship between these concepts is complex. *Development* is a broad concept that includes both economic growth and other (positive) societal changes, such as the addition of intangible value and maybe a more even distribution of income. *Growth* may mean simply an increase in GDP but ideally should be growth in true income (i.e., including various welfare-

related aspects, such as the environment). For the purposes of this discussion, the meaning approaches the more general concept of development, but the closer it gets, the harder it becomes to measure.

2. Similarly, the distribution of income, the degree of competition, the transparency of decisionmaking, the degree of corruption, and other variables are partly endogenous to economic development but can be influenced by policymaking and then, in turn, have a decisive impact on economic development.

3. Actually, more conditions are required, such as absence of indivisibilities and advantages to scale. More technically, all production and consumption sets must be convex, and all agents must have perfect foresight.

4. The complexities of society sometimes make it impossible to aggregate individual utilities to a social welfare function. If people care about only individual income, it would work. However, if social concerns include altruism, if people have preferences concerning income distribution, or if welfare depends on relative rather than absolute levels of consumption, then the mere construction of aggregate welfare functions may not be feasible.

5. *Cost-effectiveness* means achieving the given goals at least cost. *Efficiency* includes the meaning of cost-effectiveness but also requires that the goals be set optimally with respect to welfare.

# CHAPTER 3

# Public Economics and Information

*T*HE MOST CLASSICAL OF MARKET FAILURES is the failure to provide public goods that are not consumed by individuals but enjoyed by all or most citizens as a whole. Provision of public goods is one of the fundamental reasons for government. Many natural resources or ecosystem services are at least to some extent "public" in this sense. In order to discuss the implications for policy design, several related kinds of public good must be distinguished, as they are in this chapter.

One of the most important and often neglected "goods" in the economy is information. Information is vital for economic transactions and for market function. Information is sometimes a public good but is often unevenly—or asymmetrically—distributed. Together with the stochasticity of ecosystems and risk aversion among people, asymmetric information can create many serious market failures that have bearing on environmental issues.

## Public Goods, Club Goods, and Common Property

*Public goods* are goods that are used collectively by society. *Pure public goods* are characterized by *nonexcludability* (if a public good is provided for some individuals, others cannot be excluded; e.g., national defense) and *nonrivalry* (the enjoyment of a public good by one individual in no way reduces its availability to others; e.g., television broadcasts). According to public economics, the market alone cannot allocate resources optimally between public goods and private goods; nonexcludability directly invalidates the use of the price mechanism for resource allocation. Affected public goods include defense, law and order, education, and health—even "a clean environment," which can be seen as a kind of public good (and pollution, which might be considered a "public bad").

Because the market does not provide public goods (except in small quantities or special cases, such as when public goods are provided through charity or some form of sponsoring), the state or some other political body is the origin of collective action that produces them. The most common starting point for a discussion of the optimal provision of public goods is the Samuelson rule, which declares that the social value of a good is equivalent to the combined willingness to pay of (or utility to) all the consumers of that good (Samuelson 1954, 1955) (see Box 3-1). This rule is similar to the one discussed for externalities in Chapter 2. If agent $i$'s consumption somehow leads to a benefit for $j$, then this extra utility will not automatically be considered by $i$—but it should be if social welfare is to be maximized. If everyone enjoys a given public good equally, then the benefit for society is the sum of all the individual utilities. This is an abstract, "first-best" rule for the provision of public goods in a world where, among other conditions, individually differentiated lump-sum taxes are possible, so income distribution is not a variable that must be considered.[1]

In more realistic models, the political and tax systems present various difficulties. For instance, the optimal "second-best" provision of public goods is lower than the first-best because of the cost of raising the tax needed to finance the public goods or the effect of the tax on labor supply. The public sector may have several goals, including an even distribution of income. Such a goal typically complicates models if the distribution of consumption patterns is different for the rich and the poor. When income distribution goals cannot be met by using taxes and subsidies, and if the public good in question is particularly attractive to poor people, then it may be optimal to increase its provision. Similarly, the optimal tax structure for different goods may reflect a mixture of goals: a negative externality related to a particular good is a factor that leads to a higher tax, whereas consumption of that particular good by the poor is usually an argument for a lower tax. The construction of optimal taxes is complex, and the results depend on the model. With some forms of income tax, this contradiction need not arise.

Two other categories of goods are closely related to public goods but often have some degree of congestion (i.e., "costs" are related to the use of a good by many people) and rivalry in use. *Impure public goods* include such seemingly classical public goods as parks and roads, but the utility of one user typically is reduced by an increase in the number of other users. *Club goods* (sometimes called *mixed goods*), categorized between private goods and public goods, can be consumed by many individuals without diminishing the consumption of others (e.g., a movie). However, exclusion (of nonmembers) is possible.

For both of these categories, the first-best Samuelson rule still holds in the sense of Equation 3-2: the club good should be supplied in such a quantity that its price corresponds to the sum of the $n$ club members' marginal willingness to pay. Several interesting issues remain to be solved, such as the number and size of the clubs ($n$). As long as exclusion (of those unwilling to pay the price of the club good) is possible, the decentralized market economy can solve this problem and provide club goods in an efficient manner (Buchanan 1965). Similarly, if people are willing to make trade-offs between the quality of *local public goods* (i.e., goods that are public to a community or municipality) and taxes, then different local communities could offer mixed public goods of different quality and citizens will base their location decisions on the choice between paying higher taxes with more "free" services or lower taxes with fewer such services (Tiebout 1956).

## Box 3-1. Derivation of the Samuelson Rule for the Provision of Public Goods

To determine the optimal provision of a pure public good, the utility for household 1 ($U_1$) is maximized as $U_1(x_1, G)$, where $x_1$ is household 1's consumption of a private good and $G$ its consumption of the public good, subject to a minimum level of utility for other households ($j$) and a production function ($F$). It can be achieved by maximizing the Lagrangean function $L$:

$$\max L = U_1(x_1, G) + \Sigma_{j=2,\dots,n}[U_j(x_j, G) - U_{j0}] - \lambda F(x, G) \qquad (3\text{-}1)$$

Taking partial derivatives and solving gives the Samuelson rule:

$$\Sigma_j \frac{U'_{jG}}{U'_{jx}} = \frac{F'_G}{F'_x} \qquad (3\text{-}2)$$

where $U'_{jx}$ is the marginal utility for household $j$ of private good $x$; the other derivatives are defined analogously. The provision of public goods is optimal when the sum (over all households) of the marginal rates of substitution between the public good and any private good is equal to the marginal rate of transformation (in production) between them. If $x$ is the numeraire, then the expression in Equation 3-2 is simply the price of the public good. If lump-sum taxation is not available, then more restrictions are taken into account, and typically, additional terms (such as welfare weights for individuals with different incomes) would be added to the left side of Equation 3-2.

The ecosystem functions of the natural environment and various natural resources not only enter into the utility functions but also commonly are inputs into production. When these resources have characteristics such as mobility (e.g., fish or wildlife) or variability (e.g., of harvest due to rainfall)—resources that typically offer some technical advantage to joint management or have some difficulty with the definition of private property rights—they may be referred to as *common pool resources* (or "commons"). Some of these resources have no formal owner, but typically some form of ownership control is exercised, collectively or by private individuals. In several respects, these resources act like local public goods or club goods except that they are generally more a source of production inputs than goods for direct consumption. Common pool resources are usually characterized by costly exclusion, and there is typically rivalry in use: the resource is common, but the goods produced are private (Ostrom et al. 1999).

In the past, the concept of common pool resources has been confused with that of "free access" resources; the difference is similar to that between free fishing in the ocean and participation in a local river fishery, which is strictly regulated by village elders, respectively. However, the exclusion of outsiders is an important feature of successful common pool resources. Excluding members (or temporarily limiting harvests) is also practiced. The main difference between common pool resources and local impure public goods is that the size of the resource is determined by nature: overharvesting can reduce the productivity of the resource, and the resource is depletable. Good management may enhance the productivity of the resource, but its size cannot be decided arbitrarily. Resources that have certain features (such as fluidity, mobility, and extensiveness) are often

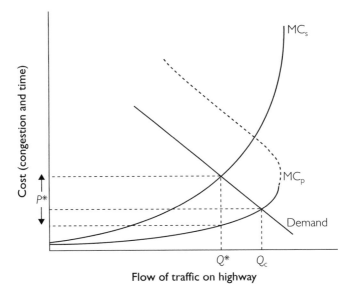

**Figure 3-1.** *Congestion Pricing for Highway Traffic*

*Notes:* $MC_s$ = social marginal cost; $P*$ = extra road charge or tax; $MC_p$ = private marginal cost (fuel and vehicle-related costs, plus the motorist's own time); $Q*$ = optimal traffic volume; $Q_c$ = congested equilibrium.

used by many people. These resources are referred to as *common pool resources*. If these resources are managed jointly, this is *common property resource* (CPR) *management*, an important policy instrument (see Chapter 10).

## Congestion

Public goods are rarely "pure"; usually, some rivalry or congestion exists. For example, at low levels of use, a highway is a pure public good that gives utility to any user, irrespective of others' use. With increasing levels of use, congestion increases, and the utility derived by each motorist is a decreasing function of the number of other motorists; each additional car implies a negative externality on all other motorists. This effect is illustrated for traffic congestion in Figure 3-1, in which the social marginal cost $(MC_s)$, which includes the sum of the time costs caused by the delays that this motorist imposes on all other motorists as a result of increased congestion,[2] diverges more and more from the private marginal cost $(MC_p)$, which consists only of fuel and vehicle-related costs plus the motorist's own time. The first-best optimum use of a given highway would be at an optimal traffic volume $(Q*)$, which could be obtained by charging an extra road charge or tax[3] $(P*)$ (Figure 3-1).

Moving from *congested equilibrium* $(Q_c)$ to $Q*$ implies a collective gain in welfare $(W)$ by the motorists. The trouble with this solution is that the motorists have to pay the entire rent[4] $(R = P*Q*)$ to get $W$. Therefore, they may be collectively better off in the congested equilibrium $Q_c$. This is a dilemma for policy-makers, who should perhaps consider spending at least a part of the revenue on costs that the motorists perceive as benefits to their group. If all the external costs were congestion costs and if only motorists were affected, then perhaps a case could be made that the state should refund at least such a share of the taxes as to leave the motorists no worse off in $Q*$ than in $Q_c$. However, in the real world, where congestion-related externalities (e.g., exhaust) also affect nonmotorists, the issue becomes more complicated (for alternative views on earmarking road funds, see Chapter 21). Some participants in the debate consider that all road charges "should" be returned in the form of road investments. In several countries, including the United States, it is mandatory to use road funds in this way. This approach is not at all in accordance with a welfare theoretic analysis, but it still may carry political weight because motorists tend to be a fairly vociferous lobby.

Congestion depends on the relationship between flow and density. If congestion is severe, what is called *hypercongestion,* then the marginal damage curve "bends backward," and congestion costs increase. In some cases, motorists gain from a charge that reduces congestion and increases traffic flow. When congestion grinds traffic to a halt, almost any measure that increases traffic flow is better than nothing (see Part Four).

Many other resources with free access—such as fishing—could be characterized by and analyzed in terms of congestion.

## Asymmetric Information and Uncertainty

Information is a special good or service that is vital for any decentralized resource allocation mechanism such as a market economy. Simple economic models assume full and free information for functioning markets, but technical information is produced by hard work and usually is not free. Economists know that there are no free lunches. Yet, for a long time, economic models have assumed that information is free. Economists have come to realize that lack of this vital tool—particularly in conjunction with risk and uncertainty—has real effects on resource allocation and on welfare.

Information is difficult to protect with property rights, although patents partially fulfill this function.[5] It is difficult to exclude noncustomers from benefiting from information; therefore, information tends to be undersupplied in a pure market economy, a well-established motivation for public funding of research, dissemination, and teaching.

*Information asymmetry* (or "agency problems") is a situation in which some agents (e.g., employees, firms, or regulators) have different information. With risk, information asymmetry is an important underlying cause of different kinds of nonoptimality in the economy. It may be directly relevant to environmental issues (e.g., the risk of chemical accidents or oil spills) or indirectly relevant, because it causes distortions (e.g., in the factor or insurance market) that may interact with environmental externalities. Information asymmetry also is likely to turn up in the design and implementation of policy instruments themselves, because firms typically do not automatically disclose the whole truth to their regulators.

Most productive activities require collaboration between agents. Advantages to scale, specialization, and other factors such as level of risk aversion result in many people contributing to one final outcome. This introduces the problem of writing contracts to define the relative contributions of the agents and their remuneration. For example, consider poor farm laborers who work for a landlord. If the harvest (outcome) depends on weather (a random variable), then the remuneration of at least one party to the contract must be stochastic. If the laborers cannot tolerate large uncertainty because they live close to the poverty line, then they may be better off when the landlord (capital owner) bears the risk and pays the laborer (agent) a fixed but low wage. This potentially good arrangement gives security to the laborer and a higher long–run (expected) payoff to the more risk-neutral landlord. The main trouble is that the incentives for the laborer to maximize harvests are gone, and the landowner may have to spend large amounts of resources in monitoring (Cheung 1969; Stiglitz 1974). This situation is characteristic of many rural settings and is necessary to incorporate in analyses of pollution and natural resource degradation in poor countries (e.g., Binswanger and Rosenzweig 1986).

In the typical principal–agent problem, in which the interests of the principal (shareholder) and agent (manager) differ, the outcome for the principal depends on the effort of the agent as well as stochastic factors (i.e., random variables such as weather), so the principal cannot in any simple way isolate the effect of the largely unobservable effort by the agent. This information asymmetry after the contract is signed is called a *moral hazard.* (It also occurs when an agent who is insured against risks takes less care in protecting himself or herself because he or she does not bear the full costs of a mishap. In such a case, the insurance company must determine whether damage was incurred as a result of carelessness or random events of the kind that insurance is intended to cover.)

A related category of problems is referred to as *adverse selection,* a situation in which the information asymmetry concerns not the agent's behavior but the agent's characteristics when the contract is signed (see Chapter 13). If the agent has private information about his or her productivity, health, or other characteristics that are relevant to the outcome and that the principal cannot observe, then first-best contracts are unobtainable. For example, healthy individuals are least likely to purchase insurance policies. Therefore, policies priced based on average health statistics will be unprofitable for the insurance company if their customers are adversely selected (i.e., their propensity to illness is higher than the average), so the insurer charges a higher premium, which reinforces the self-selection of policy buyers. It tends to indicate that a healthy person cannot purchase reasonably priced insurance, so at least some healthy people will not insure themselves at all—an undesirable state and one that may affect their behavior in other areas of relevance to the environment. Still, a healthy person who is extremely risk-averse might purchase insurance anyway.

Adverse selection also can apply to the quality of labor, products, and services. If a buyer or employer cannot distinguish between the characteristics of products or employees, then market price and quality supplied will tend to be lower than optimal. The potential top-quality producers will not be fully rewarded if buyers cannot distinguish between low-quality and high-quality products, and either high quality will not be produced at all or significant resources will be spent on signaling (i.e., methods that are used to indicate quality, such as advertising, warranties, and educational titles). In the labor market, some economists argue that education is mainly a way of signaling relative performance rather than of preparing for actual productive tasks. The same applies for the provision of environmental information (see Chapter 10).

Some cases may allow partial solutions—for instance, based on using information from the correlation of harvests between adjacent plots, the history of agents to be insured, or signaling for quality. Sometimes the information asymmetry (or the cost of monitoring) is so severe that the principal would like to make the agent the owner of the residual uncertainty. This arrangement implies the use of a completely different form of contract, such as rental, in which a laborer (agent) rents land (capital) from a landowner (principal). The laborer reaps the rewards for any extra effort and thus has incentive to increase harvest (outcome). The landowner receives a steady stream of profits (although their expected value is lower than what could be obtained if the landowner were to assume the risks and could solve the problem of monitoring and could provide adequate incentives to the agents). In this situation, however, the laborer must bear the economic risks of variable factors such as adverse weather.

One solution in farming, business, and insurance is some form of risk sharing between the principal and the agent: crop sharing in agriculture, stock options for business executives, and deductibles in insurance contracts. Risk and reward can be shared in many ways, as illustrated for farming in Figure 3-2. Because of the costs of monitoring and differences in risk aversion, contracts commonly deviate from the simple neoclassical models in which wages equal marginal productivity. Similarly, rules such as equating the environmental tax to the sum of marginal damages may have to be modified (Laffont 1994b).

Although Figure 3-2 illustrates the relationships between a landowner and a farm laborer, they could equally well be between a capital owner and an entrepreneur or a regulator and a polluter. There are similar trade-offs: agents can reduce risk, but at a cost (payoff will be less than marginal productivity). Similarly, if principals assume risk, then they make more money but also must handle monitoring and incentives to agents.

In the case of oil spills from ships, full liability would be the counterpart to a pure land rental in Figure 3-2: the agent bears the full risk. However, this situation is unacceptable for two reasons. First, oil spills are emitted by multiple mobile sources, so to some extent, there is a nonpoint-source pollution problem, and it is difficult to prove that a particular ship is responsible. Second, the potential outcome (damages) is such a large economic cost that the agent may not be able to pay. Requiring full insurance would be the closest correlation to the pure wage contract in Figure 3-2, but such insurance is not available for reasons of moral hazard. If provided, it would remove all incentives for the ship to be careful. Thus, other combinations of instruments—regulation, partial insurance, and liability—typically are observed. For example, large individual ships or particularly sensitive bodies of water could be monitored, and techniques such as satellite surveillance and "fingerprinting" of oil loads with chemical or radioactive markers could be useful (e.g., Florens and Foucher 1999; Gottinger 1998). Multiple principal–agent relationships are involved: the environmental protection agency is

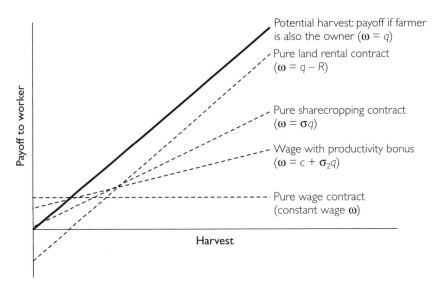

**Figure 3-2.** *Sharecropping and Other Contracts for Dealing with Risk and Uncertainty*

*Notes:* $\sigma$ = share parameter (or coefficient); $\omega$ = wage level. Solid line shows potential harvest ($q$) if the farmer works optimally; if the farmer is also the owner, then $q$ = payoff ($\omega$). Dotted lines signify payoffs to workers under four contractual schemes (e.g., a worker who pays rent [$R$] for the land receives a payoff, $\omega = q - R$).

a principal vis-à-vis its agents, the shipping companies; the companies are principals vis-à-vis their agents, the captains; and the captains are principals who must monitor their agents, the officers and sailors.

Potential problems of adverse selection and moral hazard—or inferior quality and carelessness—exist at each level. The market tends to undersupply insurance and product quality in general, producing goods of inferior quality ("lemons"). Buyers cannot separate out the lemons, and so their willingness to pay for the whole product group decreases—reinforcing the tendency for producers to supply lower quality (Akerlof 1970). Undersupply of quality is most likely to apply to environmental goods and services because of their public good characteristics. This situation in itself may be cause for policy intervention to protect safety and ensure the well-behaved function of markets (or industries, such as shipping).

Many of the problems of adverse selection and moral hazard are presumably exacerbated in developing countries, where high levels of uncertainty and greater risks of undersupplied quality and insurance exist. Low-quality product and labor markets are often the subject of casual empiricism in developing countries. This negative association can be partly remedied by increasing investment in signaling.

Adverse selection and moral hazard also have effects on insurance. The closer an individual is to the poverty line, the more detrimental variations can be, and the greater the need for insurance or other mechanisms for smoothing consumption over time (e.g., banking services). However, the administrative and informational costs of supplying these services are prohibitively high in poor countries, as are the risks associated with moral hazard and adverse selection. (Costs and risks are a factor of not only poverty but also phenomena such as mobility, morbidity, lack of public monitoring, and, in some cases, cultural heterogeneity (e.g., Narayan et al. 2000). Thus, insurance is undersupplied, which may have important environmental effects if the variation of natural systems can be controlled by technological

## Supplemental Reading

Arrow 1970
Johansson 1997b
Munk 1980
Myles 1995 (especially chapter 9)
Vickrey 1969
Walters 1961

innovations (e.g., irrigation, fertilizers, and pesticides) that imply health and ecological costs or risks. (Fluctuations may fill an important ecological role but still be quite detrimental to humans.) Consequently, environmentally hazardous technology can substitute for (unavailable) insurance, leading to the overuse of dangerous pesticides to reduce the uncertainty caused by pests, for instance. Similarly, large herds may be a nonmonetary "savings account" in response to the unavailability of conventional banking services. This response can be quite rational for individuals but severely detrimental for society (see Chapter 29).

## Notes

1. One idea behind the abstract concept of lump-sum taxes is that they are a fixed amount that does not vary with economic activity. Thus, they have no side effects on resource allocation and do not affect labor supply. Ordinary income taxes, however, imply quite severe distortions. The rule breaks down (or is at least modified) if the necessary public action cannot be financed through these ideal lump-sum taxes. A second important aspect of lump-sum taxes, in economic theory, is that if they were available, then any particular income distribution could be manipulated to give desirable welfare effects without having undesirable allocation effects. When lump-sum taxes do not exist, expenditures are inevitably financed through taxes that have distortionary and distributional effects, and thus these consequences of taxes have to be considered as an integral aspect of a program intended to provide public goods, for instance.

2. Not only does congestion cause delays and thus increased emissions per mile, but the rate of emissions itself is a function of speed and thereby of congestion (Johansson 1997a).

3. The optimization is now more complex. Utility ($U$) includes both private goods ($x$) and public goods ($G$) but also each individual's negative utility due to the externality ($g$) caused by others: $U(x, g_1, ..., g_i, ..., g_n, G)$. The Samuelson rule equates prices (transformation rates) to the sum of willingness to pay, including the sum of the (dis)utilities due to the externalities of congestion (Oakland 1972; also see Chapter 6).

4. The concept of *rent* as a payment for scarcity goes back to such classical economists as David Ricardo (1772–1823), who used "rent" to explain why land (when scarce) earns a form of income that is best measured by its marginal productivity.

5. However, the recent struggle between international drug companies and the government of South Africa over the right to import cheaper medicines for HIV (human immunodeficiency virus) shows that patents cannot always be followed. Whereas people understand that drug companies need to recuperate their research and development costs, in the face of such a massive tragedy as the spread of AIDS (acquired immune deficiency syndrome), the main concern is access to inexpensive medicines. It becomes necessary to circumvent these patents to supply enough people with whatever therapy is available.

# CHAPTER 4

# Adapting Models to Ecosystems: Ecology, Time, and Space

$E$NVIRONMENTAL AND NATURAL RESOURCES ECONOMICS applies economic theory to the often-complex world of ecosystems. In this chapter, I describe several sources of complication, starting with the biology of renewable resources such as fish and forests. The first models are for a single species, then the considerable complexities of real ecosystems are introduced. Next, I discuss dynamics—that is, development that takes time, which has both ecological and economic consequences. Finally, I introduce some of the complications that arise as a result of geographical space.

## A Simple Bioeconomic Model of a Fishery

One of the best-known bioeconomic models is that of a single-species fishery. This rudimentary model is useful primarily as a didactic and intellectual model; for empirical work it is less satisfactory because real fisheries are much more complex than the model reflects. Similar modeling may be used for other ecosystems that have some of the same general characteristics: left unmanaged, the ecosystem is only of limited use to humans; increased management can increase the yield or utility of the ecosystem; and too much management can destroy the ecosystem altogether.

Marrying ecological and economic modeling is difficult. In the simplest possible biological model, the net growth of a population is assumed to be a function of its size, which in itself gives exponential growth (i.e., more reproductive adults lead to more offspring). (*Net growth* of biomass is the aggregate growth of individuals plus recruitment of new individuals minus natural mortality. In this section, *growth* and *net growth* are used interchangeably.) However, it is also assumed that the ecosystem has a capacity to support a certain population size, and as the population approaches this carrying capacity, mechanisms such as hunger, disease, and predation tend to reduce growth, giving the logistic growth curve (see Box

## Box 4-1. Logistic Growth Curve

Growth in the stock size of a population can be modeled by the logistic growth curve. For small populations, growth is roughly exponential, because the number of offspring increases proportionally to the number of adults. However, some factor—space, food, energy, predation, or disease—always checks growth, and such factors are assumed to increase in importance as the population ($N$) approaches a carrying capacity ($K$). The result of these two tendencies is $(dN/dt)/N = g(K - N)/K$, where $dN/dt$ is the population growth rate and $g$ is the rate of growth when no restriction is imposed. When $K$ is infinite, ordinary growth is exponential ($dN/dt = gN$), but when $N$ is very close to $K$, growth is zero.

If it is assumed that fish can be sold at a constant price and that the net growth can be harvested without affecting the stock, then this growth equation can be interpreted as a curve showing the potential revenue for the resource (Figure 4-1). The growth in the fish stock is assumed to be the quantity that can be harvested without affecting the size of the stock. Therefore, every point of the curve in Figure 4-1 is an equilibrium point from the extinction equilibrium ($O*$) to the natural population equilibrium ($O$).

4-1). With appropriate assumptions, the logistic growth curve can be used to determine the potential revenue and costs of fishing for different long-run levels of fishing effort as well as stock size (as illustrated in Figure 4-1).

Increased effort generally gives higher harvest in the short run but reduces stocks (and harvest) in the long run. With no fishing effort, the natural equilibrium ($O$), yield is zero (otherwise it would not be a natural equilibrium) and stock (measured from the right origin) is at its maximum. At $O$, humanity does not receive any material benefits because no fish are caught (although there may be aesthetic, altruistic, existence, and other values). At the other end of the scale, a long-run maximum effort leads to stock depletion, or an extinction equilibrium ($O*$), and thus yield is again zero. The model contains many other equilibriums; in fact, each point along the curve is a fishing equilibrium. However, some of them (such as $O*$) may not be acceptable to future generations.

Noneconomists sometimes take for granted that the natural target of policymaking is maximum sustainable yield (MSY). The most important reason why MSY may not be a suitable target is that the costs of fishing typically increase with increasing scarcity (Figure 4-1).[1] Thus, a stock larger than at MSY would mean almost the same yield but at a considerably lower cost, giving a net profit equal to the difference between harvest and costs ($\pi*$); assuming a zero discount rate, this optimal policy is the maximum economic yield (MEY).

The logistic growth model shows that pure "hard-nosed" profit maximization can lead us to conclusions that ultimately are more, not less, conservationist than an intuitive model would be. It would be nice if the market worked this way; in fact, this is how the market would work if there were private rights to fish stock. In a private lake, the owner would manage the ecosystem according to this model if maximizing profits were the only goal. However, in most fishing grounds, the waters have no ownership and the fish are too mobile to make property rights practical. Therefore, there is no scarcity rent, and each individual fisherman continues to fish up to the point at which marginal costs of fishing are covered by average harvest, open-access equilibrium.[2] This is how open access leads to over-

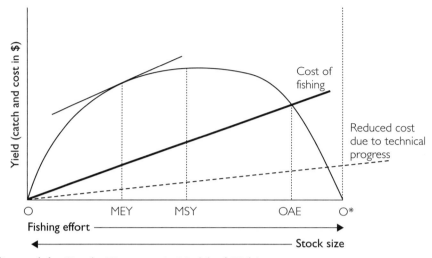

**Figure 4-1.** *Simple Bioeconomic Model of Fishing*

Horizontal axis shows fishing effort and stock size measured from the extinction equilibrium (*O\**). Vertical axis shows growth for each stock level; managing the fishery by harvesting the exact quantity that corresponds to this growth would be an *equilibrium strategy*. A simple cost of fishing function is also shown.

*Notes:* MEY = maximum economic yield; MSY = maximum sustainable yield; OAE = open-access equilibrium.

fishing, at which point, there is no rent at all (total costs are equal to total revenue). The rents are completely dissipated, which translates to a loss of economic potential to the economy.

It is not reasonable to call the simple logistic growth curve for one species a biological—or an "ecological"—model, which, by definition, would have to account for interactions between species. However, combining it with an economic model helps us understand what fertile intellectual ground this kind of cross-breeding between disciplines can create. This combination was the foundation of the first fishery economics models used by Gordon (1954) and Schaefer (1954) and thus is often referred to as the Gordon–Schaefer model. Modifications to this model abound. The dynamics are complicated (discussed in Scott 1955), but later applications of optimal control theory have greatly enhanced the analysis (Cohen 1987). For instance, the growth rate of the stock can be made to vary nonlinearly with stock size, as in $(dN/dt)/N = g(K - N^\alpha)/K$. The growth rate also can be modeled in a more complex way, with discrete time, with harvest and growth in different seasons, and possibly with different age or size groups or cohorts. Some values of the parameter $\alpha$ will imply that overfishing easily leads to extinction (if fish stocks fall below a critical threshold).

For some fish species, the relationship between recruitment (and thus growth) and stock may be much weaker, and in some models, recruitment (of "young" fish) is even modeled as exogenous or constant. For some species, the larvae float considerable distances in the ocean; thus, at least locally, recruitment is exogenous. Yet, some authors point to the strongly increasing relationship between egg production and size in some species (e.g., a 12.5-kilogram red snapper female can have more than 200 times as many eggs as a 1.1-kilogram red snapper [Pauly 1997]). Therefore, recruitment might diminish drastically with fishing stock and be an argument for marine reserves where large fish would find a haven.

Similar reasoning underlies the metapopulation models (e.g., Brown and Roughgarden 1997), which explicitly model various adult populations of a species that are connected by a common larvae pool, such as lobster. The larvae pool may be pelagic (drift in the open sea) and thus spatially separated from the various subpopulations of adults in different sites that contribute larvae to and receive new recruits from it. In metapopulation models, economists typically see strongly increasing returns to scale, and the optimal policy turns out to be strict specialization between sites that are suitable for harvesting (low costs because of distance to market and so forth) and sites that are best suited for reproduction. Taking the model literally, optimal harvest would entail harvesting all the adults in all sites except the one that is best suited to producing larvae. The conclusion for local fishing policies is thus different from the conclusions in the Gordon–Schaefer model: by harvesting all the adults, more harvest may be available in future periods because if there are excess larvae, then space, food, and suitable habitat are the only restricting parameters.

For demersal fish that live at the bottom of the ocean (e.g., flounder and haddock), metapopulation models indicate that recruitment is constant for a wide range of stock levels.

Fishing effort and the cost of fishing function may be quite complex. Some fish species swim together in large schools, so fishing costs do not rise as a simple function of increasing scarcity in stock. Much work has been done in the development of year–class models, which examine the effect of varying technology such as mesh size on the catches of different species and the age composition of the catch (which has effects on future growth and recruitment).

Recent research has grappled with the integration of some of the complexities of real ecosystems. A natural extension is to consider several species of fish (some of which compete for food) and predation. The Lotka–Volterra biological model studies interactions such as predation by specifying dynamic equations for population: growth in the population of one species depends on its own numbers as well as the number of individuals in another population (Lotka 1932).[3]

Even strong believers in the market recognize that the fishing industry is one area in which policymaking is sorely needed. Although politicians have been active, the policies adopted often are the exact opposite of those needed. A prime example is subsidies to help fishermen purchase more equipment (e.g., boats, nets, and technology) when catches decline. The trouble is that more-efficient equipment speeds up stock depletion and thus adds policy failure to market failure. The design of optimal policy is far from simple. Each fisherman creates externalities for other fishermen. Decreasing stocks increase the costs of fishing and lower future catches.

Clearly, fishing could be reduced through taxation; the understanding that fish are capital and should have a scarcity premium in situ also provides strong rationale for a tax. However, as illustrated in the congestion model (Chapter 3), implementing a tax leaves the immediate "beneficiaries" of the program worse off than without any policy. Tax solutions are not acceptable to fishermen because they take all the rent they help capture. Presumably for this reason, individual transferable fishing quotas (ITQs) have become the main instrument of fishing policy. ITQs provide a scarcity signal but still leave the rent with the fishermen.

Policies must be designed with regard to technical progress (e.g., better ships, nets, logistics, and search equipment such as sonar). Such progress lowers the marginal cost of fishing (Figure 4-1). It implies greater potential profits at the optimum but it also brings the open-access equilibrium dangerously close to the extinction equilibrium and can lead to a negative attitude toward new technology; in fact, policies often have been designed to stop various technologies from being used. From a purely efficiency-oriented viewpoint, the optimal policy would be to use the new technology but reduce effort to a point corresponding to MEY. Common property resource management and ITQs, which appear to be the most promising instruments for providing appropriate trade-offs among efficiency, monitoring costs, and social acceptability in many cases, are discussed in Chapter 28.

## Bioeconomics and the Management of Ecosystems[4]

To capture the dynamics of two species is difficult enough, but true ecosystems often contain dozens of species, even if only the most vital ones at each trophic level are counted. Ecosystems are highly nonlinear complex adaptive systems (Levin 1998, 1999) that have many features bound to create market failures. This complexity leads to the existence of multiple domains of attraction (a counterpart to equilibriums in dynamic and chaotic systems) and to the possibility of sudden threshold changes in dynamics. For instance, long periods of drought in tropical drylands can lead to soil erosion followed by loss of tree cover and eventual desertification. Dry rangelands can alternate between two states of nature: with and without trees (Perrings and Walker 1997; Perrings and Stern 2000). The question is whether such changes are reversible within the time perspective of human decisionmakers.

Ecologists have shown that the resilience of ecosystems is an important factor in preventing irreversible change. Resilience depends on such things as the ecological memory of a system—its seed banks, migratory animals, and "patchiness"(i.e., the existence of heterogeneity in local conditions that may allow species to survive and recolonize a territory when conditions improve after a period of disturbance). The diversity of species at various levels, including the genetic level, is another mechanism that may enhance resilience in a way that is somewhat analogous to the way in which diversification reduces risk in a stock portfolio.

At the same time, the ecosystem is a source (from the human viewpoint) of joint production, where some of the products have characteristics of public goods or common pool resources. All these features make management difficult and argue in favor of adaptive management that relies heavily on the precautionary principle. Table 4-1 lists some of the ecosystem services that might be provided by a typical mangrove forest, common in coastal areas in tropical countries.

Table 4-1. *Ecosystem Services from a Mangrove Forest*

| Use no. | Use value(s) |
|---|---|
| 1 | Habitat, nutrients, and breeding of shrimp, crustaceans, and mollusks |
| 2 | Spawning and breeding of demersal fish species at the bottom of the ocean |
| 3 | Spawning and breeding of pelagic species of fish in the open sea |
| 4 | Provision of fuel wood, construction wood, and so forth |
| 5 | Source of plants, herbs, and small game used locally |
| 6 | Production of salt and shrimp |
| 7 | Control of storms, floods, and erosion—extension of current land into the sea |
| 8 | Sequestration of carbon, nitrogen, phosphorus, and other nutrients |
| 9 | Protection of other marine and coastal ecosystems (e.g., coral reefs) |
| 10 | Sequestration of toxins |
| 11 | Protection of habitat for birds—amenity value of bird watching (i.e., tourism potential) |
| 12 | Option values related to future uses such as biotechnology and genetics |
| 13 | Existence (and bequest) value of ecosystem |
| 14 | Other exploitation |

*Direct-use values* accrue as a result of direct consumption. *Indirect-use values* might accrue for ecosystem services that indirectly give consumption goods (e.g., rain). *Nonuse values* include intangible goods, such as scenic beauty. In Table 4-1, Uses 1–14 correspond roughly to direct-use values, indirect-use values, then nonuse values and one "alternative" use (discussed below). The sheer number of outputs might at first appear to make ecosystem management difficult; however, an ecosystem can be compared to a firm that produces many products. From a management viewpoint, quantifying the ecosystem characteristics and identifying the different beneficiaries (some of whom cannot be expected to pay for the services provided) for the many products is more difficult. The list therefore is also intended to differentiate between uses that are relevant for different categories of user, some of whom are poor.

Uses 1–3 (breeding) are related to fisheries in the broadest possible sense of the word, and the biology and economics of optimal yield could be analyzed using the models discussed in the previous section. However, modeling is difficult for biological and economic reasons. The marine ecology is a multispecies system with complex dynamics that are poorly understood; the economics are complicated by the social structures of the societies that benefit from the system's resources. Typically, Uses 1 and 2 benefit many members of local communities by providing food, employment, and income. Use 3 typically benefits fishermen using large modern boats who originate in distant cities, even foreign countries.

Most marine and coastal bioresources are not owned by anyone and therefore are subject to overexploitation (through open access) and thus to rent dissipation as resources become scarce. With judicious ecosystem management, rents could be restored (making the aggregate of users better off), but the level of coordination required to restrict overexploitation is difficult to achieve, even under ideal circumstances. Population growth and social mobility (sometimes coupled with famine, natural disasters, and war) can lead to large population influxes to coastal areas, with increasingly heterogeneous groups. Analysts of common property emphasize how exclusive control of a common pool resource and the continuous, coherent, and evolutionary social structure of groups dependent on a common pool resource are important attributes for the success of such populations (see Chapter 10). The fact that several distinct groups enjoy or use different resources provided jointly by the same ecosystem adds a layer of complexity.

The chance of these user groups interacting is small. The Japanese fishing vessels that trawl for fish in the open seas of the Gulf of Thailand benefit from the "spawning services" of the mangrove forest but probably would not interact with the local communities to help them protect the mangroves. In fact, anecdotal evidence indicates that many Japanese fishermen invest in shrimp farming, an industry that creates large ponds by clearing mangroves and thereby eliminates the spawning grounds on which their open-sea fishing depends.

Uses 4 and 5 (provision of fuel wood and other plant products) are typically straightforward goods. Some of these goods may be difficult to value because they are used only locally and lack market prices. Data must be collected on quantities and other use characteristics. If the products are used by poor local people and are essential proteins, medicines, or products similarly essential to human life, their loss may be considered a greater loss of welfare than would be suggested by a simple valuation of close commercial substitutes. Another complicating factor is that these goods could be collected by social groups other than those who use the aquatic resources. Use 6 (salt and shrimp production) may be categorized with Uses 4 and 5 if done by small-scale operations.

Uses 7–9 (protection against storms and erosion, sequestration of some nutrients, and the associated protection of other marine ecosystems) are ecosystem services par excellence because they have true public good or common pool resource characteristics that benefit all local inhabitants. The users may or may not be the same as those of Uses 1–6. In some areas (such as off the coast of Kenya and Tanzania), the mangroves along river estuaries and coasts may protect offshore coral islands with a separate fishing population, again raising difficult issues of collaboration. The trapping of nutrients may have external benefits, such as protecting coral reefs (which are sensitive to large amounts of nutrients). Compensation in this case might seem to be a remote possibility; the fishermen on the coral reefs cannot easily compensate or interact with those who cut down the mangroves. However, the global community is starting to create mechanisms to compensate for additional positive global externalities associated with local projects. Two examples are the Global Environment Facility (GEF) and the Carbon Fund (see Chapter 17).

Use 10 (sequestering toxins to prevent damage farther downstream) is applicable only to mangrove forests near industries or subject to pesticide runoff from agriculture or effluents from cities. This use precludes many of the other uses, but it may be a cost-effective alternative to sewage treatment. For Uses 9 and 11 (protection of the coastal areas and amenity values for tourists), local decisionmakers are less concerned with consumer surplus than with revenue generation. However, income from tourism generally benefits players other than the local fishermen.

Uses 12 and 13 are nonuse values: option value (Use 12) relates to the value of keeping an "option" open (e.g., to use a resource in a new way in the future), whereas existence value (Use 13) is an intrinsic value that one might experience from just knowing that something exists, even if one does not expect to see it (e.g., a whale). Nonuse values can be local or global and are generally difficult to capture in practice, even though there are some exceptional contracts between biomedical companies and local communities to conduct joint research to look for species that can help produce new drugs and so forth. With this kind of contract, these benefits can be judged in the same commercial way as tourist income.

Some uses can be combined, whereas others are exclusive. A debate or policy issue often is triggered by a proposal from an entrepreneur who is interested in large-scale industry, a port, a refinery, salt production, aquaculture (shrimp production), or conversion to agriculture. Such projects appear to promise large revenues and massive job opportunities. Examples abound, from the Rufiji Delta in Tanzania, where the African Fish Company was initially given a lease of 10,000 hectares but later stopped in response to protests against this large-scale destruction of the mangroves, to Florida, where the expansion of Miami, Orlando, and other cities has pushed the Everglades to the verge of collapse through the demand for land, water, and flood control. In the Florida Everglades, 90% of the wading bird population has become extinct, and $8 billion will be spent over the next 20 years to restore natural water flow and thereby vegetation and ecosystem functions.

Considerable uncertainty surrounds ecosystem function and the valuations of the many services that ecosystems provide. Because there are no market values to observe, it is crucial to model the systems to obtain some good estimates. Necessity dictates the use of drastic simplifications, but the appropriate simplifications must be chosen for the type of problem at hand. For a mangrove swamp, the variables might be the area covered ($A$), the biomass ($M$), fish stock ($N$), and nutrients in the soil ($S$). By denoting the harvest of fish and wood ($H_N$ and $H_M$, respectively), the growth rates of fish and wood can be modeled by using equations that are reminiscent of those in the Gordon–Schaefer model. Denoting unimpeded growth rates ($g$) and the carrying capacity ($K$) with appropriate subscripts, the equation for fish is

$$\frac{dN}{dt} = g_N(A,S)N[K_N(A,S) - N] - H_N \tag{4-1}$$

The equation for wood is analogous.

Equation 4-1 does not account for species, age composition, and many details of recruitment and mortality, but it may be a reasonable approximation for focusing on the dynamics of the interaction among forest area, fish stock, and nutrients. For other types of ecosystems, it may be reasonable to concentrate on the variables whose dynamics are applicable in a shorter time span and abstract the dynamics of "slow" variables such as forest area and biomass (see Levin and Pacala 1997 for theories of model simplification). Still, estimating dynamic equations

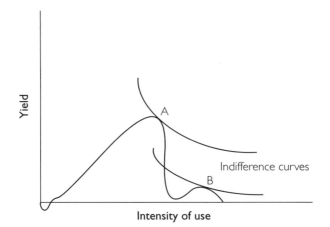

Figure 4-2. *Managing an Ecosystem with Threshold Effects*

such as Equation 4-1 is a daunting task. Waiting for time series data is hardly attractive, and using cross-sections may be unreliable because ecosystems are so idiosyncratic; the relationships governing one swamp may be different from those in a superficially similar one (Cole, Lovett, and Findlay 1991).

Still, thresholds and surprises may be extremely hard to foresee in modeling an ecosystem. Another feature of ecosystems that has not received enough attention is that they typically involve strong thresholds and may "flip" quickly from one state to another or exhibit strong irreversibilities. This feature is illustrated in Figure 4-2. The curve has a shape similar to the one for single-species fisheries in Figure 4-1, except that one section with intense use is extremely steep (or nonexistent), reflecting a threshold or irreversibility. Eutrophication in a shallow lake may have this characteristic. A small inflow of phosphorus increases algae growth, reducing light and oxygen at the lake bottom. If plants on the bottom are killed, then the sediment may be destabilized, leading to release of much more phosphorus in an almost irreversible spiral. In such cases, ordinary optimization (as represented by adaptation to the utility curve shown) can easily drive the system from a desirable state (point A) to an undesirable suboptimum state (point B), from which recovery can be extremely expensive. With true irreversibility, recovery might be impossible, but truly irreversible cases (luckily) appear to be relatively rare.

Oversimplification does have its pitfalls and is the basis for the so-called precautionary principle, which suggests that if the optimum is not known, then one should err on the side of caution. This suggestion sounds straightforward but can be too stifling for development. A reasonable strategy may be a form of careful adaptive management. By allowing small changes and continuously monitoring the ecosystem, not only are the interests of various stakeholders catered to but also valuable information is gained for modifying the management strategy itself. Thus, if shrimp farming is being proposed, it would seem wise to start with a small-scale activity rather a large-scale project like the Rufiji Delta.

A diversity of services are produced jointly by the same ecosystem but used by different users. The ecosystem is complex, dynamic, and often poorly understood. The definition of property rights is incomplete; sometimes, conflict arises from claims between local communities (which exercise traditional and culturally based rights) and outside society (represented by new immigrants, new classes of entrepreneur, the state, or foreign users). The high costs of enforcement and the lack of clear property rights may create situations of almost open access at the local level. Even if local users can collaborate, it is difficult for them to handle external effects that have repercussions on other groups. The agents using the various resources or ecosystem functions have little experience in negotiating with each other. Compounding the problem for the central policymaker are tricky issues of market structure, uncertainty, aggregation, and discounting.

The problems may sound insurmountable, yet they must be resolved. About two-thirds of the world's current population lives in the coastal areas; this proportion is estimated to increase to three-quarters (more than 6 billion) by 2025 (Lindén and Granlund 1998). In addition, fish is the main source of protein for the majority of poor people. The first step in analyzing these problems must be to isolate the most crucial conflicts of interest. When overfishing is the main problem and fishing effort needs to be restricted, severe conflicts may arise concerning who is to be allowed to continue fishing. Also, defining fishing rights or quotas may be necessary policy; when the most crucial conflict of interest is between local fishermen and the expansion of aquaculture or industry, types of regulation such as pollution control or zoning may be more appropriate. Most experience to date suggests that integrated coastal zone management (ICZM) is a promising policy instrument for this type of situation. ICZM is a constant process of learning at the scientific and institutional levels that entails bringing together all concerned stakeholders, carefully monitoring the local fisheries, conducting research, promoting local learning and capacity building, strengthening local institutions, providing feedback to the stakeholders, and, ultimately, designing new improved programs. Local management, incentive compatibility, involvement, property rights, and public information are key elements of ICZM.

## Management in an Intertemporal Setting

The temporal dimension discussed here concerns the long run versus the short run. However, other temporal dimensions exist, such as cyclical variations over time, rather than totals. For example, the distribution of rainfall and hours of sunshine across days and seasons affects agricultural production and human well-being more than aggregate totals do. Similarly, noise may be worse at night than during the day, and certain air emissions may be worse during thermal inversion or when the weather is prone to smog formation than at other times. These examples suggest that values and policies may need temporal adjustment.

Many problems related to natural resources or the environment involve stocks as well as flows; thus, the time dimension is an essential component. The archetypal economic solution to resource scarcity is to increase prices, which, in principle, can solve the scarcity problem as long as substitutability in production or consumption is sufficient. Economic treatment of scarcity and the allocation of scarce resources over time owes much to Hotelling (1931). The essence of Hotelling's argument is that scarcity is a source of value. With increasing scarcity of a good, the value of that good must rise at the same rate as for other financial instruments; otherwise, owners would not hold the good but sell, thus lowering the price. This argument is sometimes referred to as the arbitrage rule (Dasgupta and Heal 1979).

If it is assumed that the resource is finite, demand and extraction costs are constant, futures markets are perfect, and owning the resource presents no other benefits, then it follows that a scarcity rent must rise exponentially at the rate of interest of the economy. Consequently, the market price of the resource increases, and rates of extraction gradually decline so that the resource is typically not depleted in finite time (see Box 4-2 and Figure 4-3). The resource typically will go to zero, although it does so in an "optimal" fashion. For instance, the annual extraction may be a fixed percentage of the remaining stock, so the reserve-to-use ratio is constant, even though the physical quantity diminishes.

Numerous other factors affect resource prices, and other models emphasize changes in demand, technology, costs, and so forth. In such models, the cost may rise due to increasing depletion, and scarcity rents may be constant or follow another pattern. In complex models, the cost of production may depend for instance on the rate of production, the level of depletion, and the development of technology. Still, the basic message remains similar to that provided by the simple Hotelling model and has been extremely influential. For instance, the forecasters of world energy prices appear to believe strongly in it: forecasts of future oil prices are typically equal to current prices plus a couple percent per year until the date of the forecast (Manne and Schrattenholzer 1992). However, in real mineral, fuel, and other resource markets, scarcity rents and prices do not increase smoothly. This discrepancy does not disprove the Hotelling model, because it can be explained by changes in extraction costs, technological progress market imperfections, and short-run disturbances such as political upheaval. Furthermore, there is very little scarcity (yet); or, to be more precise, the resources are so large relative to use that the scarcity rent component is smaller than other factors.

For renewable resources, the interaction between harvest and biological rates of reproduction or growth affect the results of economic optimization over time considerably. Still, the effect of introducing a discount rate is analogous to that in the Hotelling model for a finite stock resource. For a simple bioeconomic model of fishing, for instance, the optimum typically would not be the same point as maximum economic yield (MEY) in Figure 4-1. Instead, the optimum would be moved to the right, to a somewhat higher level of fishing effort and lower stock size. This shift would be accompanied by a short-term increase in catch that would lower returns in the future; the proceeds of this drawing down the natural stock are invested to yield a higher rate of return elsewhere in the economy. In this model, the result may be a steady state (with constant catch each year), even though resource rents rise at the rate of discount.

Policymaking for forests must consider that there can be several *rotations* (i.e., cycles of planting, growth, and harvest) on the same acreage. This fact leads to

## Box 4-2. A Simple Derivation of the Hotelling Mechanism

Consider the intertemporal allocation of a finite source ($Q$) of a mineral—say, oil—to be allocated, and assume that this oil can be produced (pumped, refined, and so forth) at a constant marginal cost ($c$). Assume a yearly demand ($P_t$) given by the inverse equation $P_t = a - bq_t$ (where $q_t$ is yearly production of oil), which implies that total annual benefits $B_t$ will be

$$B_t = \int_0^{q_t} (a - bq)dq = aq_t - \tfrac{1}{2}bq_t^2 \qquad (4\text{-}2)$$

Maximizing discounted benefits subject to the resource constraint gives

$$\max L = \sum_t \frac{\left(aq_t - \dfrac{bq_t^2}{2} - cq_t\right)}{(1+r)^{t-1}} - \lambda\left(Q - \sum_t q_t\right) \qquad (4\text{-}3)$$

where $r$ is the discount rate and $\lambda$ is the scarcity value for the resource. Taking partial derivatives gives

$$\frac{\partial L}{\partial q_t} = \frac{a - bq_t - c}{(1+r)^{t-1}} - \lambda = 0 \qquad (4\text{-}4)$$

which, after substitution, gives the Hotelling rule—that is, that the scarcity rent, ($P_t - c$) or ($a - bq_t - c$), must grow at the following rate:

$$(P_t - c) = a - bq_t - c = \lambda(1 + r)^{t-1} \qquad (4\text{-}5)$$

The oil needs to be rationed, and thus the price of oil should equal the production cost plus a scarcity rent, which, in simple cases, grows exponentially at a rate equal to the discount rate. In reality, there are many confounding factors that blur the picture.

*Note:* An equivalent continuous time formulation using a current-value Hamiltonian formulation ($H$) for Equation 4-2 is $H = (aq_t - bq_t^2/2 - cq_t) - \mu_t q_t$, where $\mu_t$ is the current value multiplier for the flow constraint that relates extraction to the resource stock. In an internal solution (where some but not all of the resource is used), the conditions for optimal control can be used to give the marginal current value of the resource ($\mu_t = a - bq_t - c$). The costate equation for $\mu_t$ can be used to show that $\mu_t$ must grow at the rate of interest. (For a pedagogical comparison of optimal conditions for Lagrangean and Hamiltonian formulations, see Fisher 1981, especially table 2.1).

somewhat different models that are based on the Faustman equation (Faustman 1849). By including the expected discounted value of future forest rotations, Faustman correctly showed that forest rotations would be shorter with than without discounting (i.e., advancing the harvest not only brings in cash faster from current timber but also shortens the time period by which future harvest will have to be discounted). Interestingly, this point was overlooked by well-known economists of the late 1800s through the mid-1900s, including W. Stanley Jevons, J.B. Clark, and Irving Fisher, who perhaps did not know much forestry or forestry literature. The intertemporal management of complicated systems such as natural forests, which provide ecosystem and recreational services as well as timber, adds yet another layer of complexity to the task of modeling.

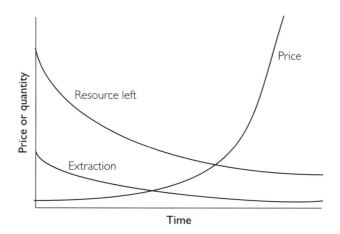

**Figure 4-3.** *Sustainable Resource Depletion According to Hotelling*

Although many applications in this area concern natural resources, the theory applies equally to stock pollutants. Many chemicals are persistent and accumulate in the environment through various mechanisms. The capacity of an ecosystem to function despite an increasing stock of a certain pollutant may be limited, and all the economic models of how to "harvest" or "mine" a natural resource stock are equally applicable to the use of the assimilative capacity of the ecosystem with respect to stock pollutants.

Historically, fossil fuels have been analyzed as a resource, but the reaction involves other resources besides fuel. Oxygen ($O_2$) is a resource, even though it is abundant. When fossil fuel is burned, oxygen from the atmosphere is replaced with carbon dioxide ($CO_2$) and water ($H_2O$) according to the (approximate) chemical formula $HC + O_2 \rightarrow CO_2 + H_2O$, where HC is hydrocarbon. In a more general sense, the assimilative capacity of the atmosphere for $CO_2$ can be considered a resource. In fact, it is a good example of the subtle ecosystem resources that are now becoming scarcer. With current trends, the carbon content of the atmosphere will soon reach concentrations that may lead to drastic changes in climate.

Furthermore, fossil fuel combustion causes the emissions of large amounts of metals, volatile organic compounds, methane, sulfates, and nitrogen oxides. Again, the assimilative capacity of the environment for these various emissions can be seen as a resource. Just because HC is not scarce does not necessarily mean that the other parts of this equation do not involve scarcities. The vital aspect is that although the sources of fossil fuel are privately owned, the ecosystem services are not. Instead, they are characterized as public goods or common property resources. The fact that the atmosphere has no ownership implies that disposal is free and that there is no scarcity rent for its use. The same applies to the biosphere's capacity to assimilate mercury, radon, and many other elements released through the processing of fossil fuels. Fossil fuel combustion is the largest source of many toxic metals and other elements from the geosphere that find their way into the biosphere (Azar, Holmberg, and Lindgren 1996).

## Spatial Heterogeneity and Land Use[5]

One essential aspect of ecosystems is spatial heterogeneity or variation. To build good models for ecological economics, economists must account for spatial dimensions—something they are somewhat unaccustomed to doing. Metapopulation models for a marine ecosystem (in which groups of adults and larvae are spatially separated) illustrate how spatial dimensions lead to conclusions different from those produced by the simple Gordon–Schaefer fishery models. Particularly in the areas of biodiversity preservation, parks management, and the protection of genetic diversity, it is crucial to consider the patchiness of nature.

A model of spatial heterogeneity can be used to analyze land use and deforestation. To understand changes in land use in general and deforestation in particular, and to suggest policy instruments, it is necessary to first understand what a *forest* is. It is not just a collection of trees but a whole system that includes the soil and its flow of water and nutrients to animals (including insects and microorganisms), which typically play important ecological roles in forest development. In this section, I distinguish *natural forests,* defined by high biodiversity and low human density and interference, from *plantation forests,* which are more or less simple monocultures. Much confusion can be avoided by carefully distinguishing these two categories.

Consider a stylized pattern of forest development. The first phase of a new settlement involves rural development and deforestation. In later stages, forest products become increasingly scarce, and prices increase. Eventually, forest investment could limit or even reverse deforestation. This pattern follows the concepts of economic geography first proposed by von Thünen in 1826 (see Samuelson 1983 for a review). The basic elements of this pattern are illustrated in Figure 4-4. Similar analyses can be applied to the allocation between more-intensive and less-intensive land uses, such as ranching, agriculture, and even fishing versus aquaculture.

The assumptions made concerning the determinants of agricultural development are necessarily simplified for the scope of this book. Agriculture is a particularly risky and stochastic enterprise, and expensive monitoring creates special problems for the labor and insurance markets (discussed in Chapters 3, 13, and 29). Nevertheless, Figure 4-4 illustrates a simple linear landscape of agriculture and forests. The origin (O) identifies the location of a new settlement and market. Land extends out into virgin forest at some distant point Z. The value of land around the settlement (A–A) is a function of the net agricultural prices at the farm gate—which are greatest when the farm is located near the market, partly because of transport costs and partly because the settlement is located in an area where the land is most productive. The value of agricultural land declines with increased distance to the market until it eventually reaches zero at Point A.

A natural cover of mature forest occupies the landscape at the time of initial settlement. The forest closest to settlement may have a negative value because it interferes with agriculture. Settlers remove the trees wherever the agricultural value of converted forest plus the value of harvesting the trees (e.g., for fuel or construction) exceeds removal costs.[6] They may leave some trees standing and plant around them. Forest value is depicted as a series of curves F–F that shift outward over time. Initially, the values for forest land are much lower than for

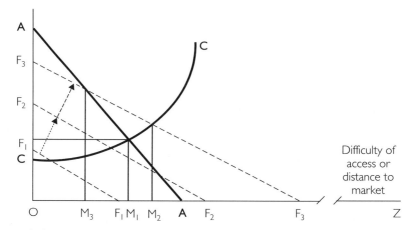

**Figure 4-4.** *Economic Determination of the Margins of Forestry vis-à-vis Agriculture in Different Stages of Development*

*Notes:* AA = value of agricultural land; FF (with subscripts) = a series of curves that depict the increasing value of forestland; CC = cost of protecting private property rights; $M_1$, $M_2$, and $M_3$ = margins of agriculture or forestry.

agricultural land (because food is more essential to survival than trees are) and also somewhat "flatter," because transport costs are assumed to be less important (because timber is less perishable than food).

Landowners must pay (C–C) to claim and defend property rights. These costs are assumed to increase as the level of public presence and thus effective control declines farther from the market center. (The cost of protection may vary depending on whether the land is used for forestry or agriculture; this aspect is simplified for the purposes of this presentation.) In the first phase, the extent of agricultural land is decided by the intersection of the value of agricultural land and the cost of property rights. Thus, farming will extend from O to $M_1$. Farther out (from $M_1$ to Z), the net value of the forest timber is assumed to be negative, and the natural forest will be managed as an open-access resource to be exploited for its native fruits and nuts, for forage and fodder.

Notice that in this first phase, deforestation is caused by the demand for agricultural land. The market price of forest products must cover the costs of production. However, growing the forest entails no costs, because the forest is already mature. The only costs of forest production are harvest and delivery costs. In this kind of forest, such policy instruments as forest taxes or subsidies (such as free seedlings) will have little or no effect because deforestation is caused by the need for agricultural land and not by the profitability of forestry.

As incomes rise, technology and tastes change, and infrastructural improvements reduce transport and other costs, the forest value function may rise to $F_2$–$F_2$. This increase would imply deforestation from $M_1$ to $F_2$. In this second phase, the agricultural frontier would be at $M_1$ and a degraded forest area from $M_1$ to $F_2$. This area is still an open-access natural forest, with no costs for management or property rights, because values are lower than the cost of protecting property rights. This area would be logged at least once and then perhaps abandoned unless it were profitable for agriculture. Deforestation in this situation would be explained by timber prices and other parameters relevant to forestry. Agricultural land conversion would no longer be a source of deforestation, because farming would take place primarily in the already deforested area, $M_1$–$F_2$. In this case, forestry policy instruments that affect the cost of harvesting would limit deforestation, but agricultural policies and free seedlings (which only affect plantation costs) would not.

It is possible that wages would eventually rise, making agriculture relatively less profitable than forestry. However, technological progress counteracts the effect of rising wages, and experience in the United States and Scandinavia indicates that rising wages favor forestry in the long run because it requires less labor. As timber scarcity increases, the frontier of economic forest activity may extend to $F_3$–$F_3$. Forestry would become more profitable than agriculture, even on some prime land that was formerly farmed.[7] Agricultural land would be limited to O–$M_3$, whereas new managed plantation forests would cover $M_3$–$M_2$. The degraded open-access region would extend from $M_2$ to $F_3$, and timber would have two distinct sources: the plantation forests, protected by secure property rights, and the natural forest at the frontier. (If property rights are secure, then according to the Faustman formula, forestry management on plantations will be optimized over time [Johansson and Löfgren 1985].) For the usual reasons, the marginal cost of growing, harvesting, and delivering products from the managed forest is expected to be equalized to the marginal cost of harvesting and delivering from the more distant natural forest. Different agricultural and silvicultural policies would have different effects on the various margins or regions (e.g., the provision of free seedlings might increase forest growth in a plantation forest but not in a natural forest).

Perhaps the most interesting characteristic of forests for policy analysis is their economic margins. The first two stages of forest development have only one economic margin (at points $M_1$ for the first stage and $F_2$ for the second), whereas the third stage has the truly unusual feature of three margins: intensive and extensive managed forest margins at $M_3$ and $M_2$, respectively, and the margin of harvests from the mature natural forest at $F_3$. The margins of the third stage in particular create some unusual policy implications. (The importance of these issues for forest policy is discussed in Chapter 30.)

Many countries have regions that can be described by one of the three possible stages of forest development discussed in this chapter. However, some might require different models, because forests and social settings are heterogeneous. Some forests have valuable tree species for which the timber value is far greater than any agricultural value the land might have. In such instances, deforestation might be of primary importance, and agriculture or ranching would be a secondary activity. Some countries have open access to resources, whereas others enforce private or state property rights to some extent. Some have large populations that hunger for productive land, whereas others have small populations and less scarcity of land.

Most forest products are either bulky or perishable and cannot be transported cheaply. Therefore, their primary markets are often geographically contained, and primary processing facilities are located near the source. Large regions of mature natural forest still exist in many parts of the world, but other parts of the world support managed forests. In some areas, such as Scandinavia and the northern United States, managed forests have been growing consistently for several decades, implying—to the surprise of some—that this (spontaneous) reforestation is a significant source of carbon sequestration in some of the most industrialized countries. However, even the southern United States (often thought to

## Supplemental Reading

Binswanger and Deininger 1997
Clark 1990
Conrad 1999
Eggert 1998
Hannesson 1998
Munro and Scott 1985
Scott 1983
Stiglitz 1999

be the most dynamic region of the world for commercial timber production) still harvests more than half of its timber from unmanaged stands.

## Notes

1. Other reasons why MSY may not be a suitable policymaking target are uncertainty and discounting.

2. *Scarcity rent* is the rent that accrues to the owner of a natural resource just because it is scarce. Throughout most of history, land has been scarce and thus has had value. Scarcity did not used to be a problem for ecosystem services such as clean air and water; however, as scarcity becomes more widespread, it must be internalized.

3. Some models analyze the difference between harvest rates and harvest quotas in multi-species models (e.g., Conrad and Adu-Asamoah 1986; Flaaten 1988; Azar, Holmberg, and Lindgren 1995). The "dove–hawk" model, also relevant, uses game theory to study the evolutionary effects of having two different strategies for two groups within a population. The doves are peaceful feeders that avoid conflicts, whereas the hawks always fight food competitors. Hawk aggressivity brings advantages but also costs, because of the risk of injury or death. This simple model creates interesting dynamic patterns of interaction with free riding and explains many biological (and economic) phenomena (Smith 1982).

4. This section draws on Arrow et al. 2000.

5. This section was written in collaboration with Bill Hyde.

6. In other cases, some exceptional hardwoods may have timber value. The focus here is on the settlers, who are in fact the cause of deforestation. In addition, "untouched" forests may be home to indigenous peoples who typically survive by hunting and gathering, perhaps with some shifting agriculture. Inevitably, this form of natural resource use (or lifestyle) requires large tracts of land per capita, and the inhabitants also usually lack formal "titles" to the land off which they live.

7. This model is simplified, but land is always heterogeneous, and some "centrally located" land may nevertheless be inferior for agriculture (because of slope, nutrient deficiency, rocky soil, poor drainage, or inaccessibility). In a more complex model, the protection of property rights to forestry may be either *cheaper or more expensive* than the corresponding protection of rights to agricultural land. Culturally and politically, it appears to be more difficult to defend property rights if land is not "used" (e.g., natural forest). Popular tradition is on the side of the squatters who work the land productively (Binswanger 1991); colonists in the Amazon often acquired both rights and subsidies through the act of deforestation.

# CHAPTER 5

# *The Evolution of Rights*

*A*S EVEN ARDENT PROPONENTS of the market economy recognize, some public institutions are absolutely necessary, such as the institutions that govern markets and property. The Pigovian approach of most economists entails comparing the "natural" or "untouched" (laissez-faire) state of markets and property with the "ideal" or "optimal" state when it comes to ownership, liability, and similar responsibilities. However, in stark contrast, Coasian thought proposes that neither state exists (Coase 1960). The natural state is an odd fiction; the current state of the economy and of any particular resource allocation is the result of a long development of property rights—not necessarily natural at all. State intervention, on the other hand, is far from optimal. Public choice research has shown how policymaking is influenced by the private interests of the policymakers.

*Property* and *ownership* are formal expressions of the relationship between humans and the natural environment (Bromley 1991). Some economists would argue that the term *property* is a construct of the modern economy and that many people who live close to nature think of humans as "belonging to the land" or that the relationship is a divine one (e.g., Hanna and Jentoft 1996). In modern parlance, property consists of a bundle of rights. To properly analyze different forms of private or common property, the following typology is recommended (Ostrom and Schlager 1996)[1]:

1. right of access: the right to use or enjoy the property for direct utility;
2. right of withdrawal: the right to use the property productively for profit;
3. right of management: the right to set up and modify rules for the use of a property;
4. right of exclusion: the right to exclude some users and set rules for access to a property; and
5. right of alienation: the right to sell, lease, or inherit property with the Rights 1–4.

---

This chapter builds on excellent research assistance by John McClanahan.

Any of these rights might be partial. For example, the right to inherit user rights but not sell those rights on a market is common. Other rights may be to move, change or adapt, and even destroy or dispose of property. According to Coase, these rights should be considered "factors of production" because they determine how much useful profit an owner can earn from a property (Coase 1960). Other important aspects of ownership are legal, cultural, and psychological—for example, the psychological satisfaction of ownership (as opposed to weaker rights to compensation) as embodied in the right to refuse to sell property rights (Rachlinski and Jordan 1998).

Bundles of rights may differ from one society to another. They typically have evolved through the interaction of neighbors and citizens living under specific circumstances. Common practice is to categorize the law broadly as private law or public law. *Private law* (or "common law") refers to the old (particularly English) legal process of making and modifying the law on a case-by-case basis. This law was made by the courts, and under the natural law doctrine, judges considered themselves "discoverers of the law." (Popular notion was that a natural law governed the design of good relationships between people living together in a society, and these natural laws could be discovered through the observation of individual cases.) *Public law* is the more formalized process of creating statutes and constitutions through legislative bodies. It could, to varying extents, be considered the expression of popular will. The implied role of government within each kind of system is very different. Under public law, the legislature plays a dominant role in the definition of property rights, whereas under private law, the courts play the leading role.

Because adjacent properties inevitably share boundary lines and because noise, smell, energy, and materials easily flow across such boundaries, interdependencies in use usually exist. However boundary lines are drawn up and however property rights are allocated, some kinds of rights that an individual exercises on his or her property almost inevitably infringe on the exercise of rights by other individuals on their properties. For example, one person's right to shoot a gun in any direction on his property limits a neighbor's right to enjoy a bullet-free environment on her property. Similarly, one person's choice of agricultural methods may have considerable effects on the aesthetics, air quality, water quality, and noise levels of adjacent neighbors. The right of one person to an unimpeded view is in conflict with the right of another person to erect high-rise buildings. The right to completely move or destroy property—such as, for instance, by removing all soil and turning a field into an open-pit quarry—is often restricted. The definition of *property rights* is closely related to the concept of *externality*, and the development of these rights over time is related to the transition from common property to private property. This process is often referred to as "enclosure of the commons."

Property rights have evolved to gradually include more kinds of rights and property. An economist would consider this evolution as a response to human need to minimize uncertainty and transaction costs as well as a reflection of increased scarcity as population grows in a finite space. Notions of property rights may reasonably have stemmed from behavior relative to immediate possessions (e.g., food, shelter, and tools).[2] However, the formal history of property rights is closely related to the development of property rights to land. Therefore, the discussion of rights starts with private land, then moves to the somewhat more complicated rights (to water, minerals, and various forms of common pool resources). These discussions provide a basis for understanding the legal underpinnings of the rights to other natural and environmental resources, including less tangible objects such as biodiversity and clean air.

## Real Property

The variations in the bundles of rights applicable to land property probably are greater than expected by those who are accustomed to any one set of laws. One example is the rights that people have on other people's land: rights to mining, water, and other resources automatically follow land ownership in some countries, whereas they are separate in others. In the United States, trespassing is highly illegal and asocial, whereas in Sweden, individuals have the right to walk; pick flowers, mushrooms, and wild berries; and camp (one night at a time) in private forests and other lands (not gardens or fields, however). Furthermore, these activities—and the associated rights of access and some partial rights of withdrawal—are not only legal but also accepted as normal. Similarly, in many developing countries, harvested fields are often used as a common property resource for grazing.

Differences in property law are a function of history. For example, the common law version of land law stems from the feudal relationships that developed after the Norman invasion of England in the eleventh century. Conquerors such as William I made way for a hierarchical system of property rights. The original inhabitants, the Saxons, were forced to give up land that was then appropriated by the Norman kings and their underlords. The king would grant use and occupation rights to lords in return for their loyalty. In turn, the lords could grant lesser rights to smaller parcels of land to tenants in exchange for some form of "payment," which could be an obligation to provide armed men or agricultural produce to the king or the lord, for instance. In modern parlance, these payments were forms of tax or rent for the right to use land. The feudal world was a hierarchical structure with the king at the top and many levels of lords and minor tenants below.

For the purposes of this discussion, the most central fact is that all the land essentially belonged to royalty. In fact, the word *real* in *real estate* refers to land owned by the king (*real* has the same roots as *regal* and *royal,* from the Latin *rex* or *regis* ["king"]). As the feudal structure developed, lords and their tenants inevitably bargained with each other. The lords wanted to extract as much labor, products, and services as possible from their land. The tenants wanted as many sticks as possible in their bundles of rights, especially the right to pass on their land to their children. This practice led to the process of land inheritance and a degree of certainty in land tenure.

Gradually, feudalism gave way to more democratic government. Representative governments took over from the king not only power but a whole system of government, property rights, and laws. By around 1500, many interests in land were recognized. The notion that the right to grant property essentially is vested in the king (or whatever government effectively rules the land) was widespread and can be illustrated by a famous American case, *Johnson v. McIntosh* (8 Wheaton U.S. 543, 1823), in which one party claimed ownership to a certain parcel of land in Illinois by virtue of purchase from an Indian tribe before the Revolutionary War. The other party claimed ownership to the same parcel based on a later grant from the U.S. government. The first party sued based on the fact that its title was older. Justice Marshall argued that the dominant political power must have the ultimate authority to govern over property and issue land titles:

> As the right of society to prescribe those rules by which property may be acquired and preserved is not, and cannot, be drawn into question; as the title to lands, especially, is, and must be admitted to depend entirely on the law of the nation in which they lie; it will be necessary, in pursing this inquiry, to examine, not simply those principles of abstract justice ... which are admitted to regulate, in a great degree the rights of civilized nations ... but those principles which our own government has adopted in the particular case, and given us as the rule for our decision.

Justice Marshall therefore ruled that the title from the U.S. government was valid even though it was issued after the earlier sale.

The McIntosh case highlights an important fact: that a well-functioning property system requires certainty. Investors would be reluctant to invest money to build residential areas, set up businesses, and explore for natural resources if they were not somewhat certain that the law would enforce ownership rights. The McIntosh verdict hinged more on what was a good and practical institution for society than on what might be considered "fair"; according to many popular conceptions of "fairness," the first owner might have won by virtue of having older rights. This dilemma plagues the transitional countries of eastern Europe, particularly Russia, where many conflicting rights to the same land remain unresolved. The failure of government to properly establish the system of rights and liabilities has created an uncertainty that is detrimental to business. Government must somehow act as the ultimate "title" holder or at least guarantee property rights.

Still, the degree of government involvement and thus the extent of the private rights can vary extensively between countries, as illustrated in the example of trespassing on private property. Another major difference concerns a landowner's rights to water and subterranean resources (such as oil or minerals) on a property. Some of the difference in economic dynamics between the United States and Mexico, for example, stems from historically diverging ideas about property rights to subterranean resources (Meyer and Sherman 1979). In Mexico, according to Spanish colonial tradition, all such rights are retained by the state (the place originally held by the Spanish crown), which gave landowners little incentive to mine. In the United States, mining rights are part of land rights, which was an essential element in creating a dynamic search for minerals and oil. On the whole, the U.S. system of property rights appears to have been crafted expressly to give more rights to individual landowners than to central government. This system presumably was created in reaction to the overly centralized systems in Europe with its vestiges of feudalism. Even today, this distinction appears to be one of the underlying differences between culture and politics in the United States and Europe.

Outside these two regions, others have had government and ownership structures somewhat similar to feudalism, particularly Japan and China. Countries that have been colonized typically have layers of legal systems with different perceptions and definitions of property rights. In policy design, considering the structure of all kinds of property rights is crucial, and in many developing countries, common or communal property is particularly important. Intermediate forms of property that have private user rights but are owned by the local community also are fairly common. In Ethiopia, in fact, many forms of traditional land ownership were structured in this way (Alemu 1999).

In northern Ethiopia, the *rist* tenure system has predominated. Traditionally, everyone in a village had a right to land to support his or her family. People who left the village would temporarily lose their rights to use the fields but could reclaim them on their return. Today, these rights have been modified. The council (known as the Derg) that seized power in a 1974 socialist revolution nationalized all land. Even though the Derg collapsed in 1991, the new Ethiopian Constitution from 1994 still affirms that all land (and natural resources) belongs to the state. However, the new regime has strengthened customary user rights to land, and in many parts of the country, the old traditions continue to dominate everyday life. Even distant descendants who can prove their lineage to village elders can come back and claim land rights. Because some families have many children and some have none, making the average holding per family uneven, and because villagers have the right to reclaim land rights, even after several generations, there is a tradition of periodic *land reallocation*. These reallocations are different from the modern notion of individual property rights, and risk of reallocation might be expected to diminish incentives for investment significantly. However, reallocations solve the problem of combining equity with hereditary rights as perceived, and they appear to have been fairly common in many African countries. And surprisingly, African systems of land tenure with reallocation generally do not have a negative influence on productivity (Place and Hazell 1993; Migot-Adholla et al. 1993).[3]

The system of land reallocation had significant repercussions on investments during the Derg regime (the so-called socialist era), when reallocations appear to have been frequent, brutal, and made partly based on allegiance to the ruling party. The original system was not necessarily illogical, but in a new historic context, frequent and erratic reallocations created a general lack of tenure security. Now that technology and investments have become so important in financing modern agriculture and increasing productivity, the absence of tenure security and thus of investment incentives has had a significant effect on economic growth (Alemu 1999).

Peasant associations are ensnared in a seemingly impossible trap because they are responsible for conflicting tasks: managing the common pool resources of the village as well as allocating "individual" land lots as equitably as possible. Compounding this difficulty, young adults increasingly want to set up their own households, and many demobilized soldiers and other settlers are returning to their villages. Because individual plots are already small, taking land from existing community members it is extremely difficult, so the village association is essentially forced to use common pool resources, which thus effectively become privatized (Kebede 2001).

In Papua New Guinea, some successful examples of communal land rights do not allow for free access. Families have individual and indefinite farming rights, but the rights of sale belong to the clan. Experience with the privatization of land has been bad. Rather than clarify rights, privatization created uncertainty and confusion, because the concepts of private land ownership are not well integrated into local culture. Many lands that had been converted to freeholds were later disputed and, on many occasions, returned to communal ownership (Panayotou 1993).

## Common Property Resources

In the history of property rights, more and more objects and attributes of objects are gradually defined as "property." Beyond the realms of personal belongings (e.g., food, clothing, tools, and housing), land, and mineral resources remain ecosystem resources, which include waterways, forests, and wild animals. Roman law called these kinds of property *res communes* ("things owned in common"). *Res communes* could not be owned in their natural state by anyone. They were thought of as belonging to the "public" as a whole: "By natural law these things are common to all: air, running water, the sea and as a consequence, the shores of the sea" (from Institutes of Justinian, quoted by Wiel [1918], who adds "fish, wild animals and game, the light and heat of the sun, and the like"). This sentiment is still often expressed, partly in opposition to one prevalent aspect of economic development, the "enclosure of the commons."

In England as elsewhere, the commons historically were areas of lands that were less productive and therefore did not justify the transaction costs (including draining and fencing) necessary for agriculture. These lands provided important ecological, recreational, and other services to everyone; they were particularly important to the poor as sources of fodder and pasture (for livestock), firewood, water, fishing, and many other uses. With increased population density and changes in social structure, technology, and other factors, more and more of the commons were privatized or "enclosed" (historically, private land had to be enclosed by some form of fence or boundary). The political struggles over the enclosure of the commons lasted for centuries in England and were sometimes fierce. In one respect, they can be seen as a process of modernization that increased productivity, but in another respect, they were a struggle in which the poor attempted to protect their rights to pasture against the landlords' desire to extend private holdings.

Today, similar struggles still take place all over the world—especially in poor countries—and the distribution of land between kinds of property (e.g., private, state, and common property) has significant importance not only for the ecosystems as such but also for the distribution of wealth and welfare in society. A study of 75 villages in India found that 30–50% of common pool resource areas had been lost between 1950 and 1982. In 1950, 70 of those 75 villages had rules for managing common pool resources, but in 1982, only 8 of them still did. Similarly, whereas 55 villages used to collect (formal or informal) taxes to maintain the common pool resources, all of them had stopped the collection by 1982 (Jodha 1992).

In addition, new "items"—biodiversity, radio frequencies, the high seas, the Arctic regions, outer space, and genetic and other information—are being made into "objects" of property rights in a way that is structurally reminiscent of the enclosure of the commons. A striking example is the "enclosure" of the seas implied by U.N. Third Conference on the Law of the Sea in 1982. A main result of the conference was an extension of the exclusive economic zone to 200 miles, which implies jurisdiction by coastal states over fishing in this area and an appropriation of most of the value from the oceans by the coastal states of the world.

Another pertinent example comes from Monsanto Company, a producer of agricultural products. Because Monsanto patents the genetic code of the seeds it genetically modifies, farmers that use it must sign a contract that prohibits seed saving (i.e., using seeds from harvested crops to plant future crops). Planting harvested seeds is well established in agriculture and could in fact be considered the foundation of human civilization. Yet biotechnology is now trying to outlaw this practice.

Some ecologists see this and similar forms of enclosure as the last frontier in a losing battle between the ecosystem and humanity on one side and capitalism on the other. Others take the opposite view, that the existence of externalities and the "tragedy of the commons" depends on the lack of appropriate definitions for property rights. With better technology—both legal and physical (for fencing, supervision, detection, communication, and enforcement)—more detailed and precise property rights could be defined, which, according to this view, would not only enhance efficiency but also minimize the extent of negative externalities and degradation to natural resources. Even in the instances where this is true, the poor might be at a disadvantage. By definition, poor people have no capital and few productive assets. For this reason, access to commons or even to degraded open-access areas may constitute a significant, even crucial, contribution to their welfare. Thus, in some instances, enclosure can raise overall productivity while further impoverishing the people in the poorest segment of society, who benefited from the natural resources in the area of concern.

Although a historic trend toward enclosure seems to be apparent, common property is not necessarily an inferior kind of property. In several respects, a well-designed, well-functioning common property resource (CPR) is like private property (Ostrom and Schlager 1996; Murty 1994). Notably, CPRs imply the exclusion of outsiders and thereby make it possible to avoid the "tragedy of open access."[4] In some ecological and social contexts (when the costs of protecting private property are high or when the yields are low and very variable), a CPR may simply have lower transaction and other costs and thus be more efficient than private property. If well-adjusted and flexible mechanisms to deal with resource allocation decisions continue to be developed, then CPRs may continue to play an important role in the future.

## Water Law

Water is the ecosystem resource that, after land, probably has the most fully developed law. It also has many attributes that help us understand property rights for other ecosystems, such as mobility, variability, and unpredictability. Two fundamental doctrines in water law are riparian and prior appropriation.

The riparian doctrine, the more common of the two concepts, developed in Europe and has its roots in Roman law (Teclaff 1985).[5] This influence is especially clear in the French Napoleonic Code. The essence of the riparian water doctrine is that the right to use water depends on access to the water through ownership of adjacent land. The owner of property adjacent to a river or stream has certain rights to the flow of that water by virtue of land ownership. Irrigation, a chief concern in developing French law, is incorporated in the riparian concept:

> He whose property borders on a running water may serve himself from it in its passage for the watering of his property. He whose estate such water crosses is at liberty to use it within the space which it crosses, but on condition of restoring it, at its departure from his land, to its ordinary course. (Wiel 1918, 258)

The principle of extending land ownership to adjacent ecosystems or geophysical objects also arises in the allocation of mineral rights and in the appropriation of the seas by coastal nations. An early American Supreme Court Justice borrowed from the civil law system (especially the Napoleonic Code) to develop the modern U.S. riparian doctrine (Wiel 1918, 245–247).

The prior appropriation doctrine of "first come, first served" began in California gold-mining camps in the mid-1800s and was developed most fully in the United States, mostly in the West. This doctrine gives the first rights to the first user in time; in other words, it protects the first person who uses a stream's water from subsequent appropriations; the first person to put water to a beneficial use has a right superior to that of subsequent appropriators.[6] Subsequent appropriators are allowed to use water only to the extent that they do not impede the use of those who have prior rights. The doctrine establishes a priority system in which the senior appropriators can "call" the river in dry years to prevent appropriations by water users with more junior rights until the senior users have received their full shares of water. Unlike the riparian doctrine, it is not necessary to own land that is adjacent to a river or other body of water to obtain a water right. All that is needed is to be first *and to use the water beneficially.*

Another category of rights is the right of state governments. Many states declare, often in their constitutions, that all waters within their boundaries are the property of the state. The state may then allocate water rights through an administrative system that is commonly divided between watersheds. In the United States, for example, a person who wants to use water submits an application to the appropriate agency, stating the source, method of diversion, and intended use of the water. Normally, there are no (or low) charges for the water. The actual water right is not vested until the water is diverted and put to a beneficial use—tying in with the prior appropriation concept. After the water right is recognized, the state issues a title for a certain quantity of water for a given purpose—say, irrigation. The quality of the water right depends on the priority date (i.e., the date of first beneficial use or the date of application). In the dry summer months, some owners may get no water at all because the users who have older water rights are entitled to take their water first. A water engineer will shut down a water supply head gate (drinking water excepted) to first supply senior water rights. It is perhaps typical that this prior appropriation system developed in a region where water is (at least was at the time) much more scarce or valuable than land. No strong traditional landowners were around to speak for riparian-style rights.

Spanish water law has had various influences and shows the interplay among the various doctrines. One of the earliest written laws from the seventh century, the *Fuero Juzgo,* had elements of both Hispano-Roman and Germanic law. The common and usufruct rather than private nature of water was emphasized: "No one shall for his own private benefit, and against the interests of the community, obstruct any stream of importance; that is to say, one in which salmon and other sea-fish enter, or into which nets may be cast, or vessels may come for the purpose of commerce" (*Visigothian Water Code,* Law 29 of Book 8, Title 4, quoted in Wiel 1918).

Moorish rule (from the early 700s to 1492) later imparted the practice of communal management, particularly in Valencia. The Tribunal of Waters had considerable elements of CPR management, and land typically was not sold without its associated water rights (Ostrom 1990; see also Chapter 26). By the thirteenth century, Roman law again became important, as Castile kings such as Alfonso X sought to centralize water law in order to be able to grant water rights and possession of canals to landowners. The dilemma was that making water common had to include rights to cross private lands with canals and pipes, but it also made more irrigation possible.

Roman law was codified around 1260, establishing once again that water was *res communes,* not susceptible to ownership. All rivers were public property, except that feudal systems influenced Spanish water law, giving property rights to kings and nobles. The effect was moderated by the communities, *pueblos,* that protected water sources for their members. Throughout the centuries, the struggle between legal principles and among groups of users continued. Gradually, feudal influences were minimized, and in 1831, *A Broad Compilation of Water Law* declared that rivers that follow their course to the sea "cannot be private property" (quoted in Dobkins 1959, 78–79). This declaration was necessary partly because of the importance of navigation. However, for nonnavigable streams, irrigation rights were liberalized.

Not surprisingly, Spanish and French laws heavily influenced laws in their colonies, especially in the Americas. The system of local control appears to have been the aspect that was most easily transferred.

## Lessons for Environmental Externalities and Commons

The riparian concept is founded on (land) property rights and also has something of an ecosystem approach, whereas prior appropriation emphasizes beneficial use and is "first come, first served." One way to think about new rights to various ecosystem and environmental resources is by extrapolation from the land and water examples. In this section, I briefly comment on cases concerning air pollution, noise, nuisance, and risks as well as unclaimed property, such as some beaches, shallow fishing grounds, and virgin forest.

If an individual or a firm can acquire water rights by prior appropriation, then it might be argued that a factory that has been openly emitting smoke for several years for productive purposes should have rights to the assimilative capacity of the atmosphere according to the principle of prior appropriation. Some people consider this argument supportive of the allocation of emissions permits by *grandfathering*, that is, in proportion to historic use.

The courts often perceive their role as one of balancing legal principles, and in so doing, they may make reference to what is a "reasonable use" or to which arrangement gives the greatest total benefit to society. These concepts are illustrated by the erosion of the principle of "nuisance" or "trespass." For example, in the case of *Hay v. Cohoes Co.*, Hay uses dynamite to build a canal. Consequently, rocks fly into adjacent property. The Court of Appeals (New York, 1848) held that "(a) man may prosecute such business as he chooses upon his premises, but he cannot erect a nuisance or annoyance of the adjoining proprietor, *even for the purpose of a lawful trade.*" At the time, the neighbor's right to a safe environment dominated. However, a few decades later, in *Booth v. Rome, Watertown & Ogdensburg Terminal Railroad* (1893), the blaster was granted the right. Even though the plaintiff's house had been hit by rocks, the New York Court of Appeals held that the blasting did not create liability if carried out with "reasonable care." The court explicitly incorporated a utilitarian consideration: a city could never be built without dynamite (Green 1997).

According to the Coasian tradition, the externalities imposed on a victim by a polluter and the costs imposed on a polluter by a victim are completely symmetric (Coase 1960). In the case of *Sturges v. Bridgman,* for example, a confectioner has been making candies in the same locality for more than 60 years. A physician builds a practice directly adjacent to the confectioner and then discovers that the pounding noise of the confectioner's mortar and pestle makes him unable to use a stethoscope and even affects his ability to think. Even though the doctor is the newcomer and reasonably knows about the confectioner's business, he is awarded an injunction, and the confectioner has to cease his business.

In some cases, externalities are completely symmetric (e.g., in congestion, overgrazing, and overfishing). Many observers might argue that there is unidirectionality in the case of *Sturges v. Bridgman*; after all, only the confectioner makes noise. However, "imposing silence" creates just as much of an externality as making noise: either the confectioner works and the doctor cannot, or the doctor works and the confectioner cannot (Coase 1960). One might still find this verdict odd because the confectioner has prior rights, but the court followed what it perceived to be the maximum social utility.

Coase emphasizes that marginal analysis can lead to the wrong conclusion. An aggregate profit function for the confectioner and the physician might be non-convex and have several maxima. A little more noise might increase profit for the confectioner without affecting the doctor, who already cannot work at all because of the noise. The absolute social optimum might be no noise and no confectioner. Thus, both total and marginal analyses are needed. Also, if there were no transaction costs or wealth effects due to the reallocation of rights, then the outcome would be the same if the doctor had paid the confectioner to be quiet or if the confectioner had paid the doctor as compensation for the noise.

However, if there were transaction costs, then the outcomes might be different. Coase believes that in such cases, society should allocate rights so that the maximum social utility is achieved; a property allocation that is intended to maximize social utility would depend on the details of a particular case.[7]

In traditional (i.e., sparsely populated and resource-rich) societies, relatively few rights are formally defined; in modern societies, more and more resource rights are narrowly defined. The institutions of property appear to have evolved in a way that is roughly proportional to the economic advantage (scarcity rent) involved. The exploitation of natural resources can be seen as a race for property rights, much like the great move to the U.S. West in the 1800s, the California gold rush in the mid-1800s, or the current colonization of the Amazon and other developing-world ecosystems.

When concepts of property rights clash—particularly when modern codified concepts take over from the more informal, culturally rooted, usufruct systems—the result can be an uncertainty that leads to a wasteful abuse of resources and, in many cases, to solutions that are perceived as unfair by the original users. For example, in the traditional island society of Mafia (close to Zanzibar), ownership rights are defined not to land (the so-called "coral rag," because it has low productivity) but to fishing sites and to coconut trees, which are valuable and belong to whoever planted them. When hotels want to buy water, land, beach rights, and so forth, negotiations tend to be uneven; hotel owners often have been successful in securing "titles" that local residents do not understand or respect (see Chapter 31).[8]

Many nation-states have taken it upon themselves to nationalize ecosystem properties—making the lands government property instead of common pool resources—allegedly to protect those resources from degradation. The consequences usually are swift and harsh, as in the central hills of Nepal, where nationalizing the forests in 1957 quickly led to deforestation because villagers no longer perceived themselves as the owners or beneficiaries of the forests. During the past couple of decades, forest management has reverted back to village communities, and reforestation has been fairly rapid.

Although government plays a vital role as an ultimate guarantee of property rights per se, it has often failed abysmally as a direct owner and manager of natural resources, partly because of the general factors behind policy failure. Several factors suggest why governments fail to turn claims to ownership into effective control and management (Panayotou and Ashton 1992):

- enormous size of areas nationalized (in many countries, more than one-half of total area),
- fast rate of this transfer from local control to national control,
- failure to recognize and respect local customary rights,
- limited budget and administrative capacity to manage the natural resources,
- increasing population pressure, and
- failure of rural development to provide alternative employment.

At the other end of the scale, it is sometimes asserted that all environmental problems could be solved if property rights to all relevant resources (including waterways, the atmosphere, and various attributes of land—biodiversity, minerals, water, radio frequencies, and air space) were to be separately (and privately) allocated. As soon as an externality was created between two rights—say, one person's right to build a second story on his house and his neighbor's right to enjoy an unobstructed view—new "rights" to views or height would be created to remove the externality by converting it into a market transaction.

Private allocation of property rights to natural resources may be useful in some instances but does not always provide appropriate solutions. The more rights are divided, the more boundaries exist between them and the greater the likelihood of creating new externalities, because externalities are essentially physical connections between attributes of different properties. This is particularly the case when underlying production functions are strongly nonlinear, like when there are strong synergistic effects in the environment. For example, assigning responsibility, liability, or rights becomes increasingly difficult in a context where hazards are caused as the joint effect of pollution from several separate but neighboring entities.

Sometimes it is preferable to assign (or maintain) some form of collective property rights. Another aspect of privatization that is important for social welfare is that poor people have little property other than common property resources or open-access properties. In this perspective, privatization or enclosure is a disenfranchisement of the poor and likely to exacerbate their poverty and thus be violently opposed. One of the challenges for environmental economists is to design systems for property rights that maximize welfare (taking into account the costs of externalities and the "overgrazing" that results when exclusion is ineffective as well as the distributional and cultural aspects). During the past couple of decades, countries have begun to come together to create mechanisms to build up an international governance of the global commons (e.g., such as the Kyoto Protocol and the Montreal Protocol, in which cost distribution is a particularly sensitive issue; see Chapter 10).

## Supplemental Reading

Baca 1993
Cohen 1954
Oxman 1997
Pigou 1932
Platteau 2000
Stone 1995
Toulmin and Quan 2000
Wolf 2001

## Notes

1. Using the same notation (Ostrom and Schlager 1996), the salmon and herring fishermen of Alaska are "authorized users" with Rights 1 and 2 because they can fish only with permission from a commission. The cod trap fishermen of Newfoundland are "proprietors" but not "owners" because they have Rights 1–4 but not 5.

2. Compare this behavior with that of a fox or a wolf defending its "property": the intensity of its aggression is a function of how close it is to the center of its territory (Lorenz 1966).

3. Thanks to Hans Binswanger for discussions on this point.

4. This is a conscious paraphrase of Garrett Hardin's famous (but somewhat mistaken) "tragedy of the commons," the point being precisely that commons may be well managed, whereas open-access resources may not be.

5. This section draws heavily on Wiel 1914 and 1918 and Dobkins 1959.

6. *Beneficial use* appears to stem from a deeply rooted feeling that owners have to use their property to defend their rights. It may apply even to such secure forms of property as a house: in some societies (such as England), if a house is abandoned and squatters move in, the owner may lose partial rights to it. One perverse example of beneficial use took place in the Amazon, where forest clearing was a criterion of beneficial use, and deforestation thus became a means of securing titles (Binswanger 1991) (see Chapter 26).

7. Ownership rights also encompass cultural and psychological aspects. For example, in *Boomer v. Atlantic Cement,* eight neighboring property owners were granted compensation but not injunction (not "property" rights—i.e., the right to stop the plant's pollution). Despite compensation, the plaintiffs continued to litigate. A Coasian analysis would be that they were merely trying to improve their bargaining position to get even more compensation; however, modern analysts conclude that there is some psychological attachment to ownership, which questions a central element of Coasian thought (Rachlinski and Jordan 1998).

8. The diversity of laws, customs, and attitudes with respect to ecosystems in different countries cannot be addressed here, but many non-Western civilizations have interpreted their relationship with nature in divine rather than material, legal, or economic terms. In the ancient Indian religious literature (Vedas, Upanishads, smriti, and dharmas), for example, the sun, air, water, earth, and forests are personified as gods and goddesses (Sankar 1998).

# PART TWO

# *Review of Policy Instruments*

I N THIS SECTION, I PRESENT THE MAIN CATEGORIES of policy instruments used for environmental and natural resources policy as well as short descriptions of how each instrument works. The purpose is to provide background and context for the chapters on criteria and policy selection in Part Three.

Policy does not function in a vacuum; it is heavily dependent on the overall policy environment. If the economy is not competitive and if the bureaucracies are not honest, well-informed, and sufficiently well funded to carry out their responsibilities, then no policy instrument will work perfectly—although some will work better than others.

Information plays a special role in policymaking, and in fact, information provision can be considered an instrument in its own right. On a general level, all policy depends on information; that is, policymakers must understand the technology and ecology of the issues under consideration. Before the ozone-depleting action of chlorofluorocarbons (CFCs) in the stratosphere was understood, for example, no CFC-regulating policy existed. A general policy based on the precautionary principle could have been formulated, but before the scientific connections are suspected, serious restrictions are usually considered overly cautious.

Much discussion on policy instruments is conducted as if there were only two instruments, standards and taxes; however, many instruments exist, with many characteristics. Policy instruments are often classified as "market based" versus "command and control," but this classification is poor. Markets involve prices and quantities, regulations often are backed by economic sanctions, and even economic theory suggests that quantitative instruments such as standards, emissions targets, or permits may be optimal in many cases.

Some political scientists insist that there are only three basic categories of policy instruments, aptly nicknamed "carrots, sticks, and sermons" to symbolize economic incentives, legal instruments, and informative instruments, respectively (Bemelmans-Videc, Rist, and Vedung 1998). This classification can be further

Table II-1. *Classification of Instruments in the Policy Matrix*

| Using markets | Creating markets | Environmental regulations | Engaging the public |
|---|---|---|---|
| Subsidy reduction | Property rights and decentralization | Standards | Public participation |
| Environmental taxes and charges | Tradable permits and rights | Bans | Information disclosure |
| User charges | International offset systems | Permits and quotas | |
| Deposit–refund systems | | Zoning | |
| Targeted subsidies | | Liability | |

*Source:* Adapted from World Bank 1997.

differentiated into physical, organizational, legal, economic, and informative categories (Lundqvist 2000). No single taxonomy is necessarily preferable, but each can be useful in a different context.

Many policy matrices have been proposed as organizing principles for the systematic collection and comparison of policy experience. One useful typology (based on World Bank 1997) for organizing the rich diversity of actual experiences in the field is presented in Table II-1 and Table II-2. The policy instruments are divided into four categories: using markets, creating markets, environmental regulations, and engaging the public. The various kinds of policy instruments listed are applicable in the area of natural resource management (water, fisheries, land, forests, agriculture, biodiversity, and minerals) or in pollution control (air, water, and solid and hazardous wastes).

The first category of instruments, "using markets," includes subsidy reduction; environmental charges on emissions, inputs, or products; user charges (taxes or fees) (see Chapter 8), performance bonds, deposit–refund systems, and targeted subsidies. It also includes instruments such as refunded emissions payments and subsidized credits (see Chapter 9).

The next category of instruments, "creating markets," consists of mechanisms for delineating rights. The most fundamental of these mechanisms has particular relevance in developing and transitional economies: the creation of private property rights for land and other natural resources. A mechanism that is relevant at the local level is common property resource management. Special kinds of property rights in environmental or natural resource management are emissions permits and catch permits (see Chapter 7). In an international context, such mechanisms are often referred to as "international offset systems."

The category "environmental regulations" includes standards, bans, (nontradable) permits or quotas, and regulations that concern the temporal or spatial extent of an activity (zoning) (see Chapter 6). Licenses and liability rules also belong in this category, connecting it to a large area of lawmaking and to the politics of enforcement. Such instruments as liability bonds, performance bonds, and (more generally) enforcement policies and penalties are all part of the instrument arsenal.

**Table II-2.** *The Policy Matrix: Instruments and Sample Applications*

| Policy instrument | Natural resource management Water, fisheries, agriculture, forestry, minerals, and biodiversity | Pollution control Air pollution, water pollution, solid waste, and hazardous waste |
|---|---|---|
| Direct provision (6) | Provision of parks (31) | Waste management (27) |
| Detailed regulation (6) | Zoning (29, 31) Regulation of fishing (e.g., dates and equipment) (28) Bans on ivory trade to protect biodiversity (31) | Catalytic converters, traffic regulations, etc. (22) Ban on chemicals (24) |
| Flexible regulation (6) | Water quality standards (26) | Fuel quality (22) CAFE (22) |
| Tradable quotas or rights (7) | Individually tradable fishing quotas (28) Transferable rights for land development, forestry, or agriculture | Emissions permits (24) |
| Taxes, fees, or charges (8) | Water tariffs (26) Park fees (31) Fishing licenses (28) Stumpage fees (30) | Waste fees (27) Congestion (road) pricing (3, 20) Gas taxes (21) Industrial pollution fees (24) |
| Subsidies and subsidy reduction (9) | Water (26) Fisheries (28) Reduced agricultural subsidies (29) | Energy taxes (24) Reduced energy subsidies (24) |
| Deposit–refund schemes (9) | Reforestation deposits or performance bonds in forestry (30) | Waste management (27) Sulfur, used vehicles (13) Vehicle inspection (22) |
| Refunded emissions payments (9) | | $NO_x$ abatement in Sweden (24) |
| Creation of property rights (10) | Private national parks (31) Property rights and deforestation (30) | |
| Common property resources (10) | CPR management (28, 30, 31) | |
| Legal mechanisms, liability (10) | Liability bonds for mining or hazardous waste (24) | |
| Voluntary agreements (10) | Forest products (30) | Toxic chemicals (6, 24) |
| Information provision, labels (10) | Labeling of food, forest products (24, 30) | PROPER and other labeling schemes (24, 25, 27) |
| International treaties (10) | International treaties for protection of ozone layer, seas, climate, etc. (10, 12) | |
| Macroeconomic policies (10) | Environmental effects of policy reform and economic policy in general (2) | |

*Notes:* Numbers in parentheses indicate chapters in this book. Columns have been "collapsed" to save space. Chapter numbers in the second column indicate where the general explanation is given; other chapter numbers refer to empirical examples. Some examples have no chapter numbers because they are not discussed at any length in this book. These and other new examples of policy instruments may be found at two World Bank websites: Environmental Economics and Indicators (http://www.worldbank.org/environmentaleconomics) and New Ideas in Pollution Regulation (http://www.worldbank.org/nipr/commun.htm).

The last category, "engaging the public," includes such mechanisms as information disclosure, labeling, and community participation in environmental or natural resources management (see Chapter 10). Dialogue and collaboration among the environmental protection agency, the public, and polluters may lead to voluntary agreements, which have become a fairly popular instrument recently.

Not included in Table II-1 but still potentially important in various contexts are four mechanisms: direct provision of environmental services (such as municipal waste disposal); international agreements (which are only a policy at the multinational level); environmental auditing and certification, which is mainly a policy instrument at the firm level (often used in conjunction with labeling and information provision); and macro policies in general (all fiscal, monetary, and trade policies have implications for the whole economy and thereby for the environment).

Because of the wealth of information available, policy matrices easily can become unwieldy. Table II-2 is a simplified example of a matrix that illustrates the policy matrix approach and simultaneously provides a "map" of this book by listing most of the instruments discussed. Chapter 11 concludes with some comparisons and reflections on national differences in policy selection.

# CHAPTER 6

# *Direct Regulation of the Environment*

$T$HIS CHAPTER STARTS WITH a short section on how to judge the optimality of policy instruments. Then, I discuss three different kinds of physical instruments in turn: direct provision by the state of public goods, the direct regulation of technology, and the regulation of performance. Although the last two categories are sometimes grouped together, they differ considerably in the freedom given to individual firms.

## Optimality and Policy Instruments

The policymaker's situation is illustrated simply in Figure 6-1, which shows the results of quantity-type policies to cut back pollution from current emissions ($E_0$) to an optimal level of emissions ($E*$), and price-type policies, which impose a tax or fee ($T*$) on polluters, thus motivating them to cut back pollution to $E*$. The marginal cost of abatement (i.e., the extra cost incurred in achieving a lower level of pollution) is derived mainly from the costs of clean production methods. In this simple model, both kinds of policies lead to the same result. In reality, the decisionmaker has many more instruments to choose from, but there is generally less certainty concerning their effects.

The basic purpose of policy instruments is to correct for market (and institutional) failures such as externalities. In Figure 2-3 (Chapter 2), it is assumed that rates of pollution are fixed, and thus, emissions can be limited to $E*$ only by limiting production to quantity $Q*$. Two ways of achieving these limitations are by regulating emissions or production quantity, or by internalizing the externalities by a Pigovian tax ($T$) equal to the marginal cost of the environmental damage done by dirty production ($MC_e$). If emissions factors are assumed to be fixed, then an emissions tax can be replaced by a product tax, but such an assumption is restrictive—the amount of emissions generally cannot be considered proportional to the amount of product produced. Many environmental cases are "cleaned up"

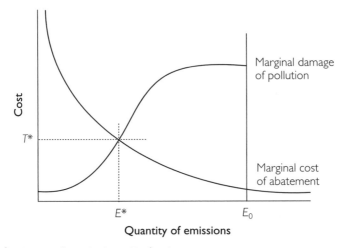

**Figure 6-1.** *Optimal Emissions Reduction*

*Note: T\** = optimal tax or fee; *E\** = optimal emissions level; $E_0$ = current or initial emissions.

not by reducing output but through abatement, and an important function of policy instruments is to encourage such abatement efforts.

To analyze abatement, the simple model is inadequate; the possibility of cleaner modes of production must be introduced, as in Figure 6-2. If there are two modes of production, one dirty with a marginal cost of $MC_D$ and one completely clean mode of production with a marginal cost of clean production ($MC_C$), and if $MC_C < T + MC_D$, then the product price will not rise to $P*$ but only to $P**$, and the decrease in output also will be smaller (see Chapter 14). The main influence on price is the adoption of clean technology. The clean technology might also have smaller external effects that lead to residual tax payments, but this possibility is ignored for the sake of clarity.

This section of the book is dedicated to describing the function of policy instruments. The standard against which to compare other instruments is provided by considering the maximization of welfare (Box 6-1), illustrated in the simplest possible case of a flow pollutant (for which there is no accumulation and thus no intertemporal aspect) and only one damage function—which means that there is only one pollutant and "perfect mixing" (i.e., no spatial dimension is taken into account). Box 6-1 shows two important requirements for optimality in this model. First, the marginal abatement costs in production should be set equal to the marginal benefit of the abatements as measured by a reduction in environmental damage. Second, the price of the product produced should reflect not only conventional production costs but also the scarcity rent that corresponds to the environmental damage that is caused despite the abatement costs incurred. This equation is essentially a formulation of the polluter pays principle. Some industrialists have complained that they are forced to pay twice—once for abatement, and once for pollution damage—but this scenario is analogous to the way the market sets the prices of other scarce items, such as water. For example, even though a firm purchases equipment that decreases overall water usage, the firm

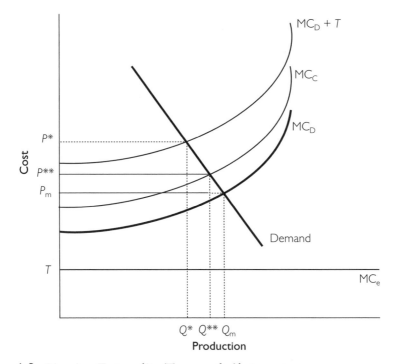

**Figure 6-2.** *Pigovian Externality Taxes and Abatement*

*Notes:* $P*$ = optimal product price in the absence of a clean production technology; $P**$ = optimal product price with clean production technology; $P_m$ = market product price in the absence of policy; $T$ = environmental tax level; MC = marginal cost of clean (C) or dirty (D) production; $MC_D + T$ = social cost of dirty production; $MC_e$ = marginal environmental damage caused by dirty production; $Q*$ = optimal quantity in the absence of a clean production technology; $Q**$ = optimal quantity with clean production technology; $Q_m$ = market determined quantity in the absence of any policy.

still is obliged to pay for water consumed. Thus, a firm that releases pollutants to the environment should pay for the assimilative action of the atmosphere.

If the assumption of perfect mixing is relaxed and each source is assumed to have a specific transfer coefficient (perhaps inversely related to the distance from the source to a particular receptor point, such as a town), then each source will contribute a different amount to each of several receptor points. Pollution levels usually vary between these receptor points, and damage functions must be modeled individually for each of them.

Two conditions are modified to reflect the sum of damages done by each individual source to all the receptor points. Emissions from source $i(e_i)$ thus cause damage at receptor $j$ of $D_j(e_i z_{ij})$; total damage at this receptor point will be $D'_{j\,i}(e_i z_{ij})$. Total damage at all receptor points will be $\Sigma_j\, D'_{j\,i}(e_i z_{ij})$, giving the following alternative conditions:

$$P = c'_q + \Sigma_j D'_j z_{ij} e'_{iq} \qquad (6\text{-}2^*)$$

## Box 6-1. Optimal Abatement

Consider a social planner who maximizes welfare ($W$), a function of net profits minus environmental damage of emissions, both of which depend on output ($q$) and abatement ($a$). Emissions are assumed to be perfectly mixed so that the damage ($D$) is equal to $D(E)$ is equal to $D(\Sigma e_i)$, where $E$ is total emissions and $e_i$ is the emissions of firm $i$:

$$W = \Sigma_i[Pq_i - c_i(q_i, a_i)] - D[\Sigma_i e_i(q_i, a_i)] \tag{6-1}$$

where $P$ is product price and $c_i$ is production costs for firm $i$. The necessary first-order conditions (dropping index $i$) are

$$P = c_q' + D'e_q' \tag{6-2}$$

$$c_a' = -D'e_a' \tag{6-3}$$

$D'$, $c'$, and $e'$ are derivatives of the damage, cost, and emissions factors, respectively. Subscripts indicate derivatives with respect to variables $q$ and $a$. The necessary conditions imply that optimal price should internalize damage costs, $D'e_q'$, which means adding them to conventional production costs $c_q'$, as in Condition 6-2. The choice of abatement technology should reflect marginal damage costs $D'e_a'$, as in Condition 6-3.

$$c_a' = -\Sigma_j D_j' z_{ij} e_{ia}' \tag{6-3*}$$

where $z_{ij}$ is the transfer coefficient; $D'$, $c'$, and $e'$ are derivatives of the damage, cost, and emissions factors, respectively; and subscripts $q$ and $a$ reflect output and abatement, respectively.

For stock pollutants, the rule implies equality of the marginal abatement cost not to the environmental damage today but to the discounted sum of the damages caused by current emissions in all future periods, taking into account their natural processes of decay or assimilation.[1] If, on the other hand, the damage function of a flow pollutant varies with the temporal dimension (e.g., seasons or time of day, as mentioned in Chapter 4), then optimal marginal abatement costs also should be adjusted accordingly.

## Direct Provision of Public Goods

The most straightforward "policy" an environmental protection agency can apply is to use its own personnel, know-how, and resources to solve a given problem. In the environmental arena, this mechanism is essentially the provision of public goods; whether the term *policy instrument* is appropriate is not clear (some economists would reserve "instrument" for policies that influence other agents), but it is important to start here.

Cleaning up public streets is a prime example of public goods provision; other examples include taking responsibility for major environmental threats, providing and maintaining natural parks, and managing certain kinds of research and control functions typically undertaken by environmental protection agencies. The U.S. Environmental Protection Agency, for instance, sponsors much research on environmentally friendly technologies. In some countries, ordinary sewage treatment or municipal waste management is provided as a public good, although the state or municipality usually tries to cover costs by charging some kind of user fees. Such fees may be part of property taxes or other taxes, or, as is becoming increasingly common, they may be user fees tied more specifically to the service provided (in the form of price-type instruments [see Chapter 8]).

During the past couple of decades, in most countries, the state has started to refrain from acting as direct producer of goods and services. The role of the state can be broken down into several components: financing, administration, provision, and control. If the state finances an activity, it does not necessarily have to manage its provision. Several activities that were formerly thought of as natural state monopolies have been organized in such a way that the government agency merely retains a control function, and private entrepreneurs are hired to provide the services. Considering that the financing and provision of activities can be more or less decentralized means that public goods—including various environmental services—can be organized in many ways, of which direct public provision is one end of the spectrum.

The starting point for a discussion of the level of provision is the Samuelson rule, which focuses on the sum over all households of the marginal willingness to pay for a public good (see Chapter 3). The provision of roads, harbors, airports, railways, and telecommunications infrastructure has considerable impact on environmental quality. The decision criteria for these kinds of investments include issues of discounting, equity, and efficiency in the tax structure and public sector in general. In many countries, cost–benefit analyses are required for all major public investments, such as roads, and even on general reforms. The rules for such analyses may include a specified discount rate, time horizon, depreciation rates, accounting value of public funds, shadow prices for unemployed labor, foreign exchange, and similar crucial variables, including how to treat issues of income distribution. The way the rules of project evaluation are specified has a considerable effect as a policy instrument on the general path of development of the economy.

## Regulation of Technology

One way of regulating the behavior of firms, households, agencies, and other agents in the economy is by prescribing the technology to be used or the conditions (through zoning or timing). The reasons why technology standards, restrictions, and zoning are still the most commonly used instruments are their intuitive simplicity and perhaps the short time perspective of many policy decisions. These kinds of regulation also may suit the interests of both regulators and polluters.

Typically, complaints are made against a polluter that is using "unnecessarily" polluting technology. The proof usually is not only the pollution per se but also a comparison with other industries or to some proven (usually end-of-pipe, or chimney) technology. This approach has many names: best practical technology, best available technology not entailing excessive cost, best available control technology, or simply best available technology. This kind of instrument is often derisively referred to as "command and control," which conjures up a rather Stalinist image, but these policies do allow for some trade-off. As the names suggest, technology must achieve a significant reduction in pollution, but at reasonable costs. The U.K. 1956 Clean Air Act used the term *practicable* and defined it (somewhat tautologically) as "reasonably practicable having regard, amongst other things, to local conditions and circumstances, to the financial implications and the current state of technology." Regulations that restrict location or timing are bans and zoning: a *ban* is a form of technology regulation in which a specific process or product is not allowed; *zoning* is a kind of regulation whereby certain methods or technology (such as certain vehicles) are banned in or are limited to a certain area (see examples in Chapters 22, 24, 25, and 31).

The trade-off between costs and benefits is a sensible requirement. However, leaving this trade-off to be determined by a bureaucracy may not be so desirable. More flexible instruments allow enterprises to make the trade-off. With mandatory technology, enterprises have little choice and are not encouraged to explore cost-efficient ways of achieving pollution control. They cannot trade reductions between sources, and they are not given any incentive to develop cleaner technology. Instead, the "best available technology" concept tends to encourage end-of-pipe solutions. However, command-and-control mechanisms or standards are not necessarily always inappropriate. In some instances (such as mandatory catalytic converters), considerations other than efficiency (such as ease of inspection) may be an overriding concern.

To understand the operation of technology regulation as an instrument, consider a firm $i$ maximizing its profit $(\pi_i)$ as $Pq_i - c_i(q_i, a_i)$. The firm's emissions [$e_i = e_i(q_i, a_i)$] are assumed to be perfectly mixed, and the regulator maximizes social benefit as in Box 6-1. With full information about abatement and damage costs, the regulator could specify the necessary individual technologies to achieve maximum welfare. In principle, both production technologies and abatement technologies could be used, but because end-of-pipe solutions are common in emissions regulations, the analysis is simplified by focusing only on abatement technologies and assuming that the regulator chooses the optimal individual level of abatement technology for each firm $(\hat{a}_i)$ as its policy instrument[2] (as in Box 6-2).

Under certain conditions, it is possible to achieve desired emissions levels through regulations on individual abatement (see Box 6-2). However, even with an optimal choice of abatement technologies, environmental damage is not reflected in the price, and thus output is not reduced as one of the mechanisms to reduce pollution. An individual firm cannot choose to meet the environmental target by searching for new (abatement or production) technologies or by reducing output.

Furthermore, it typically is not feasible for a regulator to have knowledge about individual abatement levels or technologies for each firm. The information requirements and administrative costs are prohibitive. Typically, an environmental protection agency wants a standard end-of-pipe technology that is the same for all and easy to monitor (e.g., catalytic converters, filters, or chimneys). In this case, abatement and emissions levels typically will not be optimal. Firms have individual emissions functions, $e_i = e_i(q_i, a_i)$, which implies that the same level of abatement typically will not give the same emissions rates—and individual marginal costs of abatement $(c_a')$ typically would not be equalized. Therefore, the basic

## Box 6-2. Design Standards for Abatement

Firm $i$ maximizes profit subject to its mandatory abatement technology ($\hat{a}_i$). The Lagrangean is

$$L = Pq_i - c_i(q_i, a_i) + \lambda_i(\hat{a}_i - a_i) \tag{6-4}$$

where $P$ is product price, $c$ is production costs, $q$ is output, $a$ is abatement, $i$ is a firm, and $\lambda$ is the Lagrangean multiplier. Taking partial derivatives gives the so-called Kuhn–Tucker conditions, which imply that if there is an interior solution (i.e., one in which the border conditions are not binding) with positive ($q_i^*$, $\hat{a}_i$), then the following conditions hold:

$$P = c_q' \tag{6-5}$$

$$\lambda_i = -c_a' \tag{6-6}$$

$$a_i = \hat{a}_i \tag{6-7}$$

where the asterisk (*) denotes optimum and the circumflex (^) indicates a value given exogenously—in the case of $\hat{a}$, decided outside the firm, by the regulator.

The $\lambda_i$ can be considered the shadow costs (in this context, roughly, the implicit costs) of abatement. With perfect information and if an environmental protection agency were able to select optimal (welfare-maximizing) choices of abatement for each firm ($\hat{a}_i$), then $\lambda_i$ would be equal for all $i$ (and equal to the effect of abatement on damages, $D'e_a'$, as in Condition 6-3).

Note: To include the possibility of a border solution where output or abatement is 0, equalities should be replaced with inequalities. The full set of conditions is

$$P \leq c_q' \text{ with equality if } q_i^* > 0$$

$$\hat{a}_i - a_i \geq 0, \ \lambda_i (\hat{a}_i - a_i) = 0, \ \lambda_i \geq 0$$

$$\lambda_i \leq c_a' \text{ with equality if } a_i^* > 0$$

Again, some subindices were dropped for convenience. For simplicity, only the equations for an interior solution with equality for all the other instruments are provided. They are simply referred to as the "optimality conditions."

economic condition for allocative efficiency would not be fulfilled. If the marginal cost of abatement varies between firms, then money could be saved by moving some of the total abatement effort to the firms where abatement is (at the margin) cheaper.

The type of regulation analyzed above is not a regulation of emissions but of (abatement) technology. The design standard is a mandatory technology and leaves the firm almost no choice and is therefore bound to perform poorly with respect to flexibility. It typically also will fail to meet abatement and emissions targets because there is no regulation of output (see Chapter 14). Under certain conditions, the advantages of technology standards may still dominate. For instance, some of the following conditions might motivate the use of design standards:

1. Technical and ecological information is complex.
2. Crucial knowledge is available at the central level of authorities rather than at the firm.
3. Firms are unresponsive to price signals (e.g., because of a noncompetitive, transitional setting) and investments will have long-run irreversible effects.

4. The standardization of technology holds major advantages.
5. Of only a few competing technologies available, one is superior.
6. Monitoring costs are high: monitoring emissions is difficult, but monitoring technology is easy.

In real life, all these criteria will not be fully met, but in many situations, some of them are important. Presumably, this is why technology standards are still frequently used. One important example is mandatory catalytic converters on new cars, a regulation that is now almost universal (see Chapters 4 and 13). This example meets Criteria 5 and 6, and maybe Criteria 4 and 1. Catalytic converters have been "successful" in reducing emissions, and their almost universal adoption has lowered costs. Still, it is possible that some other combination of engine modification, alternative fuels, and traffic management would have been more cost-effective, at least in some settings.

Another example of specific mandated technology is the regulation of nuclear power plants. In case of an accident, damages would be high; thus, the goal is zero incidents. Plants are regulated not only by being given maximum emissions levels; they also usually are given considerably more detailed and specific technology requirements, such as multiple separate control systems and certain kinds of containment. It can probably be best thought of in terms of Criteria 1, 2, 3, and 6 as well as the fact that society has absolved the firms from a large share of the risk. Cleanup after nuclear accidents usually is the responsibility not of the power companies but the federal government (see Chapter 13).

Technologies that are completely banned provide an example where the clarity of decisionmaking, economy of control, and ease of monitoring coincide. Examples are bans on certain kinds of chemicals (see Chapter 24), fuels or energy technologies, and vehicle types. Some of the issues concerning the potential advantages of being able to restrict bans to certain zones as well as the pitfalls of income distributional effects are illustrated in Box 22-1 in Chapter 22 (see also Chapter 23).

In natural resources management, the equivalent to technology regulation is mandatory technology (or restrictions on technology) for management, catch, hunting, farming, and so forth. It is (perhaps unfortunately) a fairly common kind of policy instrument. In fishing, the use of cyanide and dynamite are banned, which most people agree is reasonable (because these methods are so dramatically destructive to the actual habitat). However, techniques that are not necessarily destructive but perhaps simply more effective, such as those that involve enhanced nets and the use of equipment such as lights or sonar search equipment to attract or find fish (see Chapter 28), are also sometimes restricted. These restrictions are generally imposed to protect the resource but could in some cases be seen more as a way of protecting the livelihood and interests of those fishermen who use the older and more labor-intensive technology.

In agriculture and forestry, examples of technology regulation include the mandatory replanting of trees after harvest, mandatory construction of soil bunds and terraces to prevent soil erosion, and guidelines for pesticide and fertilizer use as well as other land-management practices (e.g., fallow periods or tillage technology) (see Chapter 29).

## Regulation of Performance

Under many circumstances, mandated technology might appear favorable in a superficial sense but turn out to be unsatisfactory—or even fail—when viewed with hindsight. One unfortunate characteristic of much policymaking is short-sightedness. With the time perspective of a political cycle, a given technological fix might seem adequate. Time after time, experience has shown that technology can provide better solutions than those that are immediately apparent. An important condition is to provide some reasonable incentives to develop new technology. A mandatory technology—catalytic converters or filters to stop emissions or miles of hillside terraces to prevent soil erosion—is definitely not going to help in that respect. In general, it is more efficient to target emissions rather than technology, inputs, or output.

A regulation that imposes a certain limit to harvests or to emissions instead of requiring a particular technology is called a *performance standard* (as distinguished from a *design standard*). A performance standard regulates quantities and as such is sometimes referred to as a command-and-control mechanism. This categorization is somewhat unfortunate; performance standards are significantly different from mandatory technology because they give firms considerable flexibility in the choice of abatement method by which to meet the mandated goal.[3] They also leave the firms a choice between output reduction and abatement level, and trade-offs between polluting units are possible. The formal modeling of this instrument is fairly similar to that of design standards, but the results are distinct. The regulator chooses maximum allowable emissions for each firm ($\hat{e}_i$), which optimizes within this constraint (see Box 6-3). Optimally chosen emissions limits imply not only that abatement costs will be allocated optimally but also that the product prices will reflect the costs of abatement.

---

## Box 6-3. Emissions Limits or Performance Standards

The firm maximizes profit subject to the constraint $e_i = \hat{e}_i$, and the Lagrangean is

$$L = Pq_i - c_i(q_i, a_i) + \lambda_i[\hat{e}_i - e_i(q_i, a_i)] \tag{6-8}$$

where $P$ is product price, $q$ is output, $c$ is production costs, $a$ is abatement, $e$ is emissions, $i$ is a firm, and $\lambda$ is the Lagrangean multiplier. The corresponding optimality conditions (see note to Box 6-2) are

$$P = c_q' + \lambda e_q' \tag{6-9}$$

$$c_a' = -\lambda e_a' \tag{6-10}$$

$$\hat{e}_i = e_i \tag{6-11}$$

The $\lambda$ can be interpreted as shadow prices of pollution. With perfect information and an optimal (welfare-maximizing) allocation of permits ($\hat{e}_i$), the $\lambda_i$ will all be equal to the marginal level of damage ($D'$) in Conditions 6-2 and 6-3 (Box 6-1), ensuring optimal abatement and allocative efficiency. Compared with the corresponding Conditions 6-5 and 6-9, the product price now reflects the opportunity cost of emissions permits.

Typically, the product price under a performance standard might be somewhat higher than with design standards and the quantity produced, in principle, somewhat lower. With performance standards, a firm has the additional flexibility to reduce emissions not only by abatement investments but also by reducing output. This flexibility usually is part of the socially optimal outcome because the cost at the margin of reducing output to lower emissions may be smaller than the costs associated with additional investments in abatement. However, this flexibility does not exist if the emissions limits are expressed relative to output (i.e., pollution intensity), which is a situation that is somewhat intermediate between design standards and performance standards. With pollution intensity regulations, there is no output effect (similar to a design standard), but there is some flexibility in technology choice (similar to a performance standard).

In addition to the differences in monitoring costs and flexibility, other subtle differences exist between design standards and performance standards. In many real-world cases, industrial pollution is controlled by licensing procedures that are a mixture of set emissions levels (total or relative to output) and mandated technology. Licensing procedures do not allow for flexibility in attaining goals through trading between sources. Furthermore, the information and resources available to parties (industries and local or national authorities) who negotiate the individual standards may be asymmetric, often leading to fairly lax environmental standards. The fact that many firms in polluting industries appear to prefer licensing to market-based instruments reinforces this impression. If well managed by knowledgeable authorities, these negotiations may give reasonable outcomes (e.g., Brännlund et al. 1996 on "command and control with a gentle hand"). However, one should be wary of the informational demands and the risks of corruption that this kind of regulation implies. The more judgement and flexibility exercised by the regulator, the better the potential result—but also the more expensive the regulation in terms of personnel. When regulators become familiar with the industries, conflict of interest is a risk. Incumbents always like licensing procedures because it gives many opportunities for rent-seeking. The opportunity to help formulate standards, requirements, and regulations also may give incumbents the opportunity to keep new entrants at bay.

One negative feature of controlling individual emissions rates (such as those set for new cars) is that they do not imply full control on total levels of pollution or ambient pollution levels. The total levels also depend on output for each agent and the number of agents. For example, with high mileages, "clean" (less-polluting) cars still lead to higher emissions even though the rate of emissions has been regulated, because the output (i.e., number of miles driven) has not. Even when all individual emissions (not the rates but the totals for each firm) are fully controlled, society does not have full control over aggregate totals. In the case of pollution, total emissions still depends on the number of polluters; thus, ambient pollution levels may be excessive.

In natural resources management, the total catch (e.g., grazing pressure, harvest levels) is the decisive factor for sustainability. Parameters are sometimes set by formulating the quantitative limits, such as harvest quotas, as an individual share of harvest, rather than in terms of rates (see Chapter 28).

Monitoring is discussed from various viewpoints—enforcement, political economy, and moral hazard—in Part Three. Monitoring a detailed vector of emissions is costly and often delegated to firms, but their data are not necessarily reliable. For ease of monitoring, regulators sometimes focus on the volume of inputs or some other factor that is readily controlled (e.g., number of fishing days or nets per person). One particular type of standard deserves special mention: level zero—that is, total bans. This particular level is particularly easy to monitor. With any level other than zero, inspectors have to pass judgement about the conditions pertaining to each individual case. Ease of monitoring may explain the popularity of bans, such as those on chlorofluorocarbons, dieldrin and other pesticides, cadmium, lead and other toxic paint pigments, arsenic, and the ivory trade.

## Supplemental Reading

Arrow and Kurz 1970
Bohm and Russell 1985, 1997
Dasgupta, Marglin, and Sen 1972
Little and Mirrlees 1969, 1974
Meade 1951

## Notes

1. Proofs appear in Kolstad 2000b, Baumol and Oates 1988, and Tietenberg 1992.

2. Abatement technology is assumed to be conveniently summarized by the variable $\alpha$; reality may be more complex. An environmental protection agency might require the use of some particular process, routine, or equipment that would need somewhat more complex modeling.

3. Flexibility is desirable because the firms are assumed to have the best overview of technical options and because these options vary between firms and over time. However, ease of monitoring is also an important goal. In the area of personal safety, for example, mandatory seatbelt use is not necessarily the best policy; it can encourage reckless driving and may have slowed the introduction of airbags and other constructional vehicle improvements. However, seatbelt use is relatively easy to monitor, and that factor might be enough to make it an optimal instrument overall.

# CHAPTER 7

# Tradable Permits

O NE LOGICAL WAY TO CONTROL aggregate levels of emissions or harvest is to set a total number of permits or quotas to adapt to the assimilative capacity of the environment or the sustainable harvest yield, respectively. Setting totals while allowing for some dynamics in the economy due to population growth, changing technology, mobility, and economic growth means that the allocated permits must be transferable. Otherwise, the allocation of all available rights would make all new activities impossible by definition. Tradability also allows the efficiency of the market mechanism to be harnessed to ensure that marginal benefits and costs are equalized.

The resulting instrument is called tradable emissions permits (TEPs) in pollution management and individual transferable quotas (ITQs) in fisheries management. Similarly, transferable grazing rights, development rights, and other mechanisms apply to other areas of natural resources management. The theoretical foundation of this instrument is Coasian (Coase 1960); however, such mechanisms are generally ascribed to Dales (1968a, 1968b), although others had similar ideas (e.g., Crocker and Wolozin 1966). Dales' suggestion was to create an authority in Ontario, Canada, that would sell "rights to pollute" water bodies. A local authority would thus decide total pollution, and the market would allocate the pollution rights among firms to reflect their demand for pollution or their abatement costs. The efficiency of this mechanism has been demonstrated (Montgomery 1972), and the instrument appears to have evolved just as much from the actual experience of regulation as from academic analysis. As discussed in Part One, trading follows naturally after property rights have reached a certain level of maturity.

The creation of tradable permits helps remove the externalities implied by the absence of property rights or the "public good" character of the environment. Essentially, this mechanism creates property rights to new resources or shares in the assimilative capacity or the sustainable rent production of ecosystems. As discussed in Chapter 5, we are witnessing a historic process of "enclosure." More and more aspects of the ecosystem are becoming scarce, and, in response, society develops new kinds of property rights. The fact that these property rights internalize externalities and create incentives for protection means that resources have a good chance of being put to their most efficient use (see Box 7-1).

Severe conceptual and practical problems must be overcome. For the permits or quotas to work, they must acquire the characteristics of property rights, such as permanence and reliability. It takes time and commitment to develop permanence and trust, and in the case of natural resources management, a lack of knowledge about the underlying ecosystems and lack of agreement about how they should be managed create additional difficulties. In at least one instance, allocated quotas had to be recalled (the Area 2 rock lobster fishery in New Zealand [Breen and Kendrick 1997]).

Many features of permits cannot be decided with only ecological and technical calculations. (The problem is analogous to setting the optimal tax level.) They include the definition, number, duration, and temporal and spatial validity of the permits as well as the proposed method for their allocation. These decisions are crucial, partly because permits involve the transfer of essential property rights and, potentially, some substantial transfers of wealth. With policy instruments such as licensing and technology standards, gradual calibration of the instrument to gain information and adaptation to changing circumstances can be imagined. Such changes will always meet some resistance from polluters, but they are possible. The creation of permits or quotas with the characteristics of property rights will exacerbate the difficulty due to their irreversibility.

Thus far, permits have taken two main forms. The early U.S. programs traded in credits that were created by overcompliance with other preexisting legislation (standards or emissions rates); later cap-and-trade programs (discussed later in this chapter) traded in rights that were created expressly for this purpose. Cap-and-trade programs have reduced administrative burdens and thus have encouraged the formation of real permit markets. Because the mechanism is so simple and builds on actual measurements and on predetermined permit allocations, the need for and the possibility of any discussion concerning baselines, exemptions, and so forth between regulators and firms is removed. Transparency and accountability reduce transaction costs and inspire confidence in the mechanism. However, modifying the overall level of permits would be difficult unless the permits are formulated as shares in a total that can be determined from year to year. (This procedure is common in fisheries, which are characterized by extreme variability in total allowable catch.)

Property rights must convey a sense of permanence to have an effect on costly investment and business decisions. However, this permanence is a problem for an agency struggling with severe information problems and desiring to keep some flexibility, for example, to adapt to changing conditions or new information. Optimal values for total emissions or harvest may vary considerably between locations as a result of ecological, technical, and socioeconomic conditions, and over time—both cyclically (depending on time of day, season, and weather) and as a result of changing patterns of economic activity, technology, and demographic patterns. They also may vary in response to underlying changes in dynamic complex ecosystems. Many industrial and agricultural processes do not

---

## Box 7-1. Permits as an Economic Instrument

With information on aggregate abatement and damage curves, regulators determine the socially optimal level of aggregate emissions ($E^* = \Sigma e^*$) and issue a corresponding number of permits. These permits are allocated so that each firm receives $e_{i0}$ permits. Each firm is free to choose whatever combination of production, abatement, and permits it wants—subject to trading constraints, because it must hold a number of permits equal to its emissions level.

The firm maximizes profit subject to this constraint, as in

$$\max Pq_i - c_i(q_i, a_i) + p_e[e_{i0} - e_i(q_i, a_i)] \tag{7-1}$$

where $P$ is product price, $q$ is output, $i$ is a firm, $c$ is marginal cost, $a$ is abatement, $p$ is the price of permits, and $e$ is emissions. The necessary and sufficient first-order conditions for a firm with positive output and abatement costs are

$$P = c_q' + p_e e_q' \tag{7-2}$$

$$c_a' = -p_e e_a' \tag{7-3}$$

A comparison with the corresponding Equations 6-5, 6-6, 6-9, and 6-10 and Conditions 6-2 and 6-3 for optimality (all in Chapter 6) shows the similarity between the sets of conditions. Conditions 7-2 and 7-3 should imply that $p_e$ is equal to the marginal damage of pollution ($D'$) (and thus the optimal Pigovian tax rate), which also implies that Condition 7-2 is identical to Condition 6-2 and that the output price under tradable permits correctly internalizes environmental damage.

*Note:* Optimality requires the correct number of permits and conventional assumptions concerning the behavior of firms (e.g., Baumol and Oates 1988). Both tradable emissions permits and taxes provide for a socially optimal long-run allocation when entry and exit are free (e.g., Spulber 1985; Xepapadeas 1997). Notice that for a new entrant, Equation 7-1 is modified, because $e_{i0} = 0$.

---

emit a permanent or predictable flow of a particular pollutant; however, they can be described as posing a certain risk. In the best of cases, the probability distribution of this risk might be known, and the parameters of that probability distribution might be the target of change.

How can these uncertainties be handled in a way that permits some flexibility to the decisionmaker while providing sufficient certainty to the permits? One way is to limit tradability over time. This approach has been used for pollutants that were expected to be reduced rapidly over time (as a result of technical progress in abatement) and in cases where there was considerable uncertainty about the damage function. In such cases, the permits may have a limited life, so permits for every period have to be acquired during that period with no (or limited) banking possibilities. The allocation of permits might still be proportional to allocation in previous years (a form of grandfathering).

Another mechanism that combines flexibility with certainty is conceptually similar: allow the total number of permits to vary depending on ecological or other conditions. Fisheries are a good example of this mechanism because the dynamics of the underlying ecosystem are particularly difficult to predict beforehand. As a result, tradable harvest permits for fish are often formulated as quotas—that is, shares in total allowable catch (see Chapter 28).

This component of design gives decisionmakers flexibility while still giving a reasonable degree of property rights to the firms. The need for this flexibility is readily apparent in fisheries, but society might be wise to ensure a similar type of flexibility in other areas. It could be used for sulfur permits, thereby making it easier for society to reduce future emissions levels. Naturally, industry needs a high degree of certainty to undertake cost-efficient investments. For this kind of pollutant, industry needs to know that permits correspond to certain physical quantities for more than one year. This need could be addressed by an environmental protection agency committing itself to revisions of total emissions that are infrequent and occur only on fixed, announced dates.

## U.S. Emissions Trading Programs

This section is dedicated to emissions trading programs that have been operated in the United States: "bubble" policies; ambient permits; the output-based allocation of permits used for the phaseout of lead from gasoline (which is of particular interest because it is a different mechanism); and the cap-and-trade programs that are currently used, for instance, to abate sulfur emissions.

### Emissions Reduction Credit Programs

The history of TEPs is instructive. Their beginning appears to have been as an extension of the system of emissions limits for air pollution within the Clean Air Act. In short, the polluters argued for flexibility and persuaded the regulatory authorities that they could achieve the same or even more abatement if they could allocate abatement efforts themselves and if they could could trade reduction at one source that was more than required for incomplete compliance at another. This approach opened up the possibility of economic growth in areas of nonattainment and was the start of what came to be known as the Emissions Trading Program.

The Emissions Trading Program, initiated by the U.S. Environmental Protection Agency in the late 1970s, built on a preexisting detailed system of source-by-source regulations and individual pollution permits or licenses (discussed in Chapter 6). The problem was that in certain areas of the country, these policies were not sufficient to attain the goals set for ambient concentrations of various pollutants. Such policies cannot be relied on to meet aggregate or ambient targets. These so-called nonattainment areas continued to have unacceptable levels of pollution, and progress toward meeting ambient goals was considered much too slow. Generally speaking, these areas had high population density and large concentrations of industry. Economic growth was a top priority, and the last policy they wanted was one by which industrial expansion—hence job opportunities—would be lost.

Yet such was public perception of the policy. In some cases, policy even appeared to give the worst possible outcome by being a hindrance to industrial output without being able to provide a clean environment. Policy reform was urgently needed, and the basic ideas behind the Emissions Trading Program were thus to

- provide flexibility to firms so that they could meet environmental restrictions in the most cost-effective way, which might mean reducing emissions at a source other than the one targeted in the individual licenses, and
- allow industries to build new plants (or expand existing facilities) without increasing total emissions.

Several somewhat separate mechanisms were used—netting, banking, offsets, and bubbles—but all stemmed from the emissions reduction credit (ERC), which was an "excess" reduction that a source would voluntarily undertake, thereby exceeding their legal abatement requirements. The sources could apply to the control authority for certification of the ERCs, which then became the legal tender of the program, so to speak. The credits could not be traded freely, but trades were allowed under several programs:

- The *offset policy* allowed new firms to start up (or old firms to expand) facilities, even in nonattainment areas where otherwise no new plants would be allowed. To do this, the firm was required to not only comply with stringent regulations but also to buy ERCs from existing firms in the area, so that the regional aggregate emissions would be no greater after the expansion or establishment than they were before.
- The *bubble policy* applied to existing emissions sources in an area and allowed them to act as one unit for the purposes of compliance with emissions regulations. Thus, a firm could reduce emissions at some location less than would normally be required by regulations if it simultaneously reduced emissions more than would normally be required at another location. Because compensating ERCs could be from other firms, this program provided room for emissions trading.
- *Netting* allowed existing firms that wanted to expand or modify the option of avoiding the stringent review process for new sources as long as the net increase in emissions (after reduction of ERCs acquired) was below a certain threshold.
- Through *banking,* firms were allowed to store certified ERCs for subsequent use in any of the above programs—i.e., for bubbles, netting, or offsets.

All of these mechanisms required buying "extra reductions" from the regulated firms in the region that were already fulfilling their legal obligations but could perhaps be motivated to achieve more abatement. The persuasive factor was the returns from selling their excess permits, which underscores the importance of the permanence of these permits as rights. A naïve critique of the program would claim that environmental improvement at one plant should be seen as only an environmental improvement and that the firm in question should not be allowed to sell extra credits to another firm, because then pollution is constant. This critique completely misses the point, because in this setting, firms are already regulated; additional abatement is costly and hence will be undertaken only if there is some possibility of profit, or at least cost recovery. The abatement effort that results from permit trading leads to a cleaner environment—at least relative to the ambient level obtained with the same industrial output and equivalent mandatory standards—which is the goal of the regulation. Alternatively, the policy instrument could "accommodate" more industrial output for any given level of ambient standards and total abatement costs.

In a situation where pollution is essentially unregulated, handing out permanent emissions rights would not be without risk. If many sources that can easily abate obtain large amounts of emissions rights, then incentives for abatement at other sources is reduced. The latter sources can simply buy excess permits instead of abating emissions. (This problem has been referred to as "hot air" in the context of global warming.) In extreme cases, it is conceivable that the disadvantages of this allocation of excess permits would be a stronger negative influence on pollution than the benefits associated with the economic incentives to abate created by the system. Particularly in developing countries, where some polluters may have essentially unregulated emissions, care should be taken in setting up trading systems to avoid giving away excessive rights permanently.

In many of the first U.S. programs, considerable cost savings were achieved, but the theoretical potential (worked out by comparing current regulation with an idealized model that results in equalization of marginal abatement costs, for instance) was seldom realized. Several programs had very little trading, and transaction costs proved to be important. Individual approval by the appropriate authorities of each trade was often a considerable practical problem. There also was insufficient confidence in the ERCs as secure property rights (see Chapter 24 for a discussion on cost savings in sulfur abatement). Issues of design that could be sensitive to encouraging entry or delaying exit also turned out to be crucial (e.g., Stavins and Whitehead 1997).

### Ambient Permit Trading

One problem of permit trading is that it may create regional or local "hot spots" (i.e., subregions with particularly high concentrations of industry and pollution). Permit trading can accommodate economic growth while maintaining total emissions within a country or specified region at a given level. This level might be constant or could decrease over time in response to expected technical progress or shifts in demand. The area covered must be large enough to allow trading between polluters with varying costs of abatement if there are to be any cost savings. On the other hand, it must not be too large (and heterogeneous), because then damage levels might vary considerably, and in some spots, pollution might be particularly bad ("hot spots"). In models with multiple receptors and multiple sources (discussed briefly in Chapter 6), the transfer coefficients from source $i$ to receptor $j$ ($z_{ij}$) must be taken into account. Otherwise, there is no guarantee that the trading program will not worsen ambient pollution levels at hot spots.

The problem of hot spots can be solved by having separate permit trading systems for each relevant area or by a spatially differentiated ambient permit system (APS). The APS is optimal if its transaction and operational costs are disregarded. APSs are complicated because polluters have to acquire different permits (in proportion to emissions and transfer coefficients $e_i$ and $z_{ij}$, respectively) for every receptor area. One simplification is to have an emissions permit system defined for the whole area but to make all individual trades subject to approval so as to guarantee that ambient levels do not deteriorate in prioritized hot spots. Alternatively, the emissions credits could be given different values in different areas, creating differentiated "exchange rates" for trades between areas or zones (depending on how sensitive they are). This approach is referred to as *zonal trading* and is the simplest of the APS schemes. The trouble is that these modifications remove much of the attraction of the system. The permits in the areas with lower concentrations of industry and pollution would fall in value if they could not be exchanged freely for other permits. As the complexity of the system increases, transaction costs increase, and the whole perception of permits as attractive property rights erodes.

The difficulties of APSs are particularly aggravated if the rules are modified ex post rather than being made explicit at the launch of the program, which severely restricts the possibilities for environmental protection agencies to adapt their regulations to new conditions. In the case of an inexperienced protection agency or a new pollutant for which uncertainty is great, tradable permits might be considerably less efficient than nontradable regulations.

The practical difficulties are aptly illustrated by the fact that full-fledged APSs have not been implemented anywhere. In fact, their counterpart—regionally varied taxes or fees—also involves many difficulties and is rare (however, see the Chinese example in Chapter 25). On the whole, it seems as if the difference in damage has to be considerable before the extra cost of spatial differentiation of the policy instruments will be justified. One of the best examples is the use of offsets in the United States, where new sources were allowed in nonattainment areas only if they could buy ERCs from existing sources in such a way that the total ambient levels in the relevant area were improved (or held constant) (Foster and Hahn 1995; see also Chapter 24 on sulfur trading).

### Output-Based Allocation and the Phaseout of Leaded Gasoline

The Emissions Trading Program was based on credits earned through voluntary reductions in excess of regulated levels. Such a system could not easily be used for the complete phaseout of a pollutant, such as the phaseout of lead from gasoline in the United States (see Chapter 22). Until recently, the U.S. lead-phaseout program was one of the main success stories of emissions trading.

The phaseout was to progress in the form of a stepwise function (decreasing allowances of lead in grams per gallon of gasoline). At an aggregate level, the steps had to be followed, but for some refineries, following the steps exactly would have been extremely expensive. Thus, the function of trading was to allow for flexibility and cost savings by encouraging refineries that had the technology to replace lead more quickly than required and then bank the extra lead credits for their own later use or sell the credits to companies that had difficulty complying with the time frame.

The lead-phaseout mechanism implied an output-based allocation of rights, that is, a regulation of pollution intensity rather than of total amount of pollution. Lead rights were allocated to all vendors in proportion to their gasoline sales at the refinery and retail levels. This feature is analogous to a standard for pollution intensity, with the important difference that the rights were tradable. It is also somewhat similar to the operation of refunded emissions payments (see Chapter 9). Output-based allocation sometimes is referred to as *benchmarking* because the firm must compete against a benchmarked rate of pollution intensity, so to speak. One effect of tying permit allocation to output rather than auctioning or grand-fathering permits is the lack of a role for output substitution. Another important difference is that the cost distribution among polluters is different and more favorable to new, clean firms than to firms that historically have emitted the larg-est share of pollution (and thus are favored by grandfathering). This difference can be decisive in securing political acceptability of the instrument.

With a fixed allocation of permits, an incentive to limit the production and consumption of the final good into which the pollution is embodied is intrinsic to the system, because the marginal cost of abatement (which, in equilibrium, is equal to the permit price) is part of the product price. However, with an output-based allocation system, no such incentive to reduce output exists, because increased output leads to more pollution rights. The system is theoretically less efficient than a cap-and-trade permit system (or a Pigovian charge on lead) in this respect. Nonetheless, it may be preferable in cases when the output effect is not desirable but decisiomakers want to affect abatement technology (see Chapter 9 for more analysis).

Interestingly, the U.S. lead-phaseout program started formally in 1982 as "inter-refinery averaging," and during the first couple of years, the credits could be used only in the quarter during which they were earned. Trading was later extended to include banking of credits. The important point is that the assign-ment and distribution of property rights was the first step of the process. Once the agents of the economy perceive property rights as valid and real, suggestions for trading and, consequently, flexibility in trading arrangements will come from the agents. However, constructing flexible trading mechanisms is pointless if the underlying property rights are insecure or lack credibility.

Tenure security, or the confidence placed in permits as property, is a vital aspect of permit trading systems. Having a constant value for the permit (whether in catches or pollution rights) enhances tenure security but is not always possible. A permit scheme might have restrictions on the direction of trade (e.g., into hot spots) or on time periods (e.g., seasonal permits, or permits with validity up to a certain date), for example. Permits could give the right to shares in a total (as in the case of fisheries) or could be set to decline at a certain rate, to give aggregate improvements in environmental quality over time, as in the phaseout of leaded gasoline in the United States. In addition, the system might be reviewed on pre-viously announced dates and then renegotiated or restructured as needed. Although these and other modifications are possible, they significantly reduce tenure security and tradability, which are the factors that lead to the equalization of marginal costs and thereby static (and dynamic) efficiency (see Part Three).

## Cap-and-Trade Programs

During the 1990s, U.S. policymaking moved gradually toward *cap-and-trade pro-grams*, which build on the earlier programs but emphasize tradability, security of property rights, and total emissions levels. The name of the programs is indicative of the procedure followed. First, regulators determine the total quantity of allowable emissions or resource harvest (the "cap"). Next, rights to those emissions are allocated among polluters (by auction, grandfathering, or some other mechanism but not relative to current output, because the total amount of emissions is capped). Then, the allocated rights become subject to free trading among the polluters that own the rights. Cap-and-trade programs are the most flexible programs for emissions trading to date, and they best correspond to the idealized TEP described in Equation 7-1 and in Condition 7-2 and Condition 7-3 because they also regulate total emissions.

In a somewhat Coasian way, the initial distribution of permits in a cap-and-trade program (whether by auction, grandfathering or some other mechanism) is thought to have little effect on allocative outcomes. Shares, once distributed, can be seen as a windfall to the companies that receive them—or to their shareholders.[1] Because the permits are tradable, the owner has to decide whether to sell (or buy) or retain permits; at the margin, an extra unit of pollution implies a cost to the firm. In this sense, the marginal damage of pollution (the opportunity cost of the permits) becomes part of the marginal cost of production to the firm. Hence, once allocated, the permits act much like property rights to the environment. In fact, the effect of permits on product prices is much like that of a Pigovian charge, because the final product price reflects the cost of environmental "rent." This observation assumes that firms treat permits as liquid assets with an opportunity cost and that they do not hoard them strategically to keep competitors or neighbors away, as they might do in monopolistic competition. Furthermore, there is the important issue of whether the allocation of permits is permanent, or if not, how often and in what way the system will be updated. If the system is updated based on performance indicators (such as emissions or output), then incentives for performance may be affected more or less strongly depending on the exact construction chosen. Similarly, issues such as how to allow newcomers into the system may have incentive effects on output levels or on the entry and exit of firms.

The most important cap-and-trade programs to date are the sulfur oxide ($SO_x$) and nitrogen oxide ($NO_x$) emissions trading programs in the United States. The sulfur trading program has generally been considered a great success; permit trading has been lively, and the price of permits has fallen below initial forecasts, indicating that comparative advantages in abatement have been significant and exploited and thereby lowered the costs of compliance (see Chapter 24).

RECLAIM is a program that targets $NO_x$ or $SO_x$ emissions (the main source of smog in the Los Angeles area) from permitted equipment. Participating firms receive trading credits based on past peak production and according to existing rules and control measures. Credits are assigned each year and can be traded for use within that year. RECLAIM has become known for some considerable price volatility in permit prices, which soared from some US$500 to US$60,000 in the third quarter of 2000, calling program design into question. However, it was later revealed that the increase was partly due to the exceptional scarcity of energy in California, which led to the startup of some old polluting units. In January 2001, the permit price was capped at US$15,000.

## Other Emissions Trading Programs

The concept of trading permits is spreading. Some countries outside the United States, such as Chile and Poland, have used permit trading to reduce pollution (Stavins and Zylicz 1995), and several countries within the European Union are taking a greater interest. This interest is partly in anticipation of expected climate change policies, which many policymakers assume will build on some form of permit trading.

In the international arena, plans for international offset schemes for trading of global climate permits (such as the Clean Development Mechanism within the Kyoto Protocol) are making progress.

### Chile

Chile is home to presumably the most advanced example of permit trading in a developing country. Its permit scheme applies to industrial point sources of particulate matter in the metropolitan region of Santiago, the nation's capital. The legal foundation for this system was set in 1992 by a decree that applies to sources over a certain size (i.e., that release emissions to air at a rate in excess of 1,000 cubic meters per hour) and freezes total regulated emissions. It also stipulates a maximum emissions standard of 112 milligrams of particulate matter per cubic meter. Sources are allowed to trade emissions reductions that result in lower rates of emissions than the standard, and new sources are obliged to buy credits from existing sources as a precondition to their establishment. Emissions reductions have been achieved by reducing the credits in the system over time.

To keep it simple, only particulate matter larger than 10 micrometers ($>PM_{10}$) is regulated. Temporal aspects (e.g., seasonal variation, geographical position) are not considered, even though they are important.

The success of the Chilean program has been mixed. Trading has been fairly limited and mostly within firms, partly because of the rather high degree of ownership concentration and partly because of the somewhat complicated design that requires the involvement of authorities in the trades. In this respect, the Chilean experience appears to mirror that of some early U.S. programs that were built on emissions reduction credits. Among the other limitations of the program are the resentment on the part of some members of the industrial sector that transport and other sectors were not included. To address this issue, limited proposals have been suggested whereby emissions reductions in other sectors would be recognized and made tradable. On the other hand, a considerable and important advantage of the Chilean experience was the incentive to declare emissions. Because industrial sources were allocated credits for free, based on their actual emissions, many new sources (and incorrect data for old sources) were "discovered."

### Europe

The European Commission launched the European Climate Change Programme in June 1990, which recommends emissions trading within the European Union

(European Commission 2002). Several European countries have started to set up various kinds of trading schemes. At the forefront is the United Kingdom, which already has a tradable permit scheme for packaging (see Chapter 27).[2] The United Kingdom is also in the process of introducing tradable permit schemes for landfills, $NO_x$, and carbon.

The U.K. carbon trading scheme is fairly complicated, because energy-intensive industries are treated differently from others. The Emissions Trading Scheme (ETS) is a cap-and-trade system with a fixed total of carbon emissions that are allocated freely, with grandfathering. The Climate Change Levy Agreement (CCLA), for energy-intensive firms, operates on the basis of relative targets for energy efficiency. Thus, emissions are capped, per se; firms have the option of trading permits to meet benchmarked obligations.

Because a fee (the climate change levy) is charged and agreement is voluntary, a special mechanism has been designed to entice firms to participate in the CCLA. Firms bid to sell emissions reductions (relative to the benchmark) in return for a share of the fixed funds (£215 million) that the government has made available for this program. Finally, a fairly complex mechanism ("the Gateway") regulates a limited form of exchange between the CCLA and ETS systems. This mechanism is the result of trying to balance several difficult aspects, including design efficiency, fairness, and feasibility with respect to competitiveness. The energy-intensive sectors are typically distinct from the rest of industry and often have the power to demand special treatment (see Chapter 24 concerning energy taxation in Sweden).

Several European countries are starting to trade "green certificates" based on the obligation, by decree or law, of energy companies to use a certain percentage of renewable or nonfossil sources in their electricity production. Electricity producers that exceed their shares are allowed to sell excess certificates to producers that have difficulty meeting their obligations.

In addition, several other countries have operational trading schemes or advanced plans, such as the Czech and Slovak Republics (for $SO_x$) and the Netherlands (for $NO_x$). The Dutch program puts considerable emphasis on institutional design and development and was first in Europe in setting up a specialized Environmental Tax Unit to systematically review and develop green taxes. In a similar manner, they also have set up an Emissions Authority to monitor emissions and oversee the development of emissions trading.

## Trading Programs for Other Resources

Another major and generally successful set of tradable permit programs is the ITQ program for fisheries (see Chapter 28). Many of these programs have dramatically reduced excessive fishing effort and thus restored profits and saved fish stocks. Despite this success, the process has not been without problems. Fisheries are a valuable resource, and the concentration of shares in the hands of a relative minority has created considerable social tension in some fishing communities. Also, some problems remain, such as the discarding of juvenile fish. Still, the overall picture is positive, and one of the contributing innovations is that the quotas are not for fixed harvests of fish but for fixed percentages of a total allowable catch. The total allowable catch, in turn, can be changed at short notice in response to variations in the stock.

Other examples of tradable permit programs in natural resources management include grazing rights, water rights, and transferable development rights (TDRs) for land planning. TDRs allow land planners to overcome many of the shortcomings associated with traditional zoning practices. A TDR program restricts development in one zone, for instance, to allow for the creation of a park. In exchange, the landowner is given the right to transfer a "development right" to another zone where development is permitted, with the help of TDRs purchased from the first zone. To create a green zone around a city usually entails the problem and cost of expropriating properties from landowners. Using TDRs, these landowners are partially compensated, and a large group of landowners shares the burden. However, numerous legal issues still surround this instrument (Miller 1999).

## Supplemental Reading

**Emissions Trading Programs (Early U.S.)**
Hahn 1989
Krupnick 1986
McGartland and Oates 1985
Montero 2000
Portney 1990
Tietenberg 1985, 1990

**Emissions Trading Programs (Non-U.S.)**
Bohm 1999, 2000
Convery 2001
Convery and Katz 2001
EEA 2000
Grubb 2001
O'Ryan 1996
Stavins 2001

**Zonal Trading and Ambient Permit Trading Schemes**
Kolstad 2000b
Tietenberg 1992

## Notes

1. There may still be some difference between grandfathering and auctioning due to wealth effects, and there could be some effects on entry and exit—and thus on industry size and structure. The major difference is likely to be in the political feasibility of a given permit scheme. If there is no difference, then the only arguments in favor of grandfathering would appear to be either a "prior appropriation" concept of rights or simply the bargaining power of the polluters.

2. Initially, the United Kingdom was somewhat of a laggard in environmental policy, but since this situation changed, permit trading is apparently becoming a favored instrument. By contrast, Germany has long had an effective series of policies that include an active use of "best available" technology, environmental taxes, and voluntary agreements. Germany is one of the few European countries that appears to be on track to meet its Kyoto Protocol targets. Thus, whether there is a real need and a role for a new policy instrument is more questioned in Germany than in other E.U. countries.

# CHAPTER 8

# *Taxes*

*E*CONOMISTS OFTEN VIEW ENVIRONMENTAL CHARGES as the most expected instrument for environmental and natural resources policy and tend to use them as a point of reference for other instruments. A pure environmental charge is referred to as a Pigovian tax if it is set equal to marginal social damage (e.g., of some pollution) (see Box 8-1). At least under several classical assumptions (including fully informed, honest, welfare-maximizing regulators and appropriate concepts of property rights), they have certain optimality properties. However, in many cases, the pure environmental tax is hard to use (e.g., if pollution is unobservable) and the available proxies or substitutes (such as input or output taxes) are more or less suitable.

In this chapter, I discuss some of the differences among these alternatives.

## Pigovian Taxes

One of the main conditions for optimality is that a Pigovian tax is set equal to marginal damages, but they are difficult to estimate. The reasons for this difficulty include lack of understanding of the multiservice and public good characteristics of ecosystems (market prices commonly are missing, and a large subfield of environmental economics is dedicated to studying various methods of nonmarket valuation). Another complication is the fact that the damage curve may have a large slope. It is important that an environmental protection agency set the tax or charge equal to marginal damages at the optimal pollution level (i.e., at the intersection of the marginal damage and cost curves), which may be different from marginal damages at the time of the decision itself (e.g., see Figure 6-2 in Chapter 6). However, analogous problems exist for the setting of standards.

Particularly for economies with rapid growth (such as in many fast-growing developing countries and many of the formerly planned economies), the current or observed marginal damages may be difficult to calculate; evaluating their cost at

## Box 8-1. Pigovian Taxes

Consider a representative firm that seeks to maximize its profit after abatement costs and taxes. It solves

$$\max Pq_i - c_i(q_i, a_i) - Te_i(q_i, a_i) \tag{8-1}$$

where $P$ is product price, $q$ is output, $i$ is a firm, $c$ is production costs, $a$ is abatement, $e$ is emissions, and $T$ is the Pigovian tax. The necessary and sufficient first-order conditions for a firm with positive output and abatement costs are

$$P = c_q' + Te_q' \tag{8-2}$$

$$c_a' = -Te_a' \tag{8-3}$$

where $c_q'$ is the derivative of production costs with respect to output.

A comparison of Conditions 8-2 and 8-3 with the conditions for optimality in Conditions 6-2 and 6-3 shows that the conditions are identical as long as the Pigovian tax is correctly set equal to the level of marginal damages ($T = D'$).

---

a hypothetical optimum that society hopes to move toward is even more difficult. In a situation with high historic levels of pollution (as in many eastern European countries), the marginal damage may be expected to fall significantly as the worst excesses of pollution are cleaned up. Not overestimating the Pigovian charge may be crucial because in the long run, it might turn out to be too high, and in the meantime, high taxes may cause unnecessary firm closures, unemployment, and opposition to the environmental charge instrument per se (see Chapter 25).

Timing is crucial in the context of policy instrument implementation because environmental charges can cause liquidity problems for firms; taxes for firms are highest while their corresponding environmental investment costs are highest. One solution to this cash-flow problem is to levy a two-tiered tax on polluters: a zero- or low-level tax up to a certain "permissible" level, then a higher tax rate for "excess pollution." Proponents of this scheme suggest that it makes polluters pay the appropriate tax at the margin while leaving firms sufficient funds to finance abatement investments (and avoid bankruptcy). A two-tiered tax would be appropriate for an intermediate concept of property rights to the environment. For example, a firm that has rights to some minimal level of ecosystem services would correspond to the part with zero or low payment. A firm that needs to pollute or use resources in excess of this level would have to pay the full rent for the resource use to which it does not have any rights. This principle is related to refunded emission payments and tax subsidies (see Chapters 9 and 15).

Other ways of avoiding liquidity problems include waiving taxes in return for substantial abatement investments, making the cost of certain investments deductible from the tax, or giving firms sufficient time to adapt by announcing the tax a few years in advance. The United Kingdom did something like this by announcing the so-called fuel tax escalator (see Chapter 21).

It is important to emphasize that a Pigovian tax (i.e., equal to marginal damages) is, under a range of circumstances and from a broad perspective, a full solution to the problem of how to internalize externalities. Such a tax is not only necessary but also sufficient, at least if appropriately adapted so that taxes vary with geographical and temporal variations in the damage function. The victims of pollution do not have to be compensated to arrive at an efficient resource allocation (Baumol and Oates 1988, chapter 4; Oates 1983). Although (ethics- or rights-based) welfare reasons for compensating victims might exist, the practice may provide an undesirable incentive for the victims to locate next to polluters and intentionally not take steps to protect themselves against the effects of the pollution so they can collect compensation.

Taxing victims could, under some circumstances, be used to "keep the victims away" from such locations, thus reducing environmental damage, at least as conventionally measured by disutility to victims (Coase 1960). In general, this policy is not perceived as attractive or fair, but it does illustrate the nature of the problem. Consider, for instance, people who have intentionally located near major flood-prone rivers and then regularly collected compensation when the rivers flooded. The "tax equals damages" rule is enough to keep damages at the socially optimal level, and from the viewpoint of allocation at least, additional instruments (such as compensation) only distort the incentives away from the optimal level. However, in some cases, the victims are few and may not be able to do anything to increase or decrease the effect (e.g., through self-protection or through perverse effects of moving relative to the polluter's location)—say, if the community decides to implement a technology that will disturb a small minority of particularly sensitive people. In such cases, compensation probably will be motivated but should be designed in a lump-sum fashion to avoid perverse incentives for the victims.

Setting the level of tax is far from trivial; in fact, it is one of the main difficulties with the tax or charge approach. In addition to the purely technical difficulty of estimating slopes of damage and abatement cost curves (and their intersection), numerous informational and political issues must be addressed. The firms on which an environmental protection agency will rely for information usually have a tactical reason for either withholding or distorting information, and polluters (businesses, their trade unions, or even other public bodies such as municipalities, which are among the worst offenders) may exert considerable pressure on an environmental protection agency for lower (or no) taxes. In addition, because municipalities typically lack funds and play the dual role of polluter and regulator, they sometimes can grant themselves some form of immunity.

In many real-world situations, the charge is adjusted by trial and error. A form of tâtonnement (using trial and error to obtain an optimum solution; in this context, not only end-of-pipe investments but all investments where abatement may be only one small but perhaps inseparable component) can be emulated. However, this approach may be unsatisfactory, particularly when abatement investments are expensive, and long-run stability is important for any policy that is to affect investments without undue cost. Introducing a given tax $(T)$ in small, previously announced steps may help reduce resistance to the tax. Most important, this approach may leave the firms with sufficient funds to avoid the liquidity problems commonly associated with high environmental charges. However, "path dependency" may develop if low tax levels lead the firm to invest in minor end-of-pipe investments that later prove to be insufficient and thus unnecessary because the process has to be fundamentally changed as a result of higher environmental taxes. A fee or charge that is partly refunded (see Chapter 9)—or another instrument that implies smaller total (but the same marginal) costs to the polluter—might generate less resistance from polluters and could be set higher than would have been possible for a pure tax.

When damages are difficult to estimate, the abatement costs are sometimes used as a proxy. Although the two marginal costs are supposed to be the same in equilibrium, they both have slopes, and so a point estimate of one is unlikely to be the optimal value. However, if only the abatement costs can be estimated, this approach has the advantage of giving regulators a feel for the reasonable range of values and an idea of the financial consequences for the firms.

In some settings where regulation is the main instrument, fines are levied for noncompliance. Fines are commonly set equal to the "money gained" by not complying—which is roughly equivalent to abatement costs avoided. If the fine is the only punishment for noncompliance (and if detection is automatic), then this regulation can be seen and analyzed much like a charge, where the fine takes the place of the charge (see Chapter 16 for more complex cases with probabilities of monitoring and risk aversion).

Sometimes the difficulty in deciding the level of an environmental charge or charge tax is attributed only to the policy instrument, which is somewhat misleading. For instance, it may be asserted that an environmental protection agency must use trial and error to find the charge level appropriate to meet the target emission level. This assertion would be correct if the abatement target level was exogenously given and the only concern was to meet the target cost-effectively. However, the target cannot be determined without precisely the same information needed to find the charge level. One might just as well speak of an optimal charge level (which might be based on a fairly reliable estimate of how much the companies can "afford"—or will accept without migrating or fighting the tax in the courts) and trial and error with quantitative limits to achieve the corresponding abatement cost.

From the viewpoint of efficiency, the determination of the two variables (fee and quantity) is simultaneous and symmetric. However, in practical environmental protection, target levels may be formulated by ecologists or engineers (rather than economists) and gain general acceptance without being explicitly deduced from an analysis of marginal costs of abatement and environmental damage. In situations with strong irreversibilities or threshold values, this approach may in fact be reasonable (instruments used under conditions of uncertainty are compared in Part Three). It also may apply to national policies formulated to meet obligations under an international agreement.

## Taxes, Charges, and Earmarking

The terms *taxes* and *charges* are sometimes used interchangeably in this and other texts. However, *taxes* usually are reserved for politically rather than administratively decided fees and typically go to the treasury rather than being earmarked for local or sectoral use, whereas *charges* may be levied and appropriated by sectoral agencies. In principle, this distinction is important, but some hybrid or bor-derline types of fees also exist, such as politically decided fees that are earmarked for environmental funds and locally decided fees that are paid to the general treas-ury. The previous analysis applies to the classical Pigovian tax. However, taxes have a couple of disadvantages, one of which is the relatively complex legal pro-cess involved in passing and modifying tax laws, which can make the tax instru-ment somewhat blunt.

Furthermore, many politicians have encountered considerable resistance to environmental taxes, and local or sectoral charges (over which the polluter or resource user has some influence) typically are more readily accepted (see Chap-ters 14 and 16). In many countries, money that goes to the central treasury is per-ceived as being "lost," whereas money that stays within the sector or region is not. Environmentalists, too, commonly favor charges that are earmarked for pub-licly financed abatement. Some even view the main purpose of the fee as to raise money for such purposes, which misses the central points of a Pigovian tax (i.e., the incentive for abatement and the output substitution effect).

In general, economists are skeptical of earmarking taxes for special purposes, typically arguing that all tax revenues should go to the treasury and that public goods (including abatement) should compete on an equal footing for public funds. In both developed and developing countries, earmarking is considered an additional constraint in the optimization of government taxes and expenditures (OECD 1996; McLeary 1991). With such an approach, it never can be the opti-mal policy. Other approaches—such as public choice or asymmetric informa-tion—are required to allow for the possibility of earmarking as a ("second-best") strategy.

In general, the financial tradition in Europe is negative to earmarking or "hypothecation" (the United Kingdom and France are discussed in Chapter 21). However, in the United States as well as many other industrialized and developing countries, funds are commonly earmarked for such major economic undertakings as road construction. In an economy where many funds are already earmarked and where the public allocation of funds does not operate optimally, political economy arguments may favor the earmarking of some environmental fees, too. An analogous situation in environmental policy relates to water and sewage management, for which fees are charged to finance supply. Similarly, the money from fishing and hunting licenses commonly is used to preserve habitat. The U.S. system appears to be more in favor of earmarking not only fees for road construction but also environ-mental charges, as illustrated by Superfund, the Oil Spill Liability Trust Fund, and other funds to which most U.S. environmental taxes are paid. Despite the generally negative attitude toward earmarking (e.g., of fuel taxes) in France, water charges and emission taxes on sulfur, volatile organic compounds, and nitrogen oxides are ear-marked for abatement investments by the polluting firms (e.g., OECD 1995).

If one accepts some form of earmarking, then one reasonable case might be to use sulfur taxes to finance the liming of lakes because it helps redress their ability to withstand the acid rain caused by sulfur emissions. Although it would be pure coincidence if the optimal allocation for liming happened to equal the proceeds from a sulfur tax, it may still be a politically convenient arrangement. An early proponent of earmarking suggested strong public choice arguments in favor of earmarking as a flexible mechanism for allocating public funds (Buchanan 1965). Distributional concerns may also be important. If all the taxes are raised in one urban area, for example, then it seems realistic to expect local pressure to keep the proceeds rather than spread them across the whole nation (see Chapters 25 and 31). Similarly, if a certain group of people (such as motorists) pays more taxes, then that group will pressure for spending that benefits the group. In the real world, the government poorly distinguishes between agents of the economy. Particularly when environmental taxes have large potential proceeds (such as those resulting from a carbon tax), the distributional effects of taxation and spending may be significant. In such cases, the general equilibrium effects of the proceeds must also be considered (Chapter 14). In the absence of optimal tax and transfer instruments, earmarking may be a second-best strategy (Pirttilä 1998).

## Taxes on Inputs and Outputs

When emissions monitoring is impossible, difficult, or costly, taxes may be levied on some input or output that is more easily monitored and a good indicator of (or proxy for) the pollution to be regulated. For the sake of simplicity, taxes on inputs or outputs are sometimes included among the environmental taxes, but they are actually somewhat different. They sometimes are referred to as "presumptive taxes" because, in the absence of direct monitoring, the agent that uses a certain input or produces a certain output is presumed to be polluting. Because monitoring individual emissions is often impossible in developing countries, some researchers suggest that presumptive taxes should be used—for instance, taxes based on inputs or outputs that can be presumed to be close complements to the pollution. A polluter that demonstrates abatement or clean technology can be exempted or refunded, but the burden of proof is moved from the environmental protection agency to the firm (Eskeland and Devarajan 1996). The importance of this factor is presumably greatest in poor developing countries that have little experience with environmental control. Excise taxes on products such as cars, fuel, and cigarettes already make up a large share of taxation in these countries (Sterner 1996). Although environmental motives may have played some role in creating these taxes, the main reason is probably the ease of tax administration; another factor may be the relative attractiveness in taxing these goods as luxury goods, or for moral or ethical reasons.

If the output of a firm is closely correlated to the pollution to be regulated, then the product tax may be a good proxy. For example, consider a firm that produces hydrochlorofluorocarbons (HCFCs; substitutes for chlorofluorocarbons that are somewhat less damaging to the ozone layer). An output tax would be about the same as a tax on ozone depletion, although there should be a deduction for HCFC recycled or collected in closed systems. The workings of an output tax are illustrated in Box 8-2. A similar approach applies to the use of inputs that are directly related to the amount of pollution. Thus, if a firm uses HCFCs as an input, then an input tax is almost the same as a tax on ozone depletion (unless the firm collects and disposes of the chemicals appropriately, in which case the tax ideally would be rebated).

## Box 8-2. Environmentally Motivated Product or Output Taxes

Consider a firm as in Box 8-1, with the difference that there are no abatement measures available. This can be modeled by replacing $e_i(q_i, a_i)$ from Equation 8-1 with $e_i = \xi q_i$, where e is emissions and $\xi$ is a constant. This substitution implies that there is no abatement ($a$), giving

$$\max Pq_i - c_i(q_i) - T\xi q_i \qquad (8\text{-}1')$$

where $P$ is product price, $q$ is output, $i$ is a firm, $c$ is production costs, and $T$ is the Pigovian tax. There is now only one first-order condition:

$$P = c_q' + T\xi \qquad (8\text{-}2')$$

where $c_q'$ is the derivative of production costs with respect to output.

The welfare function (Equation 6-1) that the social planner seeks to maximize becomes $W = \Sigma_i[Pq_i - c_i(q_i)] - D(\Sigma_i\xi q_i)$, giving the condition $P = c_q' + \xi D'$, where $D$ is marginal damage (compare this condition with Condition 6-2; Condition 6-3 does not have a counterpart because there is no abatement). As expected, this result is achieved by setting the tax equal to marginal damage $T = D'$. Because the emission coefficient is constant, the emissions tax can be replaced by an output tax of $T\xi$ per unit of output.

If absolutely no abatement technology is available (which is highly unlikely), then a product tax will be the optimal response because it incorporates the externality into the product price and thus allows the demand side to give an "output" effect (i.e., reduced output as a result of consumers choosing other goods). However, if the possibility of some type of abatement exists, then an output tax will be suboptimal. If the abatement possibilities are severely limited and monitoring is difficult, then taxes on inputs or outputs may be good second-best instruments. Gasoline taxes are one example. Because of its flammable nature, gasoline is sold subject to a certain amount of physical control in most countries, so the extra (administrative and technical) cost of taxing gasoline sales is small. The relationship between gasoline consumption and the environmental damage caused by vehicle emissions is far from simple, but valuing and monitoring emissions is so complex that gasoline taxes are in fact often chosen as a proxy. In this case, input taxes do not provide any incentives for technical progress or substitution that could help in abatement, and therefore, it may be advantageous to combine gasoline taxes with other instruments, such as mandatory catalytic converters or even subsidies for such abatement technology (see Chapter 4).

Presumptive taxes provide an incentive for clean firms to disclose their technology if they can thereby reduce their fees. One good example is the Swedish tax that is levied on the sulfur content of fuels. This sulfur is presumed to be released on combustion. However, the tax is repaid on proof of abatement (see Chapter 24). This method is somewhat akin to a deposit–refund system (see Chapter 9) or a so-called *tax expenditure*—an exception to taxation that may be seen as an implicit subsidy for some kind of activity. Tax expenditures are quite common in OECD as well as in developing countries (Sterner 1993, 1996; see also Chapter 24).

## Taxing Natural Resources

In natural resources management, the counterpart to Pigovian taxes is the levying of fees such as mining royalties, stumpage fees, user fees, and land taxes. These fees might be charged because of external effects (including stock or catchability externalities) or because of the absence of property rights and thus of payment for a scarcity rent ($\lambda$) as in the Hotelling rule (see Chapter 4). Policymaking consists of correcting for externalities and other market failures as well as imposing a scarcity rent $\lambda$ when there is no effective owner who can claim such a rent. The analysis is closely analogous to the case of depletable minerals discussed in Chapter 4; however, for many renewable resources (such as forests, water, fish, and other ecosystem resources), the dynamics are more complicated. Fees charged for natural resources are commonly a considerable source of revenue in low-income countries, and in other countries, they could be (some examples are discussed in Part Six).

# CHAPTER 9

# Subsidies, Deposit–Refund Schemes, and Refunded Emissions Payments

$C$HAPTER 8 WAS ABOUT VARIOUS FORMS of taxes or charges. The "other kind" of price instrument is often said to be a subsidy. Although these two instruments may seem to be worlds apart to a layperson, economists tend to think of subsidies as similar to taxes. They can be partly analyzed as negative taxes, and some reflection shows that they are indeed similar in many superficial respects but that there are also crucial differences between fees and subsidies.

At a deeper level, the main differences between taxes and subsidies are their implications for the questions of ownership and rights to nature. Also, there are more combinations of and hybrids between taxes and subsidies than might at first be apparent. They can be very interesting from a policy perspective because they allow decisionmakers to keep some of the positive aspects of an instrument while avoiding some of the negative ones.

## Subsidies and Subsidy Removal

There are many forms of subsidies, from tax expenditures (see Chapter 8) to more classical direct, budget-financed payments in support of certain activities (or people). Subsidies may apply to payment for certain "services," prices for certain inputs or technology, loans, or access to credit markets.

As shown in Box 9-1, subsidies work similarly to taxes, but their properties with respect to the number of firms in an industry are different. Subsidies lack the output substitution effect of taxes. Not only are the price and thus the output effect missing, but at least some subsidies (like the kind analyzed in Box 9-1) create a perverse or opposite effect because they tend to encourage the entry (or delay the exit) of polluting firms, resulting in too many firms and too much production and pollution (possibly even more pollution than when unregulated [Baumol and Oates 1988]) (see Chapter 14).

The perverse output effect of subsidies can be reduced by careful design (this is the other alternative mentioned in Box 9-1). Instead of a payment for each unit

---

## Box 9-1. Similarities and Differences between Subsidies and Taxes

A subsidy can be either a direct (partial) repayment of abatement costs or a fixed payment per unit of emissions reduction. In the latter case, the subsidy can be seen as a kind of negative tax. Suppose that the polluting firm ($i$) has an initial emissions level of $e_{i0}$ and is paid a subsidy $T$ for each unit of emissions abated beyond the initial level. The polluter then receives a subsidy $T(e_{i0} - e_i) = Te_{i0} - Te_i$, which can be seen as almost identical to the tax case in Equation 8-1 (see Chapter 8) if $T$ is set equal to the Pigovian tax.

In fact, the subsidy is a combination of a variable tax ($Te_i$) and a fixed subsidy ($Te_{i0}$). Because the subsidy is fixed, this component does not affect the first-order conditions for optimality of a representative firm, which are still the same as Conditions 8-2 and 8-3 (see Chapter 8). Thus, this firm will have the same incentives for abatement, which shows that there is indeed some similarity between a tax and a subsidy. However, differences stem from the fixed subsidy. It lowers the total and average cost to the firm, making it lower under subsidies than under taxes. A marginal firm that would exit the market (produce zero or go bankrupt) under taxes might survive under subsidies. Consequently, the whole industry would tend to have too many firms and produce too much output.

In terms of allocative efficiency, this scenario is a loss (although one can imagine settings where it may be seen as an advantage). In the long run, entry might even be encouraged, and the number of firms in the industry would differ as between a subsidy scheme and a tax regime. On the other hand, if the subsidy covers only abatement costs, then there is no such incentive for entry.

---

of abatement compared with a baseline, the subsidy is typically a (partial) repayment of verified abatement costs (often fixed capital costs for a filter, catalytic converter, or some other item). The subsidy instrument does not fulfill the polluter pays principle, but (partly for that reason) it is popular with polluters and in some cases may be feasible from a practical standpoint, particularly when other instruments are infeasible (e.g., when no polluter can be identified, as is the case with some historic cases of pollution, particularly in developing or formerly planned economies). If the polluting company is nonexistent, bankrupt, or unidentifiable (such as after some types of oil spill), then the public sector may have little choice but to finance cleanup with public funds—or subsidize other companies to do the cleanup. The same applies if, for other reasons, the public sector has insufficient power. However, making an industry collectively responsible for certain kinds of cleanup may lead to conflicts, as shown by U.S. experience with the cleanup of hazardous waste sites under Superfund.

Other situations in which subsidies might be warranted would be when the polluter expressly owns the property rights to a certain resource that society wants, or in financing research or development that has strong public good characteristics. Such research is typically underfinanced because firms cannot profit from it. In many countries, the credit market is described as "imperfect" and skeptical toward loans for the financing of investments in pure abatement. This lack of research provision is also an instance when subsidies (or public provision) may be warranted. The most practical argument against subsidies is that they are too expensive as a policy instrument—especially in developing countries, where the opportunity cost of public funds is high. In such countries, taxes are particularly hard to collect, and many pressing needs (such as primary schooling and health) need to be satisfied.

In reality, the most relevant issue for the environment is not subsidies for abatement but the prevalence of perverse subsidies for pollution. Inappropriate subsidies promote rather than prevent wasteful and environmentally destructive behavior. Well-known examples include large subsidies for energy use in many countries, particularly oil-exporting countries and the formerly planned economies (Kosmo 1987). The formerly planned economies subsidized the domestic consumption of not only energy but all natural resources (Bluffstone and Larson 1997). In the fishing industry, harmful subsidies (that increase rather than decrease effort) are common, and enormous subsidies for forest destruction in Brazil have been documented in recent decades (Binswanger 1991; see also Chapter 30). Perverse subsidies are so common that "subsidy removal" (e.g., from pesticides, fossil fuels, and land conversion in the Brazilian Amazon) is often classified as an environmental policy instrument in itself.

It is somewhat ironic that the undoing of one bad policy should be classified as a new policy instrument. However, the removal of subsidies is politically complicated, because subsidies become intertwined with vested interests. In fact, the value of subsidies (or of tax expenditures on goods and services that are implicitly subsidized because they ought to be subject to Pigovian taxation) typically is capitalized in property values. If an individual buys a house with electric heating in a cold climate, then the dependability of the heating system is one of the most important attributes of the house. If the government changes the value of this attribute after the house is bought by abolishing nuclear power or taxing fossil power, then the value of the house may plummet in expectation of future energy bills. Properties acquired just before a policy change can suffer particularly serious losses in value. In general, the political economy of subsidies is based on the fact that those who benefit are often few and well organized, whereas those who stand to lose are many and have a small, diffuse interest (see Chapter 16).

## Deposit–Refund, Tax–Subsidy, and Other Two-Part Tariff Systems

In addition to the basic policy instrument types already mentioned, several more complex instruments exist, many of which are combinations of the other instruments. In fact, combined instruments is a fertile area. Quantity restrictions, for instance, can be combined with a (small) subsidy for overcompliance or with (high) fees for pollution emitted above a certain level. Such combinations may serve as a "safety valve" if regulators are not sure of the optimal pollution level (see Chapter 13). They also may provide a way to collect information on abatement costs.

Some combinations are so well established that they are considered policy instruments in their own right. One such instrument is the deposit–refund system, which encompasses a charge on some particular item and a subsidy for its return. This instrument can be used to encourage environmentally appropriate recycling. Assuming disposal is inappropriate for ecological reasons, the deposit–

## Box 9-2. The Technical Operation of Deposit–Refund Systems

If the output ($q$) in our model is identical to the pollution ($e$) to be abated, as in the case of cadmium batteries, and abatement ($a$) entails the collection of used product for recycling, then an equation such as Equation 8-1 can model a simple damage function for firm $i$: $e_i(q_i, a_i) = q_i - a_i$. The first-order conditions are then simply

$$P = c_q' + T \tag{9-1}$$

$$c_a' = T \tag{9-2}$$

where $P$ is product price, $c$ is production costs, and $T$ is the value of the refund. The damage cost of inappropriate disposal is included in the price, and there is an incentive for "abatement" (collection and return) up to the value of the refund.

refund combination may be categorized as a tax expenditure or a presumptive tax on inappropriate disposal. The polluters (i.e., those who do not return the item) pay a charge, whereas those who return the item collect a refund and thus pay nothing (see Box 9-2). The distinguishing feature of the deposit–refund system is that it has a clever disclosure mechanism: the refund is paid when the potential polluter demonstrates compliance by returning the item that carries the refund, thus making the monitoring of illegal disposal unnecessary. Recent developments in deposit–refund systems show that the deposit and the refund need not be the same amount. In fact, a revenue-neutral construction would have higher refunds in general if the rate of return were incomplete (Mrozek 2000).[1]

Usually, deposit–refund systems are used for certain final outputs (beverage cans and bottles are the classic examples), and abating environmental pollution has been far from the only (or even main) motivation. However, the concept is spreading, and Sweden has instituted a deposit–refund scheme on scrap vehicles to prevent such vehicles from being left to rust in the woods. It is conceivable that a similar instrument could be used for other polluting inputs, such as cadmium or mercury, but practical complications might arise (e.g., rogue imports from other countries for the sole purpose of collecting refunds).

Wastes generally can be disposed of by appropriate (municipal) waste management or by irresponsible "dumping." If there were no monitoring problems or irresponsible dumping, then the appropriate "first-best" instrument would be a Pigovian charge on dumping. If this charge is not feasible (because monitoring is impossible), then the charge must be applied to all goods at purchase and thus becomes a general consumption tax; if the good is properly returned, then the tax is refunded (Fullerton and Kinnaman 1995). The new waste disposal "charge" is thus composed of two parts: a standard waste disposal charge (to cover the cost of municipal disposal) and a refund. This amount is equal to the difference in externality between dumping waste illegally and disposing of it through the proper systems. The latter is assumed to be better from an environmental viewpoint, and thus this whole term is negative. Whether the amount is large enough to outweigh the first term is an empirical matter, but if it is, the optimal fee to the consumer for ordinary waste management might become zero, or even negative. Thus, a type of deposit–refund system has been created in which a deposit is paid on all purchases and later refunded in connection with recycling or (municipal) waste collection, which also is subsidized (applications are discussed in Chapters 21, 23, and 27).

Even though deposit–refund systems have been used mainly for waste management and recycling, the same concept can be applied in other two-tiered instruments by combining a general charge with a subsidy to encourage an environmentally beneficial behavior. In cases of asymmetric information (when the regulator has difficulty in collecting information about pollution), it may be effective for the regulator to design contracts or menus of instruments from which the agents can choose. For instance, a high tax on pollution will discourage firms with high abatement costs from disclosing their pollution levels. However, if high taxes are combined with subsidies for truthful disclosure (e.g., for measurement equipment and technical help, or certain types of abatement investments), then the combination can be made attractive even to high-level polluters.

An especially interesting two-part instrument is the tax–subsidy scheme, in which polluters are assigned property rights to a "baseline" emissions level set by regulators (Pezzey 1992; Farrow 1995). Polluters pay a charge per unit of emissions above the baseline and receive a subsidy per emission "saved" whenever emissions are below the baseline. New firms have no property rights and must pay for their emissions. This instrument has some attractive features. It provides a clear link between the extent of a polluter's property rights to the environment and the size of the baseline on the one hand, and hence, the extent to which society or the polluters pay the scarcity rent on the other hand. It also is general enough to encompass both subsidies and Pigovian taxes as special cases. If the baseline is zero, then the instrument is a tax; if the baseline is equal to, say, current (unregulated) emissions, then the instrument becomes a generous subsidy.

## Refunded Emissions Payments

Another two-part instrument with some interesting properties is the refunded emissions payment (REP; Sterner and Höglund 2000). An REP is a charge, the revenues of which are returned to the aggregate of taxed firms. Thus, polluters pay a charge on pollution, and the revenues are returned to the same group of polluters, not in proportion to payments made (which would be nonsensical because it would make all net payments zero) but in proportion to another measure, such as output. Thus, the individual polluter pays a tax on its emissions ($Te_i$) and receives back a share of the total fees collected [$(q_i/Q)TE$] based on output ($q_i$), where $Q$ and $E$ are aggregate output and emissions, respectively (see Box 9-3). The aggregate of all firms pays for abatement costs, but it does not pay anything more to society. Instead, the net effect of the payment and refund is that the firms with above-average emissions make net payments to the cleaner-than-average firms.

The incentives for abatement with an REP are essentially the same as with a Pigovian tax, as long as the number of firms is large enough that the market

---

## Box 9-3. The Refunded Emissions Payment

Each company ($i$) seeks to maximize profit as in

$$\max Pq_i - c_i(q_i, a_i) - Te_i(q_i, a_i) + q_i/(\Sigma_i q_i)T[\Sigma_i e_i(q_i, a_i)] \tag{9-3}$$

where $q$ is output, $c$ is production costs, $a$ is abatement, $Te_i$ is the charge, and the last term is the refund. The first-order conditions for a firm with positive output and abatement are

$$P = c_q' + Te_q'(1 - \sigma_i) - T(E/Q)(1 - \sigma_i) \tag{9-4}$$

$$c_a' = -Te_a'(1 - \sigma_i) \tag{9-5}$$

where $\sigma_i = q_i/Q$ is the firm's market share, $Q$ (gross output) $= \Sigma_i q_i$, and $E$ (total emissions) $= \Sigma_i e_i(q_i, a_i)$.

The social planner is assumed to maximize welfare ($W$, see Conditions 6-2 and 6-3 in Chapter 6). Comparison with these conditions and with Conditions 8-2 and 8-3 (Chapter 8) for Pigovian taxes shows two differences between refunded emissions payments and a pure tax instrument. First is a factor $(1 - \sigma_i)$ in Condition 9-4 for price and output and in Condition 9-5 for abatement. This factor is close to unity as long as a competitive economy with many firms is studied, because the market share of each firm will then be small. The second and main difference is in the supply price of Condition 9-4, which has the refund term, $T(E/Q)(1 - \sigma_i)$. This term shows the output effect: it results in a lower price and thus larger output than with the tax instrument. However, the price will still typically be higher—and output lower—than in a situation without any instrument at all.

---

shares are small and the REP instrument thus provides adequate incentives for cleaner technology (see Box 9-3). However, the effects of REPs on output price are different from those of taxes (and auctioned or grandfathered emissions permits). The average cost will be lower with REPs than with taxes (because of the refund) but higher than if unregulated (because of abatement costs). This comparison assumes that the fee in the REP would be of the same magnitude as the Pigovian tax. (However, this assumption is not realistic, because a tax will meet much greater political resistance than the REP, which results in lower net costs to firms.) In fact, output prices with REPs are similar to those under perfect physical regulation (see Chapters 6 and 14) because industry, as a whole, pays no net fees to society.[2]

These features are both the strength and the weakness of the REP instrument. An REP provides regulators with a price-type policy that can be used for firms that have the right or the power to stop environmental taxes or other instruments that are too onerous. As long as there are significant technical abatement possibilities, the instrument will work well. On the other hand, if no abatement technology is available, the REP will not act as an output or product tax because the output substitution effect is missing in the same way as with, for example, design standards.

An REP scheme in which refunds are proportional to output is analogous to an output-based allocation of emissions permits (Sterner and Höglund 2000; Fischer 2001).[3] This comparison reflects the function and character of the instrument and shows that policymakers are free to choose a price- or a quantity-type instrument irrespective of who (the polluter or society) has the property rights to the environment; even if polluters have the political power to stop pure taxes, regulators do not automatically have to choose a quantity-type instrument such as grandfathered permits (see Chapter 15 as well as examples in Chapters 5 and 6).

## Supplemental Reading

Bohm 1981
Farrow 1999
Pezzey 1992
Yohe and MacAvoy 1987

The main advantages of REPs—and indeed of tax–subsidy schemes—over taxes relates to the distribution of costs and thus the political economy of the instrument. Because all firms will pay less and some firms even make money in a REP scheme, it will not create the same kind of compact resistance (and lobbying) as taxes. The REP charge also can be set higher, giving stronger abatement incentives than a tax. The fact that the REP is cost neutral also means that it can be used for a subgroup of polluters without greatly affecting their competitivity vis-à-vis those who are not included. In some cases, this feature might provide valuable degrees of freedom to the decisionmaker.

## Notes

1. Suppose that the deposit for a can is $x$, the fraction of cans returned is $y$, and the refund is $z$. The system would be "revenue neutral" if the refund were larger than the deposit ($z = x/y$). Regulation can be analyzed as a two-stage process in which the regulator applies a charge $T$ on an emission that can be avoided by the customers' return of the product to the firm. The firm chooses a refund (which need not be the same level as the charge) to entice the customers to hand in the product. The product might have some reuse value (or the offer to the customers might in practice be part of a deal to buy a new model, as is common with vehicles), and the fact that it is returned helps the firm reduce its environmental charges to the state. Firms tend to offer refunds to customers as long as the tax savings plus reuse and other values are larger than the costs involved (Bohm and Russell 1985).

2. This concept can be neatly illustrated for an average enterprise with a fixed rate of emissions ($e = \xi q$), $de/dq = e/q = \xi = E/Q$: the new refund term $-T(E/Q)(1 - \sigma_i)$ would be equal to the tax term $Te'q(1 - \sigma_i)$, which it would cancel, leaving $P = c'_q$.

3. In this case, assume that the firm needs $e_i$ permits but permits are allocated depending on the firm's current output, which means $(q_i/Q)E$ permits. Compared with the zero-revenue auctions of permits (Hahn and Noll 1983), the permits are auctioned but the revenues reallocated to the polluters in proportion to a fixed algorithm (e.g., historic pollution share).

# CHAPTER 10

# Property Rights, Legal Instruments, and Informational Policies

IN THE PREVIOUS CHAPTERS IN PART TWO, I discussed many classic policy instruments, such as taxes, zoning, and permits. In this chapter, I turn to a more subtle group of policy instruments that goes to the heart of the structure of the economy and society. I start with the definition of property rights, which may be either private, collective, or other. The actual bundle of rights may be defined or restricted in different ways. I dwell particularly on common property resource (CPR) management, because it is important to the management of ecosystem resources in many countries. The chapter then continues with a section on liability and other legal instruments, which can be seen as restrictions on the bundle of rights mentioned. Legal systems and enforcement vary considerably from one country to another, and this creates diverse conditions for different instruments. I end this chapter by considering environmental (so-called voluntary) agreements and the role of information provision and environmental labeling.

## Creation of Property Rights

The foundation of individual negotiation, bargaining, and all contracts within an economy is the creation or definition of property rights (Amacher and Malik 1996). The evolution of property rights is closely linked to the development of types of scarcity (see Chapter 5); with increased scarcity, value increases and pressure builds to create property rights to the benefits of natural resources (Hanna, Folke, and Mäler 1996). The definition of property rights also can be used consciously as an instrument.

In many countries, a considerable degree of skepticism and uncertainty surround the concept of creating new property rights. The defining feature of property rights is that they are perceived to be permanent; it is this inalienability that gives owners the confidence and incentive to make long-term and costly productive investments in their properties. However, lack of flexibility may entail considerable risk: if a government incorrectly assesses the carrying capacity of an aquifer, a fishery, or a forest area, then it may find that it has given away key national assets (such as water supplies) inappropriately.

Government agencies must strike a fine balance between stability and flexibility. Typically, excessive hesitancy in creating and distributing property titles leads to lost opportunities for positive development. Whereas governments may be considered the best protection against valuable assets falling into "the wrong hands," they have turned out to be bad owners in many cases. They may be such bad owners because they are too distant to provide constructive management or perhaps prone to corruption. The most common problem is a government's inability to develop resources efficiently.

Another aspect of "bad ownership" is that government often is perceived as having taken rights claimed by local communities (particularly of tribal or indigenous peoples). In many cases where such rights were informal (i.e., unwritten), government has used its power to assert ownership. Large social movements have formed in developing countries to protest state ownership of forest and other resources. One famous example is the Chipko Movement in India. *Chipko,* which means "tree-huggers" in Hindi, is a pacifist movement whereby villagers have often triumphed against logging companies simply through passive resistance (i.e., hugging trees).

Despite many examples of bad ownership, the government has reason to maintain ultimate responsibility of natural resources, even if some rights are privatized. Bundles of rights (e.g., of exclusion) are not and should not necessarily be identical across social contexts (see Chapter 5). In some poor countries, for instance, complex webs of local culture and other social norms dictate that poor people have the right to graze their animals on richer farmers' land after the harvest (Chopra 1991, 1998) (see Box 10-1).

In general, the process of modifying rights is fairly slow, and environmental policy instruments (e.g., tradable emissions permits, transferable grazing or fishing rights) evolve relatively quickly. A permit or quota should for instance be seen as a real title to property.[1] In this sense, the development or assignment of new kinds of property rights is a definitive policy instrument of natural resources or environmental management.

During the twentieth century, particularly since the end of World War II, international political organizations such as the United Nations played an active role in defining rights (Wiener 1999). The process started with the U.N. Universal Declaration of Human Rights (United Nations 1948). Although the declaration perhaps was not originally intended to have binding force, its provisions reflect customary international law and have gained binding character as customary international law. This document might be considered the foundation for environmental rights because it guarantees everyone "the right to a standard of living adequate for the health and well-being of himself and of his family" (Article 25:1).

The Universal Declaration of Human Rights was followed by many more declarations, treaties, conventions, agreements, and protocols that span the spectrum of environmental and natural resource issues (e.g., desertification, migratory species, biological diversity, wetlands endangered species, and hazardous wastes and disposal). Some of the more important environment-related agreements that have been adopted in recent years include

---

### Box 10-1. Shaping Property Rights: A New and Sensitive Form of Foreign Assistance

Well-established property rights to land and other resources are a decisive asset for many western countries. In some other countries, these rights either have never existed or were destroyed by colonialism. Some aid agencies (such as Sida) try to help countries with processes such as land surveying and property delineation, which are essential steps in the development of private enterprise (which requires private property rights).

The lack of clarity concerning rights to property is the most serious obstacle to business in many parts of the world, from Africa to the former Soviet Union. However, the definition of *property rights* is difficult, and in some cases, such projects have been abandoned because powerful groups were acquiring an unfair share of the rights.

---

- Rio Declaration on Environment and Development (June 1992),
- U.N. Framework Convention on Climate Change (May 1992), and
- Montreal Protocol on Substances that Deplete the Ozone Layer (adopted Sept. 16, 1987, and subsequently adjusted and amended).

Another important environmental agreement is the 1982 United Nations Convention on the Law of the Sea (mentioned in Chapter 5), through which coastal countries received extended property rights to adjacent portions of the ocean. Some countries found the process unfair, but on the whole, the definition of *property rights* in this and the other treaties is generally acknowledged as being a necessary prerequisite to whatever other attempts at management the countries have undertaken.

The U.N. Environment Programme publishes various environmental law materials with a view to promoting a wider appreciation of environmental law. Organizations such as the World Bank and OECD also have helped shape international environmental law. Within the OECD, the Environment Committee and the Environment Directorate were established in 1971, and the "polluter pays principle" was adopted by the OECD Council in 1972: "The polluter should bear the expenses of carrying out environmental protection measures decided by public authorities to ensure that the environment is in an acceptable state. In other words, the cost of these measures should be reflected in the cost of goods and services which cause pollution in production and/or consumption. Such measures should not be accompanied by subsidies that would create significant distortions in international trade and investment" (OECD 1975). Worded in this way, the principle rules out subsidies but does not necessarily mean that the polluter has to pay the full "rent" on the environmental resource used.

## Common Property Resource Management

Rights to common property are particularly important for poor people in many developing countries. The development of property rights in general is a gradual process that clarifies the definition, boundaries, and defense of property (see Chapter 5). As more and more resources become scarce, common property rights are replaced by private property rights (this privatization is often called "the enclosure of the commons" [see Chapter 5]).

Some researchers maintain that common property resources (CPRs) may be more primitive in an administrative sense but more advanced in a social sense, and they maintain that common property may be a superior institution under certain conditions (Balland and Plateau 1996), primarily for common-pool resources (Ostrom 1990; Stevenson 1991; for an analysis of various African cases in which the introduction of formal property titles failed to increase land productivity, see Migot-Adholla et al. 1993). For instance, when the "services" provided by a certain ecosystem are erratic (e.g., rainfall in some areas) or mobile (e.g., the availability of fish or game), then people may find it in their interest to collaborate intensely. When the "services" provided by a certain ecosystem are meager (e.g., pasture in extremely dry areas), the productivity of the land may be too low to cover the basic costs of enforcing private property rights (e.g., by installing fences). However, this kind of relationship is not defined indefinitely; with new technology, productivity and the costs of enforcement can change dramatically.

One basic criterion for determining the need for CPR management is whether the profitability of a private property rights regime would be lower than that of a common property rights regime as a result of either the excessive cost of private property rights on marginal lands (see Chapter 4) or technological factors that make CPRs more productive (see Hanna and Jentoft 1996 for some fishery examples).

Many social scientists have questioned the sustainability and optimality of CPR institutions, arguing that they ultimately will break down because of the temptation to free ride. Individuals vary their behavior depending on circumstances. Some adopt a narrow self-interested perspective and will always be free riders, whereas others will behave selfishly only in certain situations, and still others rely on reciprocity and are thus able to overcome the temptation to free ride. A good deal of experimental work has been done on this topic, and results indicate that norms to support sustainable CPR management may evolve easily, particularly if the proportion of permanent free riders is not too large. CPRs are important in many developing countries, particularly for the welfare of the poorest individuals (Ostrom 1990, 1997, 1999). This logic follows largely from the fact that the poor have little or no other property. CPRs also are usually necessary for the protection of sensitive and marginal ecosystems (Dasgupta 1993).

The problems of cooperation, free riding and free access, and CPRs can be illustrated by the "Prisoner's Dilemma" game (see Table 10-1). Known among political scientists as a "Social Dilemma," this game is commonly used to illustrate the problems of free riding in CPR management. It offers an explanation to the overgrazing of resources where ownership control is insufficient. The two players choose one of two strategies: "cooperate" or "cheat" in the management of some form of public good or common property. For both public goods and CPRs, players' selfish actions (free riding) create negative externalities. In this context, collaboration is beneficial. (However, collaboration is not usually a good thing, as illustrated, for instance, by collusion on a market.)

Both parties have a common interest in protecting the resource as long as the other one does; each party receives 10 points if they both collaborate. This "payoff" can be thought of as the reward for modest harvesting with respect to the

Table 10-1. *Prisoner's Dilemma*

|  | A plays a1 ("cheat") | A plays a2 ("collaborate") |
|---|---|---|
| B plays b1 ("cheat") | (0, 0) | (−1, 11) |
| B plays b2 ("collaborate") | (11, −1) | (10, 10) |

*Note:* Values are the payoffs received by Player A, then by Player B.

Table 10-2. *Prisoner's Dilemma: Solution through State Intervention*

|  | A plays a1 ("cheat") | A plays a2 ("collaborate") |
|---|---|---|
| B plays b1 ("cheat") | (−2, −2) | (−1, 9) |
| B plays b2 ("collaborate") | (9, −1) | (10, 10) |

*Notes:* The state puts a "tax" penalty of 2 points on cheating. Values are the payoffs received by Player A, then by Player B.

carrying capacity of the ecosystem. Individually, a free rider can gain 11 points as long as the other player continues to collaborate—but the collaborator will get a catastrophic payoff of −1. Therefore, the collaborator is not likely to collaborate further, bringing both players to cheat and receive 0. The point of the game can best be seen by recognizing the uncertainty for each player about the other player's actions. Player B's actions are unknown to Player A. Whatever Player B does, Player A actually appears to do better by cheating. If Player B cheats, then Player A's payoff increases from −1 to 0 by cheating; if Player B collaborates, then Player A's payoff improves from 10 to 11 by cheating. This effect can be analyzed as a reciprocal externality: the actions of one player affect the payoff of the other.

If Prisoner's Dilemma is analyzed as a static game, then it appears to have no desirable solution. This analysis is often done by looking for Pareto and Nash equilibria.[2] There is only one Nash equilibrium (top left, where both cheat), and Pareto equilibria are the other three outcomes. In this simple static formulation, the game has no solution, which is both a Nash and a Pareto outcome—which is a technical way of saying that the spontaneous outcome of this game does not maximize overall welfare.

Several ways out of the prisoner's dilemma have been discussed. One key factor is the distinction between playing the game once and playing it repeatedly. Another important factor in overcoming the dilemma is the buildup of institutions for control and punishment. Table 10-2 illustrates the "perfect" government control policy: by putting a fine of 2 points on cheating, the incentive for cheating is removed and the payoffs are changed so that the game only has one "natural" solution. (The technical term is a "core" solution, which is both optimal—from a purely selfish, individual, or "market" viewpoint—in the Nash sense and collectively beneficial in the Pareto sense.)

The trouble with assuming a perfect state control policy is that state control entails monitoring, information collection, and enforcement that are neither costless nor perfect. An unfortunate but likely outcome of state control is that the state controller starts fining the wrong people at the wrong time, creating havoc with the irrigation system, fishery, or other CPR under analysis. Local people then will be worse off than before; they not only will have to bear the costs of state control but also will have to contend with the consequences of its failures, which may lead to the deterioration of local resources management as well as quality.

Some observers might conclude that the only sustainable solution is privatization of resources. However, privatization may cause problems related to distributional or ecological concerns or to the reoccurrence of negative externalities. For example, if all the rich people (perhaps of one ethnic group) manage to buy all the resource shares, social tensions may quickly become unbearable. Also, in some ecological settings, the transaction costs associated with assigning private property rights are greater than the value of potential benefits (e.g., the cost of putting up fences may be higher than the value of the harvest produced on the land). In such cases, a third alternative may be superior: cooperative management of CPRs with local enforcement, the cost of which is often lower than for state enforcement for several reasons.

First, locally selected monitors have better knowledge of local social and ecological conditions and are more motivated than state monitors. They also may have lower salaries and often can combine monitoring with everyday resource management tasks. In addition, control by local peers commonly is perceived as more legitimate than control exercised by state authorities. The cultural setting of CPR management is much better modeled as a repeated game. The dynamic, repeated game turns out to be quite different from the static, single game. In a repeated game, it is profitable for the players to "invest" in such "assets" as reputation, good relationships with neighbors, and the building of collective institutions. Seven conditions are considered essential for stable CPR management (Ostrom 1990):

1. Boundaries are clear, and outsiders can be excluded.
2. Rules of provision and appropriation are adapted to site-specific conditions.
3. Decisionmaking is participatory (democratic).
4. Locally designated agents monitor resources.
5. A local court or other arena is available to resolve conflicts.
6. Graduated sanctions are used to punish infringements.
7. Outside government respects the CPR institutions.

Condition 1, clear boundaries and exclusion of outsiders, is a general prerequisite for any kind of property. Condition 2 concerns adaptation of the rules of provision and appropriation to local ecological conditions; rules (e.g., concerning rights to harvest) that are appropriate for one setting may be inappropriate in another. Conditions 3–7 concern the internal "sociology" of decisionmaking. Rules and processes must be democratic, legitimate, efficient, and effective. (Some of these aspects are illustrated in the examples in Chapter 6.)

This form of CPR management is a policy instrument that operates primarily at the local level, but central (or local) government also plays an important role. If government is prepared to accept the autonomy of CPRs, it can benefit from their good management. Central authorities can aid CPR management by providing the necessary legitimacy and by not interfering too much, as suggested by Condition 7, respect for CPR institutions. In situations without CPRs or where the underlying culture has broken down, the central government might try to revive or recreate communal institutions, but rebuilding generally is much more difficult than sustaining institutions that have already evolved.

Because an essential element of CPRs is the gradual evolution of democratically established principles (Conditions 2–4) that have coevolved to fit a cultural and an ecological context, no "quick fix" can restore institutions that have eroded. This is not to say that such situations are hopeless; arrangements can be made, perhaps partly on the basis of historical precedent but in a way that suits modern society, technology, and changing ecological conditions. In Turkish Alanja, for example, institutions had broken down, but fisheries were rebuilt over a period of more than 10 years to become successful again (Ostrom 1990).

## Liability and Other Legal Instruments

The set of rules that ensures the enforcement of other rules is essential to mechanism design. Some instruments are specifically intended to ensure compliance. Such instruments—criminal penalties, fines, liability, and performance bonds—are commonly called "legal" instruments even though other instruments are also based on laws and definitely are not "illegal." The distinction between legal and other policy instruments is often related to the degree of responsibility required of individuals who cause injury or economic damage to others (e.g., through their businesses).

In Roman law, a distinction was made between degrees of causation and intent, which range from *causa,* which means "cause" (with unawareness or possibly negligence), to *culpa,* which means "causality and fault or responsibility," to *dolus,* which is "causality with intent." When two cars are involved in a traffic accident, for example, both drivers are inherently part of the cause, regardless of who is at fault; however, the consequences (e.g., insurance payments and legal action) depend crucially on the assignment and degree of fault or intent. The determination of fault and intention could affect whether a driver faces criminal charges of reckless driving, manslaughter, or even murder.

The character of penalties or sanctions applied to environmental problems depends on the degree of causality. For new environmental problems, when the polluter could not have known about a hazard, the natural response is to provide information about the threats and possible technical solutions. When the causal relations become more established, the main policy concern focuses on the exact level of abatement or the timing, location, and character of emissions. Next, administrative rules and regulations typically become the main instrument, and economists argue for the use of market-based instruments as their complement or substitute. Regulations (and economic instruments) must be backed up with some ultimate threat in case of noncompliance, and these threats are usually financial (e.g., fines). When a firm has knowingly broken environmental or health laws to make a profit, fines are commonly imposed (partly to remove the competitive advantage that the company gained through noncompliance). When the offense is so serious that human health is impaired or people die, administrative and financial sanctions are generally not considered sufficient; in such instances, the responsible managers or employees may by punished under the criminal law system.

Because risks are complex, policies are often used in combination. Good illustrations of combined policy are found in the field of medicine, in which regulations and rules assign liability and require that doctors and drug companies be correctly insured against malpractice. The appropriate progression and mix of civil (economic) and criminal law depends on the character of the society, the problem at hand, and many other factors. In practice, fines and charges may not be so different from a firm's point of view; corporations may "calculate" losses from "expected fines" (factoring in the extent of monitoring, the probability of being apprehended for breaking the rules, and the probability of losing suits).

Whether companies should be strictly or only partially liable for damages has important implications for their behavior. Negligence or fault-based liability implies that the person injured has a right to compensation only if the party causing the injury has been negligent and has taken less than due precautions. Strict liability means a right to compensation, irrespective of precautions. Both principles should lead to the same level of "care," but the business that causes damage often knows that many people who suffer injury never litigate. Suing an employer or a doctor—not to mention government agencies or large firms—is risky and expensive, and many individuals do not have the necessary resources to hire a lawyer and proceed with a lawsuit. Strict liability gives more rights to injured individuals in particular (and perhaps to the weaker parties in society in general). Strict liability could be viewed as the ultimate policy instrument because it should lead to the internalization of all environmental damages and risks. However, its main drawbacks are that it increases the number of cases of litigation in the courts and perhaps generally hampers all economic activities that entail any risk.

In several countries, legal policy appears to be moving toward strict liability. One applied example of a liability instrument is producer responsibility for vehicles in the European Union, modeled on the German experience. In the future, all E.U. auto manufacturers will be forced to accept, manage, and reuse increasing percentages of their vehicles at the end of their useful lives. In the United States, many companies (from the chemical industry to general industry, and manufacturers of guns and cigarettes) have been sued for negligence as a form of product liability (whereas in Europe, firms probably would not be sued, but their pollution would be regulated or taxed by the government). One perverse element of the U.S. suits is that the companies (and CEOs) find it in their interest to prove that they did not know that the products were dangerous. This approach reduces interest in detailed environmental audits, because audits remove the convenient excuse of ignorance. This problem is more prominent under negligence than under strict liability.

Environmental liability typically involves multiple layers of principle–agent relationships; it starts with the relationship between the environmental protection agency and a company and then continues with the relationship between different levels of management within the company's hierarchy. Many serious environmental outcomes not only are the result of poor management practices but also have a stochastic element that introduces the problem of moral hazard.[3] Also, in some cases, there is considerable distance in time or space between the "cost-saving" in the plant today and ensuing (risk of) injury. This distance may reduce management's attention to environmental cost issues. Companies commonly are analyzed as if they had eternal life and "goals" of maximizing expected profit, but in fact, they are run by managers who have short-run goals, and this discrepancy may also introduce a bias against environmental caution.

To complicate matters further, the worst-case scenario is not the loss of value equal to maximum environmental damage but bankruptcy of the firm (which typically is worth less than the value of the environmental damage). This is the implication of limited company liability, an institutional innovation that makes large risks easier to handle and is often seen as one of the main factors behind economic growth in market economies. However, limited company liability can be abused by entrepreneurs who repeatedly and systematically use bankruptcy to get rid of debts. In the case of large hazards, the situation could have more serious repercussions. The owners of plants that represent particularly large hazards can divide them into separate subsidiaries—legally distinct companies. Such separation allows the parent company to reap the profits but avoid the risks because, in the event of an accident, only one subsidiary goes bankrupt, and damages to the parent company are thus minimized. Swedish nuclear power reactors are one example of this approach. Each reactor can be a separate "company" even if it is 100% owned by larger electricity companies.

Environmental problems commonly involve a fairly long chain of causality that precedes injury. For instance, the waste, effluents, or emissions of several firms may be mixed; chemicals may react in the atmosphere; there may be synergies so that the joint effect of two chemicals may be much worse than the sum of their separate effects; and toxins may bioaccumulate in the food chain. Damage to an ecosystem or to human health may not become apparent until years after the hazardous emissions, and the links to a certain productive activity may be tenuous and difficult to prove. Several companies could have been involved, and some may no longer exist. Emissions may have been released legally, and their hazards might have been unknown at the time of the release. In such cases, strict liability may not be a sufficiently forceful instrument. Liability must also be retroactive, and to be really effective, all the parties involved must be liable. Such strict, retroactive, and "joint and several" liability is in fact enforced in the context of the U.S. Superfund program (see Chapter 24).

Literature abounds on these issues. In a comparison of four programs under imperfect information (effluent taxes, physical controls, and two types of liability rules—negligence and strict liability), the strict liability rule was found to be superior to quantity and price control (White and Wittman 1982). Negligence has a potential advantage over strict liability in that fewer cases are litigated but also implies the disadvantage of higher expected deadweight costs. However, the underlying assumption of the strict liability rule is that the firms or polluters have sufficient assets to cover the costs or losses incurred by liability lawsuits. Similarly, individuals can alter risks privately in many ways, and therefore, it has been suggested that strict liability be retained only for extremely hazardous chemicals (because of the expected severe damages from human injuries, and because victims are presumed unable to take precautions) (Shogren and Crocker 1999). For less dangerous chemicals, negligence liability might be sufficient. A modified strict liability standard has been suggested for chemicals to reduce the moral hazard problem of the producers. The modification would consist of victim precaution requirements, including knowledge of contamination.

Citizens can sue firms for damage, but doing so may be risky. Firms can fight back because they believe that they need protection against unwarranted and harassing suits by overly zealous environmentalists (and others). Firms may, for instance, respond with a Strategic Lawsuit against Public Participation (SLAPP) to stifle environmental activism. Corporate SLAPPs, which amount to more than US$7 million a year, are intended to intimidate. Fearing the loss of the constitutional right to petition, several states have or are considering explicitly restricting SLAPP regulations. However, some analysts think that restricting SLAPP regulations will not increase the efficiency of the dispute resolution (Hurley and Shogren 1997).

Whereas the typical CEO would not desire bankruptcy, it is not clear whether the threat of such an event is serious enough to make a CEO internalize the appropriate social cost of decisions made on behalf of the firm. Therefore, policymakers are faced with several additional policy options, such as criminal responsibility for employees, mandatory insurance against environmental liabilities, and liability of the parent company and even lenders. However, insurance is a two-edged sword, because it introduces the risk of moral hazard. Insurance itself may guarantee funds in case of damages but does not necessarily provide incentives to guard against damage.

The use of liabilities also changes the game for the company, bringing in more players and presumably raising the costs of capital. New players are faced with the same information asymmetries as already mentioned for the regulator, and the result is not necessarily a simple and clear improvement. Another possible instrument when risks are considerable is to demand a liability bond, which is a deposit of money to be used for repairs in the event of an accident. This instrument is effective in providing protection but can be expensive for a firm. A firm will enjoy the interest earned on the bond, but depending on the characteristics of the capital market, it may have difficulty raising sufficient capital, and the policy would then be a strong discouragement to investments. Liability bonds have not been used extensively.

For firms whose operations involve a risk of such large environmental accidents that they generally do not have sufficient assets to cover the costs, the main moral hazard problem is that such accidents depend on the firms' efforts to prevent them, which is not observable to the government (Strand 1994). Many governments are considering introducing laws to recover cleaning costs caused by pollution damages, which may cause many banks as well as regulators to suffer adverse selection and moral hazard problems. In particular, banks that finance the firms that cause environmental damages may be considered liable for the cleanup costs. Hence, these banks play a role as additional insurers (Boyer and Laffont 1997). The conventional wisdom about the lender liability is that penalizing lenders for environmental damages caused by their borrowers will drive up interest rates, thus increasing the cost of capital (Segerson 1993); however, this assumption has been challenged (Heyes 1996; see also Jin and Kite-Powell 1995).

A risk-neutral firm, even if it is held liable for its environmental damage, may prefer to run the risk of bankruptcy rather than become fully insured, even for a fair insurance premium (because the value of the firm is less than the potential value of maximum damage). Moreover, the firm may exert insufficient effort to decrease the risk of an accident. These factors will lead to excessive investment by the bank and inadequate prevention of pollution. In the presence of adverse selection, the firm's bank does not observe the level of profits but does observe the firm's level of effort to prevent an accident, whereas both profits and effort are unknown to the insurer. Full liability of the bank allows for full internalization of the negative externality but leads to insufficient lending. When the cost of alleviating the asymmetric information is too high, partial responsibility might be better than full responsibility, because it can balance the need to internalize the externality and the reluctance of the banks to lend. Moral hazard exists on two levels: one between the government and the firms, and another between the firms and their workers; it follows that adverse selection also exists on two levels (Strand 1996). More importantly, the liability rules of firms also apply to their own workers, especially for long-term environmental risk or hazard.

To end on a practical note, government should perhaps consider selecting contractors and licensees in risky industries on the basis of firms that have sufficiently large assets of their own, because this approach might provide the best guarantee of payment to victims in the event of damage as well as the strongest incentive to avoid such damages (e.g., Lewis 2001).

## Environmental Agreements

In recent years, "voluntary agreement" or "voluntary approach" (VA) has become a catchphrase, and VAs are promoted as a new instrument of policymaking. According to the U.S. Environmental Protection Agency (U.S. EPA 1998), more than 13,000 U.S. firms, nongovernmental organizations (NGOs), and local agencies are estimated to be involved in voluntary U.S. EPA initiatives in 2000. The word *voluntary* implies that the polluters are not coerced; on the other hand, if abatement is purely voluntary, then this method can hardly be described as an instrument. Presumably, there is always some voluntary level of abatement; an economist would surmise that firms at least clean up to the point at which marginal abatement costs are equated with private environmental damages, which may include secondary loss effects to reputation as well as relationships with neighbors, workers, and customers.

To call a VA an instrument would presume a level of abatement beyond that achieved voluntarily. The word *agreement* suggests more than a purely autonomous decision by the polluter. In fact, the term VA appears to be used mainly for a form of negotiated (and verifiable) contract between environmental regulators and polluting firms.[4] A firm agrees to invest, clean up, or undergo changes to reduce negative environmental effects. In exchange, the firm may receive some subsidies or perhaps some other favor, such as positive publicity, a good relationship with the environmental protection agency, and perhaps speedier and less formal treatment of other environmental controls. This agreement is formalized in a model in which the polluter agrees to adopt cleaner technology in exchange for more lenient regulation (Amacher and Malik 1998a). The potential gain from VAs, for both parties, can be greater under taxes than under quantity regulation (Harford 2000).

What distinguishes this kind of policy from an ordinary command-and-control kind of licensing or regulation may not be immediately apparent. The main difference may be a cultural and psychological one. For example, the "covenanting process"—that is, the dialogue itself, rather than the formal agreement—has been touted as the feature that makes VAs successful (Glasbergen 1999). Today, most companies have environmental expertise of their own and are conscious of image and public relations issues. They may prefer a new label on what is essentially the same old negotiation with an environmental protection agency. The proactive, voluntary approach may be a good way not only of building public image but also of preempting effort by the agency. By taking the initiative in some areas, a firm may be able to divert attention from other areas and be able to set a level of environmental standard closer to its preference (Maxwell, Lyon and Hackett 2000). By winning the public relations war, a firm may be able to focus on issues and solutions of its own choosing. In this sense, VAs are closely related to labeling schemes (see later).

In general, one might say that VAs build on the provision of information, as illustrated by one of the most successful VA programs in the United States, the U.S. EPA's 33/50 Program. A voluntary "add-on" to the Toxics Release Inventory, the program achieved voluntary reductions of the release and transfer of 17 priority chemicals, 33% by 1992 and 50% by 1994 (one year ahead of schedule) (see Chapter 25). Furthermore, the positive, voluntary approach may motivate the staff of a giant organization, in which it is not sufficient for the highest echelons of management to be convinced but the whole organization must become more environmentally aware.

According to Coase (1960), Pigou recommended subsidies for factory chimneys. To modern-day observers, chimneys are rather obvious. However, almost a century ago, in a world where factory smoke was taken completely for granted and the installation of a chimney or filter was seen as an extremely unusual and benevolent expense for the public good, even Pigou seemed to consider that such efforts warranted gratitude from society. Today, VAs are popular in cases where firms are in some way rewarded for engaging in more abatement than normally would be required. It is important to note that the definition of an environmental investment is context specific. Chimneys are no longer considered a specifically environmental investment, and brakes are not considered an extra safety item in cars (in some countries, though, safety belts may be). After a period of constant use, former "add-ons" eventually become ordinary operating costs.

VAs give (up to a point) property rights to the firm. Firms have to pay some abatement costs, but VAs may be a good way for the firm to avoid instruments that require payment for unabated pollution, such as charges and auctioned permits. Formally, this policy could be modeled as a design standard (Chapter 6). The incentives for abatement under VAs may be correct, but the output substitution effect is missing. VAs are most promising when the opportunity for technical abatement is good while imperfections (in the product or technology markets) make the use of conventional instruments, such as taxes, difficult (Carraro and

Siniscalco 1997; Carraro and Leveque 1999). In addition, VAs are an alternative to taxes when emissions verification is problematic (Nyborg 2000).

Although collaboration concerning the science of common abatement problems might be desirable, economic collusion is not. In fact, avoiding such collusion is usually one of a policymaker's main goals. A policy instrument that encourages cooperation between competitors may be undesirable from this viewpoint.

VAs might appear attractive in cases when the environmental protection agency does not have sufficient power to coerce the polluter. In contrast, VAs probably lead to the most abatement in the opposite case—that is, when the state does have sufficient power—because then, firms are strongly motivated (for a theoretical model, see Ingram 1999 or Segerson and Miceli 1999). The notion that VAs work best against a backdrop of tougher instruments is illustrated by the case of the voluntary banning of chlorine from paper bleaching in Sweden. The result has been successful in the sense that companies have stopped using chlorine, but their response was a reaction to plans to introduce an environmental tax on chlorine. (Presumably, the demand for "green" labels on paper products also contributed to this shift.) As the laws were being drafted by the Swedish Environmental Protection Agency and parliament, industry decided to be a "first mover" and voluntarily ban chlorine. In this way, they avoided the tax payment, gained good publicity, and set a precedent for future battles concerning environmental legislation.

The performance of VAs is difficult to evaluate, but they very probably cannot be used in place of other instruments.[5] Anecdotal evidence is both favorable and critical, typically concerning the behavior of large multinational firms. However, the value of such case studies is limited by "success bias." One of the main reasons that a company undertakes environmental protection activities is to improve its public image. Thus, a good deal of resources are spent on publishing and disseminating the "good news" (e.g., GEMI 1998).[6] The Global Environmental Management Initiative is an organization that does just that.

On the other hand, a natural strategy for individuals affected by pollution is to attack the companies where they are most vulnerable, usually in the public relations arena. One of the first attempts to systematically survey the VAs in the United States concentrated on four major U.S. EPA programs: 33/50, Green Lights, Responsible Care, and Project XL (Mazurek 1999a, 1999b). The 33/50 Program has already been mentioned and is further discussed in Chapter 24. The Green Lights Program encourages energy-efficient lighting technologies. Several thousand participating institutions are required to upgrade lighting "where it's profitable and where lighting quality is maintained or improved" (U.S. EPA 2002b, 2002d). Average returns are said to be 25% on investments, so it appears that this program is not really causing "costs" to firms in the economic sense. The Responsible Care Program, started by the Chemical Manufacturers Association in 1988, is the most prominent industry-led initiative in the United States. Companies that choose to participate in Responsible Care (known as "members") voluntarily adopt six codes of conduct intended to promote safety and avoid pollution.

The U.S. EPA has introduced a great deal of flexibility in its licensing and control of the Project XL Program in exchange for voluntary commitments to abate over and above the required norm. For Intel, which produces processors for computers, the saying "time is money" is almost an understatement. Every time Intel comes up with a slightly faster processor, having to wait to put in on the market would cost the company a million dollars a day. A new processor may well require a "change of production process" that automatically requires new licensing from the U.S. EPA. For such a company, meeting environmental standards and abatement costs are not a big burden, but delays are costly. Thus, the company is prepared to make significant investments in abatement (which also gives the company a good image) in exchange for fast-track treatment on regulatory approval (see Boyd, Krupnick, and Mazurek 1998).

Several of the claimed success stories for VAs appear to be overstated (Mazurek 1999a). Although VAs can motivate and mobilize employees, they do not necessarily lead to expensive abatement programs. Furthermore, VAs raise numerous legal problems vis-à-vis regulatory and anti-trust legislation.

The findings concerning VAs generally also apply to "proactive" or "green" companies. Some green companies may be in sectors that have an inherent interest in environmental awareness (e.g., health products and ecotourism), but some may be green partly because of the preference or foresight of managers or as a preemptive business strategy. The World Business Council for Sustainable Development (WBCSD) is a "coalition of 150 international companies united by a shared commitment to sustainable development ... [that] pursues this goal via the three pillars of economic growth, environmental protection, and social equity" whose members represent more than 30 countries and 20 major industrial sectors. WBCSD has succeeded in attracting the CEOs of some of the largest (and, in some cases, heaviest polluting) industries.

WBCSD has some striking corporate policies or experiments to point to, which seems to validate the claim that the council truly is serious about the environment.[7] Critics still see these examples as a form of "greenwash," and a true empirical verification is a difficult task. However, the main strength of VAs lies in the creativity generated by employee involvement. By involving employees, the many thousands of small, complex everyday issues that are hard to regulate or tax might be addressed. Even though VAs cannot replace other instruments when it comes to large-scale, costly environmental issues such as global warming or acidification, the motivation and proactivity of some of these companies' employees might be beneficial.

## Provision of Information

All policy instruments require information to function, and public disclosure of information has come to be seen as an instrument in its own right. Public disclosure is a hot topic in the search for effective policy instruments in settings where, for one reason or another (e.g., complex technology and ecology, unequal balance of power between polluters and victims, or differences in ideology), traditional policies fail. Information disclosure can take any of several forms, depending on the degree of interpretation and aggregation of information as well as on the character of the organization that is responsible for certification: labeling, public disclosure, or rating and certification.

Labeling schemes are categorized as Type 1, 2, or 3. Type 1 is a voluntary certification that companies apply for; independent agencies set criteria and evaluate products. Type 2 certification takes place in house, without fixed criteria or independent outside review. Type 3 certification is the provision of raw data, without interpretation or judgement, but sometimes in the form of life-cycle analysis (LCA). As such, Type 3 labeling provides qualified product information—a detailed description of individual effects—but without outside evaluation. Schemes also may be categorized by the item certified: products, firms, processes, or management procedures.

The number of product labeling schemes has been increasing rapidly; in 1999, about 20 national OECD programs (plus an E.U. program) and several developing country proposals were already in existence (Nadaï 1999). "Organic" certification of food is widespread and probably one of the oldest schemes. Type 1 "green" labeling of products has become popular in northern Europe; the German Blue Angel, started in 1977, was the first national eco-labeling program. In Scandinavia, the Nordic Council of Ministers started the so-called Nordic Swan in 1989, and in Sweden, the Swedish Society for Nature Conservation (an NGO), runs the Environmental Choice independent labeling scheme (see Chapter 27). Other examples of certification programs include the Canadian Environmental Choice, the U.S. Green Seal, the Japanese ECO MARK, and the French NF Environnement (OECD 1997).

Somewhat similar to product labeling are public disclosure, ranking, and rating schemes such as the PROPER initiative in Indonesia (see Chapter 25). However, such schemes rate firms or plants rather than products, and the rating or certifying agent is a ministry rather than an NGO.

Another form of disclosure is environmental certification of firms by ISO 14000 or EMAS standards.[8] This kind of policy is oriented toward management, because the administrative routines and administrative structures of firms (or other organizations) are certified, not the environmental standards or performance per se. ISO 14000 requires implementation of an environmental management system and specifies requirements for establishing an environmental policy, determining impacts of products or services, and planning and meeting environmental objectives through measurable targets. Companies are required to organize a credible procedure for environmental management and aim to continually improve in environmental management. In return, companies are certified, which (at least in some markets) adds value to the firm by boosting its credibility. Certification is not only a labeling scheme; it is also an instrument within the organization. There is a great deal of inertia in large organizations, and management may have to fight to get its policies implemented throughout an organization. In this respect, ISO 14000 is akin to quality control (ISO 9000), and the two standards seem to be becoming closely integrated.

Type 2 and 3 labels are fairly common in industry. Volvo, for instance, provides detailed environmental data and evaluates its performance according to several criteria and its own internal goals, and the results are published in environmental reports. A scheme for public disclosure of information in which industry provides raw data to public authorities is the U.S. Toxics Release Inventory (TRI). One criticism of the TRI is that the public cannot interpret such information, but experience has shown that, building on such information, other rating and evaluation schemes have sprung up, such as Scorecard (see Chapter 24), which provides digested information to NGOs, investors, neighbors of emitting firms, and others.

The various forms of information provision are becoming increasingly popular among theorists and in real-world applications. Information provision has been referred to as the "third wave" of environmental policymaking (after legal regulation and market-based instruments), and its popularity may be explained by the changing costs of providing, processing, and disseminating relevant information (Tietenberg 1998). In this area, there have been some interesting experiences from natural resources management in both industrialized and less developed countries (e.g., organic farming, forest certification programs). In some countries, labeling is being applied to pollution issues and may be particularly promising in situations where the information and other administration costs of more traditional policies are excessive.

The northern European labeling schemes are presumably among the most successful. However, few if any schemes have been evaluated scientifically in a satisfactory manner; thus, *successful* refers in a general sense to the range of use and to the market share and popularity attained. The two most common labels in Sweden (Environmental Choice and Nordic Swan) together had the following approximate market shares in 1996:

| | |
|---|---|
| facial and bathroom tissue | 100% |
| laundry detergent | 65% |
| other detergents and cleaning products | 50–70% |
| soap and shampoo | 1–10% |

In addition, a range of products from public transportation systems to green electricity producers have been certified. It is not obvious how large market shares should be viewed: if zero, then not much progress has been made (yet); if high, the criteria may be too lax. In the case of "green electricity," the homogeneity of the good implies that the price differential provides a good measure of willingness to pay for the "green" attribute, which has no direct, practical importance for the consumer. In Sweden, the extra value of the green label was somewhat above US$0.001/kWh or some 5% of the electricity price in 2000–2001 (data were actual offers on the market [Kraftbörsen 2001]). The market share grew from 0 in 1996 (deregulation) to 5% in 1999.

One reason cited for the success of eco-labeling in Sweden is the presence of strong environmental NGOs and thus consumer demand for green products (OECD 1997). At least two strong rival labeling programs exist in Sweden. The European Commission originally hoped to include all European labeling schemes within the E.U. scheme, but evidence now indicates that competition between rival schemes is good rather than bad, and thus the drive for harmonization has slowed (Karl and Orwat 1999). If this assumption is true, then part of the success in Sweden might be due to competition. Certainly, competition encourages debate on certification criteria. Because the technical and ecological issues are complex, several sources of independent valuation might provide the optimal information input for the consumer.

On the other hand, industry takes a negative stance toward a plethora of standards. Industry also has some reservations concerning Type 1 labeling (ICC 1996):

- The existence of too many labeling schemes may hamper international trade, particularly if there are many individual national labels, and may thus affect competitiveness.
- Labeling schemes should be based on sound science, especially the use of LCA, to avoid creating overly simple criteria. However, LCA is so complicated that it is sometimes impracticable.
- The adoption of labeling schemes should be voluntary, and the criteria should be transparent. Business representatives should be involved in the formulation of labeling schemes.

Interestingly, opinions on the role of external auditors are divided on opposite sides of the Atlantic. The International Chamber of Commerce (ICC) as a whole states that the belief in the necessity of third-party eco-labeling is misguided because there is no reason to distrust the companies' own labeling; taken together, the companies' voluntary efforts and national regulations provide sufficient environmental protection. ICC therefore warns manufacturers and governments to be "cautious" when contemplating the use of third-party labeling. The U.S. Council for International Business, a chapter of the ICC, is much more emphatic in its disapproval. In a disclaimer to the ICC document, the U.S. chapter "cannot endorse this paper because it does not adequately reflect their view that multi criteria-based environmental labeling schemes are based on an inherently flawed concept that results in protectionist trade practices, barriers to environmental progress, and deceptive and misleading information to consumers" (ICC 1996).

Others who might be concerned in principle are the exporters of goods from developing countries who might perceive eco-labeling as a nontariff barrier to trade. This concern was investigated empirically by interviewing North American firms that had either the U.S. Green Seal or the Canadian Environmental Choice EcoLogo (Chua and Fredriksson 1998). At the time of the interviews, the green labels did not appear to be important on the U.S. market, and thus, their existence was hardly a barrier to exports from other (developing) countries to the United States.

The demand for information is increasing, and the costs of supply are decreasing. One of the main functions of policymaking in this area seems to be to provide a structure for information flow. In Part Three (particularly Chapter 13), I analyze several conditions and criteria for selecting policy instruments in situations that have uncertainties or information asymmetries. Information provision through labeling, disclosure, or other mechanisms is particularly important in these cases. However, the mere provision of information per se is insufficient, whereas the simultaneous use of information as well as incentives that build on that information improves firm performance (Sinclair-Desgagné and Gabel 1997).

In Europe, there has been some political discussion of officially encouraging labeling schemes by requiring public authorities to give preferential treatment to labeled products. E.U. Commissioner for the Environment Margot Wallström has suggested one of the more drastic schemes: differentiated value-added taxes (VATs), with lower rates for labeled products. The feasibility of this scheme is being investigated by various national governments within the European Union. That such a relatively strong policy would be considered at all in light of the difficulties often encountered in passing more conventional environmental taxes might seem odd. Sometimes the logic of political feasibility is complex; thus, no possibility should be ruled out completely.

## Supplemental Reading

Amacher 1998
Arora and Cason 1994
Bizer and Jülich 1999
Laffont 1989b
We the Peoples 1995

A differentiated VAT would be a strong factor in promoting the eco-labeled goods but also would entail several complications. It would completely change the political nature of the labeling schemes by exposing them to much stronger lobbying and would raise trade-related problems by discriminating against imports from countries where the producers may not have had the opportunity to obtain certification. In addition, the environmental benefits achieved by "sanctioned" firms would have to be proportional to the tax advantages received; how to measure these values would have to be determined. The instrument could turn out to be quite costly without being very precise in targeting environmental effects.

## Notes

1. Sometimes new technology can be vital in establishing property rights. In the Amazon Basin, for example, Brazilian authorities recently started to use geographic information system (GIS) maps to monitor annual development. This approach has been effective in allowing the enforcement of land-use regulations (i.e., finding forest clearers so they can be brought to court). Similar technology is needed to make property rights secure (CPTEC 2002; Chomitz 2000).

2. The Nash equilibrium is one in which neither player can improve payoff by unilateral action, that is, without the possible reaction of the other player. A Pareto equilibrium is one in which no player could get a higher payoff without a simultaneous reduction in the payoff of some other player.

3. In 1984, the Union Carbide plant in Bhopal, India, leaked a massive amount of methyl isocyanate, a poisonous chemical gas. An estimated 1,500–4,000 people were killed and many others permanently disabled. The chain of events leading to the catastrophe is complex, and whether the incident depended on minor operating faults and sloppy standards or possibly sabotage from a disgruntled employee is apparently disputed (Fischer 1996). At any rate, the behavior of employees in this context illustrates the moral hazard issue.

4. Some authors categorize VAs as purely unilateral, publicly run voluntary systems and negotiated settlements.

5. Voluntary agreements (VAs) should be used in conjunction with command-and-control or market-based instruments. In this way, and with independent monitoring and public diffusion of results, VAs can be successful. However, using VAs as a substitute for another policy in a deregulated environment creates considerable risks for insufficient abatement. Taxes are a more

effective threat than regulation when it comes to coercing companies into VAs (Bjoerner and Jensen 2000). With taxes, firms have to not only abate but also pay for unabated pollution.

6. A research network aimed at the analysis of voluntary agreements is Concerted Action on Voluntary Approaches (CAVA), the European Research Network on Voluntary Approaches for Environmental Protection (CAVA 2002).

7. Two examples are of interest in this book. The Deutsche Bank is one of the world's largest financial institutions, and as a member of the WBCSD, it also supports the U.N. Global Compact, the U.N. Environment Programme Financial Services Initiative, and the environmental management system ISO 14001. One example of Deutsche Bank's commitment to promoting sustainable development is the Deutsche Bank Microcredit Development Fund, which is said to be modeled on the Grameen Bank (see Chapter 29). Another WBCSD member has affirmed support for the Kyoto Protocol: Shell unilaterally fixed a target reduction for itself of 10% by 2002 (compared with 1990 levels) and is testing internal carbon trading, with prices fixed at US$5/metric ton of carbon until 2010 and US$20/metric ton thereafter (Shell 2001). However, the emissions reductions discussed by Shell (and other oil companies, such as BP, which has a similar system) are process emissions—not the ultimate total emissions related to the burning of their fuels, which are much larger.

8. ISO 14000 is run by the International Organization for Standardization (ISO), and the Eco-Management and Audit Scheme (EMAS) is run by the European Commission. EMAS has higher demands on public disclosure than ISO.

# CHAPTER 11

# *National Policy and Planning*

T HE ORGANIZATION OF THIS BOOK might give the impression that real governments objectively select policy instruments. However, more often than not, factors such as tradition, culture, habit, and expectations enter into the selection process. As a case in point, patterns of policy tend to characterize geographical areas or sectors. These patterns give insight into the underlying ecological, technical, cultural, political, and legal factors that shape policymaking. According to the principles of subsidiarity or fiscal federalism, decisions should be delegated to the appropriate level of government, and local policy should vary in order to adapt to local conditions. On the other hand, countries or regions that compete to attract firms by offering lax environmental (or other) standards may be doing themselves a disservice in running a "race to the bottom," especially in matters of transboundary or global pollution.

The structure of environmental planning and policymaking along with administrative structures and institutions (such as ministries and environmental protection agencies) can be seen as an "instrument" in itself. Some political scientists emphasize this aspect and study the process of innovation and diffusion of experience within the planning process across countries (Jänicke and Jörgens 2000; Jänicke, Carius, and Jörgens 1997; Jänicke 2000; Kern, Jörgens, and Jänicke 2001). The first step appears to have been the establishment of autonomous environmental ministries or agencies, starting in the late 1960s in Sweden (1967), the United Kingdom, and the United States (1970). Another nine countries followed suit in 1971, mostly in industrialized OECD countries. Today, most countries in the world have some form of environmental protection agency or environmental ministry.

Agenda 21, adopted at the Rio Earth Summit in 1992, called on its signatories to strengthen institutional capacity for sustainable development. To speed the implementation of Agenda 21, many countries formulated and adopted national environmental action plans (NEAPs). The forerunners were Denmark, Sweden, and Norway in 1988, and a large majority of OECD countries now have such plans in place. Furthermore, the World Bank has been actively promoting, and even requiring, NEAPs from its client countries in the developing world.

Similar trends emerge in legislation and concrete policies for carbon taxes and soil conservation laws. Small, proactive countries tend to create policy innovations that spread, but prominent countries such as the United Kingdom and the United States serve as role models for many others in the areas of institutional setup and overall policy culture (Kern, Jörgens, and Jänicke 2001). The variations in policy structure and instrument choice across nations is impressive.

Tradable permits are common in the United States but rare elsewhere; the only important applications are in fisheries (as transferable quotas). Applications outside the fisheries sector are relatively few, save some minor examples in other natural resources sectors (such as grazing) and the use of tradable permits for ozone-depleting substances in Singapore, where permits are auctioned every quarter but 50% of allocation is grandfathered (Markandya and Shibli 1995). Similarly limited in dissemination are environmental taxes, which have been rarely used and often resisted in the United States but are common in northern Europe and in some developing countries, notably, China (see Chapters 5 and 6).

Total U.S. excise taxes in 1997 were $59 billion, of which $14 billion were for alcohol and tobacco, around $30 billion were for fuel, and $9 billion were for air traffic and telephones. Strictly environmental taxes were only about $100 million, down from about $1.5 billion (mainly on ozone-depleting substances and Superfund) in 1995. If *environmental* is defined rather liberally (excluding telephones and alcohol, but including fuels and tobacco), one might say that U.S. environmental excise taxes were about $48 billion in 1997 (IRS 2002). This amount was only six times the corresponding taxes in Sweden, even though the population of Sweden is only about 3% that of the United States (SOU 1997). Using OECD statistics, energy- and environment-related taxes have been compared as a share of gross domestic product (SOU 1997). In 1993, the U.S. share was 0.8%, whereas the average share in European OECD countries was 2.5%. In the same comparison, other shares were as follows: Canada, 1.3%; Japan, 1.7%; Germany, 2.4%; Sweden, 3.2%; the Netherlands, 3.5%; and Greece, almost 5% (the highest of the countries compared).

In general, the United States prefers permits (particularly freely allocated, grandfathered permits), whereas European countries prefer taxation. These trends seem to reflect underlying beliefs in individual rights: the prior appropriation doctrine in the United States, and the benign and paternalistic state as a representative of societal interests in Europe. There is probably a good deal of "path dependency" in the development of national policies. Once a country starts with one set of policies, it leads to certain patterns of lobbying, learning effects, and experiences that shape the future acceptance of policy instruments. Originally, taxes in some European countries were intended mainly to finance water supply, sewage treatment, road construction, and other municipal expenses. However, as policymakers and others observed that such taxes also led to energy and resource savings, they were motivated to gradually increase taxes. Society adapted to the high taxes, and some groups may even have realized that they benefited from paying taxes (see Chapter 21).

Germany is one of the strongest economies in the European Union and places considerable attention on high technical standards in many areas of environmental concern. Significant environmental progress has been achieved in Germany, mainly through the ambitious use of standards combined with a strong belief in technical progress and good engineering. Only recently has Germany started to use taxation for environmental protection. On April 1, 1999, a law was enacted that was one of the main projects of the new German government (since October 1998), which is a coalition of the Social Democrats and the Green Party (Koalitionsvereinbarung 1998): "First step toward an ecological tax reform" (Deutscher Bundestag 1999). The law, which is supposed to be the first in a series of steps, has already been sharply criticized by businesses as well as labor organizations, economists, and environmentalists. The ecological profile of the law is not clear, although it does deal with carbon dioxide emissions. Numerous institutional factors (e.g., E.U. legislation) make national environmental taxes difficult to implement (see Chapter 24).

In Japan, consensus building is considered the main instrument for environmental policy; government decisionmakers meet with representatives from industrial conglomerates to reach a consensus. This approach may be similar to the mechanism referred to as voluntary agreements in the United States or voluntary approaches in the United Kingdom. However, it is difficult to know how much of the cultural differences in policy instruments are functions of different terminology, rather than in the way decisions are actually made.[1]

Environmental taxation appears to be rare in all Japanese sectors except the pricing of water and energy (resources that have been carefully managed to encourage conservation) and sulfur emissions. "Lean production" is the key strategy for many Japanese industries, which encourage managers to work hand in hand with other employees to avoid wasting material, energy, space, or time. This concept seems to naturally extend to the environment. However, consensus seems to include some important elements of subsidy—for example, generous tax incentives for environmental abatement investments, extra depreciation allowances against company tax, waivers on local property taxes for abatement-related purposes, and other exemptions from land and urban planning taxes as well as ordinary business taxes.

Another policy instrument that has come to the forefront in Japan is the assertion of liability and the role of the courts. With its high population density and rapid (and somewhat special) postwar economic growth, Japan had been shaken by several serious environmental diseases, such as Itai-itai, a bone disease caused by exposure to cadmium in polluted water, first reported in 1946 (Adams and Motarjemi 1999); mercury intoxication, caused by poisoned fish from Minimata Bay, which occurred for the first time in 1956 (Mineta 2002); and "Yokkaichi asthma," an array of respiratory disorders believed to be caused by emissions from a petrochemical complex in Yokkaichi, which became apparent around 1960 (WHO 2001). In 1970, the Japanese supreme court presented new guidelines that essentially reversed the burden of proof: by allowing the submission of epidemiological evidence, courts forced polluters to prove their innocence in litigation. As a result, in the early 1970s, plaintiffs won all their cases related to these environmental diseases. These rulings sent a clear signal to polluters that compensation was costlier than abatement, for the firms as well as for society. In 1973, a law was passed that gives people who have certain diseases the right to compensation for medical expenses and loss of earnings (but not pain and suffering). The funds are raised through taxes on sulfur and automobiles.

One common trait among all the industrialized countries mentioned in this chapter is that basic regulation and licensing form the backbone of environmental protection. The market-based and other policy instruments are not construed as alternatives so much as complements to the traditional policy instruments. Market-based and other instruments cannot replace the traditional instruments because market-based instruments generally require a considerable amount of monitoring and control.

Many of the largest economies in the world have a colonial past (e.g., Brazil, Canada, Indonesia, and the United States). Fairly large portions of their legal systems were created under colonial rule and thus were modeled on western legal concepts, but the systems have had to adapt and evolve to respond to local demands and conditions. One common local condition is a lack of financial and administrative resources. In India, for instance, environmental legislation and standards generally are not lacking, but enforcement of such rules is. The Indian court system has also been slow to try cases; almost 3,500 of 6,200 cases submitted under the Water and Air Acts were still pending in 1998. To speed up environmental progress, public interest litigation was introduced in 1982 as a means by which the common people (or a third party on their behalf) could directly petition the courts for the enforcement of rights to clean air and water. This process has been successful in several cases (Sankar 1998). Public interest litigation has made the courts an important player in Indian environmental policymaking (see Chapter 24).

Several formerly planned economies of eastern Europe had already implemented a form of environmental charges under the old planning regime. They did not operate as Pigovian taxes because the firms had "soft budget constraints"; they did not earn their revenues on a market but were allocated ministerial funds relative to "needs." However, the charges did raise revenue for funds that were used for environmental abatement and restoration while raising environmental awareness. The charges also facilitated the introduction of some real (but still low) taxes as the economies embarked on their transition toward market economies.

The weakness of institutions, monitoring, and all the other prerequisites for secure property rights in the transitional economies—and in many developing countries—is a considerable obstacle to any environmental policy. However, the positive experiences of environmental taxes in Colombia and China (see Chapter 25) are inspiring. In other countries, information provision, contests, voluntary agreements, and some green labeling have been successful, and the tradable quota concept has been used (mainly for natural resources management).

Two other traits are particularly striking about many developing countries. First, common property resource (CPR) management is important to social welfare, particularly of poor people. CPRs may be useful ways of managing sensitive ecosystems that perform important functions for society. Second, many developing countries use a tax structure different from that used in developed countries. Most developing countries have much smaller income taxes (as a share of total tax revenues) (Sterner 1996), so they are forced to rely on excise, trade, property, and profit taxes, many of which are seriously distorting. In recent years, pressure has mounted to streamline taxes, but the fact remains that income taxes are more difficult to collect in some countries than in others. For this reason, several environment-related taxes (e.g., on roads, fuel, energy, cigarettes, and alcohol) already play a central role in many poor countries' budgets. It may be a way of financing environmental protection or general revenue. To ensure acceptance of policy for environmental or natural resources, the cost of protection must be distributed among different agents, polluters, and taxpayers in a way that no party is overburdened. Many countries set up environmental funds, whereby environmental charges collected are earmarked for environmental protection expenses. Another instrument that is gaining in popularity in developing countries (as well as in others) is information disclosure.

Strong welfare arguments favor subsidiarity as an effective policy instrument for developing countries (Oates 1998). Considerable gains can be reaped by allowing local legislatures to adapt policies to local conditions if the legislatures are large enough to internalize most of the pollution and have the appropriate instruments to implement an optimal policy. For transboundary pollution or the management of global commons, however, international policies (or coordinated policies) are called for (see Chapter 17).

During the past few years, environmental policymaking has started to move out of the national arena into a more global one. This change is partly a reflection of the internationalization of all policymaking and partly an attempt at leveling of the playing field between competitors. An industry in one country that has been forced to adopt stringent process or product requirements tends to want to impose the same restrictions on all its competitors, even if it originally was not supportive of the regulation as such. Once adapted to the regulation, each industry finds it in its interest to spread the same standards to other industries. Thus, industries in developed countries become the inadvertent allies of the green parties in promoting environmental standards in developing countries.

Other driving forces behind the internationalization of policymaking include the pervasive nature of many pollutants; the need for standardization of management in multinational companies; and the generally prominent role of such international institutions as the World Bank, the Japan Economic Foundation, the North American Free Trade Agreement Secretariat, the European Union, and the United Nations.

## Note

1. Informal consensus approaches may be quite important in the United States and Europe, too. In Germany and France, issues concerning environmental regulation probably are discussed and decided among colleagues and former schoolmates from the elite *grandes écoles,* such as the French Ecole des Mines (see Chapter 10 on the "covenanting process" created by voluntary agreements). Consensus building is also considered a fundamental trait of Finnish policymaking (Sairinen and Teittinen 1999).

# PART THREE

# *Selection of Policy Instruments*

T HE CHOICE OF OPTIMAL ABATEMENT LEVEL IS conventionally given as the intersection of two curves that depict marginal damage and marginal cost of abatement (see Figure 6-1 in Chapter 6). However, the smooth marginal cost curves are an unlikely simplification, for several reasons. First, the abatement technologies are not perfectly known, even to the primary decisionmakers—the agents themselves and the regulators. In fact, regulators must rely on partial and possibly inaccurate information supplied by the agents. It may not be possible to know beforehand because the ultimate abatement technologies—particularly fundamentally clean production methods—have yet to be developed, and sometimes the basic science has yet to be researched. Typically, only a couple of point estimates that represent technology blueprints are available, with some approximate estimate of emissions reductions and costs. If these estimates were used as the basis of a cost of abatement curve, then the curve would be very "fuzzy" and have considerable uncertainty.

Considerable uncertainty is also assumed for the damage costs; they are difficult to estimate because they depend on ambient concentrations rather than emissions. The total pollution load on a given population or biotype determines ecosystem and health effects. Damage is probably also nonlinear and presumably varies strongly with location, time, population density, and interaction with other pollutants. Combining the two kinds of uncertainty creates an even greater uncertainty about the optimal regulation or tax levels. Some instruments may yield smaller expected costs of error than others, depending on the circumstances. The mere existence of uncertainty concerning the optimal level of abatement or the optimal charge probably will encourage lobbying. The actual policy used will always be the result of negotiations and thus, in comparing instruments, one cannot take for granted that, if used, the instruments would necessarily aim for the same optimum.

The chapters in Part Three review several special cases in which theory can suggest how to choose policy instruments. In applied work, the problems of policy selection and design are often seriously complicated by the interaction of many of these factors, which are discussed in isolation. Although the facts in any one case may be highly uncertain, there is a great need for stability and credibility in policymaking. Any policy instrument that lacks credible, long-run commitment will be resisted by both judicial and political means, and if it is not expected to last, then it probably will have little if any effect on investment and management decisions by firms, public agencies, and individuals. This situation is particularly applicable to instruments that entail the transfer of property rights and to tax laws—two of the main mechanisms for potentially effective policymaking. Faced with uncertainty, policymakers often like to experiment with policies and adjust them gradually. This approach is legitimate and understandable but may be difficult, because the adjustment itself may be perceived as a weakness and an invitation to lobbying.

The diversity of policy instruments is much richer than the simple "taxes versus regulation" scenario presented in elementary textbooks (see also Chapter 11). In general, no policy instrument is optimal. Given the complexities of real-life situations, there is always a trade-off between several desirable attributes or criteria such as efficiency, distributional effects, administrative costs, and political feasibility. The kinds of sterile debates in which economists typically argue in favor of taxes and other market-based instruments while lawyers and engineers argue in favor of regulations are clearly not productive. A more constructive way of depicting policy selection is to recognize that several instruments are available and that theories from various disciplines can help determine the choice and design of instruments.

The policy matrix, introduced in Part Two, can be used to identify and compare various countries' or sectors' experiences with policy instruments. To further analyze policy instrument selection, it is necessary to relate experience to theory, and the discussion would have to be structured in a policy selection matrix. Such a matrix would supplement the policy matrix in two important dimensions: the criteria for policy selection, and the states of the world or conditions that may be relevant. The criteria include efficiency (in various forms, such as static and dynamic allocative efficiency, but also including efficiency with respect to the use of public funds and their transaction costs), effectiveness, fairness, effects on income distribution, and other aspects related to the distribution of welfare, incentive compatibility, and political feasibility. Among the relevant conditions, some pertain to the technology and ecology at stake, such as the existence of nonlinearities, threshold effects, interactions, or temporal effects or the spatial distribution of effects. Other relevant conditions concern the economy: market forms (e.g., competition, oligopoly, or monopoly), the structure and availability of information, and the preexisting degree of inequality in distribution. Sociopolitical factors (e.g., the extent of corruption, democracy, or openness) also may be crucial.

Combining all this information into a matrix would make it three-dimensional (instruments, conditions, and criteria). To simplify their presentation, conditions and criteria can be joined; thus, the rows would be combinations of criteria and conditions that are relevant for the choice between the various policy instruments, which could be columns in the matrix. Because of the overwhelming array of policy options, criteria, and relevant conditions, a complete policy selection matrix cannot easily be presented on a single page. However, one attempt at summarizing some of the main points is Table 18-1 (see Chapter 18). In some instances, a subset of selected policies might be compared according to several relevant criteria. Given this impracticality, the policy selection matrix is thus more of a "virtual" matrix or an organizational principle. In that sense, all of Part Three is an attempt to describe some of the general traits of the choice a policymaker faces, and the different issues raised all represent rows in such a virtual matrix.

The first criteria to be studied are the classical static and dynamic efficiencies, often economists' core arguments in support of market-based instruments, at least given heterogeneous or rapidly changing abatement costs. However, heterogeneous damage costs can also be a powerful argument for quantitative regulations. In some instances, exact goal fulfillment can be important, which makes effectiveness a crucial criterion. Chapter 12 covers a range of conditions that may affect static and dynamic efficiency, including technical progress, economic growth, inflation, and ecosystem properties that affect the way rules are designed, giving agents the correct incentives to act in the common interest.

Chapter 13 is concerned with uncertainty (e.g., about the development of abatement technologies or about costs of abatement and damage) or asymmetric information. In Chapter 14, the importance of market structure, the general economic context within which the policy instruments operate, and general equilibrium effects are discussed. Chapter 15 focuses on the distribution of costs associated with the implementation of different policy instruments. This leads the way to the discussion of the political economy of policy instrument choice in Chapter 16, which also includes issues related to enforcement and monitoring. Chapter 17 deals with international aspects of instrument choice: transboundary and global pollution as well as trade and competitiveness issues (as well as the funding of abatement in poor countries). Chapter 18 summarizes the most important conclusions concerning the selection and design of policy instruments.

# CHAPTER 12

# Efficiency of Policy Instruments

CURRENT ENVIRONMENTAL CHALLENGES are considerable, and some will be quite costly to address. Hence, it is imperative to select goals that balance costs and benefits reasonably in order to reach those goals at least cost; this concept is the same as the everyday meaning of *efficiency*. Policy instruments must be used because otherwise market failures impede the market from operating efficiently.

In this chapter, I begin by discussing static efficiency (essentially, profit or utility maximization in a static model) under various conditions—notably, the heterogeneity of abatement costs or environmental damage. I then pass to dynamic efficiency, which is the intertemporal efficiency in the use of a finite and depletable stock, and end with a general discussion of efficiency in the presence of such macroeconomic factors as technological progress, economic growth, and inflation.

## Heterogeneous Abatement Costs

Choosing a policy instrument depends on many factors, and one prominent point of contention among policymakers is whether to use market-based policy instruments, such as charges and tradable permits. Market-based mechanisms are generally more efficient than other instruments when pollution is uniformly mixed and marginal abatement costs are heterogeneous—that is, when the pollution and damages are the same but the ease with which firms can reduce emissions differs. When differences in abatement costs (or in resource management efficiency) are large, the companies that hold a comparative advantage (i.e., can abate at a lower cost) should be responsible for the lion's share of reparation, and the market is the best instrument to allocate the appropriate tasks. This idea is obvious to most economists. However, because it is an essential component of environmental and natural resources policy, the topic warrants further discussion here.

Figure 6-1 (see Chapter 6) is a classic diagram in environmental economics, but its simplicity is misleading. It illustrates two curves: marginal cost of abatement and marginal cost of environmental damage. These cost functions are drawn as curves or sloping lines rather than fixed cost levels (flat, horizontal lines) because the variation in costs with emissions is important. The marginal approach is an important contribution from economics that also is relevant to the understanding of policy design. There is no such thing as "the cost" of abatement or of environmental damage; both values depend on the level of emissions or emissions reduction, respectively. Thus, the (implicit) price of pollution and pollution prevention vary—often considerably—with the intensity of pollution. Typically, the marginal damage from pollution increases with emissions (an upward slope). However, after attaining a certain level of damage, the curve may flatten or fall toward zero. This means that a point has been reached at which additional emissions cause little additional damage (e.g., a river may cease to be "a river" in the traditional sense and become "a sewer"). Similarly, the marginal cost of abatement typically increases with increased abatement, which is the same as decreasing with increased emissions (a downward slope). Indivisibilities, varying returns to scale, and technological development may modify these typical curves in individual cases.

The point at which these two curves intersect is the optimum. With more emissions, the marginal damage is higher than the marginal costs of abatement, and so a cleaner environment is preferred. To the left of the optimum is the opposite situation: the marginal costs of abatement are higher, and thus the cost of continuing to clean up the environment is more than the benefits are worth. Environmentalists might find it difficult to accept, but the resources that would be required to further decrease emissions could be used more efficiently if spent elsewhere in the economy, which includes spending on other environmental issues.

If environmental damages are external (e.g., because ownership rights are not properly defined or enforced), then the market will not arrive at this optimal level on its own, and policymaking may have a role ("may" because such a policy makes sense only if it does not introduce new distortions or costs that are worse than the externality it is intended to correct). Under many overly simple conditions, it would then not make any difference if a charge ($T*$) or a quantitative restriction ($e*$) on emissions is implemented. However, determining the optimal level of abatement is not so simple if you consider that this abatement has to be allocated among many firms. One of the main reasons why the curve for the marginal cost of abatement slopes upward might not be that each firm finds successive abatement more expensive (although this certainly can be the case) but that different firms have different abatement costs.

The first step in modeling this scenario is to introduce heterogeneity into the abatement equation. To do this in a simple manner, consider two firms that have abatement functions of the form $MC_i = h_i a_i$, where MC is marginal cost, $h$ is heterogeneity, and $a$ is abatement as defined in some appropriate physical measure. For clarity, normalize $h_1$ to 1 and set $h_2 = h$ as the key heterogeneity variable to indicate how much more expensive abatement is at Firm 2 than at Firm 1, thus assuming that $MC_1 = a_1$ and $MC_2 = ha_2$. The (optimal) aggregate abatement is

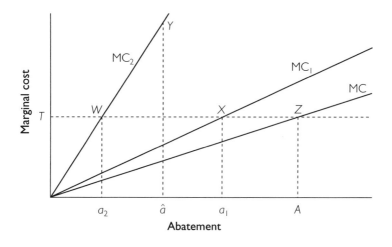

**Figure 12-1.** *Static Efficiency with Heterogeneity in Abatement for Firms 1 and 2*

*Notes:* $MC_1$ and $MC_2$ = marginal costs for Firms 1 and 2, respectively; $MC$ = aggregate marginal cost; $T$ = tax level; $a_1$ and $a_2$ = levels of abatement that equalize marginal costs for Firms 1 and 2, respectively; $â$ = equal abatement for both firms; $A$ = aggregate abatement ($a_1 + a_2$, or $2 \times â$).

derived from these individual abatement curves and labeled as MC in Figure 12-1. Emissions are assumed to be uniformly mixed, which means that the sum of the emissions causes the environmental damage; otherwise, emissions could not simply be traded between sources (the case with heterogeneous damages is discussed below). The intersection of the curves for the marginal costs of pollution damage ($MC_e$) and aggregate abatement costs ($MC_a$) determines the optimal levels of emissions and their shadow cost, $T^*$ (see Figure 6-1 in Chapter 6).

Figure 12-1 shows the allocation of abatement between the two firms. Either abatement could be mandated at a target level of abatement ($â$) for each plant (abatement levels are assumed to be equal here, which might be a reasonable approximation of regulation if the firms were of equal size[1]), or a market-based policy instrument such as a charge ($T$) could be used, giving abatement of $a_1$ and $a_2$, respectively. This solution is the most cost-efficient, and the same result can (under ideal assumptions) be achieved by allowing the firms to trade their rights to emissions reductions. Under the trading scenario, Firm 1 (with lower abatement costs) will voluntarily undertake more abatement to create and sell surplus emissions credits or rights to Firm 2, which finds that it is cheaper to buy these on the market than to incur the costs of abatement.

The total abatement costs to the companies are the areas under the MC curves in Figure 12-1. When reductions in abatement are equal, costs are the sum of two triangles with base $Oâ$, of which one ($OâY$, costs to the company for which abatement is difficult) would be large. Under a market-based allocation scenario, the company with lower abatement costs is responsible for most of the abatement, and thus aggregate costs are significantly lower ($Oa_1X + Oa_2W$). If there are big differences in the abatement cost functions of the two firms, then the equalization of MC of abatement can reduce aggregate costs significantly.

Table 12-1. *Cost Heterogeneity and Savings due to Efficiency*

| Heterogeneity | Cost | | Savings by MBI (%) |
| --- | --- | --- | --- |
| | Equal reduction | Efficient reduction | |
| 1 | 2 | 2 | 0 |
| 1.5 | 2.5 | 2.4 | 4 |
| 2 | 3 | 2.67 | ~11 |
| 3 | 4 | 3 | 25 |
| 4 | 5 | 3.2 | 36 |
| 9 | 10 | 3.6 | 64 |
| 99 | 100 | 3.96 | ~96 |

*Notes:* MBI = market-based instruments. Cost values are indices of cost, which must be multiplied by $â^2/2$ to get actual costs. Assume a target reduction in abatement of $2â$. An equal abatement reduction plan with $a_1 = a_2 = â$ would cost $â^2/2$ for Firm 1 and $hâ^2/2$ for Firm 2, giving a total of $(â^2/2)(1 + h)$, where $h$ is heterogeneity. Equalization of MC would imply $MC_1 = MC_2$ and thus $a_1 = ha_2$. Because $a_1 + a_2 = 2â$, the emission reductions would be $a_1 = 2hâ/(1 + h)$ for Firm 1 and $a_2 = 2â/(1 + h)$ for Firm 2. The costs for the two firms would be $2h^2â^2/(1 + h)^2$ and $2hâ^2/(1 + h)^2$, respectively. The total cost with equalized marginal costs can be written as $(â^2/2)$ $(4h)/(1 + h)$. A comparison of total costs indicates that equal emission reduction is more expensive than the least cost (i.e., equalized MC) by a factor of $(h + 1)^2/4h$, which gives the cost savings for different values of $h$ listed in the table (see Box 15-1 in Chapter 15 for more details).

## Savings from Market-Based Instruments

Although the exact savings depends on several factors, including the number of firms and the form of the abatement functions, the general principle can be illustrated with a simple example. For the function presented above, in which the marginal cost of abatement increases linearly with emissions reductions, assume that the regulator has chosen a fixed target for emissions reductions. The potential cost savings from market-based instruments (MBIs) compared with an equal reduction are listed in Table 12-1.

Thus, if $h = 1$ and the abatement costs of the firms are identical, then the costs are the same for the two policies and equalization of marginal cost will not yield any savings. With small differences in abatement costs, the savings are minor. Even with 50% differences in the firms' abatement costs ($h = 1.5$), the potential savings from MBIs would only be 4% compared to that with an insensitive regulation that required equal abatement. However, when differences in abatement costs are substantial (e.g., one or two orders of magnitude), the savings can be significant. Although such results depend on a special set of assumptions (i.e., two firms with linear marginal abatement cost and a fixed abatement target), they illustrate how strongly the savings from the static efficiency of MBIs is related to the underlying heterogeneity of the abatement cost functions.

Because the so-called static efficiency property of MBIs is one of the strongest arguments for their use, it is important to consider when and why abatement costs might differ. Some fundamental reasons are related to the timing of abatement investments. The price of abatement technology may decrease over time or

## Box 12-1. Cost and Timing of Abatement Investments

The relationship between cost and timing is nicely illustrated by vapor recovery systems for gas stations. The cost of equipment for gasoline vapor recovery is almost negligibly low, and the gasoline saved (as condensed vapor) as a consequence of its use provides some long-term cost savings for its owner. The net cost of installation of such a system in a new gas station or in a station already undergoing major renovation is estimated to be quite negligible (around 0.01 cents/liter) (Katz and Sterner 1990). The cost of closing a station for a week and tearing up the asphalt, however, is sizeable—particularly for a small station—and could be a hundred times higher, around 1 cent/liter. Consequently, retrofitting a gas station to comply with a regulation on short notice may be prohibitively expensive.

with increasing experience. If the abatement technology is developed by one of the firms in the industry, incentives for sharing technology with competitors may be complicated, depending on the policy instrument used (Höglund 2000). The cost of abatement may depend on capacity use in the abatement equipment industry; the cost may be artificially (and temporarily) high if many firms are making similar investments.

If abatement investments are complementary with other investments or if the abatement technology is embedded in other capital, then abatement investments can be timed to coincide with other investments the company is planning, thus lowering costs of abatement. The cost of complying with an environmental regulation in connection with a new investment or a major reinvestment is often a fraction of the cost of retrofitting a plant at any other time (which probably would require closing the plant and halting production; see Box 12-1). A similar difference is found in the costs of reducing emissions from transport. Reducing emissions at the moment of purchase (or design) of a new car is relatively inexpensive, as is retiring old and polluting cars. However, reducing the level of pollution emitted from the average five-year-old car is an expensive undertaking (see Part Four).

A related issue about timing is that the costs of an abatement investment depend heavily on whether there is slack capacity in management, engineering, and production resources of the firm. A large share of the cost is opportunity cost (i.e., productive work foregone); if management or technological experts in the firm are fully employed in productive activities, then the time spent on compliance with a new regulation may be a considerable opportunity cost. These costs could have been avoided if the company had had the leisure of choosing the time for compliance. On the other hand, polluters cannot be allowed to choose their own timing either, because they would then have too strong an incentive to postpone action.

Cost heterogeneity may result from plants of various ages in an industry, or from the emission of the same pollutants (greenhouse gases, lead, chlorofluorocarbons [CFCs], and solvents) by several sectors. The inherent differences in scale, type of technology, and other factors increase the chances that abatement costs will differ. For example, it was easier to phase out the use of CFCs as propellant gases in spray cans than in some areas of refrigeration and cleaning. In such cases, MBIs may be more efficient, although a common-sense flexibility in the use of regulations—such as banning CFCs or lead in one area of application long before they are banned in another—can capture a large share of benefits, too. The speed of technological development and thereby cost reductions in different areas also may be affected by the choice of instrument.

Empirical evidence shows that abatement costs are heterogeneous in many cases, thus suggesting that MBIs are warranted. For instance, variations in actual abatement costs in the United States for various classes of air pollutant reportedly vary by a factor of at least 5–10 (Hartman, Wheeler, and Singh 1994).

## Heterogeneous Damage Costs

The counterpart to heterogeneity in abatement costs is heterogeneity in damage costs. In the model of heterogeneous abatement costs, pollutants are assumed to be uniformly mixed. When damage costs are heterogeneous, the damage of a certain pollutant varies depending on the timing, location, or other circumstances of its emissions. For simplicity's sake, in this section, abatement costs are assumed to be homogeneous. Damage costs could be heterogeneous for two reasons: either essential characteristics of the pollutant or emission are different, or decisive variations in the ecosystem characteristics of the environment determine the level of damage caused by a given disturbance.

If pollution is purely local and each pollutant unique, then the argument that one enterprise may be better at abatement than another is no longer relevant. If gold mining leads to mercury emissions that affect one river and the disposal of bagasse from sugarcane production affects another river in a completely different way, then there is no point in discussing the relative ease of abatement or of trading rights between the two cases. Reducing pollution of one type in one locality simply will not have any effect on the welfare of individuals in the other locality.

For health costs, all else equal, the number of people affected will increase the estimated external costs of emissions. This assumption applies to any industrial or transportation pollution whose damage costs are related to population density (discussed in Part Four). This heterogeneity is different from heterogeneity in abatement costs because the optimal charge itself varies with the cost of emissions (Figure 12-2). The optimal level varies because there is not one single market for pollution or abatement but several geographically separated ones. A familiar example is the availability and price of parking, which has considerable spatial variability. An early and environmentally important example was estimated gains from spatially differentiated environmental charges for managing the estuary of the Delaware River (Kneese and Bower 1968). Equal taxes may be relevant for global (mixed) pollutants whose only effect is global warming (e.g., carbon dioxide). However, most other pollutants are, to varying extents, both local and regional in effect, and their tax ideally should be differentiated.

Although differentiated taxes may be feasible in some cases, they may be difficult to implement in others. Regulations also would have to be differentiated rather than homogeneous. However, differentiated regulation (or zoning) is often considered to be much more acceptable than differentiated taxes, which bring differences in not only the degree of abatement demanded (hence, in abatement

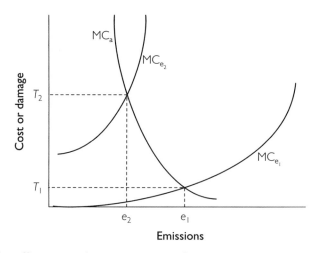

**Figure 12-2.** *Efficiency with Heterogeneity of Damage Functions*

*Notes: T* = tax level (for Firms 1 and 2); MC$_e$ = marginal damage due to emissions (of Firms 1 and 2); MC$_a$ = marginal cost of abatement.

costs) but also tax payments for unabated pollution. These taxes may be much more important than the difference in abatement costs, as illustrated in Figure 12-2 (see also Chapter 15 on the politics of cost distributions).

The concept of "equal taxes" is deeply engrained in culture; it often is protected as a principle by law and, in some countries, in the constitution. Inevitably, urban polluters confronted with a differentiated tax would complain that their rural counterparts are unduly advantaged if the urbanites are more heavily taxed "only" for living in a densely populated region. (The counterargument is that this approach is akin to other market prices: land is, for example, more expensive in cities without being considered unfair.) Similarly, area pricing fees that charge motorists entering an (urban) zone will inevitably be thought of as an unfair "entry fee" by motorists from rural areas.

Taxing a polluter for certain emissions and charging rates that reflect differences in location, time, and other variables requires sophisticated monitoring. In many cases, actual emissions cannot be monitored or taxed; instead, the tax is levied on an input or output that is assumed to be a close complement or cause of the pollution (e.g., leaded gasoline), which makes it more difficult to differentiate the tax with respect to timing or geography of emissions (see Chapter 8). Such is the case if transport costs of the taxed good are small and no other monitoring of use or emissions is possible. For instance, differentiating the price of gas to reflect external effects inside and outside a city is difficult because consumers are mobile. However, water prices are increasingly being differentiated. As more sophisticated meters become less expensive, decisionmakers are realizing that this approach may be one way of internalizing the costs of pollution and sewage treatment associated with water consumption (see Chapter 26).

When might damages be expected to be heterogeneous? One example has already been mentioned: the health effects of local air pollution. Instruments such as zoning are applicable to many local pollution (emissions, noise) problems that depend on the number of people affected. Zoning can significantly reduce many externalities at an administrative cost much lower than regionally differentiated taxation or ambient permit trading. Zoning (or, more generally, geographical separation) of economic activities that have external effects on each other or on third parties can substitute for or complement other policies to reduce the external effects. For instance, if the emissions from a paper mill damage fisheries, agriculture, or some other industry (and possibly vice versa), then conceptually, three kinds of policy might improve the situation: relocating the firm that causes emissions, relocating the activities that suffer from the emissions, or abatement. In such situations, a partial analysis usually is not sufficient to determine the optimal policy combination; a total analysis of all the possible policies may be necessary.

Damage costs also commonly vary with other factors, such as time of day and weather. The time of day has obvious importance for a valuation of noise and may affect the cost of air pollution, because the number of people affected typically changes throughout the day. The damage curve is often nonlinear, so the marginal cost of an extra unit of pollution is higher in heavily polluted areas (or perhaps lower—pollution over a certain threshold may have little additional effect).

This nonlinearity, together with the resilience of the ecosystem affected, forms the basis of the concept of *critical loads* of pollution. Whereas some areas may be able to tolerate additional acid rain or runoff (e.g., of pesticides or metals) without incurring significant damage, the damage might be considerable in other areas. For example, the buffering capacity of underlying rock varies so much that the damage from acid rain in much of Scandinavia is several orders of magnitude worse than in areas such as England, where bedrock contains primarily clay and chalk. Similarly, some weather patterns (e.g., inversion) cause pollutants to accumulate, and thus additional pollution may have a greater impact on those days. Another reason for time differentiation is the daily rhythm of the labor market and urban transportation, which cause congestion and the rapid buildup of pollutants just when the concentration of people affected is at its highest. As a result, the environmental damage cost of driving a car in an urban area typically varies strongly depending on the time of day; therefore, restrictions and congestion fees should be designed to reflect this variance (see Part Four).

Another reason for differentiated policy is the plethora of pollutants. Innumerable chemicals have different properties depending on the ecosystems into which they are emitted. Thus, the number of damage curves to consider is large. Given that knowledge and resources are inherently limited, considerable uncertainty will always surround the determination of appropriate levels of emissions and taxes. Synergies between pollutants are not uncommon, so the total damage from all pollutants is nonlinear. Thus, the estimate of the damage curves will vary with changes in knowledge and changes in pollutant concentrations.

As long as pollution permits are not tradable, differentiating the permits by geographical area is simple in principle. However, to reap any of the gains from specialization or trade mentioned in the previous section, permits trading must be allowed. If pollutants are perfectly mixed, then trading is a natural solution. If each instance of pollution is completely separate, then trading makes no sense. The intermediate case—that is, when there is some mixing of pollutants, or when each emission has both local and regional or global effects—is the most common. The more complex realm of ambient trading schemes, or differentiated taxes, is discussed in Chapters 6 and 7.

Damage curves may be fairly steep and shift frequently, even stochastically, with an exogenous variable such as time of day, weather, or other social or ecological factors. Examples of such damage include atmospheric emissions, for example, from vehicles. The level of damage may depend strongly on population density and weather. Similarly, the effect of a certain level of fishing activity may depend on the current fish stock. The fish stocks vary stochastically—in some cases, chaotically—and thus, a level of fishing that may be acceptable when stocks are high might lead to rapid depletion when stocks are low.

These examples illustrate situations in which the appropriate level of an activity should change often. The chosen instrument must be flexible. The authorities that have the relevant knowledge must also have the power to fine tune the policy instrument frequently. This thinking goes against the traditional kind of license, but many precedents exist for constructing rules that vary according to certain predetermined characteristics: allowing noise only at specific times of the day, differentiating speed limits by route, and tailoring emissions levels to specific chemicals. In the fishing industry, a common policy solution is to denominate the transferable catch rights not in tons but in percentages of a total allowable catch, which then can be altered by authorities without affecting the relative competitive positions of the participants (see Chapter 28).

Price instruments also can be varied. Because taxes have to pass through a complex legal and political procedure, they are generally thought to be less suitable when fast and frequent changes are required. Fees or charges do not have to go through the same kind of procedure and should be much more flexible; however, in many countries, the right of local authorities to change tariffs and charges as they please is severely restricted.

The treatment of heterogeneity in damage costs is particularly interesting in China, where pollution charges are supposedly equal across the country (for the sake of fairness). In fact, implementation is uneven, so the effective fee turns out to be significantly higher in some parts of the country than in others. The more densely populated and affluent areas tend to pay more than other areas, which may well correspond to a reasonable de facto differentiation of the fees to reflect actual damage costs (Wang and Wheeler 1996) (see Chapter 25).

When both damage costs and abatement costs are heterogeneous, policymakers must base their choice on the relative strength of the two sources of heterogeneity as well as the degree to which each is observable (see Chapter 13). Sometimes both issues can be addressed simultaneously, for example, if input prices can be differentiated locally or if a permit trading system can be supplemented with local restrictions in the areas where pollution is worst. If the damage costs are high in cities, then the regulations would be strict there, and no trading of permits would be allowed between sources in the city and sources outside the city unless a special "exchange rate" were used. If abatement costs vary among the plants in the city, then large savings in abatement costs can still be achieved by trading permits among those plants. If polluting inputs have high transportation costs, then they, too, can be priced higher in the cities (see Table 12-2).

Table 12-2. *Policy Selection with Heterogeneity in Both Abatement and Damage Costs*

|  | Homogeneous damage costs | Heterogeneous damage costs |
| --- | --- | --- |
| Homogeneous abatement costs | (Either policy) | Individual permits, zoning, and other restrictions |
| Heterogeneous abatement costs | Charges, taxes, or tradable permits | Differentiated input prices; local trading of permits in hotspots |

## Efficiency in an Intertemporal Sense

Natural resources economics focuses on the management of stocks. For pollutants that accumulate, the stock rather than the flow of pollutants causes damage. Examples of such pollution range from the buildup of toxic metals or persistent organic chemicals (such as DDT and PCBs) in the food chain to the emission of greenhouse gases (such as carbon dioxide) that cause climate change.

Most pollutants cannot be categorized as "pure stock" or "pure flow"; they have elements of both types of effect. Noise is a good example. Technically, it is a pure-flow pollutant because noise is momentary and does not accumulate. However, medical health effects such as hearing loss and stress are linked to cumulative exposure; therefore, the problems caused by noise pollution may have some stock characteristics. Also, one problem can have several causes (e.g., global climate change is caused by several gases that persist in the atmosphere for different lengths of time). To a large extent, carbon dioxide has stock characteristics: an increase in carbon dioxide leads to some increased assimilation; thus, the long-term increase is not equal to the short-term increase, but decay takes a long time. For water vapor, ozone, or nitrogen oxides ($NO_x$), the assimilative processes are much more rapid and thus have more of a "flow" character. Pure stock pollutants essentially use up a finite resource (i.e., the assimilative capacity) and can thus be analyzed analogously to the optimal use of natural resources. Regulators might impose or create some kind of scarcity rent to compensate for the fact that no owner or market mechanism does this automatically (see Chapter 4).

A scarcity rent can be captured either by gradually raising charges or by implementing physical constraints that become increasingly binding over time. Both options have the potential to work, but they bring about different consequences. A third option is to create property rights, as has been done in some fisheries (see Chapter 28). Some stock pollutants (such as DDT, which accumulates in the food chain) have been banned completely in some countries because the hazards are large and substitutes exist.

MBIs must be used for pollutants whose costs of abatement vary significantly (e.g., atmospheric carbon). Making a distinction between stock and flow externalities can shed some light on the choice between price-type instruments and regulation for global warming (Pizer 1999). The general starting point of the analysis is that because climate change has a stock character, the benefit curve for emissions reductions is fairly flat, which favors price-type instruments.[2] In addition, the optimal control path for emissions depends on the relative slopes that tend to further favor price controls. However, this result can be modified and counteracted by several factors, including low rates of discount, low rates of stock decay, and positive correlation of cost shocks across time.

In some cases, other factors such as administrative simplicity, scientific criteria, or international agreements have led to the explicit adoption of quantitative goals. The existence of such quantitative targets may in itself be an argument for regulation or tradable permits over taxes, for instance, in application to climate gases where complex international bargaining appears to focus on percentage reduction targets for each country. Such targets suggest an advantage for permit trading over taxes also at the level of national policymaking. However, the experience of congestion pricing in Singapore indicates that iterative changes in tax may sometimes be chosen, even if a quantitative target is being pursued (see Chapter 20).

To choose policy for ecosystems with more complex characteristics (e.g., those discussed in Chapter 4), not only stock aspects and intertemporal optimization but also risk and uncertainties related to irreversibilities and interacting non-linear effects must be considered.

## Technological Progress, Growth, and Inflation

Changes in population, income level, available technology, and economic variables such as prices can warrant the adjustment or redesign of environmental or natural resources policies. The ease of this task depends on the kind of instrument used.

Substitution possibilities can change as a result of technological progress. For example, 10 years ago, it was easier to replace the cadmium in paint pigments than in rechargeable batteries. Now, cadmium can be replaced easily and affordably in batteries, too. Removing cadmium from some phosphate fertilizers remains an expensive process, but that situation may change in the future.

Technological progress is a very strong force. People who have worked with or even casually observed environmental protection over time have observed many instances in which polluters first claim that they cannot reduce emissions and then, a short time later, proudly announce that they have found a way to cut the same emissions drastically.[3] At the micro level, pollutants from vehicles are commonly reduced by one order of magnitude per decade as a result of technological improvements. At the macro level, pollution intensities are declining dramatically in most OECD countries for most pollutants.

Some values for pollution reduction in Sweden are listed in Table 12-3. For several of the pollutants that have experienced a rapid reduction in emissions (e.g., mercury, cadmium), the main instruments providing an incentive for the development and application of technological progress have been information and regulation. Environmental taxes have also been part of many cases, sometimes playing an important role (e.g., taxes on batteries containing mercury, lead, or cadmium; refunded emission payments for $NO_x$ and various energy, carbon, and sulfur taxes [Löfgren and Hammar 2000]; taxes on lead in gasoline; taxes on nitrogen and phosphorus in fertilizer). To determine the actual relative contribution requires detailed case-by-case analysis. Because the potential for clean tech-

Table 12-3. *Emissions Reductions for Selected Pollutants in Sweden, 1970–1995*

| Substance | Reduction (%) |
|---|---|
| $CO_2$ | 52 |
| $NO_x$ | −28[a] |
| Mercury (to air) | 94 |
| Mercury (to water) | 99 |
| Cadmium (to air) | 97 |
| Cadmium (to water) | 83 |
| Lead (to air) | 88 |
| Lead (to water) | 97 |
| BOD | 73 |
| $SO_x$ | 89 |
| VOCs | 17 |
| Phosphorus (water) | 78 |

*Notes:* $CO_2$ = carbon dioxide; $NO_x$ = nitrogen oxides; BOD = biological oxygen demand; $SO_x$ = sulfur oxides; VOCs = volatile organic compounds.

[a]$NO_x$ emissions actually increased by 28%.

*Source:* SOU 1997.

nology is evident (see Table 12-3), incentives for the adoption of such technology must be offered (Kneese and Schultze 1995).

As shown in Figure 12-3, the regulation imposing a cap on emissions ($\hat{e}_0$) will have the same effect as the charge $T_0$ on the same (short-run) marginal cost of abatement curve. The difference is in the incentives to look for new technology that could lower the abatement cost curve. Regulation tends to keep the firm in the same position, and after the firm has mastered the abatement technology to meet the required standard, the standard tends to become nonbinding and may have no further effect (unless other changes are imposed, such as output expansion with a fixed emissions target). If a charge is used, management is reminded repeatedly (by the environmental tax bill) of the cost savings they could achieve by adopting innovative abatement strategies. Making the pollution permits ($\hat{e}_0$) tradable has a similar effect as long as transaction costs are negligible. Permit trade may result in less abatement than a tax in the presence of technical progress because the progress makes the permits cheaper and lowers the incentives more than a tax would. For similar reasons, permits might lead to stronger incentives in the presence or economic growth or inflation, because these factors raise the permit price above that of a fixed tax.

Given the importance of technological progress, it makes sense to study the process of creating technological innovations as well as their subsequent spread and adoption. The three steps in technological progress have been distinguished as innovation, diffusion, and policymaker response (Milliman and Prince 1989). In the first stage (innovation), all the MBIs are equal, and direct controls provide fewer incentives, as in Figure 12-3. The second stage (diffusion) is more compli-

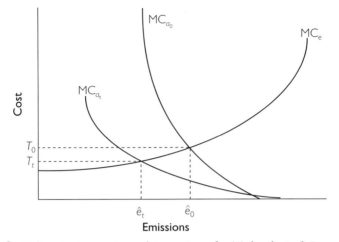

**Figure 12-3.** *Policy Instruments and Incentives for Technological Progress*

*Notes:* $MC_a$ = marginal cost of abatement at time $t$ or $0$; $MC_e$ = marginal damage caused by emissions; $T$ = optimal taxes (at time $t$ or $0$); $\hat{e}$ = optimal emissions (at time $t$ or $0$).

cated because one must analyze the interests of the innovator and of the firms that may or may not copy the innovation. Auctioned permits are the best instruments, and emissions taxes (or subsidies) are second best. Free permits (e.g., of the traditional grandfathered type) provide a smaller initial incentive for technology diffusion because they are free for some pollutants. Still, all permits have opportunity costs. With purchased (e.g., auctioned) permits based on a fixed pollution target, the diffusion of the abatement technology should cause permit prices to fall; lower prices are in the joint interest of all the polluting firms. Pollution taxes do not fall automatically, but in the third phase of technological progress (policy response), policymakers who rely on taxes may be assumed to react by lowering the tax level, which thereby would provide incentives for both innovation and diffusion.

Besides technological progress, many other variables in the economy change over time. The most obvious are overall income level and price level. The effects of economic growth and inflation differ by policy instrument. Inflation may be seen as only a change in the value of money, and as such, it would not warrant changes to quantitative instruments such as permits (whether tradable or not); however, all price-based instruments (particularly taxes but also subsidies, refunded charges, deposit–refund systems, liability bonds, and fines) would need to be adjusted.

The difficulty of making such updates also depends on the instrument. Tax changes typically have to pass through a legislative body, making them fairly complicated and somewhat unpredictable to modify. Administrative fees are simpler in this respect. Taxes and charges can be indexed to some measure of the price level, but this kind of indexing may be considered to fuel inflation and is not always popular with politicians because it is perceived as reducing their discretion. If inflation lowers inappropriate subsidies without the need for painful political fights, it may be an advantage. On the other hand, rapid inflation often is partly due to high budget deficits, making all subsidies unsuitable because they contribute to such deficits.

Economic growth (and population growth) requires more fundamental change, because such growth means that the curves for marginal cost of abatement and marginal cost of damage move; thus, the optimal level of pollution changes, too. Unless strong nonlinearities in the damage function provide a clear threshold value as a fairly stable goal, permit and tax levels would have to change as a result of economic growth. Usually, this process is slow; however, under some circumstances, local growth rates in particularly attractive or dynamic regions may be high, which would require rapid policy responses.

## Supplemental Reading

Baumol and Oates 1988 (Chapter 8)
Hoel 1998
Requate 1998

## Notes

1. Actual physical regulation may well be more subtle, and if regulators can correctly estimate marginal costs and adjust emission reductions or permits accordingly, then regulation may come close to the optimum (e.g., Brännlund et al. 1996). Individual regulation may offer advantages, particularly when damages are heterogeneous. The risks and difficulties involved in this kind of differential standard setting are notorious because they provide a strong incentive for firms to misrepresent their true abatement cost functions (see Chapter 13). Some transport examples are presented in Part 4 and industrial examples in Chapter 24.

2. See Chapter 13 for more analysis of the choice between price-type instruments and regulation. Because the rate of emissions is the time derivative of the stock level, in order for the marginal damage of emissions curve to be steep, damage would have to be a rapidly accelerating function of stock. Although this result is possible, it does not seem to be in line with the current estimate of the situation. However, for stocks that are on the verge of irreversible collapse, the implication may be a steep marginal damage (see Chapter 4).

3. One example is the U.S. sulfur emissions trading program, in which permits initially were expected to sell at $300–700/ton but ended up selling at around $100/ton; another example is the case of cleaner diesel fuels in Sweden (see Chapter 24). Comparing ex post and ex ante cost estimates for pollution control provides more empirical evidence (Harrington, Morgenstern, and Nelson 2000).

# CHAPTER 13

# Role of Uncertainty and Information Asymmetry

*I*N THE PREVIOUS CHAPTER, I FOCUSED on the relative efficiency of policy instruments under economic conditions, including cost or damage heterogeneity, technical progress, and economic growth. In this chapter, I add the condition of uncertainty. If the marginal benefits of pollution control are fairly constant (flat) and the marginal abatement costs are uncertain or change, then the tax will be approximately correct, but the number of permits would need constant adjustment (see Figure 12-3 in Chapter 12). Conversely, if the damages are insignificant below a certain threshold level and then rise steeply above that level, then quantitative permits may be a more appropriate instrument; the exact tax would be hard to estimate, and a continued pressure for more abatement under the threshold would serve no purpose.

One of the earliest and most famous articles in environmental economics focuses on formalizing this intuition on the design of policy instruments under conditions of uncertainty and asymmetric information (Weitzman 1974). It focuses on the choice between regulation through permits and regulation through taxation. This choice turns out to hinge on the slope of abatement and cost curves. Sometimes, neither permits nor taxes can be used, and the lack of information as well as agency problems can make policy design quite complicated.

First, I briefly present some general analyses of policies for adverse selection and moral hazard, then several cases in which the availability or provision of information to some agents can facilitate policy implementation. Examples include the provision of aggregate or ex post information or of information to peers and third parties.

## Uncertainty in Abatement and Damage Costs (Price vs. Quantity)

The seminal Weitzman article, entitled *Prices vs. Quantities,* concerns the choice between a price (P) instrument, which is a tax or charge, and a quantity (Q) instrument, which may be thought of as a tradable permit system (Weitzman 1974). The regulator is assumed not to have full information on abatement costs. When there is certainty, the two instruments can be seen as equivalent; both lead to the optimal outcome.

If there is uncertainty about the benefits of environmental cleanup, there will be uncertainty about the appropriate target level for the charge or regulation, but this uncertainty does not affect the comparison between charges and permits. The two instruments give the same (uncertain) result because the results of the model are determined by the firms' abatement costs. However, if these abatement costs are uncertain, then the expected outcome will be different with the different instruments.[1] If the marginal cost of abatement is mistakenly overestimated, then the regulation would be less stringent than the optimal one, whereas if charges were chosen, they would be set too high—leading, implicitly, to an excessively stringent abatement.

As shown in Figure 13-1, the price and quantity instruments do not just lead to a symmetric over- or underregulation. With overestimated marginal cost (dashed lines), the regulator will aim for $O_E$ instead of $O$, and the tax level ($P$) will be too high, leading to overabatement ($A_P$). On the other hand, regulation will be too lax (at $A_Q$), but the size of the economic welfare cost caused (as measured by deadweight losses, i.e., welfare losses due to distortions in consumption choices caused by distorted prices, for instance) depends on the relative slope of the marginal abatement and cost curves. If errors in judgement are assumed and the sum of expected consumer and producer losses minimized, then the expected losses turn out to be different with P- and Q-instruments, as shown by the relative size of the deadweight triangles; the assumption that the curves are linear makes for a simple algebraic analysis (Weitzman 1974). If the marginal benefits of abatement are flat and the marginal costs of abatement steep (see Figure 13-1), then one can predict with reasonable precision the "price" for pollution, and the loss due to taxation will be small. The exact regulatory level is difficult to give with precision, and the risk of large costs is considerable with the Q-type instruments. The opposite applies when the benefit of abatement (or pollution damage) is steep and the marginal cost of abatement flat: the deadweight loss due to an excessively high pollution tax would be large, whereas the error caused by choosing a given target abatement level is small. Thus, a P-instrument should be used if the cost of abatement curve is steeper than the damage curve, and a Q-instrument should be used when the abatement cost curve is flatter than the damage curve.

Since Weitzman's original analysis (1974), numerous articles have explored aspects of the P-versus-Q issue. One intriguing suggestion is to introduce hybrid instruments that allow polluters or users a choice between buying permits and paying a charge. For instance, an environmental protection agency might issue several permits corresponding to its target pollution level but allow additional permits to be bought at a certain (but high) rate as a "safety valve" for companies that might otherwise face unbearably high abatement costs. The combination of the two policies in a hybrid policy should reduce the risk of large errors. Many hybrid policies are potentially available, including very high fees for noncompliance. One suggested hybrid includes permits and side payments (taxes and subsi-

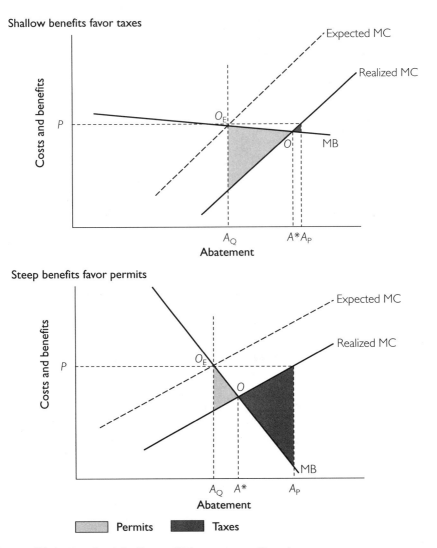

**Figure 13-1.** *Deadweight Loss of Taxes versus Permits*

*Notes:* $P$ = optimal tax level; $O$ = target for optimal policy; $O_E$ = target for policy under expected costs; MC = marginal costs; MB = marginal benefits; $A_Q$ = abatement under regulation; $A*$ = optimal abatement; and $A_p$ = abatement under a tax.

dies) to better emulate a nonlinear (increasing) function such as the environmental damage of increasing pollution levels (Roberts and Spence 1979). Another suggestion is that the regulator should simply adjust the number of permits by open-market operations (Collinge and Oates 1980).

A hybrid mechanism could combine a limited number of permits that are freely allocated with a menu of call options for additional permits (Requate and Unold 2001). Different sets of permits would have different prices, thus allowing

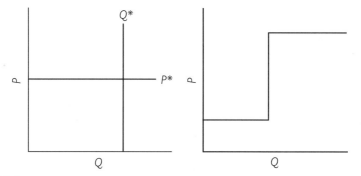

**Figure 13-2.** *Conventional (left) and Step-Shaped Hybrid Policies (right)*

*Notes:* $P$ = optimal tax level; $Q$ = quantity; $P*$ = optimal tax; and $Q*$ = optimal quantity regulation.

regulators to reflect the rising costs of environmental damage. Diagrammatically, these policies can be thought of as step-shaped functions as opposed to the vertical and horizontal functions that symbolize Q- and P-instruments, respectively (see Figure 13-2). This kind of policy might have a low fee (or free or cheap permits) up to a certain level and then high charges or fines (or more expensive permits) in excess of that level.

It may not be appropriate to treat uncertainties as random.[2] Although a measure of true uncertainty (that can be modeled as random) exists, there also are many strategic behavior and information asymmetries. Different instruments provide different incentives for reporting by the firms. A simple way of illustrating these dynamics is to compare taxes and standards as regulatory tools in a setting that includes several firms with heterogeneous abatement costs (as illustrated in Chapter 12, Figure 12-3). Under standards, the polluters have an incentive to tell their environmental protection agency how difficult and expensive abatement is in order to bargain for lenient standards (such as emissions level $[\hat{e}_0]$). But because environmental protection agencies have long-standing relations with firms, this situation creates perverse incentives. After polluters have done the first round of bargaining and received the lenient standard, they have an incentive to "prove that they were right" by not engaging in any research that might lead to technological progress in abatement. This scenario can create a culture of denial rather than of research and collaboration.

Under a regime of taxes or charges, the incentive for polluters to exaggerate abatement costs in this way is weakened because it would logically lead the environmental protection agency to choose a high tax rather than a low one. By giving an honest prediction of the lower abatement cost curve, polluters might be able to get a lower tax rate ($T_t$) associated with a lower level of pollution ($e_t$). If firms thought that they could affect the environmental protection agency's estimates, then there would be an incentive to underestimate the abatement cost to get an even lower tax rate. This approach might backfire if the company does not reduce its emissions as much as suggested during the next period. This scenario allows for some feedback, and given the long-term relationship between regulators and the firm, the firm usually will avoid misrepresenting its costs if there is not some considerable gain at stake. However, the asymmetric nature of information flows often makes it necessary to design sophisticated (two-part) policy instruments to persuade companies to reveal some of their information.

Another line of research models investment decisions when several distinct (abatement or production) technologies have different externalities. With two externality-producing technologies, a Pigovian tax does not guarantee a "first-best" solution (Turvey 1963). Unless nonlinear taxes are available, detailed regulation of technology choice is needed.[3] In the context of imperfect competition, the effects of tax and regulation tend also to be distinct (see Chapter 14).

Hazardous emissions with distinct threshold values warrant a Q-type instrument, such as a ban or a regulation. The same may apply to the management of a natural resource in danger of extinction. However, steep abatement costs (and flat damages) favor the P-instrument. Steep abatement costs may be due to cost heterogeneity (see Chapter 12), which reinforces the advantages of price-type mechanisms. It is not easy to generalize about which economic conditions give rise to steep or flat cost curves. The application concerned as well as the character of the economy determine the shape of the curve. It is sometimes assumed that countries with low income and little previous pollution control have low abatement costs (because they are at the "beginning" of the abatement curve). However, such economies often have limited access to credit, technology, and skilled technical staff—factors that may make abatement difficult. They also typically have high unemployment rates and fear the flight of capital to other countries and maybe social unrest if unemployment increases. Consequently, the abatement curve can be quite steep.

Costs of abatement change rapidly with new technology and with the restructuring of industry. In many transitional (and some developing) countries that have old, polluting plants, abatement could be so expensive, at least in the short term, that a given environmental standard might bankrupt a company. In an economy with high unemployment and extensive poverty, the bankruptcy of a single firm may be such an undesirable outcome that it will be resisted at all cost, which may help explain the observed preference for low environmental fees (with revenue recycling), which place a small and predictable burden on the enterprise, thereby minimizing the risk of bankruptcy while providing some incentive for gradual cleanup.

## Uncertainty Concerning Type of Polluter or User

Incentives for cheating and costs of monitoring are two primary aspects of the same underlying problem of designing policy instruments with asymmetric information. The problems of moral hazard, adverse selection, and stochasticity of the environment as well as genuine lack of scientific knowledge or understanding can be labeled "uncertainty" in everyday language. However, in more technical parlance, it is customary to reserve the term *uncertainty* for stochasticity or randomness of the environment. It must be distinguished from a more general lack of knowledge on the one hand and from asymmetric information on the other.

Asymmetric information can be described as either "adverse selection" or "hidden information" concerning the type of polluter and moral hazard (see Uncertainty Concerning Polluter Behavior).[4]

Under some conditions, policy instruments can be designed to circumvent these difficulties such that the agents to be regulated find it in their interest to cooperate. This strategy is common in the insurance literature and is referred to as the *revelation principle*. The insurance company knows that some people are more prone to get sick than others, and it is assumed that the individuals know their own risks but that the company cannot readily find out.[5] If an insurance company asks, "Are you likely to get sick?" then an applicant has no incentive to tell the truth. However, the company can indirectly get the answer by designing two or more plans one with different combinations of premium and copayment or deductible. A low premium with a high copayment or deductible will be more attractive to individuals who have low risks, and a high premium with a low copayment or deductible will be more attractive to individuals who have high risks. Thus, individual applicants reveal their risk types in choosing the policy.

Note that this design is still not optimal in the sense that ordinary individuals cannot buy full insurance at a price that corresponds to their true risk; neither high-risk nor low-risk individuals get the optimal insurance they would have been able to get without asymmetric information. Therefore, some low-risk individuals may choose not to buy insurance at all because of adverse selection, making the premiums more expensive for those who do. A similar principle determines the structure of the contracts discussed in Chapter 3: an employer cannot judge the types of workers and thus may offer a low fixed wage; a landowner cannot judge the types of tenants and may offer some mix of wage and sharecropping. Either way, wages are not equated to marginal productivity, and incentives for effort are not the same as in a risk-free world.

A similar principle determines the structure of policy toward different kinds of polluters. In a situation where pollutants are not mixed and abatement costs vary, individual tax rates may be needed (see Chapter 12). However, if regulators do not know the abatement costs of each polluter, polluters will have an incentive to claim low costs in order to get low tax rates. To get polluters to voluntarily reveal that they have high costs of abatement, two-part contracts can be designed that combine mechanisms—for example, individual emission fees with a subsidy for certain kinds of abatement technology.

Consider an agent (individual or company) that has private information about its type (described by the variable $\tau = \{\tau_i \mid i = 1, \ldots, n\}$. This type could be productivity, risk aversion, abatement costs, or the emissions factor of an unobservable pollutant. For an emissions factor, the principal (an environmental protection agency) cannot observe the type directly but is assumed to know the set of possible types as well as their distribution. For any given individual, there is an estimated probability $(\Omega_\tau)$ that the agent's type is $\tau$. Any other information that regulators have (e.g., the production function and emissions for each type) can be used to design policies that encourage firms to cooperate. One policy will be the most attractive to each polluter, and the principal has sufficient general information about technologies to calculate the optimal alleged type $\tau$ for each actual

## Box 13-1. The Revelation Principle

Formally, the agent selects an action $a$ from a set $A = \{a \mid a_{min} \leq a \leq a_{max}\}$. The outcome (pollution) can be any one of $M$ outcomes $q_j$. The probability of outcome $j$ depends on action $a$ and the type ($\tau$), as in $\Phi_j(a, \tau)$; $a$ itself depends on $\tau$. The probability is assumed common knowledge. A contract function is a set of contingent payments $\omega = \{\omega_j\}$ from the principal to each agent for each outcome $q_j$. The utility ($U$) of the principal is a probability-weighted average of profits $u(q_j - \omega_j)$ based on an estimate of the probability distribution over types ($\Omega$).

$$U(\omega, a) = \Sigma_j \Sigma_i \Phi_j[a(\tau), \tau] \, \Omega_\tau u(q_j - \omega_j) \qquad (13\text{-}1)$$

It is not assumed that the agent will be truthful but that the agent will assert a type $\tau'$ corresponding to the contract found most profitable. The utility of an agent type $\tau$ who chooses a contract made for type $\tau'$ is

$$V\{\omega(\tau'), a \mid \tau\} = \Sigma_j \, \Phi_j[a(\tau), \tau] \, v[\omega_j(\tau'), a \mid \tau] \qquad (13\text{-}2)$$

where $v$ is utility in given outcome $j$. This is thus the payoff function the agent maximizes by reporting a type $\tau'$. However, the principal can calculate the optimal alleged type $\tau'$ for each actual type $\tau$. The principal maximizes Equation 13-1 given the following conditions:

$$V[\omega(\tau), a(\tau) \mid \tau] \geq V^*(\tau) \text{ (for all } \tau) \qquad (13\text{-}3)$$

$$[\tau, a(\tau)] \in \text{argmax}_{(\tau', a')} \, V[\omega(\tau'), a' \mid \tau] \text{(for all } \tau) \qquad (13\text{-}4)$$

Condition 13-3 is the reservation constraint that the agent utility must equal at least some baseline utility level $V^*$. Condition 13-4 is the incentive compatibility constraint that the contracts must be designed so agents find it profitable to disclose their true type $\tau$ rather than another type $\tau'$.

type $\tau$. In this sense, the choice of contract reveals the type of the agent (see Box 13-1 for a formal definition).

A concrete and detailed example of the revelation principle as pertains to the design of contracts that internalize the use of scarce water is presented in Chapter 26, which deals with the design of contracts when metering of water is not possible. The only information regulators have is the range of technologies used (i.e., the water demand functions for distinct types of farmers). Regulators can write self-revealing contracts that combine an offer to buy a certain volume of harvest with type-specific taxes on water. With similar information on the emissions coefficients of different types of producers, an environmental protection agency could design contracts with allowable production volume and fixed fees that would make polluters self-select, thereby allowing the environmental protection agency to steer production to minimize pollution. For example, French water agencies have designed contracts that give firms incentive to treat their water over and above the levels motivated by the current tax (which is too low to provide adequate incentive) (A. Thomas 1995). This kind of principle might be useful in regulating informal enterprises or other cases in which direct monitoring is impossible.

## Uncertainty Concerning Polluter or User Behavior

In the previous section, "hidden information" concerned the type of polluter or user. A somewhat similar class of problems arises when the payoff to one party (the principal) of a contract depends on the performance of another party (the agent) but that performance cannot be observed or monitored inexpensively. The essence of such a principal–agent problem is not hidden information but hidden action, although both could occur at the same time. With hidden action, the challenge is to design a contract that offers incentives for the desirable level of performance of the agents. In general, this is done by making the agent's payment a function of the outcome as in the sharecropping models discussed in Chapter 3.

In the analysis of moral hazard, it is assumed that regulators know the agent's indirect utility function ($V$), which makes it possible for them to anticipate the polluter actions and thus work out the optimal strategy. If it is a case of adverse selection, then regulators know that the producer has one of several production functions and that emissions depend on the "type" (i.e., production function). If it is a case of moral hazard, then regulators know the production function and can observe output and thus draw conclusions on emissions (see Box 13-2).

In many real-world situations, regulators must design instruments with little knowledge. They may assume that firms maximize profit and that households maximize utility, but they do not know the production or utility functions, and the uncertainty cannot be reduced to a single probability distribution for some decisive variable—the lack of information is more general. However, when some useful information is available, regulators should use the information that is correlated with the desired outcomes. In the following sections, I present some of these special cases: when ex post (actual) information is available, when only aggregate information is available, and when information is available only to peers or third parties.

### Ex Post Availability of Information

In many cases, direct monitoring is almost impossible, but ex post verification of abatement may be possible. One area in which many such examples are found is in the inappropriate disposal of hazardous waste, in which small quantities of highly toxic materials are very difficult to monitor. Monitoring the crushed rock or tailings left by a mining company or the total volume of trash from companies or households is relatively easy. However, monitoring the disposal of a few grams of arsenic, cadmium, or mercury or even residual quantities of pesticides and household chemicals can be difficult.

It is hard to envisage the kind of monitoring that would enable an environmental protection agency to trace such small amounts of individual emissions occurring anywhere at any point in time. Even for large companies that monitor their main effluents, it is difficult to devise a system that can detect a brief, maybe momentary, dumping of some hazardous chemical—which could perhaps even be timed to coincide with scheduled repairs or with breakdowns of monitoring equipment. Yet such releases of some chemicals fulfill the dual criteria of poten-

## Box 13-2. Moral Hazard

In a nonstochastic, single-agent moral hazard problem, the agent chooses an (unobservable) action $a$ from the set $A$, giving $M$ outcomes (as in Box 13-1). The probability ($\Phi$) then depends only on the action $a$. A contract between the principal and the agent is a pay-off schedule ($\Omega$) equal to a pay-off or wage $\{\omega_j \mid j = 1, ..., M\}$ for each outcome ($j$). The expected utility ($U$ and $V$) of the principal and agent are then

$$U(\Omega, a) = \Sigma_j \, \Phi_j(a) \, u(q_j - \omega_j) \tag{13-5}$$

$$V(\Omega, a) = \Sigma_j \, \Phi_j(a) \, v(\omega_j, a) \tag{13-6}$$

where $\omega$ is a set of contingent payments and $u$ is the agent utility for a given outcome. Again, the principal designs the contract to maximize expected utility $U$ subject to the following conditions:

$$V(\Omega, a) \geq V^* \tag{13-7}$$

$$a \in \text{argmax}_{(a')} \, V(\Omega, a') \tag{13-8}$$

where Condition 13-7 is the individual rationality constraint and Condition 13-8 the incentive compatibility constraint. (If the action were observable, only Condition 13-8 would be needed, but then there is no moral hazard problem.) Assuming the agent chooses $a$ to maximize the expected utility and assuming a unique interior solution, Condition 13-8 can be replaced by the following first-order condition (Laffont 1989a, 1989b):

$$\partial V(\Omega, a)/\partial a = 0 \tag{13-9}$$

tial harm and incentive: they can cause grave damage to the environment, and their illicit release could result in large cost savings for a firm (or an individual).

One solution that is sometimes appropriate is the use of mandatory equipment (e.g., filters, catalytic converters, and storage ponds) that is designed to abate emissions of the pollutant in at least a satisfactory (if not necessarily optimal) way and is highly visible. The importance of visibility is to facilitate monitoring. The potential disadvantages of this solution include the usual problems related to mandatory technology (i.e., that mandatory equipment does not necessarily lead to optimal results and that there are no incentives to develop better technology) and nonoptimal operation of the equipment (e.g., automobiles might have catalytic converters that are effectively nonoperational because of poor fuel, and factories may have filters and storage ponds but not use or maintain them properly).

Another class of instruments that goes more directly to the heart of the issue builds on a deposit–refund concept. Shifting the burden of proof to the agents, the environmental protection agency charges potential polluters as if they were polluting but then refunds the charge if they can prove compliance by handing in the physical residues (or by following an appropriate procedure for certification of disposal). Deposit–refund schemes are used mainly for bottles and cans, but one Swedish example keeps scrap cars from being dumped in the woods. All car buyers pay a deposit that is refunded when the car is turned in for demolition. Potential applications extend further. For example, any emissions of mercury or sulfur (which cannot be created or destroyed in the process of production) must come from the inputs the firm acquires. Therefore, regulators could tax the purchase of the input and then pay a refund for its return. (This policy does not work for synthetic toxins such as chlorinated organic compounds or pollutants like $NO_x$, which are manufactured and can be destroyed.)

Various pitfalls may be associated with setting the appropriate deposit. In principle, the deposit should be set at the Pigovian level to reflect the potential damage from inappropriate disposal (Equation 9-1 in Chapter 9). However, these damage costs may be very high, and the deposit cannot be higher than production costs—otherwise, there might be a risk that inputs would be produced or imported for the sole purpose of collecting the refund.[6] It might not be necessary for the deposit to be at the Pigovian level if the costs of abatement (i.e., collection or appropriate disposal) are limited. The fee needs to be only high enough to encourage collection and safe disposal. For households, it means overcoming the inconvenience costs of storage and disposal. (Some empirical examples of sulfur taxation and waste management fees in Sweden are presented in Chapters 23 and 27.)

Discussion on deposit–refund schemes tends to focus on beverage containers, but deposit–refund schemes are part of a larger class of two-tiered price schemes that can be used in a much broader set of circumstances. Similar concepts are used to entice motorists to buy less environmentally dangerous vehicles and to have cars inspected (see Chapters 19, 21, and 22). Mandatory programs such as vehicle inspection and maintenance are difficult to enforce; owners of the oldest and most polluting vehicles have an incentive to evade control. One possible response is to mimic the refund mechanism by making it attractive for vehicle owners to have cars inspected (or to buy cleaner vehicles), for instance, by raising annual registration fees but paying rebates for vehicles that somehow reveal themselves as cleaner through certification, on-board diagnostics, or inspection. Similarly, owners could receive rebates for vehicles that demonstrate (through global positioning or similar systems) that they are not driven in urban areas where environmental damage is costly.

## Availability of Aggregate Information (Nonpoint-Source Pollution)

Diffuse pollution, referred to as nonpoint or nonpoint-source pollution (NPSP), is a category of cases for which there is no individual monitoring, typically not because the levels of emissions or effluents are low but because emissions are spread over a large area. Usually, monitoring the ambient levels of pollution that depend on emissions from a group of sources is possible. The partial information that this monitoring provides makes the use of policy instruments such as ambient charges possible. However, under some circumstances, other mechanisms that build on (and foster) cooperation between the polluters are preferable.

NPSP includes many important categories of environmental problems. A prime example is water pollution from the agricultural runoff of fertilizers and pesticides, which is extremely costly to monitor by its nature (i.e., diffusely spread over large geographical areas, unevenly and unpredictably distributed over time). Agricultural runoff causes the eutrophication (accumulation of excessive biological matter) of streams and lakes, pesticide accumulation in rivers and coastal waters, and soil erosion—the consequences of which include the death of fish, other water species, or ecosystems (including coral reefs); the accumulation of toxic chemicals in the food chain; and the siltation (obstruction) of dams or power stations, respectively. These consequences have negative effects on fisheries, recreation, tourism, health, and many other sectors. Furthermore, NPSP issues are interesting at a theoretical level because most environmental issues have some degree of NPSP character (i.e., some degree of monitoring cost), and analyses developed for the pure NPSP cases may throw additional light on other cases of industrial or transport pollution with difficult monitoring.

Consider a simple setting in which $n$ polluters share a watershed and total pollution is to be regulated. If "perfect mixing" is assumed, then all the transfer coefficients are identical (and equal to 1), so the only concern is aggregate emissions $(E)$, which are equal to $\Sigma e_i$. If the equilibrium charge level is $T$, then an ambient charge paid by every polluter $(TE)$ would have the right incentive properties at the margin, because $TE = Te_i + Te_{-i}$, where $Te_i$ is a standard Pigovian charge on emissions for agent $i$ and $Te_{-i}$ is a fixed transfer payment from the viewpoint of agent $i$ $(e_{-i}$ is the sum of all $e$ except for agent $i$). In reality there are numerous problems with such a tax. It assumes that $E$ is known, even though in reality, the regulator has only imprecise estimates. The tax level is also far greater than the damage caused by each polluter. With serious externalities and large groups (large $n$), the payment $TE$ may even exceed what individual agents earn and thus what they can pay. Furthermore, an equal fee would be unfair for small producers. This scheme brings in total charge revenues $(nTE)$, which is $n$ times total damages, which may pose serious problems vis-à-vis the right for local communities to raise such taxes. The fact that the revenues are so large provides the option of modifying the instrument in some way to refund the revenues.

Several increasingly sophisticated solutions have been suggested for dealing with policy design in situations with aggregate information. An early important contributor to address the NPSP problem suggests two possible reasons for the inability to infer behavior from observed outcomes (Segerson 1988). First, given any level of abatement, the effects on environmental quality are uncertain because of stochastic (random) variation. Second, the emissions of several polluters contribute to ambient levels, and only combined effects are observable. Because of the presence of uncertainty and monitoring difficulties in an NPSP situation, mechanisms that provide incentives for compliance must be used instead of direct regulation of each polluter's discharge. In essence, this suggested solution is a combination of the ambient charge and a penalty, which leads to a first-best allocation when polluters are risk-neutral and follow Cournot–Nash (self-interested) behavior. If the ambient pollution level is less than the predetermined standard, then firms receive a positive payment. If the ambient pollution level exceeds the predetermined standard, then each polluter has to pay a fee that is proportional to the level of ambient pollution, plus a penalty for exceeding the ambient pollution standard.

If monitoring is not impossible but just costly, then it is assumed that the regulator will monitor sometimes, but not always. The expected cost to the firm would be $\Psi T$, where $\Psi$ is the probability of detection and $T$ is the fee charged for noncompliance (theoretically, $T$ is based on damages, but because damages are difficult to calculate, the value is often based on a rule of thumb or an estimate of how much the firm saves through noncompliance). If $\Psi$ is small, then the expected value of the fee is likely to be insufficient to motivate compliance, but increasing the fee might encourage compliance. The fee level that ensures compliance might be too high to be feasible, but assuming risk aversion, the fee can be set lower and still have an incentive effect that would be on a par with a certain fee of $T$. Several mechanisms have been suggested along these lines to enforce NPSP regulation.[7]

If monitoring is so difficult that regulators have no information, then they have little alternative but to pay for abatement, which is an example of the direct provision of public goods (possibly with voluntary agreements or cost sharing by the whole industry concerned). Sometimes information may be available only to the polluters' peers. In such cases, collaborative policy designs such as common property resource management may be appropriate (see Chapter 29).

## Risk, Insurance, and Environmental Policy

When economic outcomes depend largely on the random variation of nature, insurance (or other mechanisms to even out the revenue flow, such as savings) will generally be required. In the presence of adverse selection and moral hazard, the demand for insurance will be at least partly unsatisfied. For this and other reasons, many developing countries have poorly developed markets for insurance, banking, and related services. This situation may be relevant for the choice of environmental policy instruments in these areas.

Sometimes the risks are private, but the absence of insurance leads to public risk in the form of pollution. For example, individual farmers bear the private risk of whether their harvests will be reduced by pests, but they create a public risk by using excessive amounts of agricultural pesticides (see Chapter 29). Overstocking of cattle is another mechanism through which pastoralists typically attempt to compensate for the lack of formal savings institutions. Unfortunately, this kind of system often leads to overgrazing, which has negative consequences for the environment. Such situations might be addressed by charges on or regulation of the pesticides or number of cattle. However, given the potential gravity of the welfare losses, these policies generally are neither politically feasible nor desirable. Instead, the appropriate response may be to provide or facilitate the provision of insurance or banking services (or the institutions that create conditions for the market to supply these services).

At the other end of the scale are isolated large-scale events that lead to private risks for all or many citizens. Such events include accidents at nuclear power plants or chemical plants, such as the Union Carbide plant in Bhopal, India, which leaked toxic gas into the local community in 1984, killing or injuring thousands of people. Because of steep damage curves, direct regulation is generally the most appropriate and commonly used policy. Technical complexity and the inherent difficulty of dealing with small probabilities add extra layers of difficulty. The magnitude of the worst possible event may be much larger than the net worth of the responsible firm; therefore, various arrangements may be considered, such as strict liability, mandatory insurance (at specified levels), or the provision of liability bonds (which are essentially set-asides of money to be used in case the company defaults on its obligations or goes bankrupt). Attention may need to be focused on the limited liability inherent in company law to make parent companies, stockholders, creditors, and other agents share the risk. The provision of public insurance, whereby the government implicitly or explicitly assumes risks to limit an industry's liability (e.g., the Price Anderson Act, which limits the liability of the U.S. nuclear power industry), may be an incentive for business but may also worsen moral hazard and thereby require even more regulation.

In both risk scenarios, the provision of information to the producers, communities, banks, insurance firms, customers, and society is an important by-product of insurance provision. By requiring and providing insurance, the market itself is forced to calculate, disclose, and provide information about the risks in a form that will be useful to several parties in society and help them make rational decisions. A somewhat different approach to risk and uncertainty is for firms to adopt the technology recommended by an environmental protection agency or similar authority explicitly, to protect themselves against criticism (or penalty) from the same agency in the event of any incident or accident. This mechanism gives authorities considerable leverage over technology but reduces their ability to hold the firms accountable as long as they follow the technical prescriptions. This approach turns an environmental protection agency into a clearinghouse for technology, a development that has happened in some areas of the United States.

## Availability of Information to Consumers[8]

Sometimes, the relevant primary decisionmakers on environmental issues may be not government regulators but individual consumers. For instance, government regulation of the level of pesticide residues in and on food may be fairly tight, but consumers can still make an additional trade-off between price and quality by choosing certified organically grown produce to suit their tastes and budgets. Such choices cannot be made without information, and the assumption of "full information" is not very appropriate here. The natural science and medical facts concerning pesticides and other toxins are so complex that few laypeople can grasp them, even if they have the information that usually is available only to the producers. In such cases, information might be considered a public good, and government agencies should provide the institutions to aid its dissemination and interpretation.

In recent years, several technologies have been developed that make the provision of and access to information as well as its processing, assimilation, and comprehension easier (e.g., remote sensing, satellite photography and mapping, genetic and other biotechnology, advances in computing, and the ubiquitous Internet). The difficulty of transmitting information about quality is a problem for consumers as well as producers and applies not only to environmental attributes (such as clean products or processes) but also to all other quality attributes that are subtle or difficult to detect. It is sometimes referred to as "the problem of the lemons."

If inferior goods ("lemons") are difficult to distinguish from high-quality products, then consumers have little incentive to buy high-quality products and probably will avoid them because they are more expensive. Good products are nearly always expensive, but a high price is no guarantee of quality because inferior products (which are indistinguishable) may also sell for a high price. The result may be an undesirable equilibrium in which only low-quality products are sold. If there is some way to distinguish the products through signaling—through warranties, information, or labeling—then the high-quality producers typically will try to use such techniques. For environmental attributes, the issue is complicated by the fact that the "quality" attribute is not private (such as product durability) but public. This feature makes it problematic for the seller to issue warranties and may be a motivation for public investment in and support for environmental labeling schemes.

Information provision is expensive and not necessarily wholly positive. For example, providing full information about dangerous chemicals may have the desired effect, making consumers wary of and cautious about the products. However, providing detailed information about less dangerous chemicals can have a similar effect, making consumers wary about those chemicals, too. How are consumers supposed to judge the relative dangers of 10 grams of zinc, 50 grams of an enzyme, or 1 nanogram of dioxin when emitted into different recipients under different conditions? A layperson, who does not know which chemicals and interactions are truly problematic, may be frightened by the mere sight of a chemical formula. There is thus some risk that information provision, particularly of raw data, can create a "scare," but the advantage of this approach is that openness is encouraged and competing interpretations can argue in public. One might compare with the general dilemma of a democracy in which the electorate has to make many complicated choices they do not have full information about ("If we can choose governments, we should be able to choose laundry detergents!"). One alternative to full information disclosure is the provision of processed information, but then the problem becomes one of determining the party to interpret the information. The solution is far from obvious, and the choice of policy instrument and the optimal degree of decision decentralization probably depend on the complexity of the issue at hand.

When monitoring is difficult and resources for information monitoring and processing are limited, the first priority may be to elicit information. The likelihood of receiving accurate information can be increased by standardizing the data-collection method (i.e., specifying the nature of the information to be gathered as well as collection or processing procedures, such as life-cycle analysis) and by making the penalties for falsifying information sufficiently large. The Toxics Release Inventory (TRI) is a large, structured collection of raw data about the use of chemicals in the United States (see Chapter 10). Organizations such as Scorecard use these data to provide maps and other easily accessible information that allow laypeople to make their own comparisons and informed decisions. Similarly, various national programs that promote environmentally friendly practices by awarding "green labels" to products (e.g., the United States' Green Seal, Canada's Environmental Choice, the Nordic Council's White Swan, Germany's Blue Angel, and Japan's Ecomark) make an effort to interpret information for consumers (see Chapters 23, 24, and 27).

Information disclosure can be voluntary or mandatory. Farmers who use organic methods traditionally have sought organic certification voluntarily, because organic certification is considered to convey added value. (In contrast, no requirements are placed on conventional farms to list pesticides used.) On the other hand, "right-to-know" mechanisms (such as the TRI) require all firms to provide information about emissions.

The crucial issue for understanding labeling schemes is to understand the nature of the credence good, for which the value is derived from beliefs rather than actual experience (Nadaï 1999). The technical issues involved are so complicated that the consumer cannot evaluate a green label, even after purchase of the good. The credibility of the issuing parties is thus of paramount importance. Clearly, consensus among all involved parties (industry, regulators, and possibly consumers or environmental organizations) is a good way to build credibility. As a result, industry is faced with an important strategic decision: whether to collaborate and influence the design of the criteria to arrive at a labeling scheme that is credible (and thus has marketing value), or to try to kill off the whole idea by opposing and discrediting labeling schemes in general.

How industries handle labeling schemes depends on factors such as the ease of satisfying criteria, the degree of consumer awareness and interest, and the ease of collaboration (collusion) within the industry. Documented approaches include the following (Nadaï 1999):

- Industry cooperation—In the paint and varnishes industry, the main concern is to reduce volatile organic compounds (VOCs). Wanting to maintain the distinction between water-based and other paints, the industry believed that it could best maintain profits through a collaboration that allowed influence over criteria in a way that would suit its interests concerning future technological developments.
- Industry confrontation—Manufacturers of hairsprays considered that the labeling requirements were difficult to achieve and not demanded by consumers.
- Strategic groups—Industry might split into strategic groups in which one or more collaborate and others do not, as in the laundry detergent industry (see Chapter 27).

Some degree of monitoring is vital, and in most cases, it is done by the regulated entity itself. When the mandatory reports submitted to environmental protection agencies are also publicly available, transparency is greatly enhanced, and the reports can be used by private enforcers as a basis for raising noncompliance claims. The reports also act as a check on not only the polluter but also the regulator. This element of discipline is emphasized in the design of the PROPER scheme in Indonesia (see Chapter 27) and public interest litigation in India (see Chapter 11). It is most effective in countries where the legal system encourages litigation, such as the United States. Management styles vary by region. For instance, North American industries appear to be more negative toward labeling than their European counterparts.

The relationship between regulators and polluters is by necessity delicate, and it is vital not to make it overtly antagonistic. Many companies find that the transaction cost of simply communicating and complying with the environmental agency's bureaucracy is a heavier burden (e.g., the U.S. Project XL [see Chapter 10]) than the abatement costs themselves (which tend to be perceived as legitimate and to carry side benefits for the company in its relations with consumers and others). Being service-oriented and nonbureaucratic is important for environmental protection agencies but difficult when asymmetric information problems must be overcome and the temptations of corruption and nepotism must be avoided. In addition, many agencies struggle with fundamental problems such as funding of staff and equipment.

## Supplemental Reading

Adar and Griffin 1992
Amacher 1996, 1998
Dasgupta, Hammond, and Maskin 1979, 1980
Dosi and Tomasi 1994
Hoel 1998
Kwerel 1977
Roberts and Spence 1979
A. Thomas 1995

## Notes

1. For instance, uncertainty in abatement costs may arise because of technical progress combined with heterogeneity between firms and asymmetric information. Furthermore, the errors in judgment of abatement costs may be not random but strategic, because information presumably comes from the company itself.

2. Some researchers have studied the correlation between benefit uncertainty and cost uncertainty. With simultaneous uncertainty, a positive correlation tends to favor the use of quantity-type instruments and a negative correlation favors price-type instruments, and positive correlation might be more frequent (Stavins 1996).

3. The results of a modeled "game" between an environmental protection agency and firms with endogenous choice and several discrete technologies suggest that the relative merits of taxes and regulation depend on the timing of the decisions to be taken by regulators and firms (Amacher and Malik 1999).

4. The terms can be illustrated with reference to several distinct cases. "No knowledge" would perhaps be the best description of an issue that cannot be known, such as whether there will be an outbreak of a new pest. "Uncertainty" describes occurrences that can be predicted to some extent, such as next year's rainfall (the probability distribution, but not the outcome, is known). One example of asymmetric information is when an insurance company does not know whether an injury really was an accident but the injured person does. In the absence of information, no regulation is possible—except perhaps persuasion or information collection and diffusion. However, in some instances, partial information is available and can be exploited to design policy instruments in such a way as to provide incentives. I discuss some such cases in this chapter.

5. Some diseases might be related to genetic factors and are assumed to be private knowledge; this is a case of pure adverse selection. If the risk of disease is related to lifestyle that the individual not only knows about but also can influence (e.g., diet), then the case also entails moral hazard.

6. This activity could be monitored at some cost; in fact, some automatic mechanisms are already in place. Swedish can-deposit machines distinguish between Swedish and Danish beer cans so that the latter cannot be inserted to collect the Swedish refund. In the United States, the refund for the same can may differ by state.

7. Modeling transaction costs as proportional to taxes raised, Smith and Tomasi (1995) show that taxes are not always superior and that the optimal regulatory policy is generally a mixture of taxes and standards. For a case of limited information (individual polluters cannot be moni-

tored but ambient concentration is observable), Xepapadeas (1991) proposes budget-balanced contracts in which each polluter faces a random risk of an ambient fine (where the proceeds are distributed among the other polluters). Trading between point and nonpoint sources may save costs but raises issues of monitoring; Letson, Crutchfield, and Malik (1993) assess compliance by verifying technology and conclude that the optimal trading ratio depends on the relative costs of enforcing point versus nonpoint reductions, and on the uncertainty associated with NPSP. Shortle, Horan, and Abler 1998 is a good summary of NPSP research issues.

8. I am grateful to Joseph Stiglitz for enlightening discussion of some issues discussed in this section, particularly labeling.

# CHAPTER 14

# *Equilibrium Effects and Market Conditions*

$A$N ECONOMIC INSTRUMENT has several effects. Consider a tax on a polluting input, which will directly affect input mix and technology in production, and this direct effect is often the main intended effect. The tax also will have indirect effects: it makes the final product more expensive, and thus reduces demand, which is referred to as the output or output substitution effect. Finally, the tax raises revenue for the treasury.

In this chapter, I focus on the indirect effects. First, I discuss situations in which the indirect effects can be expected to be more important than the direct effects. Next, I discuss the revenue and other effects that are associated with the so-called double dividend debate. Many of these effects depend on market conditions. The existence of monopoly or cartels will affect and may completely distort the optimality of, say, taxes. The last section is therefore dedicated to a discussion of market conditions.

## Goal Fulfillment, Abatement, and Output Substitution

The emissions from an economic activity or sector typically depend on its pollution intensity or rate of emissions ($\xi$) and its level of activity or size ($q$). Symbolically, $e = \xi q$ with the appropriate summation over firms and pollutants. Typically, the focus of environmental policy is on the pollution intensity or rate of emissions, $\xi$. In addition to technological measures for abatement, emissions can be reduced by decreases in output, and one of the advantages of certain market-based policy instruments is that, if correctly designed, they will lead to changes in output composition that reduce emissions damages. This advantage applies to taxes but not to subsidies. It also applies to tradable permits, as long as they are not allocated in proportion to output.

Whether output effects are important or marginal appears to depend on the ease of abatement as well as the demand elasticity for the product. Related issues are the precision of instruments, the degree of goal fulfillment, and monitoring costs. Two contrasting examples are the use of chlorine to bleach paper products and carbon emissions, which lead to global climate change. In the case of chlorine, pollution intensity is relatively easy to reduce, and monitoring is fairly easy. In addition, the cost share of bleaching in paper production is small, and the demand for paper is fairly inelastic. All these factors suggest that if society wants to abolish the use of chlorine, it can choose an appropriate policy and reduce the use of chlorine with essentially no impact on the total use of paper.[1] In fact, Swedish authorities used "no instrument" to achieve this goal. They merely signaled a serious intent to reduce chlorine use by drafting tax laws, and the paper mills "voluntarily" stopped using chlorine as a bleaching agent. If it had not been possible to monitor the use of chlorine and the issue were a major environmental problem, then a product tax on paper might have been considered the closest proxy available. However, such a tax would have been a very blunt instrument to deal with chlorine because it uses only the output effect and has no effect on pollution intensity.

In contrast, with carbon emissions, both factors can and should be influenced: the rate of emissions of fossil carbon per unit of energy, and the total use of energy. Abatement is one option but not the only one. Nonfossil energy (at the necessary scale) probably will be more expensive, and it is efficient to include reductions in overall energy use among the options chosen. Chosen instruments must have the correct output effect, which means that subsidies and voluntary agreements are not sufficient instruments. Furthermore, if exact goal fulfillment is required (e.g., because of obligations in international treaties), then quantitative instruments are favored.

### Importance of Output Compared with Abatement

The importance of abatement cost and demand elasticity to environmental policy design is illustrated in Figure 14-1. In Figure 14-1A, abatement is cheap and product demand is inelastic, so technological abatement options are the best solution. In Figure 14-1B, where there are no technological abatement possibilities and no "clean technology," the only mechanism to reduce pollution is to reduce demand and production. The effectiveness of this option depends on the demand elasticity of the product.

In Figure 14-1C, both abatement and output reduction are important. The cost increase that results from new and somewhat cleaner production methods is large, and environmental taxes must be paid (or permits bought); thus, the increase in the marginal cost of production is considerable. Because demand is fairly elastic, the increase in price due to the tax will result in a large decrease in output from $q_a$ to $q_o$ in addition to the reduction from $q_m$ to $q_a$ caused by the direct increase in production costs due to abatement. Algebraically, the output and abatement effects of an instrument such as tax or permits ($T$) can be compared as follows: $T$ tends to reduce output $q(T)$ and rate of emissions $\xi(T)$, as in $e(T) = q(T)\xi(T)$. Differentiating gives $de/dT = qd\xi/dT + \xi dq/dT$; multiplying by $T/e$ and substituting $e = q\xi$ gives the elasticity of emissions with respect to the instrument ($\varepsilon_{e_T}$) as the sum of the elasticities for output ($\varepsilon_{q_T}$) and abatement ($\varepsilon_{\xi_T}$):

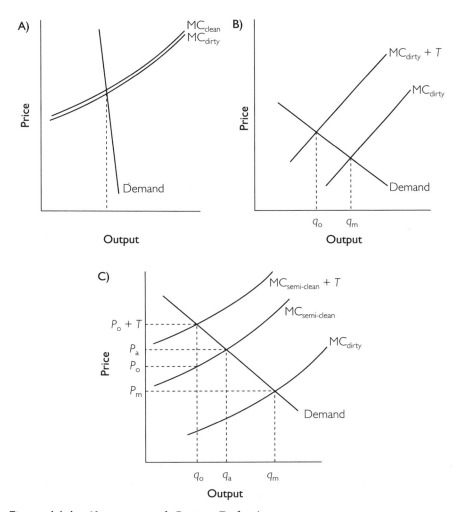

**Figure 14-1.** *Abatement and Output Reduction*

*A:* Inelastic demand, cheap abatement. *B:* Elastic demand, no clean technology. *C:* Elastic demand, clean technology.

*Notes:* MC = marginal cost of production (for clean, semiclean, or dirty technology); $P_o$ = optimal production cost; $T$ = tax; $P_a$ = price if the optimal (semiclean) technology is used but no taxes levied; $P_m$ = market price in the absence of any instrument; $q_o$ = product quantity (with $q_o$, $q_a$, and $q_m$ are defined analogously to $P$).

$$\varepsilon_{e_T} = \varepsilon_{q_T} + \varepsilon_{\xi_T} \qquad (14\text{-}1)$$

The extent of the abatement response will depend on the sum of technological abatement and output substitution (i.e., the partial and general equilibrium effects). The output effect can be expressed in Equation 14-1 as an interaction of the demand slope, $q = q(P)$, and a shift in the supply curve, $P = P(T)$. Combining gives $q = q[P(T)]$ or

$$\varepsilon_{q_\mathrm{T}} = \varepsilon_{q_\mathrm{P}} \varepsilon_{P_\mathrm{T}} \qquad (14\text{-}2)$$

and substituting into Equation 14-1 gives

$$\varepsilon_{e_\mathrm{T}} = \varepsilon_{\xi_\mathrm{T}} + \varepsilon_{q_\mathrm{P}} \varepsilon_{P_\mathrm{T}} \qquad (14\text{-}3)$$

The total response in emissions depends on the possibilities of reducing the rate of emissions as well as the product of the price elasticity of demand for the final product and the tax elasticity of the marginal cost of supply ($\varepsilon_{P_\mathrm{T}}$). As already mentioned, it depends on how much more expensive clean production methods are or the abatement elasticity.

This relationship is nicely illustrated for the simple special case of linear additivity in Box 14-1. The tax elasticity of supply is equal to the share in total marginal costs of the marginal environmental tax ($\xi_\mathrm{T}$); these terms must be compared to the response in abatement technology or the rate of emissions, $\varepsilon_{\xi_\mathrm{T}}$. If the output effect is a large share of the total effect of the instrument, then there will be a big difference between using the tax and any other instrument that does not place the full cost of abatement and tax payment on the polluter. On the other hand, if the costs of abatement are small in relation to production costs, then the difference will not be as large. In the latter case, the use of other instruments or other ways of refunding taxes might be preferable. In small open economies or economies whose main goals are alleviating poverty and creating employment, the idea of addressing environmental issues through reduced output is bound to generate solid resistance (see Chapters 15–17).

## Output Effects of Subsidies and Other Policy Instruments

The output effect—reduced emissions due to a higher product price—may be an essential part of the economic response to a policy instrument. (There may also be cases in which emissions increase, despite abatement reducing emissions per unit; but because output grows, this response is sometimes called the *rebound effect*.) The importance of the output effect was highlighted in an early analysis that focused on the perverse effects of abatement subsidies (Baumol and Oates 1988). In that model, the payments for emission reductions could exceed abatement costs and thereby lower total costs and increase output enough to counteract, and even reverse, the abatement effect. When marginal abatement costs are constant, the subsidy is simply a repayment of abatement costs. The marginal and average costs (calculated in Box 14-2) are illustrated in Figure 14-2. Because subsidies cover the abatement costs exactly, the output produced under subsidies is the same as (but not larger than) that in the unregulated case.

Figure 14-2A shows the short-run marginal costs (MC) and average costs (AC) for a representative firm without abatement as well as the corresponding curves with abatement induced by subsidies, charges, or performance-standard or regulated emission. With the subsidy construction chosen, the MC and AC curves are identical for a subsidy and the unregulated case.[2] The tax causes the largest increase in MC and AC, whereas regulated abatement or emissions rates bring about intermediate MC and AC because the regulation affects abatement costs

## Box 14-1. Tax Elasticity of Supply (Linear and Separable Abatement Costs)

Consider a firm maximizing profits equal to $Pq - C(q) - c(\xi_o q - e) - Te$. Production costs are here additively separable and linear in abatement. Production ($q$) entails production costs $C(q)$ and a "normal" level of emissions ($\xi_o q$) that can be abated to any desired level of emissions ($e$) by a linear cost function $c(\xi_o q - e)$. The firm undertakes abatement to reduce environmental taxes ($Te$). Differentiating with respect to $q$ and $e$ gives first-order conditions $c = T$ and $P = C_q' + c\xi_o$ and, by substitution, $P = C_q' + T\xi_o$, where $P$ is product price.

In this case, the increase in the supply price is the tax $T$ times the original rate of emissions $\xi_o$. Differentiating with respect to $T$ gives $dP/dT = \xi_o$, and expressing this as an elasticity gives

$$\varepsilon_{P_T} = \frac{T}{P} \cdot \frac{dP}{dT} = \frac{T\xi_o}{C_q' + T\xi_o} \tag{14-4}$$

Thus, in this simple linear case, the tax elasticity of supply is the relative share of the marginal environmental tax in marginal production costs. Equation 14-4 can be inserted into Equation 14-3, allowing for an exact determination of the relative size of abatement and output effects in this model.

---

but not the costs for the unabated pollution. As illustrated in Figure 14-2A, the higher costs of taxes—and, to a smaller extent, the regulation—create a higher market price, which may or may not have an effect on firm output in the short term. (There is an effect, but it is small.) In the long run, the price is bound to reduce demand. In addition to firm output, the number of firms is endogenous, and the higher market price caused by abatement costs and taxes will lead to exit from the industry. A regulation of emission factors would not have the full output effect, even if the individual emissions rates were optimal. The subsidy instead will lead to a price reduction, delaying exit (and possibly encouraging entry).[3]

## General Equilibrium, Taxation, and the Double Dividend[4]

In recent years, there has been a great deal of interest—perhaps excessive, in that it distracts from the task of managing the environment—in the so-called double dividend issue. Although taxes are optimal policy instruments in terms of allocative efficiency and incentive structure, they are often strongly resisted by those who have to pay. For this reason, "green tax reforms" were proposed, whereby the revenues from environmental taxation were to be used to reduce other (distorting) taxes. This idea can indeed be a good one, but some enthusiasts got carried away, suggesting that these proposals would thereby give double or even triple dividends of environmental as well as economic improvement (manifested as increased employment and growth).

To understand these issues properly, one needs a thorough knowledge of the theory of taxation. Only a few of the most basic principles of taxation are outlined here. If a particular consumption good is taxed, then its relative price will rise and consumption will decline. The consumer's pattern of choices will

## Box 14-2. Marginal and Average Costs under Different Instruments

Assuming the same linear, additively separable cost functions as in Box 14-1 [total costs (TC) = $C(q) + c(\xi_o q - e)$], the total costs for various instruments would be as follows:

| | |
|---|---|
| without regulation: | $TC_m = C(q)$ |
| with taxes: | $TC_T = C(q) + c(\xi_o q - i) + Te$ |
| with subsidies that exactly cover abatement costs: | $TC_s = C(q) + c(\xi_o q - e) + T(e - \xi_o q) = C(q)$ |
| with individual regulation of the emission rate ($\xi_R$): | $TC_r = C(q) + c(\xi_o - \xi_R)q$ |

Taking derivatives with respect to output ($q$), observing that $c = T$ gives marginal costs (MC) under taxes and subsidies, and dividing by $q$ gives average costs (AC) as follows:

$$MC_m = MC_s = C_q'$$
$$MC_T = C_q' + c\xi_o$$
$$MC_r = C_q' + c(\xi_o - \xi_R)$$

$$AC_m = AC_s = C(q)/q$$
$$AC_T = C(q)/q + c\xi_o$$
$$AC_r = C(q)/q + c(\xi_o - \xi_R)$$

thereby be distorted. This distortion gives rise to a loss in consumer surplus or excess burden of the tax, usually measured by the so-called Harberger triangles under the demand curve. Under certain conditions, the *inverse elasticity rule* states that only goods with inelastic demand should be taxed because the consumption of these goods is hardly affected and thus the distortion mentioned is small. (More precisely, it is the relative degree of substitution with leisure. Even if the demand curve is vertical, a product tax could reduce labor supply by increasing product prices and reducing real factor returns.)

The inverse elasticity rule applies literally only in simple models. The tax system has to take many factors into account; one of the more important is effects on income distribution. Because the demand for perfume, jewelry, and other luxury goods is fairly elastic, it may not be effective to tax these goods; however, they are consumed by rich individuals, and thus there may be equity reasons for taxing them. Taxation of luxury goods is particularly effective when it is not possible to tax wealth directly or to use nonlinear income taxes. In practice, it is fairly difficult to find goods with inelastic demand; one example is food (particularly among low-income groups), but taxing food more heavily than other goods is not acceptable from an ethical, distributional, or political viewpoint. In addition, differential taxation of goods entails many problems of practical administration, rent-seeking, and income distribution. Thus, the inverse elasticity rule is not always a very practical policy.

One special "good" is labor itself, and income tax is commonly used. In simplified models (without foreign sector, savings, and so forth), taxing income and taxing all consumption goods (a value-added tax) are equivalent. One might

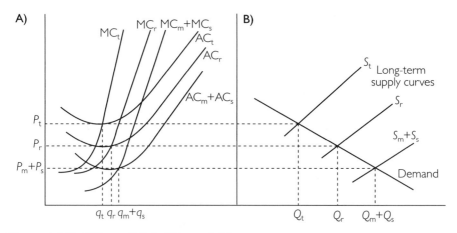

**Figure 14-2.** *Effects of Subsidies and Taxes*

A: Short-term costs for firms. B: Long-term market equilibrium for industry. Marginal abatement costs are constant, and the subsidy is equal to abatement costs.

*Notes:* $P$ = price ; $MC$ = marginal cost; $AC$ = average cost; $q$ = firm output; $Q$ = total output; and $S$ = supply or marginal cost curve. Subscripts s, t, r, and m indicate subsidies, taxes, regulated (or performance-standard) emissions rates, and "market" without any instrument, respectively.

think that this approach avoids distortions, but it does not, because it may distort the allocation of time between labor and leisure. With high taxes, the incentives for (market) labor are diminished, and people may choose (excessive) levels of leisure (or nonmarket labor). There is a strong contradiction between wanting high and progressive taxation to fulfill goals related to income distribution, equity, the provision of public goods, and fighting poverty on the one hand and the disincentive effects that such taxation may have on the high–income earners on the other.

A few additional "rules" of taxation are commonly discussed. One is to tax factors or goods that are not easily mobile (to reduce tax evasion); examples are land and forestry taxes and, to some extent, property taxes (although the size of housing may also respond in the long run). Another concerns the treatment of intermediate goods. It is generally not desirable to tax intermediate goods because certain inputs are taxed twice. Consider a market for home heating, which can be satisfied by using oil or electricity. If both are taxed by a uniform tax on consumption and the oil used to produce electricity is also taxed, then the home owner who buys this electricity would be taxed twice, thus distorting the choice between the two sources. In practice, having different tax levels for goods that can be either consumed directly or used as an input can be difficult (see the discussion on diesel taxes in Chapter 20).

The simple double dividend argument has been shown wrong in the following sense: if the environmental benefit (i.e., the first dividend) is disregarded, then an optimal tax structure can hardly be improved by lowering one of the tax rates—say, the rate on wages—and levying a supplemental tax (i.e., beyond what would be motivated by externalities or other market failures) that will give the same additional revenue on an environmental bad (Bovenberg and Goulder 1996). First, it is an illusion that the environmental tax is not a tax on labor; it is indeed a tax on wages because people use their wages to pay whatever good generates the environmental disturbance (such as gasoline), and so, disregarding the externality, it is an inefficient tax on labor. The inefficiency that arises from lost consumer surplus due to a nonoptimal distortion of the consumption basket makes the tax effectively bigger in welfare terms than the labor tax it was supposed to replace. Thus, there is a direct welfare loss rather than a gain to this tax swap. To compound this injury, the decrease in effective real salary creates secondary effects, often assumed to reduce labor supply.[5] Furthermore, the environmental benefit (as opposed to the loss in private consumer surplus) is a kind of public good, and thus, considering the cost of public funds, its provision would be lower in a "second-best" optimum than the "first-best" reasoning would suggest.

It is thus incorrect to say that an environmental tax would be good, even if there were no environmental problem. This response has sometimes been used as a last line of defense for environmental tax proposals in which it is scientifically difficult to exactly measure the value of the ecological or health benefits. Interactions between the environmental tax and the preexisting taxes may be counterintuitive. Thus, the advantage of the dividends from the environmental tax are not greater in economies with high preexisting levels of (marginal) labor tax; rather, the high degree of distortion results in greater costs to public funds and bigger deadweight losses.

This attack on the double dividend neglects, however, to address the environmental externality. If it is addressed through regulations, then even these environmental regulations will have tax interaction effects; if labor taxes are high, then environmental regulations distorting consumption choices will lead to a large welfare loss and a significant social cost through reductions in labor supply. Taking this effect into account changes the picture somewhat. The relevant comparison is not only between an environmental tax and a general tax, because a general tax leaves the environmental problem unresolved; instead, an environmental tax should be compared with a combination of general tax and an environmental regulation. In this comparison, the environmental tax proceeds are a positive rather than a negative factor.

An optimal environmental tax operates through five effects (Goulder et al. 1999):

- the *abatement effect*, which is fairly self-evident;
- the *input substitution effect*, which refers to the substitution of "cleaner" inputs for "dirtier" ones in production;
- the *output substitution effect* (discussed earlier);
- a *revenue-recycling effect,* a positive effect of the environmental tax that confirms what many people mean by a double dividend (i.e., if there is a cost to public funds, then an instrument that just happens to collect such funds must have some advantage due to this fact); and
- a *tax interaction effect*, which is related to the loss of consumer surplus and real wages that are due to distortion of the preferred consumption (or input) basket and reduces labor supply and tax incomes, leading to further losses.

Between the last two related effects, the tax interaction effect may dominate over the revenue-recycling effect, but regulation and other instruments cause tax interactions, too. For command-and-control regulations, the tax interaction effect is a welfare loss, although small, because the price of the polluting good is not raised as much as with taxes. Overall, command-and-control policies lead to higher primary efficiency costs and higher second-best costs than other kinds of policies because the output substitution effect is not fully used and the sum of revenue recycling and tax interaction costs turns out to be higher (Goulder et al. 1999).

The earmarking of funds also must be discussed in this context. If environmental taxes are earmarked for use in an environmental fund, then presumably, part of the revenue-recycling effect is lost (unless these funds happen to be the best use of the tax proceeds in that particular situation). The negative tax interaction could, for related reasons, be worsened.

One practical goal of using environmental taxes is the desire for a stable tax base, sometimes referred to as a *revenue goal*. Clearly, there is a contradiction of goals here: when abatement is easy, the pollutant emissions may go to zero and thus the tax base would be gone (see Figure 14-3B). If the prospect of revenues from such a tax had been used to lower other taxes, there would be a shortfall for the treasury. Because changing the tax system is disruptive and expensive in itself, it probably is not optimal to apply taxes (as opposed to charges) to environmental problems that are amenable to abatement. Administrative charges or simple information and persuasion (because abatement costs are assumed to be small) may be a more appropriate instrument.

On the other hand, when demand for the polluting good is inelastic, the tax base is potentially stable (but there is little reduction in pollution; see Figure 14-3A). In this case, abatement is difficult, and an environmental tax might appear to be a failed policy. However, because the demand curve is steep, any other policy would also encounter difficulties in achieving significant abatement. At least the tax provides a continuous incentive for innovators as well as the output substitution and revenue-recycling effects. Presumably, some pollutants fall into an intermediate category (see Figure 14-3C), where the tax instrument performs well: first, it provides incentives for abatement through demand reduction; then (as demand becomes more inelastic), it provides a stable tax base, the revenues of which can be recycled or used for other purposes.

## Adapting to Market Conditions

Up to this point, the discussion of policy instruments has assumed perfect competition, but such a description of the market structure is hardly exact in many markets—particularly not in developing countries. The problems of policy design under imperfect competition were reported more than three decades ago, when it was shown that the optimal charge for an externality produced by a monopoly would be lower than the standard Pigovian level.

The reason for this effect on optimal taxes is the existence of two imperfections: the monopoly (or oligopoly or cartel) that tends to indicate that optimal

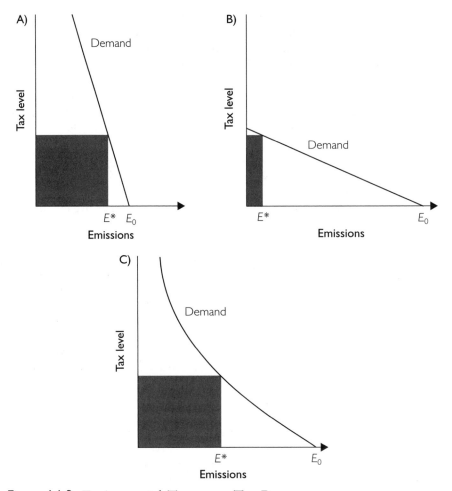

**Figure 14-3.** *Environmental Taxes as a Tax Base*

*A:* Steep demand, little effect on emissions, large tax base. *B:* Flat demand, large effect on emissions, small tax base. *C:* Nonlinear demand, large effect on emissions, large tax base.

*Notes:* $E_0$ = original emissions level, $E*$ = optimal emissions level.

output would be higher than the output observed, and the negative externalities that pull in the opposite direction (see Figure 14-4).[6] The monopoly produces output $Q_m$ at the monopoly price ($P_m$), and the social optimum would be a production of $Q_o$. If a tax is set according to the Pigovian rule (equal to the sum of marginal damages), then the monopoly would further reduce output (that is already too low) down to the level $Q_{m_T}$ instead of increasing it.

A second-best charge can be calculated (as $T_m = D'(e*) - (P/|\varepsilon|)q_e'$, where $|\varepsilon|$ is the absolute value of the elasticity of demand) but is not always feasible or even desirable. In the case illustrated in Figure 14-4, the second-best tax would be strongly negative. One might also directly discuss a combination of policy instruments: a subsidy to increase supply (to overcome the "monopoly" effect) and a

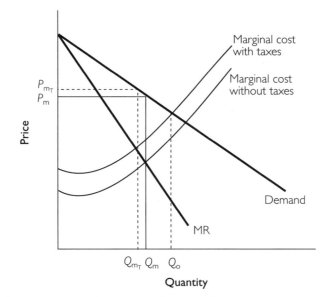

**Figure 14-4.** *Environmental Taxes and Monopoly*

*Notes:* $P_m$ = monopoly price; $P_{m_T}$ = monopoly price if a Pigovian tax is (wrongly) applied; MR = marginal revenue; $Q_{m_T}$ = quantity sold under monopoly with tax; $Q_m$ = quantity sold under monopoly; $Q_o$ = optimal quantity.

tax on emissions (to overcome the externality problem). In general, this approach will not be practicable, either. Subsidizing monopolists is hardly attractive. The way the problem is typically stated, environmental policymakers are about to implement charges when they realize that the polluter is a monopolist. This begs the question why policymakers do not address the existence of the monopoly itself (e.g., by deregulation). When monopolies cannot be avoided (e.g., in some sectors of small economies), trade liberalization is the only viable alternative. However, it may be difficult in the presence of barriers to trade or high transport costs. In many countries, signs indicate that the classical monopolies are declining in importance as a result of globalization.[7]

When monopolies cannot be deregulated or broken up, regulators may have to choose between the inefficiency caused by a charge on a monopolist and the inefficiency caused by the externality, especially when many different industries produce the same environmental externality (e.g., carbon or sulfur emissions). If the tax works well in the competitive industries, it seems unlikely that policymakers would want to reduce the tax because of possible perverse effects in one oligopolistic industry. The relative importance of these two inefficiencies depends on the relative importance of abatement effects and output substitution effects. With inelastic demand and a relatively low cost of clean technology (e.g., as in Figure 14-1A), a tax instrument might be satisfactory; a monopoly would involve too low a production level. An environmental tax would in this case have only a negligible effect in further distorting output. However, the tax would give incentives for cleaner technology. Finally, policy packages with combined instruments can be used. In a context with imperfect information and the possibilities of both moral hazard and adverse selection, the optimal policy might be a combination of an emissions tax, a subsidy per unit of production, and a lump-sum tax on profits (Laffont 1994a, 1994b).

Most industries do not belong to either of the ideal types (monopoly or "perfect competition") but operate in intermediate markets with some degree of imperfect competition. In duopolies and other market structures, it is natural to frame the analysis in a game-theoretical framework. Results might be expected to be midway between the monopoly and competition cases, but numerous complex cases are possible. Complications arise, especially if industries make separate choices about technology and operation (e.g., Simpson 1995; Carlsson 1999, 2000; Katsoulacos and Xepapadeas 1995; Ebert and von dem Hagen 1998).

Furthermore, additional aspects must be considered, such as those related to the distribution of the burdens, transaction costs, political economy of decision-making, and power of the polluters. One elementary observation is that designing complex instruments such as tax laws that have to be approved by a legislative body may be costly if regulators only want to regulate one firm (or a few firms). In such cases, direct negotiations are perhaps more appropriate. This approach appears to have been confirmed by the phaseout of leaded gasoline in El Salvador, a small country with only one refinery (see Chapter 22).

Another important consideration when choosing a policy instrument in situations where competition is limited is to avoid instruments that may be used (or abused) by companies to increase market power, build collusive agreements, or impede the entry of newcomers. Several instruments convey such risks: subsidies for abatement, voluntary agreements, the creation of property rights (as with grandfathered permits—unless the regulator specifically designs the instrument to provide for new entrants), and standards (particularly if the standards for old and new plants differ considerably).

> ## Supplemental Reading
>
> Bovenberg and Goulder 1996
> Carlsson 2000
> Mayeres and Proost 2001
> Mirrlees 1971
> Myles 1995 (Chapters 4 and 5)
> Parry and Robertson 1999
> Repetto et al. 1992
> Xepapadeas 1997

## Notes

1. Whether other bleaching agents (such as ozone) also cause problems is beyond the scope of this discussion. In principle, the prices of and markets for paper would be affected somewhat if the new processes were more expensive than the chlorine process, but this effect would be very small.

2. Because of the definition of subsidy chosen and the constant marginal cost of abatement, the implications here are not the same as in the main case presented in Box 9-1 (see Chapter 9) or in Baumol and Oates 1988.

3. Any subsidy will make average prices lower than in the case with taxes. The generous subsidy analyzed in Baumol and Oates 1988 makes them even lower than in the unregulated case and may lead to a perverse effect: that total pollution increases due to the subsidy. The Pigovian charge provides the correct incentive for the socially optimal number of firms in the

long run (Spulber 1985). In a simplified form, for $n$ identical and optimal firms, welfare can be written as $W = nPq - nc(q,a) - D[ne(q,a)] - nC_F$, where $C_F$ is the fixed cost of entry.

Other instruments can be analyzed analogously. Output-based allocation of permits or refunded emission payments will lead to results similar to that of regulated emissions rates: the marginal cost will be close to $MC_T$, and the average cost is close to $AC_r$. The long-run price will be close to $P_r$, reflecting the loss of some of the output effect of a tax.

4. Many thanks to Ian Parry for comments on this section.

5. If reducing the externality itself (e.g., congestion when driving to work) is complementary to labor supply, then the labor supply effects may actually be positive rather than negative (Parry and Bento 2001).

6. In a wide range of natural resource extraction models, this intuitive rule fails to apply (Brown 2000). In determining the optimal steady-state stock, prices do not enter the equation. The stock is decided by the intrinsic growth rate of the stock and the discount rate alone; market structure has no effect on optimal stock.

7. On the other hand, new firms with considerable market power are emerging, but they do not necessarily make their profits by reducing output and keeping prices up—as shown by Microsoft. Also, environmental policy instruments may be designed to speed up the process of deregulation. By taxing emissions rather than products, import competition would normally be strengthened.

# CHAPTER 15

# *Distribution of Costs*

$P$OLICY INSTRUMENTS NOT ONLY CAUSE different allocative effects but also impose different financial burdens for polluters, victims, and society. The distribution of costs is distinct and thus linked to concepts of rights (see Chapters 5 and 10) and to the politics of policymaking. In this chapter, I focus on two kinds of cost distribution—among polluters, victims, and society and among the polluters themselves—and then address the effects on income distribution in general.

First, consider the victims of pollution. The Coasian perspective emphasizes that parties should be left to negotiate with minimum support from the state (see Chapter 5). The size of the damages depends on many variables, including the number of victims, and in a sense, a polluter may become "swamped" with "victims" who choose to locate near the source of pollution (which commonly is also a source of employment). A company may have built a plant with small externalities as measured conventionally, but when people move in, the external effects become larger. If environmental charges are based on the total value of such externalities, then it can be argued that the victims impose an externality on the polluter. However, in practice, negotiations are not effective with large numbers of victims. One exception is the health-related claims brought against U.S. tobacco firms, but the number of agents was not as large as the number of smokers, because the states sued. In this kind of setting it may be reasonable to assume that victims are represented by society (see Chapter 18), providing a rationale for general policymaking on behalf of the victims.

Several factors influence regulators' choice of policy instrument. Typically, the focus is mostly on efficiency, and insufficient thought is given to distributional and political economy aspects. Different instruments can lead to significant differences in the distribution of costs and thereby in the support or opposition that policy proposals engender. Efficiency addresses the total cost of abatement, which is the relevant variable for aggregate welfare analysis, but small differences in the distribution of the costs may often be politically more sensitive than fairly large differences in total cost. The managers of large firms today are often primarily

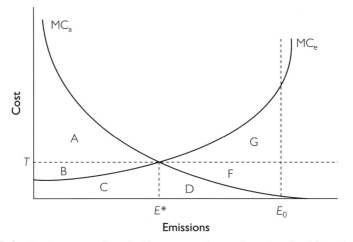

**Figure 15-1.** *Environmental and Abatement Costs Associated with a Reduction in Emissions*

*Notes:* $MC_a$ = marginal cost of abatement; $MC_e$ = marginal cost of damage due to emissions; $E*$ = optimal emissions level; $E_0$ = original emissions level, $T$ = Pigovian tax level; A–G = areas that represent different cost components discussed in the text.

judged by and concerned with their relative performance vis-à-vis competitors or vis-à-vis the market average. This relative performance is an important measure of their success. The typical polluter might thus be concerned with two distinct aspects of the distribution of the costs: its impact on the whole industry and (generally more important) its relative impact, that is, compared with competitors. With market-based instruments (MBIs), firms with the lowest marginal abatement costs typically do more abatement and thus end up having to bear higher total abatement costs. Because the level of abatement is chosen voluntarily by the firm, these costs must be more than compensated by the side payments (the difference in taxes or purchase of permits) involved in an MBI.

## Distribution of Costs and Rights between Polluters and Society

The distribution of costs and benefits of instruments between the aggregate of polluters and the aggregate of victims and society can be discussed with the help of a simple illustration (see Figure 15-1). The regulator wants to reduce aggregate emissions from an initial level $E_0$ to a socially optimal level $E*$, where marginal abatement cost ($MC_a$) equals the marginal cost of damage to the environment ($MC_e$). The benefit of environmental improvement from the emissions reduction is equal to the sum of areas D (which also represents abatement costs), F, and G. The net gain to society is thus always F + G, but the distribution of costs between parties varies with different instruments. Area C is the cost of unabated pollution; B + C is the classical Pigovian tax or scarcity rent, and B is thus a financial transfer from polluters to society when using taxes or auctioned tradable emissions

**Table 15-1.** *Distribution of Costs with Different Instruments and Environmental Rights*

| | Ownership rights to the environment | | | | |
|---|---|---|---|---|---|
| | Polluter (absolute) | | Polluter (relative) | Mixed | Victim (polluter pays principle) |
| Instrument | (1) | (2) | (3) | (4) | (5) |
| *Distribution of costs and benefits* | | | | | |
| Environmental benefit | D + F + G | D + F + G | D + F + G | D + F + G | D + F + G |
| Polluter costs | F | 0 | −D | −(C + D) | −(B + C + D) |
| Costs to society | −(D + F) | −D | 0 | C | B + C |
| *Type of instrument* | | | | | |
| Quantity-type | Public cleanup | | CAC, VA, TEP (free) | TEP, partly auctioned | TEP, auctioned |
| Mixed | | | Hybrids | Hybrids (e.g., TEP + tax) | Hybrids |
| Price-type | Subsidies | | REP, Tax–subsidy | Partly REP | Tax, DRS |

*Notes:* CAC = control-and-command policy; VA = voluntary agreement; TEPs = tradable emissions permits; REPs = refundable emissions permits; DRS = deposit–refund scheme. Single letters refer to areas in Figure 15-1.

permits (TEPs) as instruments. Similarly, $D + F$ would be the size of a subsidy if the state paid the polluter a fixed subsidy $T$ for each unit of abatement. Only area D corresponds to costs, whereas area F is a financial transfer from society to polluters if a fixed subsidy per unit of abatement is paid. However, subsidies are often designed to cover only D or only a part of D. The transfers B and F are a consequence of using a single market price on the environment even though the marginal costs are not constant. Figure 15-1 can also be interpreted to cover users of ecosystem resources such as water or air or even jointly owned resources such as a fish stock.

Depending on the choice of instrument, the various costs illustrated in Figure 15-1 are distributed differently between the polluters and the victims of environmental damage. The top part of Table 15-1 shows the different distributions of costs and benefits that achieve the same abatement, that is, bringing the level of emissions from the initial $E_0$ to the optimal $E^*$. The environmental benefit is the same for all columns. The costs (or payment) borne by polluters and society include physical abatement costs as well as tax or subsidy payments. For the sake of simplicity, *society* represents the victims of pollution. (In a fuller treatment, I could distinguish between situations in which some victims of pollution are compensated and some are not.) The bottom half of Table 15-1 presents the instruments best suited for meeting each definition of environmental ownership.

Whether the polluter or society should bear the costs of pollution is a question with aspects of efficiency, welfare, and ethics. In the spirit of Coase (1960), the question can be approached as a problem of how to define property rights to the environment. The columns in Table 15-1 refer to different ways of distributing the cost burden between the polluter and the victim (i.e., corresponding to an equivalent set of concepts of ownership rights). When the polluter has exclusive and absolute ownership rights to the environment, a victim (or society) who wants a clean environment will have to pay for it—either through abatement subsidies (if a price-type instrument is chosen) or through a publicly financed cleanup (if a quantity-type instrument is chosen) such as municipal trash pickup. In Column 1, subsidies are paid for every unit of pollution abated. Thus, society pays the "extra" subsidy (F) to the polluter, who in this case does not pay anything but earns money. In Column 2, abatement subsidies only cover actual abatement costs (D; in practice, it may only be a part of D, e.g., when credits for abatement investments are partially subsidized by low interest rates).

Column 5 of Table 15-1 represents the most far-reaching interpretation of the polluter pays principle. The ownership right resides squarely with society (perhaps in representation of the victims of pollution), and polluters will have to pay the equivalent of a market price for the right to use "environmental services." The price-type instrument to reflect this type of ownership would be a conventional (Pigovian) tax, and the quantity-type instrument would be an auctioned TEP scheme.

In Column 3, the polluter has limited, intermediate rights—the polluter may use the environment or resource in question freely but has to maintain it in some acceptable condition. The polluter must pay for some reasonable level of abatement (D) but not for the pollution caused at that level. This distribution of costs may be law-based (the polluter or resource user has the rights), or it may be a recognition of the relative clout of the different parties. Either way, the regulators have the option of choosing ordinary regulation (design or performance standards), voluntary agreements (VAs), a TEP scheme in which the permits are allocated free of charge, a tax–subsidy scheme, or a refunded emissions payment (REP) scheme. Intermediate instruments (e.g., TEPs with a refundable payment for excessive pollution) also are conceivable.

Column 4 represents an intermediate concept of ownership and shares the cost burden between the polluter and the victim. In this interpretation of rights, polluters must pay the costs of abatement and unabated pollution but not the "extra" financial transfer (area B) associated with a single market price on environmental services. It could be approximated by a TEP scheme in which permits are grandfathered up to a certain total emissions level and auctioned thereafter; or, permits could be grandfathered but carry a charge or tax. This approach has been used in the United States for allocating permits for chlorofluorocarbons and halons, which were taxed by the U.S. Congress when it became apparent that this program conferred large monetary benefits to the permit recipients (Tietenberg 1995; see also Chapter 24). Similarly, emissions payments could be partially refunded and, for example, a share of the revenues set aside to pay the costs of unabated pollution. This scheme is similar to some of the fee systems used in formerly planned economies and in developing countries, where fees are paid only in excess of certain levels and then paid into various environmental funds with different (regional or other) scope and mandate (see Chapter 25).

A TEP scheme could allocate permits to current polluters (in proportion to output) or to historic polluters. The output allocation corresponds more closely to the property rights implied by the REP. One interpretation of rights is that they ultimately belong to society but that the polluter earns these rights by producing something useful or desirable for society. The grandfathered rights correspond better to a classical "prior appropriation" concept of rights because they are given to the polluters who had them in the past. For polluters, REPs are preferable to a tax or charge. Similarly, the exact allocation mechanism for TEPs (auctioned, grandfathered, or output allocated) as well as the finer details of how the mechanisms are implemented—such as the choice of baselines and starting years—make a big difference to the firms involved, and these aspects tend to dominate practical attention.

Instrument choice is commonly discussed as if the main choice were between grandfathered permits and taxes. These mechanisms lead to different consequences for the distribution of payments. Thus, the distributional issue can get mixed up with the choice between price- and quantity-type instruments. The point of Table 15-1 is to show that regardless of who has ownership rights and who should bear a certain share of costs, there is still a broad choice between price- and quantity-type instruments. Other instruments also are related to the material presented in Table 15-1. Deposit–refund systems belong with taxes in Column 5. VAs are listed in Column 3 but also can be applicable in Column 2, because the firm only has to pay partial costs of abatement. Liability is more complex because it introduces the probability of damage (and detection). However, strict liability appears closest to the concept of rights in Column 5, whereas negligence liability would correspond to the more intermediate concept of rights in Columns 3 and 4. Informational instruments and labeling do not have such a clear position but are presumably most applicable with concepts of intermediate rights. They definitely do not fit (as the sole policy instrument) with the polluter pays principle in Column 5.

## Allocation of Rights

When a permit system is used, decisionmakers must decide how to allocate the permits. For REPs, the corresponding problem is how to refund the charges. Because the permits can have considerable value, a great deal of interest is focused on their allocation. In the 1999 allocation process for summertime U.S. nitrogen oxide permits in 22 eastern states, the total estimated value of the permits was US$1 billion to US$2 billion. The efficiency gains that tend to be the main focus of economics may be larger in absolute magnitude, but their incidence is spread across many producers and consumers over a long period of time and are thus to a large extent "invisible." However, the permit values are to be distributed among a small number of firms and are highly visible. Thus, they inevitably tend to attract most of the attention.

Pollution "rights" (explicit for permits, implicit for refunds) can be allocated in several ways. First, it must be decided how large these rights are: do they cover the entire environmental "rent" (B + C in Figure 15-1), part thereof, or even more (corresponding to areas D and F)? Auctioned permits or taxes build on the idea that society, as represented by government, originally owns the rights. This method has several positive properties but tends to be resisted by polluters—especially in cases of congestion, in which the polluters or resource users (e.g., motorists, fishermen) are (almost) identical to the victims. Even though congestion can be solved in an allocatively optimal manner through auctioned permits, the agents are worse off with a tax or auctioned permits than with no policy; more rent is available to society, but it is all appropriated by the government, leaving nothing for those agents it was supposed to benefit.

Two basic choices must be made in designing the allocation principle: whether to use historic or current variables, and whether to base the allocation of rights (or refunds) on output, emissions, or some other variable. To enhance long-run efficiency (with respect to output and entry effects), polluting firms must include the opportunity cost of pollution rights in their calculations. However the rights are allocated, the allocation must be perceived as definitive. Renegotiation with respect to current variables should not be possible. The independence of current variables guarantees the absence of perverse effects on business decisions; it is one reason for the success of cap-and-trade permit programs, even though it reduces the flexibility of the system from the viewpoint of politicians and environmental protection agencies.

The historic variable most commonly used as a basis for allocation of permits is levels of emissions or, for natural resources policy, harvest levels, which ties in well with the legal doctrine of prior appropriation. Examples include the grandfathering of free permits and the tax–subsidy scheme (proposed in Pezzey 1992 and discussed in Chapter 9). The latter is instructive in this connection because the polluters have to pay a tax on emissions except that they are given a certain baseline level (related to their historic levels) to which they have a right; they have to pay the charge on emissions only in excess of this level. If their emissions are below the baseline, the instrument effectively becomes a subsidy.

This principle is naturally popular with polluters, but one disadvantage is that it does not encourage "proactive" behavior; as such, its use is undesirable. In fact, if all polluters knew that regulators would always grandfather the allocation of permits, then they would never have an incentive to clean up problems in anticipation of future legislation and policy. On the contrary, polluters would have a strong incentive to emit as much as possible in order to get a large allocation of permits or a "high" baseline against which to measure future abatement. If all firms in an industry acted similarly, it would be difficult to find demonstrated clean technology, and the development of clean technology would be slow.

The undesirable feature of historic variables can be partly compensated for by choosing baselines many years prior to the decision, but this method has other practical disadvantages: measurements may be inaccurate, old values tend to be thought of as outdated, and newcomers will be highly critical. Newcomers tend to be the most progressive (in terms of efficiency and environment), and a policy instrument that supports the oldest and most polluting industries to the detriment of the newer ones is unsuitable, at least in that respect.

In the context of global warming, the concept of prior appropriation is fiercely contested by countries with large populations and small (but rapidly growing) levels of pollution, which would be more favored by a per capita or needs-based allocation (which might also be favored by countries with cold climates). Some have even emphasized the stock characteristics of the pollutants and have argued for a per capita allocation of total emissions (historic, current, and future), thereby leaving little allocation in the future for the industrialized nations.

When the carrying capacity of an ecosystem varies rapidly and unpredictably and is threatened by a market failure such as overuse, then policy instruments must meet two distinct goals: first, they must be efficient, which might mean that there is a strong need for permanent and verifiable property rights (permits or quotas); second, ecological factors require some degree of flexibility. A good example of this is found in the case of fisheries. Fishermen need quotas that are private property in order to plan their economy (e.g., take loans). At the same time, stock dynamics of fisheries are such that total allowable catch must be varied each year. This has been solved (as already mentioned) by denominating the individual transferable quotas in terms of shares in total harvest. The fishermen have fixed shares (which can be sold, inherited, or used as collateral), but the total catch can be set by a board of fisheries experts based on recent assessments of stock (see Chapter 28). If the world had to implement a full allocation of permits for carbon emissions into the atmosphere for the whole of this century, it would be wise to include some similar mechanism; if it were later determined that the amount of emissions to be reduced was more than allocated, then society would face an extremely costly burden of buying back the allocated permits.

To circumvent the somewhat arbitrary nature of an allocation based on historic variables, current variables can be used. Current emissions are not a good measure to use. However, two natural ways of allocating permits are in proportion to a key input (such as fuel) or to output, which is the same as regulating emissions rates rather than total emissions (see Chapter 22 on the U.S. lead rights program or Chapters 9 and 24 on the Swedish policy of refunding charges on $NO_x$).[1] One disadvantage of these approaches is that they provide an effective "output subsidy" to the firms: if the refund or permit allocation depends on a variable that the firms can influence (such as output), then firms may be perversely encouraged to overproduce. (For monopolies with suboptimally low supply, an output subsidy might actually be more appropriate.) The ordinary command-and-control approach (e.g., best-available technology) has similar properties.

## Incidence of Costs between Polluters

Another important area in which the distribution of costs may be important is between polluters. Although the sum and the distribution of polluters' abatement costs should be the same with any MBI (at least to a first order of magnitude; "wealth" and entry and exit effects or effects of strategic behavior might be distinct), the distributions of transfer payments among the polluters are usually different depending on the permit distribution or refund mechanism chosen. The distribution between polluters is important not only for reasons of fairness but also because it affects political support for and resistance against certain kinds of instruments.

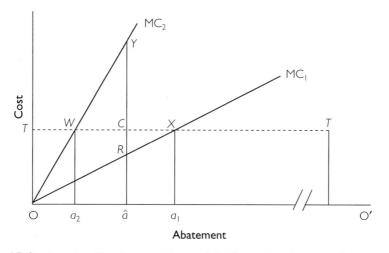

**Figure 15-2.** *Cost Implications to Firms of Different Market-Based Instruments*

*Notes:* MC = marginal cost of abatement for Firms 1 and 2; $T$ = Pigovian tax level; $a$ = optimal level of abatement for Firms 1 and 2; $\hat{a}$ = equal level of abatement for both firms; O = origin.

Grandfathered permit allocation favors firms that have been large polluters in the past and disfavors newcomers, which are likely to use more modern techniques and be more efficient. Taxes, auctioned permits, and REPs are relatively more favorable to firms with low emissions rates, and such firms may conceivably approve of these instruments despite their cost, because competitors are faced with even higher costs. In Figure 15-2, $a_2WY\hat{a}$ is the net savings in abatement for Firm 2, which has higher abatement costs than Firm 1; $\hat{a}RXa_1$ is the increased abatement cost for Firm 1; and the difference between these costs is the net savings to society in reduced abatement costs.

To realize this cost savings, Firm 1 must be motivated to bear the extra abatement cost and MBIs must be used—that is, firms must receive subsidies, pay taxes, or trade permits. Whichever mechanism is used, the difference in payments between the two firms is $a_2WXa_1$; if an environmental tax is used, the taxes are higher for Firm 2 by the same amount; and if permits are used, then the value of the permits are half of $a_2WXa_1$.[2] The amount of the cost savings depends on the heterogeneity in abatement costs (see Chapter 12), and the level of the taxes depends on the amount of the emissions ($O'a_1$ and $O'a_2$). The relative sizes of these amounts can vary considerably. For some parameter values (low heterogeneity in abatement costs [see Box 15-1]), the transfer payments caused by the MBIs between agents will be much larger than the net gain in reduced savings. In these cases, distributional issues will be thought of as more important than overall allocative efficiency.

Additional complexities enter into the picture in empirical cases. For environmental taxes, there may be exemptions; for subsidies, baselines have to be formulated; for permits, an allocation mechanism must be formulated and implemented. The exact allocation of permits is often the result of a long and complicated political process and is difficult to foresee exactly in advance (Hahn

## Box 15-1. An Example of Abatement and Permit Costs for Two Firms with and without Trading of Permits

Taking the linear marginal cost functions ($MC_1 = a_1$ and $MC_2 = ha_2$) from Chapter 12 as an example, the costs for each firm can be calculated. With equal abatement reductions ($∂$), Firm 1 pays $½ × ∂^2$ and Firm 2 pays $½ × h∂^2$ (see column 3 in the table below). Total abatement costs are $½ × ∂^2(h + 1)$. If trade is allowed, then Firm 1 will abate $2h∂/(1 + h)$ and Firm 2 will abate $2∂/(1 + h)$ (see column 5). The MC of abatement is $2h∂/(1 + h)$, so total costs of abatement are $2h^2∂^2/(1 + h)^2$ and $2h∂^2/(1 + h)^2$, respectively ($OWa_2$ and $OXa_1$ in Figure 15-2). Total abatement cost savings are $∂^2(h − 1)^2/2(h + 1)$. Total costs for each firm consist of abatement and permit costs (columns 5 and 6). The total difference in cost to the two firms is $2h∂^2(h − 1)/(1 + h)^2$, which is bigger than the total resource savings in abatement costs even for quite large values of $h$ (roughly, under 5).

Abatement and Permit Costs for Two Firms with and without Trading of Permits

| | Equal reductions | | | Equalized MC after trading permits | | |
|---|---|---|---|---|---|---|
| Firm no. (1) | Abate- ment (2) | MC (3) | Cost (4) | Abatement (5) | Abatement cost + permit cost (6) | Total cost (7) |
| 1 | $∂$ | $∂$ | $∂^2/2$ | $2h∂/(1 + h)$ | $2h^2∂^2/(1 + h)^2$ $− 2h∂^2(h − 1)/(1 + h)^2$ | $2h∂^2/(1 + h)^2$ |
| 2 | $∂$ | $h∂$ | $h∂^2/2$ | $2∂/(1 + h)$ | $2h∂^2/(1 + h)^2$ $+ 2h∂^2(h − 1)/(1 + h)^2$ | $2h^2∂^2/(1 + h)^2$ |

Notes: MC = marginal cost; $∂$ = abatement reductions; $h$ = heterogeneity.

1990; Stavins 1998). Subsidies, output-allocated permits, and REPs entail some distributional consequences of the output effects discussed earlier. Grandfathering tends to benefit (the shareholders of) older, heavily polluting firms, whereas output allocation (or REPs) is more beneficial to cleaner newcomers.

Efficiency dictates that marginal costs be equalized, so firms with low marginal costs of abatement bear high total abatement costs. How does an environmental protection agency persuade one firm to bear higher costs than its competitors if not through side payments (as in the MBIs)? If marginal abatement costs differ because Firm 1 has historically done much less abatement, then Firm 1's higher requirements might be thought of as "reasonable." However, if that is not the case and the difference in marginal abatement costs is due to other factors, then the firm probably will not voluntarily agree to undertake more abatement than Firm 2 without receiving some kind of compensation. Regulation would be much less costly to society if it were possible for an environmental protection agency to make informed, reasonable judgements and arrive at an approximately optimal allocation of abatement. This goal might be achieved with VAs, which are sometimes suggested as an alternative to MBIs, although the likelihood is very low.

Instruments that are sensitive to the distribution of the costs of abatement and transfer payments appear to be at least as relevant for developing and transitional economies as for developed economies (Chapters 5 and 6). The reasons for this relevance include higher emissions levels, the less competitive nature of the economy (which increases interest in strategic effects), and inadequately developed credit markets (which can make abatement difficult to finance).

## Income Distributional Effects and Poverty

Three kinds of costs are involved in pollution: environmental damage, abatement costs, and expenses related to the instrument chosen (e.g., transfer payments, information, and administration). Each of these categories has effects on distribution. Many of the most important effects of policy instruments on income distribution are related to the cost factors discussed in the previous sections. The distribution of costs has effects on employment, rents, profits, and taxes.

Environmental programs are sometimes accused of putting firms out of business and creating unemployment for the sake of protecting national parks and other ecosystems that provide amenities and existence values to the already privileged. This claim may, in a few cases, be true; because privileged individuals tend to be more vociferous, their environmental projects may be more likely to be funded. In principle, some other projects might support ecosystems that poor individuals rely on or enjoy, and a particular allocation of costs might disproportionately affect rich individuals. Distributional issues with respect to individuals builds heavily on the effects already analyzed in the previous section because most individuals are affected as workers, consumers, or owners of property or shares.

In general, little can be said about distributional impacts except that they must be analyzed on a case-by-case basis. The distributional impacts of environmental damages are completely separate from the corresponding impacts of abatement or instrument costs. Any of these components might be progressive, neutral, or regressive in general terms, and any kind of cost may or may not affect the welfare (in benefits or damages) of specific groups.

The provision or enhancement of environmental amenities that are luxury goods tends to have a regressive profile. Although it may apply to some cases (epitomized by scenic beauty or recreation), it is far from being a general truth. The poorest segments of society generally are the most dependent on intact common property resources for their livelihood and survival in terms of food, fuel, building materials, and medicines. A program that seeks to maintain mangrove swamps, for example, will benefit the poor individuals who gather various products there. The same program may increase costs to companies producing shrimp in shrimp ponds, but those companies are likely to be wealthier than the individuals who are dependent on the resource. (However, shrimp farms may employ low-income people, so the issue is not clear.)

Similarly, it is the poor who are most affected by bad water quality because they cannot afford bottled water, filtering systems, inoculations, and other medical treatment. Projects that protect sources of drinking water thus have a progressive distribution profile.

Despite these counterexamples, many environmental projects may have regressively distributed benefits. In a similar way, abatement costs may be distributed progressively or regressively. Generally, the agents who produce the pollution are the ones who will be most affected by abatement costs. The choice of instrument also can make a big difference: an instrument such as an environmental tax or auctioned permits puts the burden explicitly on the polluter. However, instruments such as grandfathered and output-allocated permits, REPs, and subsidies have different properties with respect to the cost distribution. The structure of a market determines whether a polluting company can pass on the costs of abatement and regulation to its consumers or back to its workers and suppliers earlier in the production chain. With information on the distribution of costs and benefits, this information must be considered when choosing and designing instruments. It also is important to distinguish between equilibrium costs and adjustment costs. Closing down old, polluting plants may be beneficial for the environment and, in many cases, for the economy overall. Closures still might cause considerable harm to employees and others in the form of short-run adaptation costs. The distribution of these damages between categories of people or between legal entities may vary just as much as the incidence of the environmental damages.

The cost distribution may have many dimensions—not only between owners of capital and laborers or between rich people and poor people, but also between generations, between healthy people and sick people, between urban communities and rural communities, and among many other categories that may be difficult to enumerate or predict. Concern should be for the welfare of all citizens, and it is problematic that some groups are adversely affected because a particular characteristic is correlated with consumption or production patterns. People who live in a certain area; have a certain kind of family structure, job, culture, or habits; commute by a certain route or vehicle; or in some other way use certain goods may cause environmental damage, and the policy instruments to correct this damage may end up causing costs to these same people. To a large extent, this effect is inevitable, but the extent of the costs borne by certain groups may sometimes cause problems.

Each case must be examined with an understanding that considerable welfare effects may be related to the exact distribution of costs and benefits and that these effects might even overshadow the aggregate effects. In some cases, the use of social welfare weights might be called for in the analysis, design, and evaluation of policy instruments. Particular care needs to be taken to avoid unnecessary effects related to unemployment and relocation. On the other hand, attempting to soften all the blows of economic restructuring can make policies very expensive. Good, detailed analyses of the income distribution effects are rare. The Swedish Green Tax Commission is (at least partially) an exception (see Box 15-2).

One of the main purposes of the tax system is to affect (directly or indirectly) the distribution of goods and welfare. Public economics explains that it requires considerable information and runs the risk of creating incentive problems and control costs. Because of the complexity of tax systems, adding new environmental goals exacerbates interaction effects and inevitably increases the overall complexity. However, not only environmental tax instruments but also regulations have such interaction effects (see Chapter 10).

In developing countries, the poorest of the poor often suffer a disproportionately large burden of child mortality and disease. According to one study, the same people lack education as well as access to physical infrastructure such as

## Box 15-2. Distributional Effects of Raising Carbon Taxes in Sweden

| | Loss | |
|---|---|---|
| Income group | SKr/cap/yr | % of total consumption |
| Poorest 20% | 888 | 1.24 |
| Richest 20% | 1,026 | 0.78 |
| Urban areas | 1,261 | 0.88 |
| Rural north | 1,392 | 1.16 |

*Notes:* Carbon taxes were increased from 0.37 to 0.74 Swedish krona [SKr] per kilogram (from about US$0.04 to US$0.07 per kilogram, where 10 SKr = US$1). Loss is essentially measured as the compensation necessary to stay at the same utility level despite the tax. "Urban areas" refers to Stockholm and the two next largest towns.

*Source:* SOU 1997.

The results show distribution effects of increasing carbon taxes from 0.37 to 0.74 SKr/kilogram according to two criteria: income, and a combination of geography and rural/urban location. The richer consumers are more affected in absolute terms but less in relative terms. Judging this welfare effect is difficult because it depends on the curvature of the aggregate welfare function. The rural consumers of the cold and sparsely populated north who need energy for heat and mobility are, as expected, on average more affected than the inhabitants of the cities. However, it does not apply to all subcategories: many rural farmers, pensioners, and others do not commute and use biofuels for heating. Thus, the issues are typically very complex. Total effects are quite small.

water and sanitation: of all households, <10% of the third and fourth deciles, 30% of the second decile, and 70% of the poorest decile lack sanitary facilities (Bonilla-Chacin and Hammer 1999). Similar disparities applied to several indices, such as access to safe water (only 15% of the poorest decile). The difficulties are compounded because the poorest people often have low "social and human capital" resources. People with physical, mental, or social handicaps (including India's "untouchables" and refugees of natural disasters) are in this group. They may be unable to take advantage of programs intended to help them.

One natural conclusion from this information is that poor populations should be better targeted, particularly in times of budget cuts and leaner government. However, it is far from easy to do because the "slightly less poor" have appreciable power to protect themselves from cuts in the benefits they receive, and it is possible (in some cases, even likely) that poor populations will be more than proportionately hurt by budget cuts (Ravallion 1999). A certain "leakage" to nontargeted (or less-targeted) groups may be part of the political economy cost of reforms; perfect targeting is simply not realistic considering the characteristics of these groups.[3]

At least some groups among poor segments of populations may be more than proportionately hurt by abatement costs and environmental programs; however, many poor people suffer severely from the consequences of natural resource degradation and certain types of pollution. This issue concerns not only the welfare but also the efficiency of poor households, inasmuch as their (potential) resources or abilities cannot develop given the serious effects of a poor environment on their health and thus opportunity.

## Supplemental Reading

Grafton and Devlin 1996
Kriström and Riera 1996

## Notes

1. In the context of the REPs, it would be hypothetically possible to emulate grandfathering by refunding the charges not to the current polluters but to polluters of some base year. However, it would be an odd policy to search out the largest polluters in some historic "base year" (even if they have closed down since then) and then pay what are essentially tax dividends from today's polluters to them.

2. With taxes, Firm 1 will pay taxes $O'a_1XT$ and Firm 2 will pay $O'a_2WT$ (assuming $O'$ is the origin from which emissions are measured). With subsidies per unit abated, Firm 1 will receive $Oa_1XT$, and Firm 2 will receive $Oa_2WT$ (disregarding any output effects of the subsidies). With equal permit allocations of $O\hat{a}$, Firm 1 will sell $a_1XC\hat{a}$ permits to Firm 2, which is just sufficient for that firm (i.e., $a_2WC\hat{a}$).

3. This observation should not be taken as condoning complacency. In examples of seemingly successful reforms, general subsidies were cut and more targeted programs introduced; for example, in Tunisia, universal food subsidies were replaced with a food subsidy program that targeted poor individuals (Tuck and Lindert 1996). In Chile, strict marginal cost pricing was combined with means-tested subsidies for water pricing (see Chapter 26).

# CHAPTER 16

# *Politics and Psychology of Policy Instruments*

A COMMON ASSUMPTION IS THAT POLICYMAKERS design policies to maximize something like a social welfare function. Although this assumption is not very realistic, it is often made because it allows for the use of tractable mathematical models.[1] Even if policymakers did try to maximize such a function, and even if the welfare function were a simple utilitarian one, the design of policy might be complicated by market failures, information asymmetries, and other problems. If the social welfare function is more complex (e.g., including "equity" as an independent element), then the design of policy instruments must meet several goals. Also, complex individual utilities—for example, including altruism or relative (rather than absolute) consumption—have profound consequences for the design and implementation of policy instruments.[2]

Some environmentally relevant goods may also be positional (e.g., car owners may get utility from having a better car than others), which will influence optimal taxation (e.g., Frank 1985). If there is habit formation, myopia, or even addiction, then the complexity of the utility function increases more (Rabin 1998; Becker and Murphy 1988), which may or may not have consequences for policy instrument design. If policymakers maximize a function completely different from welfare (e.g., chance of being reelected) or if they follow some other rationale for decisionmaking (e.g., psychological factors, ideology, or norms), then it must be incorporated into the analysis, as is traditionally done in the public choice literature—particularly if the purpose of the analysis is to be positive rather than normative. In this area, economists tend to recommend "optimal" policies (mainly based on efficiency criteria) and thus slip into a rather normative role. Equally as important as designing policies should be analyzing and understanding why certain policies are selected or designed differently under different circumstances. Psychological, cultural, and political factors may dominate over economic ones.

## Politics of Policy Instrument Selection

To understand the politics behind the selection and design of policy instruments, one must realize that governance is a complex phenomenon. It is an area in which more collaboration is needed among economists, political scientists, psychologists, and other social scientists. Some political scientists categorize policy instruments as normative, suasive, or informative (or "carrots," "sticks," or "sermons," according to Bemelmans-Videc, Rist, and Vedung [1998], who state that in some situations, the government must use "sticks" or "carrots," but the tendency to use the more paternalistic informative instruments is increasing) (see Chapter 6).

Political scientists have different views concerning the degree to which central government truly governs in the everyday sense of the word. Some contend that central government is being reduced to a position in which it coordinates and bargains with other centers of power, ranging from local government to supranational and intergovernmental organizations on the one hand, to networks of other actors in the private and public sectors on the other (Rhodes 1996). In Europe, the historically rapid process of integration through the European Union appears to be coupled with a strengthening of regional and municipal government in what sometimes seems to be a joint effort to reduce the power of the traditional national governments. The opposite view is that the central governments maintain crucial control over the important decisions (Pierre and Peters 2000). The deregulation of capital and other markets has created greater flexibility for companies and may increase their negotiating power. Lobbying groups form to influence and determine the actual structure and implementation of policies. Economic interest groups can be successful by lobbying—that is, investing money to influence the political process in their favor (e.g., Grossman and Helpman 1994). Lobbying groups essentially seek to get advantageous trade policies that tilt the relative prices in their favor. Competition among rival lobbies may be a way of selecting efficient policy instruments (Becker 1983). Similar models have specifically analyzed lobbying in the area of environmental policymaking (Fredriksson 1997, 1998). Once it is accepted that policymaking is not only the result of a neutral effort by the state to promote welfare but also a reflection of the self-interest of some groups, it is not difficult to understand that the most powerful, well-established, and concentrated groups tend to have an advantage over other groups. A few industrial polluters often have more opportunity to band together as lobbyists than the much more numerous, dispersed, and unorganized victims of industrial pollution. In contrast, many polluters are unorganized and relatively powerless, and the capacity for nongovernmental organizations to capture and represent the interests of victims should not be underestimated.

An important force in building policy is the formation of political lobbies to protect the status quo. An established industry has many strings to pull, not only through the traditional (direct) lobbying mechanisms but also through its relationships with employees, local communities, customers, and suppliers. Subsidies tend to be self-perpetuating in that their existence will be strongly defended by the beneficiaries, who may be prepared to spend a large share of the subsidy in lobbying for their continuation. Individual licensing of plants is a process that can easily degenerate into a mechanism for limiting competition. For example, because technology improves constantly and because it is easier to incorporate new technology into new plants than to retrofit old ones, setting higher emissions standards for new plants might be considered a reasonable approach to regulation. When the old plants have lobbying power and succeed in lobbying for such a system, they effectively create a barrier to entry; this practice is referred to as *new source bias*. Even tax mechanisms may develop lobbies or groups that favor the charge. As Canard (1801) said two centuries ago, "*Que tout vieil impôt est bon, et tout nouvel impôt est mauvais*" ("Every old tax is good, and every new tax is bad").

Political scientists tend to focus their analysis of policy instruments on many other aspects in addition to the efficiency of the instrument on which economists concentrate. Other factors include the process of decisionmaking, the influence of lobbyists, the structure of political institutions, and the dispositions of the political agents (Lundqvist 2000). Some political scientists use the whole range of factors to provide a broad synthetic theory of instrument choice (Linder 1988; Linder and Peters 1990, 1991). Political institutions tend to affect the selection of policy instruments by kind of regime: democratic regimes tend to apply stronger policy instruments than authoritarian regimes as strategies for environmental abatement. Some countries have rules that restrict or prohibit lobbying, whereas others encourage or tolerate it; the prevailing approach affects the balance between lobbyists and technocrat experts arguing for efficiency, for example. The scope for local or regional instruments such as taxes or fees depends on their legality. Because many pollutants are either local or regional, policymaking ideally would take place at levels other than national. In many countries, local communities' rights of taxation are narrowly regulated, making it impossible to introduce environmentally differentiated waste charges, for instance. The state may be keen on keeping its monopoly on taxation, but more fiscal federalism and revenue sharing between levels of government often would be beneficial.

Similarly, national sovereignty sets limits to taxation and other instruments at the supranational or regional level. Another factor that influences instrument choice is the characteristics of the decisionmakers themselves. Rationality and time for decisionmaking are limited; decisionmakers cannot be expected to know everything about all relevant topics, and the time they are given to amass all relevant information on any given topic is usually short. As a result, they commonly rely on various methods to help them make their decisions: consulting experts, applying rules of thumb, learning by doing, and deferring to ideology. Ideologies are not trademarks invented to win elections; to a reasonable extent, they are belief systems that cover not only the goals of regulation but also the means, which is why some politicians prefer legal restrictions while others prefer financial incentives or the persuasive power of good examples.

In Sweden, the government has been accused of using the tide of environmental opinion for its own purposes to formulate the Local Investment Programme for Sustainable Development (Lundqvist 2000; Regeringskansliet 2001). The Swedish government allegedly strove to attract green votes and please ecologically oriented collaborating parties in an election year (1998). Instead of taxes, permits, licenses, or regulations (which tend to annoy at least some people), the policy instrument chosen for the program was large-scale public spending and subsidies. The local investment programs are grants that typically cover up to 30% of the costs for investments that are intended to create an ecologically sustainable society (according to Göran Persson, Swedish Prime Minister, Parliamentary Record, March 22, 1996). The program has been criticized for the size of the grants and the speed of selection and decision mechanisms. At least in 1998, the first year, all the conventional channels (environmental protection agencies and relevant ministries) were bypassed, and 42 municipalities received roughly US$500 million.

The Local Investment Programme has been strongly criticized (Riksrevisions-verket 1999b, 28; Riksrevisionsverket 1999a, 37; Riksdagsrevisorerna 1998/99; Kågeson and Lidmark 1998). The municipalities chosen to receive subsidies allegedly were those with the largest number of swing votes (Dahlberg and Johansson 2000). This case as well as previous Swedish public investment and subsidy programs for energy savings and renewable energy investments have been criticized for their low efficiency. This critique is not necessarily always correct but it does illustrate the peril inherent in subsidies as an instrument that can be used, or seen, as a means of buying political support.

Abatement subsidies may strengthen lobbying in favor of lower environmental taxes (Fredriksson 1997). In Part Four, I discuss how the political economy of lobbying can lead economies into distinctly different equilibria: some countries eschew fuel taxes almost completely and opt for physical regulation (e.g., the United States), whereas other countries, where the fuel taxes are much higher, only mildly oppose them and even tolerate further increases. One might expect that the political pressure to lower gas taxes would be high in countries with high fuel taxes, such as Sweden or Italy (see Chapter 21), and that it would be fairly easy to increase the modest gas tax in the United States. However, it is not: the parties who stand to lose are visible, organized, and vociferous, whereas those who potentially might gain are not organized and, in some cases, not yet in exist-ence (e.g., people who would be employed if there were more public transporta-tion options). Whereas the political resistance to a tax might be expected to increase with its magnitude, when it does not, it is indicative of a form of multi-ple equilibria or states that depend on the interaction between politics and economics.

The power of lobbying groups is another factor that in practice makes the taxa-tion of monopolies or others with market power likely to be less severe, because they typically have the resources and the incentive to engage heavily in lobbying (see Chapter 24 on energy taxation in Sweden and Germany). The choice of policy instrument should take into account potential lobbying groups. A tax has the decided disadvantage of uniting polluters, who all lose money from a tax. Other market-based instruments, such as tradable permits and refunded emissions permits, may possibly diffuse some of the potential power of such lobbies because some seg-ments of the polluting industry may gain rather than lose from the policies.

Conversely, firms that have political power typically have preferences about not only the target level of abatement but also the choice of instrument. Different instruments can have dramatically different consequences for the distribution of the total cost burden. The established firms have more influence and can shape technical regulations so that they have a heavy new source bias, which helps explain why physical regulations and licensing are so frequently used. Similarly, grandfathered permits are popular with established firms, whereas output-based allocation of permits and refunded emissions payments are less popular. Least popular, presumably, are taxes and auctioned permits. Striking a balance between what is perceived as fair while being realistic with respect to those who have power can be difficult. Political scientists often emphasize that the success of a policy depends not only on a fair distribution of costs but also on respecting due process, which means following the traditional procedures for gathering information, engaging in debate, ensuring representation, and participating in decisionmaking.

Researchers have addressed the development of environmental planning as a way to face the multiple uncertainties related to environmental policymaking (Jänicke and Jörgens 2000). A shift from "public policy" to "public management" signifies that the reliance on general rules, hierarchical decisionmaking, and fixed policy instruments has given way to a more strategic approach in which government emphasizes the process of formulating environmental visions with concrete goals and then relies on decentralized decisionmaking and flexible instruments for implementation.

## Enforcement, Monitoring, and the Psychology of Instrument Choice

The efficiency of policy instruments depends on the mechanisms through which they function. Common assumptions (e.g., the profit-maximizing firm and the utility-maximizing individual) are gross simplifications. Particularly in complicated situations (e.g., those with stochastic outcomes), psychological research has shown that individuals do not maximize expected utility in a simple statistical sense: they have difficulties perceiving the relevant parameters of a situation, may have "costs" of mental calculation, may interpret the probabilities subjectively, and may be averse or prone to risk.

Instead of calculating the (subjective) expected utility based on expected probabilities and outcomes, decisionmakers might place more weight on possible outcomes or on probabilities. Psychologists call this practice the *prominence effect,* which may be part of a mental strategy in which people seek to reduce the conflicts inherent in choice by finding a cognitive structure that shows one alternative as dominant. According to prospect theory, individuals do not calculate expected outcomes and typically care more about losses than gains (Tversky and Kahneman 1981). The nature of such mental processes has great implications for the success of different policy instruments. If agents are particularly concerned about being "caught," then frequent and efficient monitoring and enforcement may be the most important aspects of a good policy. If the size of the maximum penalty influences the behavior of the agents, then the penalty structure might be the most important design factor. These distinctions are well known in criminal law, but the details may be different for infringement of environmental regulations than for other crimes.

Aspects of enforcement and penalty not only influence strategies of compliance but also have a role in forming attitudes toward environmental infringements. The emission of chlorofluorocarbons into the air is not the obvious moral equivalent to violent crime, but by stipulating certain punishments for emissions offenders (such as prison sentences), law can create the desired values in a population. A tax on child pornography, for example, might be more efficient than certain criminal penalties in limiting the prevalence and severity of such activity; however, taxation would be considered a morally unacceptable policy in this instance because laws are important instruments for forming and communicating ethical values. Other psychological factors may also affect the choice of a policy instrument.

Usually, economic incentives are assumed to have an effect on actions (strong or weak, but definitely positive). However, economic incentives can "crowd out" moral motives (Frey 1997). In one experiment, two groups of subjects played managers who were instructed to allocate part of their budget to scrubbers that would reduce emissions (Tenbrunsel and Messick 1999). One of the groups was required to make decisions subject to a weak external sanction: if the group did not comply, there was a small risk that the group would be fined. Compliance was about 75% in the reference group but 50% in the "sanctioned" group. Without the sanction, the decision was seen as an ethical one; with the condition, it became a business decision, and because the expected costs of cheating were low, abatement was simply not profitable.[3] In a similar study that focused on common property resource management, people who used the same finite water resource were offered the opportunity to voluntarily "buy out" other users to save the resource (i.e., a smaller group would more easily think of the common good [White, Anderson, and Ford 1995]). However, the policy did not have the intended result but rather hastened depletion. In debriefing, participants indicated that the payment structure made them think in terms of maximizing their own outcome rather than solving the collective dilemma. Thus, economic incentives had an effect that was perverse or opposite to that intended.

Some research has focused on integrating the economics of enforcement into the decisionmaking process, because all instruments have transaction and implementation costs. Assumptions about enforcement range from the naive belief that compliance is automatic to a pessimistic view, that there is no compliance without strong incentives. Compliance is costly, but surprisingly, many studies find that firms usually comply even though the monitoring is far from meticulous and punishments seldom enforced (Heyes 1998).[4] This is the so-called Harrington paradox, which has several plausible explanations (Harrington 1988; see also Russell, Harrington, and Vaughan 1986). For example, it could stem from the (altruistic) preferences of managers as individuals.

One argument against firm compliance might be that the firms would be at a competitive disadvantage compared with those whose only goal was to maximize profits. One argument in favor of compliance is misjudgement of the risk for (or consequences of) being monitored and prosecuted. These risks are small, but managers may be averse to these kinds of (partly private) risks. Detection and prosecution for environmental noncompliance may carry more costs than the official fines or other penalties; the company's reputation and trademark could be tarnished, employee motivation could suffer, and shareholder interest could decrease in response to negative publicity.

Finally, some explanations focus on the nature of the enforcement process itself. The inspection, monitoring, and enforcement process can be seen as a repeated game between an environmental protection agency and a firm, and the agency may have informal means of "punishing" violators (e.g., through more frequent or rigorous monitoring, or stricter enforcement of minor regulations). Anecdotal evidence indicates that firms are often dependent on regulators' "favors" (e.g., lenient interpretation of rules or quick processing of permits) that can be discontinued at any time. An agency that agrees informally to certain forms of flexibility with firms that are cooperative acquires a great deal of power or leverage vis-à-vis the firms that are not cooperative (however, this mechanism is almost the same at the one corrupt officials use to collect bribes). In this kind of model, overcompliance may well be the optimal response (Harrington 1988). However, not all regulation is informal, and agencies including the U.S. Environmental Protection Agency (EPA) also take formal monitoring and enforcement actions (see Box 16-1). Even so, some economists consider that U.S. EPA has little real control (and presumably, that the situation is worse in many other countries) and propose considerably strengthening the legal rights for U.S. EPA to conduct unannounced inspections (Russell and Powell 1996).

There is no neat way of summarizing the above factors, but their influence on instrument choice or design can be profound. In a culture where polluters dutifully follow regulations despite insignificant penalties, inefficient monitoring, and lax enforcement, simple policies may be sufficient. In other cultures, such a policy would be overly naive. Considering the social value of compliance, policymakers should consider the long-run effect of new regulations (and policy instruments) on public attitude. Unreasonable legislation, for instance, may weaken a law-abiding culture in general.

## Policymaking in Severely Resource-Constrained Economies

Examples of the Harrington paradox also are found in developing countries. Still, in some contexts, compliance appears to be lower, but whether it depends on the weakness of the laws, public agencies, or other factors is unclear. Because monitoring institutions are weak overall, it is impossible to know for sure whether compliance is common.

In poor countries, many social, political, and other factors that typically are distinctly different from those in developed countries must be taken into account in designing policy instruments. Poverty is often related to several factors that make policy design and implementation difficult: low education, lack of finance, lack of administrative tradition, and weak institutions. With the tensions present in some of the multiethnic nations carved out by the colonial powers, these factors sometimes contribute to higher-than-usual incidence of corruption and nepotism. Corruption has been identified as a major threat to development in general (Keefer and Knack 1997) and makes the design of policy instruments difficult in particular. Judicial corruption and favoritism can lead to distortions in agents' incentives to invest in assets such as trust and reputation and thus reduces the efficiency of policy instruments (Mui 1999).

## Box 16-1. U.S. Environmental Protection Agency Sets Enforcement Records in 1999

The U.S. Environmental Protection Agency (EPA) announced record enforcement actions and penalties for FY1999 that included US$3.6 billion for environmental cleanup, pollution control equipment, and improved monitoring (an 80% increase from 1998 levels); US$166.7 million in civil penalties (60% higher than in 1998); and 3,935 civil judicial and administrative actions. Criminal defendants were sentenced to a record 208 years of prison time for committing environmental crimes.

Also during FY1999, U.S. EPA settled the largest Clean Air Act case in history against seven diesel engine manufacturers whose products were alleged to have caused millions of tons of excess emissions of nitrogen oxide ($NO_x$), a contributor to smog. Under the settlement, the companies were required to spend more than US$800 million on producing cleaner engines and to pay a penalty of US$83 million.

*Source:* U.S. EPA 2000a.

Other special circumstances typically apply in developing countries: the informal sector (e.g., small or home-based businesses) is large, making monitoring difficult; property rights may be poorly delineated (often as a result of conflicts between many layers of historical and cultural traditions); agriculture, fishing, forestry, and other industries based on often-sensitive ecosystems tend to be the mainstay of the economy (Platteau 1999); and capital and credit markets are marked by risk, uncertainty, and restrictions, making it difficult to borrow money to finance "nonproductive" (e.g., abatement) investments. Frequently, abatement costs are not annual variable costs but investment costs. Liquidity problems result because abatement equipment incurs high costs at the same time that the polluting firm is expected to pay its unusually high taxes.[5] This situation can be alleviated by using tax or charge proceeds to create environmental funds for financing loans to polluting companies for abatement-related outlays; such funds also can finance local environmental protection agencies (Panayotou 1995, 1998). Liquidity problems also can be lessened by refunding emissions payments or by using a two-tiered tax (see Chapters 7 and 15) such as the tax–subsidy scheme, whereby polluters pay taxes only on "excess pollution."

Constructing environmental and natural resources policy instruments is a complex task of tailoring mechanisms to specific situations, and it is not expected to be easy. In many instances, several goals are pursued simultaneously, and an array of instruments might need to be combined for optimal results. The challenge of policy choice and design is to overcome market failures without incurring policy failures that are worse.

The technology and ecology of an issue may be so complex that individual licensing must be part of the policy package chosen. Unfortunately, licensing creates risks of "regulatory capture": polluters dominate (and maybe corrupt) the regulators. Polluting firms typically have more information and greater resources at their disposal than regulators, who are faced with the dilemma of not only designing policy instruments but also—first and foremost—constructing their own agencies. Together with state and local inspectorates, U.S. EPA employs on the order of a hundred thousand individuals. Most agencies in developing countries have only a few employees, who sometimes have little technical training and equipment as well as limited legal authority. Managing the environment and natural resources is no trivial task for a handful of civil servants. However, training, dedication, and other factors are probably more important than sheer numbers.

Under these less-than-ideal circumstances, informational policy instruments may be a good first step in regulation. By collecting and disseminating information, an environmental protection agency can achieve several important goals:

- creating a baseline for future action;
- encouraging transparency so that individual inspectors can at least not secretly agree to unreasonable emissions;
- building relationships with the polluting industries that are not purely adversarial; and
- opening the way for pressure from customers, workers, investors, neighbors, and other involved parties.

## Supplemental Reading

**Pollution Fees in Transitional/Developing Countries**
Bluffstone and Larson 1997
Carlsson and Lundström 2000
Vincent and Farrow 1997

**Psychological Decision Theory**
Montgomery 1983

**Public Choice**
Arrow 1951
Buchanan and Tullock 1962
Tullock 1965, 1981

**Political Economy of Environmental Taxes**
Boyer and Laffont 1996
Dijkstra 1999
Keohane, Revesz, and Stavins 1998
Wallart 1999

**U.S. Environmental Regulation (empirical analyses)**
Hahn 1989, 1990
Hahn and Hester 1989
Hahn and Stavins 1991

Many (but far from all) problems can be greatly diminished by simple means, and positive partnerships must be built in which the environmental protection agency may assume the roles of not only "police" but also "facilitator" and "teacher." Local industries must acquire a reasonable perspective on the range of abatement technologies as well as the requirements posed by customers in national and international markets. Local agencies cannot perform this task or any other task (e.g., facilitating contact with credit institutions) well if they are understaffed and underfunded. Small fees on appropriately selected pollutants may help provide required financing.

A certain degree of pragmatism is important here. Finding the exact Pigovian level of a certain tax may not be important in an economy in which the level is likely to change by several orders of magnitude as a result of rapid industrialization, migration, technical progress in abatement, and other factors related to resource-based or informal-sector activities and poorly developed markets and institutions. However, information policies may be treated as a first step toward fees and other instruments, such as labeling (e.g., see Chapter 25 on PROPER). Recording information early in the process is important as a baseline against which to evaluate the success of policy measures undertaken. Without such evaluations, the local environmental protection agency will not be able to learn from its successes and failures.

The social and political (as distinct from economic) costs of bankruptcy can be high, which tends to indicate that high taxes are hard to collect. Establishing the principle that pollution is a cost to be internalized is crucial, and a low fee that is implemented and leads to payments (and some incentive effects for abatement) is better than a fee that has no chance of being implemented. If carefully designed, a charge that is partly refunded to the industries to finance abatement and partly used to finance the environmental protection agency may bring all parties into a productive collaboration. (Although risks are involved in this arrangement, most developing countries that use environmental charges also put the proceeds into environmental funds [see Chapter 25]). One of the advantages of such a fee is that it may be set considerably higher than a pure tax.

## Notes

1. Actually, ecologists also assume that plants, animals, and even genes maximize something akin to utility: "fitness." This practice does not necessarily reflect a belief in the existence of conscious purpose in nature but reflects the fact that if maximizers compete with nonmaximizers, then maximizers will tend to win and thus dominate in the long run through natural selection.

2. For an analysis of environmental taxation under different forms of altruism, see Johansson 1997b. The debate on whether utility is derived from mainly absolute or relative income has many antecedents in economic thinking. The mainstream emphasis is on absolute income, but some authors (e.g., Oswald 1998) conclude that relative income is dominant (for an overview, see Weiss and Fershtman 1998). Johansson-Stenman, Carlsson, and Daruvala (2002) carry out an empirical test and find both to be important.

3. The financial incentive was designed to be weak; a strong incentive might have reinforced cooperative decisionmaking. Some research finds that with higher financial stakes, behavior tends to be "expected" and "rational" (Smith and Waker 1993).

4. Compliance is complicated by the fact that it is generally an unobserved variable (Heyes 2001). Only after monitoring are some instances of noncompliance found, and even so, the state of the others—typically the majority, who are not found to be noncompliant—cannot be known for sure. According to a "well-known GAO study" quoted (but not referenced) by Heyes, only 3% of the firms that the U.S. Environmental Protection Agency had designated as compliant with air emissions standards were really compliant.

5. The taxes are high mainly because the firm's emissions are high. With high total emissions, the marginal damage is also high. The literature is somewhat ambiguous as to whether the appropriate Pigovian tax is the marginal damage in equilibrium or in the current situation (Baumol and Oates 1988, chapter 6.7). With the latter interpretation, the tax rate would be much higher, compounding the liquidity problems for the firm.

# CHAPTER 17

# *International Aspects*

G OVERNMENTS ARE BOUND by numerous international conventions concerning the environment and the use of natural resources in particular (see Chapter 10) and investments, political relations, and trade (e.g., the General Agreement on Tariffs and Trade [GATT], the World Trade Organization [WTO]) in general. At least two distinct sets of factors affect domestic environmental policymaking imposed by international regulations: international externalities and the trade, investments, and other economic relations between countries that may indirectly restrict local policymaking. International policymaking plays a necessary role in the provision of public goods and in dealing with problems of transboundary or global pollution. The literature on fiscal federalism indicates that the case for harmonization or international policy is not so strong when pollutants are purely local (Amacher 1998b; Murty 1996; Oates 1972; Oates and Schwab 1996).

## International Environmental Issues

Transboundary pollution is an environmental issue that may affect several nations. Externalities that originate in production activities in one country may be felt predominantly by citizens, consumers, or producers in other countries. One common example is acid rain that originates in continental Europe but kills lakes and forests in Scandinavia. Pollution and natural resources issues have no respect for the human-made borders that demarcate countries. Many major environmental issues are so pervasive and affect so many nations that they are referred to as "global pollution" or "global commons." Global environmental damage such as ozone depletion and effects such as global warming are caused by aggregate levels of global pollution—similar in concept to "perfect mixing" of nonpoint-source pollution in the international dimension except that no supranational authority acts as a regulator to impose collective penalties. The ideal policy might be Pigovian taxes or auctioned permits, but these mechanisms may not be feasible because international agreement and voluntary cooperation would be required of all the countries involved.

Treaties could be considered yet another instrument of environmental policy, but purely for international policymaking (Barrett forthcoming). It is essential to realize that some important relationships of complementarity and substitution exist between policymaking at the national level and policymaking at the international level. The structure and design of environmental and other treaties are usually analyzed with game theory. International treaties must be negotiated under the restriction of sovereignty, which means that no supernational agency can enforce policies. There is a strong element of "Prisoner's Dilemma" (see Chapter 10) in the structure of the games, because although countries jointly benefit from collaboration, there are also incentives to free ride. National sovereignty means that treaties must be written so that compliance is in the interest of each nation, because there is no possibility of enforcement. However, because the games are repeated, countries have incentives to invest in building good reputations and relations, which facilitates the achievement of collaborative equilibria. Trigger strategies may be used, and there may be various possibilities for retaliation because nations "play" many different "games" with each other that can be seen as connected (Folmer, Mouche, and Ragland 1993). For example, one country might react to unfavorable regulations in one area (e.g., environmental policy) by retaliating in another area (such as trade).

The moral hazard issue is applicable in such a context. In one study, researchers assume that a group of countries that can potentially commit to cooperate to protect the environment are environmentally conscious (Petrakis and Xepapadeas 1996). These countries can provide self-financing side payments to a second group of less environmentally conscious countries such that the two groups of countries can form a stable coalition (global or partial) that agrees to reduce emissions. However, the moral hazard problem arises because the emissions level of an individual country is unobservable by the rest of the participating countries; monitoring the emission levels of other sovereign countries is not always possible. In this case, each country has an incentive to defect from the cooperative agreement to free ride the efforts of the countries that abate emissions. Hence, a punishment mechanism is designed to make the countries that are less environmentally conscious report their emission levels truthfully (see the analyses of ambient pollution in Chapter 13). A somewhat similar mechanism is used in a study on the strategies used in coastal states with limited resources to contract other states' fishing fleets (Clarke and Munro 1987).

The coordination between the international treaties and domestic environmental policies may be complicated. Environmentalists often want to be proactive and suggest that their own countries should set a good example. Sometimes it is important to show that certain goals or technologies are feasible, and for those who believe in the Porter hypothesis, it may be advantageous to be first. However, it is also important to think tactically when playing noncooperative games (Hoel 1991). Altruistic pollution reduction in one country may easily be undone by increased production and pollution in a competing country. For example, the Scandinavian countries, the United States, and Canada all agreed to ban the use of chlorofluorocarbons in aerosols about a decade before the Montreal Protocol; no other country followed after. (However, their actions did help pave the way for the Montreal Protocol.) This effect is somewhat similar to the trade effects of a domestic tax on pollution in a small, open economy (see the next section).

Environmental treaties are mainly intended to regulate polluters as a way of minimizing environmental damage. In some special (but quite common) situations, the victim of pollution (or overexploitation of a natural resource) is prepared to contribute to abatement (or preservation), perhaps because of differences in valuation or income. In a typical case, a developed country may be prepared to pay to prevent the deforestation of tropical forests or other forms of environmental degradation or destruction in a developing country. This situation can be analyzed in terms of layers of external effects: for instance, a local forest has global as well as local positive external effects. The local effects may not be strong enough to warrant protection alone but create a strong case for protection in combination with the global ones.

*Fiscal federalism* is a branch of public finance theory focusing on the division of responsibilities (including taxation and provision of public goods) between different levels of government. Taking fiscal federalism as a point of departure, it is natural to deconstruct the "optimal" tax into a revenue tax, a tax on local air pollution, and an international carbon tax—the proceeds of which could be shared among the countries of the world (Murty 1996). This idea is illustrated in Figure 17-1, in which the marginal private benefits (i.e., benefits to the firm or owner) are small so that the "market" level of conservation is low. Adding local external benefits would, in principle, encourage a higher degree of protection. Such benefits might be in the form of local public goods (e.g., protection of microclimate, soil protection, or water purification services), which might be important but are often difficult to implement in practical terms because of imperfect institutions and other factors. Many local commons are destroyed every hour across the world because the people who depend on them do not recognize their actions as destructive or do not have the money, power, or influence to defend their interests.

The addition of international external benefits such as biodiversity, climate, and so forth clearly increases the extent of conservation (see top benefits curve in Figure 17-1). With this addition, the extent of conservation is clearly increased. Equally if not more important in practice is the fact that the likelihood of any conservation may be significantly increased. International payments can be channeled through various mechanisms that would reflect a willingness to pay to avoid (the international part of) the external effects; some of these mechanisms are the Global Environment Facility (GEF), the Clean Development Mechanism (CDM), and debt-for-nature swaps (DNSs)[1] (see Chapter 30 on the CDM and DNSs and Chapter 27 on GEF funding). Numerous problems must be addressed, such as "additionality," which asks

- How does a donor country know that the specific hectares of forest being protected are truly in addition to the ones that the country would have chosen to protect or conserve anyway?
- How does a recipient country know that the money received is really "additional"?

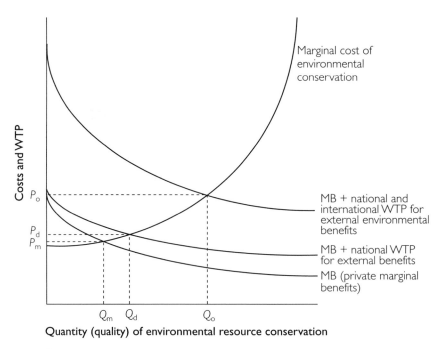

**Figure 17-1.** *Private and Social Benefits at Different Levels*

*Notes:* WTP = willingness to pay; $P$ = price; $Q$ = quantity or quality variables; MB = marginal benefits. Subscripts o, d, and m represent global optimum, domestic optimum, and domestic market with no environmental benefits, respectively.

Recipient countries may assume that the donor countries have a rather fixed budget share for international aid and that this kind of fund simply displaces funds that would have been received anyway. However, if these and other difficulties (e.g., concerning control and verification) can be resolved, then this kind of mechanism may not only complement but also help organize local willingness to pay for local benefits.

## Trade, International Relations, and Local Policymaking

Irrespective of the direct environmental links between countries, national environmental policies must respect several restrictions regarding other aspects of international relations. For instance, trade agreements are typically harsh on "nontariff" barriers, and a common complaint is that local environmental regulations are designed not to protect the environment but are a hidden agenda to support or give preference to local producers. Conversely, governments complain (or conveniently excuse themselves) that they cannot use the environmental policy instruments they prefer because they are bound by irrelevant regulations intended to protect free trade. Free trade is generally thought to increase welfare, although it need not do so in the presence of environmental distortions (Zhao 2000).

The regionalization of economic activity is increasing, as exhibited by the increased importance of such trade blocks as the European Union and the North American Free Trade Agreement, and creates restrictions on national policymaking. It has advantages and disadvantages, depending on how much emphasis is put on harmonization and how much respect is paid to the principle of subsidiarity. The GATT and WTO processes for the liberalization of trade are essential in this context. The current set of organizations is the culmination of many decades of efforts to reduce irrational trade barriers, which purport to protect local interests (employment and others) but also reduce efficiency, growth, and aggregate welfare. Negotiators are understandably wary of new kinds of (explicit or implicit) trade barriers. Large, closed economies can implement local environmental policies more or less at will, but to understand the importance of trade, it is instructive to look at what restrictions free trade imposes on the small, open economy. The basic condition in such an economy is that product prices are decided by the world market and that world market prices cannot be influenced by local policy.

This condition has somewhat different implications for production- and consumption-based taxes. If the consumption of a certain good leads to externalities, then a consumption tax is clearly called for. Such a tax will reduce consumption in a small country. When domestic demand falls, world market prices also fall, and consumption in the rest of the world increases somewhat. With conventional price elasticities, aggregate consumption still decreases, and thus the tax serves its purpose, although only partly. One classical example is a carbon tax on fuel consumption in a country with high fuel taxes. Although fuel consumption in the taxing country is reduced, theoretically, the global price of carbon resources (such as oil) falls somewhat and compensates for a part of the decline in use. This effect is often referred to as *carbon leakage*.

When production per se gives rise to externalities, input or product taxes are needed. These instruments are particularly troublesome in small, open economies because they make national production more expensive, leading to reduced exports and increased imports. This effect is illustrated in Figure 17-2, in which the left side shows how the world market sets the equilibrium price $P_w$. The right side shows the small, open economy with demand $Q_d$. The local tax has no effect on the world price or domestic demand but depresses the profit margins of local producers so that their supply decreases (from $Q_{s1}$ to $Q_{s2}$), so the small country ceases to be a net exporter (of $Q_{s1} - Q_d$) and becomes a net importer (of $Q_d - Q_{s2}$). Because the world market price is not affected, total consumption and thus production would be the same; the only effect would be to move production to other countries (that do not levy the tax in question). Relocation results in roughly the same amount of pollution being generated elsewhere. If other countries were to use less-efficient or more-polluting technology and the pollutant in question were truly global, then the net effect of the tax policy could even be to increase global pollution. Short-run costs to the local economy would be lost production, income, and employment, and long-run costs would include exchange rate adaptations and a new equilibrium with, in principle, a lower level of equilibrium salary. Similar effects would result from any policy instrument that raised the aggregate costs of production in the small, open economy. Tradable permit that were auctioned or grandfathered would

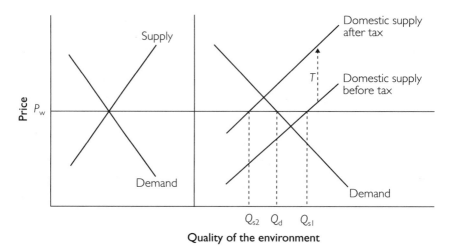

**Figure 17-2.** *Effect of Environmental Taxes in a Small, Open Economy: World Market (left) and Domestic Market (right)*

*Notes:* $P_w$ = world market price; $Q_s$ = domestic supply before (1) and after (2) tax; $Q_d$ = domestic demand; $T$ = domestic tax level.

have the same effect; the effect would be much smaller with output-based permit allocation or a refunded emissions permit scheme (or if a command-and-control mechanism were used) (see Chapter 9).

When selecting policy instruments for small, open economies, several special aspects must be considered. First, taxes on production in competitive sectors are not as effective or desirable as in a closed economy. If an instrument is needed for a certain sector (where there is a perceived risk that some companies will be run out of business as a result of competition from firms that are not subject to any control), then instruments such as regulation, output-based permit allocation, tax–subsidy schemes, or refunded emissions payments might be preferable. For protected sectors (e.g., local health care and road building, in which the pressure from foreign competitors is absent or smaller), taxes may still be preferable. An effect that might become problematic with large-scale programs such as carbon trading is that multinational companies trade permits and move abatement costs between subsidiaries in such a way as to move profits from countries with high (business) income tax to tax havens or to low-tax countries. This effect may be strongest in settings with partial trading regimes (Fischer 2000a, 2000b).

The previous analysis applies strictly to global environmental problems; local environmental problems can still be addressed with charges if their solutions warrant the cost of lost production, jobs, and exports. If this is the case, then the country simply selects the appropriate charge or other instrument and accepts the loss of production because the loss is smaller (from a welfare viewpoint) than the local externality that is involved. This trade-off is the essence of the fiscal federalism or subsidiarity approach (Oates 1972; Oates and Schwab 1988, 1996) (see Chapter 25 for a study of how effluent fees are varied within China to reflect local priorities). Also, there may still be problems concerning welfare distribution at the local level: some people may receive employment or other benefits while others may bear a disproportionate burden of the deteriorated ecosystem.

Traditionally, most interest in pollution abatement has concentrated on factory chimneys and effluent pipes. However, another class of environmental problems is related not to industrial processes but to products (and thereby to consumption). Such products (which include fluorescent lightbulbs, lead batteries, pesticides, and paint as well as cheap but unsafe foods, toys, dyes, and fuels) may or may not have locally important emissions of mercury, lead, or other toxins, but their release "through the factory gates" (i.e., in the product) is typically orders of magnitude greater. There is no trade-off between production and local or global environment but between local environment and consumption, including imports. Consumption (product) taxes are effective in such cases, and policymakers do not need to worry about employment effects or capital flight. For local pollutants, the relocation of consumption (mentioned earlier in connection with the concept of carbon leakage) is not an issue.

The ultimate example is hazardous waste, which is sometimes unscrupulously sold to firms in developing countries. This practice is difficult to regulate because it is usually covert. The waste may be sold (and labeled) as fuel or fertilizer, but the true purpose of the transaction is for the "seller" to avoid expensive treatment or disposal by having the "buyer" get rid of the waste cheaply. A combination of instruments (including information, liability, and criminal law) are most appropriate (see Chapter 27).

The small, open economy hypothesis is relevant for many developing and formerly planned economies. In fact, policy advice (from the World Bank and other international forums) to such countries is to increase their degree of openness. It is an important step toward economic growth but also results in several restrictions on local policymaking. Many developing countries must address considerable capital flows and major international ownership of important mineral and other natural resources in addition to trade relations. Consequently, external parties have great power, which could create considerable conflict concerning taxes on resource rents or other policies to promote conservation. The private interests of foreign owners must not be perceived as interfering with more general policy advice to countries that are heavily dependent on natural resources for their development. Advising countries with vast untapped natural resources to "open up" their economies is likely to create strong incentives for detrimental overexploitation of these resources if done before they have the capacity to formulate environmental and natural resources policy, enforce regulations, assign and defend property rights, and so forth.

In contrast, increased openness is urgently needed for other sectors of developing and formerly planned economies. The continued operation of inefficient plants with distorted relative prices (supported by irrational tariff structures, quotas, or subsidies) has detrimental effects on the local economy and environment. Deregulation is not wrong but should be done carefully, with respect for the inherent risks. In general, micro- and macroeconomic policies must be developed in conjunction with fundamental institutions such as property rights. Structural adjustment is necessary if an economy has been seriously derailed; if interest or exchange rates are out of line with those of other countries, then all other microeconomic policies will be ineffective. Periods of instability or transition can lead to unforeseen effects and heighten the need for good institutions at the micro level in sectors that govern vulnerable natural resources.

## Competitiveness and the Porter Hypothesis

Much of the political attention on trade issues focuses on competition. Particularly in small, open economies, there always is the question as to whether environmental (and other) policies affect competition. The relevant hypotheses are contradictory (Albrecht 1998). On the one hand, the "industrial flight and pollution haven" hypothesis claims that strict regulation will lead to industrial relocation. In principle, most economists expect this response based on comparative advantage, although the effect is likely to be small. On the other hand, the Porter hypothesis, formulated by Harvard management guru Michael Porter, states that environmental regulation will increase productivity as a result of its secondary effects on innovation (Porter 1990; Porter and van der Linde 1995). The most important underlying ideas are that

- cleaner technologies have not been explored and generally turn out to be more efficient and thus lead to cost savings and
- the effort of having to adapt to harsh regulation forces the firm to increase its productivity, which gives the firm a strong position vis-à-vis competitors.

(The latter is particularly true if the regulations spread between countries.)

The logic of the Porter hypothesis has been criticized by economists who say that if the productivity opportunities were real, then they would be exploited irrespective of legislation. However, it is possible to construct models in which some other market or regulatory imperfection could lead to a Porter effect (Bonato and Schmutzler 2000). One model confirms the logical impossibility of the pure Porter effect but points to several mechanisms that indicate that the cost of compliance with regulation may be low (Xepapadeas and de Zeeuw 1999). Compliance with regulation is not necessarily automatic but must be reasonably assumed for the Porter hypothesis to work (see Chapter 16). One model of regulation under lobbying shows that if the business community believes in the Porter effect, then the environmental protection agency should take into account Porter's views—even if it considers them wrong and misguided (Heyes and Liston-Heyes 1999).

The empirical evidence that supports these hypotheses is ambiguous. Porter does not formally test models but uses successful examples or case studies to support his hypothesis. However, it would be easy to come up with counterexamples, such as companies that have not increased profits despite environmental legislation or environmental investments (Palmer, Oates, and Portney 1995).[2] These effects are probably small because the cost share of abatement or environmental taxes in production is bound to be small (for most sectors). This kind of relationship, by its nature, is not easy to prove. Clearly, the "industrial flight and pollution haven" hypothesis would lead to more investments, industrial activity, and exports from sectors that benefited from being located in a country with lax environmental legislation, whereas the Porter hypothesis would indicate the opposite.

One stumbling block of this analysis is the difficulty in quantifying whether environmental legislation is lax or strict.[3] Another is finding a counterfactual situation. Such trends presumably would take a long time to materialize, and during that time, many other important factors are at play that are likely to be related to the laxity of legislation and the patterns of growth, and thus the results would be difficult to identify statistically.

## Supplemental Reading

Adams 1997
Barrett 1997, 2000
Carraro 1987
Hoel and Schneider 1997
Huhtala and Samakovlis 1998
Jaffe et al. 1995
Mäler and de Zeeuw 1998
Smith and Walsh 2000
Ulph 1996

## Notes

1. The Global Environment Facility (GEF) is a source of credits for environmental investments in developing countries. It was set up in 1990 after a French proposal and is run jointly by the World Bank, the U.N. Environment Programme, and the U.N. Development Programme. The clean development mechanism (CDM) funds originated in international treaties such as the Montreal Protocol, which mandated compensation to poor countries to cover their technology-related costs. They are a new mechanism, although somewhat similar to the GEF, launched in 1997 at Kyoto for global climate–related issues (Deacon and Murphy 1997; Pearce 1999). Debt-for-nature swaps are agreements by which some party (often an NGO) "buys" bad debts, thereby relieving indebted countries of some of their debts. As part of the deal, national funds are used to guarantee some ecological projects such as the protection of natural ecosystems.

2. In one example, the industrial flight hypothesis was advanced, but industries later admitted that the regulation (the Swedish ban on trichloroethylene) did not really have such an effect on competition (see Chapter 24).

3. Abatement figures declared by the companies typically are used for these studies, but using such a variable is conceptually problematic. Aside from the issue of data reliability, the abatement figures are endogenous rather than exogenous. If anything, they may reflect the change in (rather than the level of) environmental laxity.

# CHAPTER 18

# *Design of Policy Instruments*

$I$N REAL POLICYMAKING, all the aspects discussed in Part Three must be blended together and must interact with the intricacies of other policymaking. To facilitate a dialogue with the policymaker, it is desirable to summarize the modeling results of instrument design under various conditions in a policy selection matrix (PSM).

Such an exercise may be possible for a concrete issue, for which the number of choices can be limited, but at the general level of the present discussion, it is impossible to do justice to all the complexities in only one table. However, for the sake of illustration, Table 18-1 presents the outline of such a PSM.

## Environmental Policy Selection Matrix

Row 1 of Table 18-1 shows the preferred policies for static efficiency (i.e., cost efficiency in a static allocative sense). It is the most classical economic argument explained in many textbooks. The basic concept is that if the differences in marginal costs of abatement (but not in damages, where perfect mixing is assumed) are large, then considerable savings are possible by allowing the agents that have the lowest costs of abatement to do most of the abatement. The regulator usually is assumed to have imperfect knowledge about costs and may thus choose uniform reductions when using command-and-control instruments. Market-based instruments (MBIs) are preferred because they tend to equalize marginal costs and thus allocative efficiency in the use of resources for cleanup. Row 2 is a closely related situation in which the marginal cost of abatement is steep and the benefits of abatement are flat; in this case, price-type instruments are generally the best choice.

In contrast, Row 3 reflects a situation in which the costs of environmental damages vary considerably, giving individual licensing, detailed time-of-day or zoning regulations a clear advantage. The alternatives (e.g., ambient permits or differentiated taxes) may be too complex. When it is necessary to secure a definite level of goal fulfillment, as in Row 4, most MBIs rely on an estimate of responsiveness (i.e., an elasticity) to achieve the correct response level. If the country is bound by treaties or if there are considerable thresholds in the damage function (or if it is steep) and it is absolutely necessary to meet a goal exactly, then searching for the appropriate tax or charge level would be risky. Physical regulation is preferred in this case, but regulating one source at a time can cause problems for aggregate pollution and thus for ambient levels. The best instrument may thus be the tradable permit that combines the features of a market mechanism with a design that ensures aggregate goal fulfillment.

Row 5 shows how the instruments adapt to changes such as inflation; complicated changes such as economic growth and entry (presented in Chapter 12) are not discussed here. The basic principle is that quantity-based instruments give a constant emissions level that is desirable (e.g., given inflation and maybe entry). Price-based instruments give a constant price signal that is undesirable in the face of inflation; inflation will weaken the policy, which normally is undesirable. (One exception could perhaps be subsidy instruments, because permanent subsidies can have many negative effects and thus it might be good if the subsidy is gradually reduced by inflation.)

Dynamic efficiency is a measure of how well a policy instrument encourages the efficient use of resources—including environmental resources—when faced with rapid technical change (Row 6). If the marginal costs of abatement decrease by 50% per year, then this year's socially efficient level of pollution will be excessive a few years hence. Under a command-and-control policy, the regulator will be forced to either update permits often or accept that the wrong levels prevail for a long time. In principle, similar problems could occur with the appropriate level of a tax or charge, but when the marginal damage of pollution is relatively constant, these instruments are superior because they continue to provide roughly accurate incentives for abatement without adjustment. In this situation, tradable emissions permits would not adjust automatically; a more elaborate design would be needed, with frequent revisions in the number of permits.

It is difficult to make taxes sufficiently detailed for complex environmental conditions (e.g., some industrial chemicals and agricultural pesticides), as indicated in Row 7. Many chemicals are potentially hazardous, and although product taxes could be used in some contexts, creating a sufficiently detailed set of taxes that could discriminate between substances would be a difficult task. In choosing between two products or two processes, the amount of information available is so vast that it must be distilled to make the costs of information transfer and assimilation manageable. If there were individual charges on each chemical, then the market mechanism would perform this interpretative function; however, this approach is hardly practicable. Thus, in this case, other instruments have an advantage. The best policy instruments for some kinds of complex industrial processes may be traditional licensing procedures, which allow an authority to weigh many factors and specify various rules, or voluntary agreements. At the other end of the scale is information disclosure. In some cases, green labeling appears to work efficiently in condensing information for provision to consumers, shareholders, and citizens so that they can make educated decisions. In com-

**Table 18-1. Example of a Policy Selection Matrix**

| Row no. | Criteria and conditions | Taxes and charges | Two-tiered instruments | Subsidy | Tradable emissions permits | CAC instruments | Information disclosure and other |
|---|---|---|---|---|---|---|---|
| 1 | Static efficiency | *Best:* With heterogeneous abatement costs, MBIs save on costs. | | | — | Costs are high, especially with mandated technology. | — |
| 2 | Marginal costs steep, benefits flat | *Best:* With flat benefit curves or steep abatement curves, price-type policies are preferred. | | | — | *Best:* Licensing, zoning, and regulation are preferred. | Information provision and liability legislation are necessary prerequisites or substitutes. |
| 3 | Efficiency with heterogeneity in damage costs | Tax variation is perceived as unfair and impractical. Modern technology may increase feasibility. | | | Ambient permits are difficult. | | |
| 4 | Marginal costs flat, benefits steep | Exact goal fulfillment is difficult because MBIs rely on demand estimates. Low fees lead to excessive pollution. | | | *Best:* Aggregate emissions control is achieved. | Individual but not aggregate emissions are controlled. | — |
| 5 | Inflation | Price-based instruments are sensitive. | | | Unaffected by inflation. | | — |
| 6 | Dynamic efficiency | MBIs imply cost savings and intertemporal efficiency. Subsidies distort entry and exit conditions. Attention to details of permit allocation and refunding is necessary. Quantity instruments are less satisfactory with fast technological progress. | | | | CAC licenses offer no incentive. | |
| 7 | Complexity | Technical or ecological complexities put MBIs at a disadvantage. | | | — | License, liability, VAs, or information disclosure may be best. | |
| 8 | Distribution and political issues | Taxes are unpopular with polluters; charges may be preferable. | Flexibility is an advantage. | Most popular with polluters. | Flexibility is related to permit allocation mechanism. | Polluters like because they pay only for abatement. | Information disclosure is a first step. Liability may be important. |

| # | Context | | | | | | |
|---|---------|---|---|---|---|---|---|
| 9 | Asymmetric information and risk, nonpoint-source pollution | Hard to use without good monitoring. Ambient taxes are difficult. | *Best*: Self-revealing contracts are preferred. | Ultimate policy if no polluter can be found. | Hard to use without good monitoring. | Mandatory technology is easy to inspect. | Labels, liability, and CPR management are good examples. |
| 10 | Small number of polluters | ——— Unsuitable for reasons of administrative efficiency and thin markets. ——— | | | | ——— May be best for small numbers. ——— | |
| 11 | Rent-seeking | Engenders opposition. | Has good potential. | *Bad*: Leads to rent-seeking. | May be used as barrier to entry. | Individual negotiation has risks. | Subsidy removal is needed. |
| 12 | General equilibrium | Optimal due to output and revenue recycling effects. | Has no output effect. | Perverse output effects are costly for the state. | Output effect but no revenue unless permits are auctioned. | ——— Normally not applicable. ——— | |
| 13 | Developing economy | Inflation and corruption complicate price-type policies. Environmental funds are attractive. Political aspects of monitoring and enforcement need attention. | | Promising, but legal issues must be resolved. | | Information disclosure and regulation are natural starting points. CPR management may be useful. Institutions must be built. | |
| 14 | Global pollution; small, open economy | Coordination is needed, particularly in competitive sectors. | May be preferable. | Often conflict with trade rules. | | Quantity instruments may fit better into framework of international treaties. | International treaties. Money transfers to poor countries. |

*Notes*: Two-tiered mechanisms include deposit–refund schemes. MBIs = market-based instruments; CAC = command and control; CPR = common property resource; VA= voluntary agreement.

plex cases, the policies needed may be considerably different, depending on the state of the ecology. Before an ecological collapse, the precautionary principle may suggest quantity-type policies (although there is a trade-off with efficiency in case there are many polluters). After an ecological collapse, the costs of recovery and adaptation may be so considerable that subsidies are inevitable in some form.

Row 8 shows the potential for resistance to certain environmental policies by affected agents. Companies, which stand to lose large sums of money in environmental charges, and individual consumers alike may challenge the legitimacy of the policy at hand. Political attention is often focused on policies that might have a regressive effect, so that poor consumers might have larger budget shares for a particular commodity (e.g., some forms of energy). In such cases, careful design of the instrument appears to be crucial for its political acceptability. Tradable emissions permits are sometimes said to be preferable to taxes because the main polluters are compensated by being given property rights, and charges have turned out to be more palatable than taxes in many instances. If polluters share genuine concern for a particular environmental issue, then they are eager to see collected charges earmarked for that environmental cause. For emissions of nitrogen oxides ($NO_x$) from large combustion sources in Sweden, refunding charges to the polluters makes the net cost of the policy low, facilitates its acceptance, and makes it easier to apply the instrument to only one subgroup of polluters. Established polluters often feel most comfortable with conventional licensing, but licensing may involve significant new source bias. Numerous other aspects concerning the psychology and sociology of policy instruments are omitted from the table because of space constraints.

Row 9 reflects the many situations in which monitoring is almost impossible—for instance, the small "emissions" of metals and other toxic chemicals from households. These emissions take place on such a small scale and are so dispersed and intermittent that they are impossible to monitor. They make a mockery of any tax, charge, or physical regulation. The only mechanisms that can work under these circumstances are information (e.g., persuasion or education) and incentives that are self-revealing and self-enforcing (e.g., deposit–refund schemes and certain subsidies). Subsidies and public provision are the best instruments when the environmental issue has a large public good component (as with research into abatement technology or common property resource management). The great advantage of a deposit–refund scheme is that the equivalent of an emissions tax is paid when a product is bought (which is, a priori, the only moment when it may be possible to get the potential polluter's attention). This fee is refunded if there are no emissions, but the burden of proof is placed squarely on the individual, who must return the product to prove that there were no emissions or that there was no inappropriate disposal. This simple example of a revelation mechanism is central to the design of policy instruments for conditions of asymmetric information (see Chapter 13).

Rows 10 and 11 concern the number of agents and the structure of the markets in question. If there is a monopoly, then policymakers are not going to draft general tax laws to influence that one agent's behavior. It is much simpler to issue a regulation. Furthermore, taxing the emissions of a monopolist is complicated, because there are two market failures: missing markets for pollution and restricted supply by the monopolist. Also, when the agents are few, they may think and act strategically; their chief concern in so doing might not be the market for pollution per se but other markets (e.g., credit, labor, or product markets). Thus, the monopolist company might use regulators to strengthen its competitive position. This risk is obvious for direct regulation, in which more generous regulations for older firms create an indirect barrier to entry for new ones. Similarly, tradable emissions permits can be bought or withheld from sale for strategic reasons. Subsidies are the worst policy option, whereas taxes and charges presumably cause fewer such problems. When the number of victims is small, too, negotiations may be an appropriate instrument.

Row 12 refers to general equilibrium effects and the double dividend issue, which are particularly important for pollutants that are expected to generate substantial long-run revenues (when the optimal pollution level is far from zero). Row 13 is labeled "developing economy," which is too broad a category to convey real meaning. Several considerations generally apply to these countries: low overall income levels, heavy emphasis on distributional issues, and high growth rates (unfortunately, not in all of these countries). Many other special features may or may not be common: high population density, ecological complexity, high inflation rates, low effective rates of competition, issues related to corruption, lack of technical expertise, and lack of political support for environmental programs. These kinds of individual country characteristics are more decisive for policy design than the mere fact that a country is one of the many developing economies.

Similarly, Row 14 covers a wide range of issues related to global pollution and global economy issues (see Chapters 16 and 17, respectively). A small, open economy cannot tax productive inputs in the same way that a closed economy can, and this situation may lead to a "race to the bottom" or development of pollution havens unless environmental policies are coordinated in some way. Such coordination is particularly important for global pollutants, for which domestic action without the backing of international treaties is largely symbolic.

Many issues are inevitably missing from the table. With ambient pollution, some special sets of policies exist that encourage collaboration among the polluters because they can help monitor each other. In developing countries, distributional and political economy issues typically are most important, and environmental protection agencies and authorities need to be formed, informed, trained, and financed. Informational instruments and low fees (to provide funds) might add to the proper mix of instruments in the early stages of policymaking.

## Interaction between Policies

It is the nature of policymaking that new regulations and instruments tend to supplement rather than replace older ones. What are the joint effects of using several instruments? Sometimes, several instruments may be needed to address several goals. On other occasions, the use of different instruments may reflect historical development or the unintended interaction of legislation at various levels of government (local, regional, national, and international).

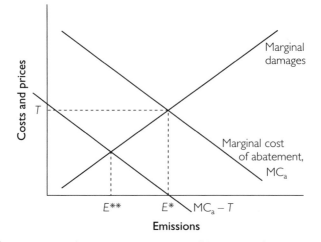

**Figure 18-1.** *Interaction between Environmental Taxes and Negotiations*

*Notes:* $T$ = environmental tax; $E**$ = suboptimal level reached through the joint use of two separate instruments (bargaining and tax), each of which would have given the optimal level if used in isolation; $E*$ = optimal emissions (and abatement) level; $MC_a$ = marginal abatement costs.

One example for the small numbers case will suffice, in which unintended interactions may be created between negotiations and other policy instruments. Figure 18-1 illustrates such a case in which the optimal level of pollution ($E*$) could be achieved (in principle) either by negotiation or through some state-initiated policy, such as taxation ($T$). However, if the state imposed a tax and the victims then succeeded in negotiating with the polluter, they might arrive at a suboptimally low level of pollution ($E**$). This combination of bargaining and tax instruments is perhaps not very common. However, many other combinations are likely.

Labeling schemes do not necessarily stop or avoid certain industries just because they are subjected to environmental taxation. Similarly, insurance and liability, in various forms, coexist with all other instruments because the payment of environmental tax does not necessarily release firms from responsibility for potential future damages. Permit systems are typically layered over local environmental regulation or other policy instruments. Information disclosure and labeling (possibly separate schemes in different countries) are used and may have market effects irrespective of whether firms pay environmental taxes.

Although the interaction of policy instruments can cause effects that are undesirable, we are likely to see much care in real policymaking. In fact, the gradual evolution of policies in a changing environment necessarily requires changes in the policy instruments used, and several instruments inevitably (and probably desirably) will be used during transition periods. Combined instruments require careful analysis of their interactions. Practical policymaking is an art of timing, combining, and sequencing instruments to meet multiple goals amidst changing circumstances.

# PART FOUR

# *Policy Instruments for Road Transportation*

$T$HE TRANSPORTATION SECTOR IS IMPORTANT FROM an environmental viewpoint: it is one of the main users of energy and a particularly prominent source of air pollutants. It also accounts for several other important external costs, such as congestion, accidents,[1] noise, and "barrier effects" (e.g., large roads create barriers to communication and movement of both humans and animals when they cut through a community, effectively making it difficult for people to shop, socialize, or work on the other side of the road). The sector has long been one of the main topics of environmental attention in industrialized countries, and in the past decade, issues of air pollution have taken priority in developing countries, too. In megacities such as Mexico City (Mexico), São Paulo (Brazil), and Santiago (Chile) in Latin America; Ibadan and Lagos (Nigeria) in Africa; and most of the large cities in Asia—Bangkok (Thailand), Bombay and Delhi (India), Jakarta (Indonesia), Manila (Philippines), Seoul (Korea), and many Chinese cities—levels of many air pollutants greatly exceed World Health Organization guidelines and generally are much worse than in the industrialized cities of richer countries. Similar situations exist in many cities of eastern Europe, the former Soviet Union, and the Middle East. Motor vehicles are a major source of this pollution.

Typical values for the vehicle share in total air pollution range from 40% to 99% for carbon monoxide, hydrocarbons, and nitrogen oxides and are somewhat lower for fine particulate matter.[2] Two of the three most important health-related problems in Bangkok are air pollution and lead contamination, both of which are, to a large extent, caused by motor vehicles (Faiz, Weaver, and Walsh 1996). The transportation sector is an important contributor to global warming, acidification, and local air pollution in megacities. In some rural areas of poor countries, runoff and soil erosion due to roads may be significant; water flowing along a hilly road in Kenya or Ethiopia, for example, can erode the soft surrounding soil

Special thanks to Per Kågeson and Winston Harrington for comments.

219

and may even cause landslides. In heavily urbanized high-income areas, valuable time is lost as a result of traffic congestion; furthermore, congestion increases other costs, such as those related to emissions.

The chapters in Part Four focus on the design of appropriate policy instruments for regulating road transportation, which entails regulating industry. Many of the transport policies actually entail regulating two of the world's largest and most powerful industries: the vehicle and petroleum industries.

By way of exposition, I begin by describing the environmental damage function for transportation. Air pollution from cars consists of many chemical substances that affect human health as well as the environment in general. Each environmental problem results from the confluence of several factors: population density, congestion, weather, type of fuel, vehicle, and driving habits. Sometimes, one main cause of a certain environmental problem is easily identifiable, for example, certain vehicles, a particular fuel, or some other factor, such as unpaved roads or climatic inversion. If so, responsibility for the damage might be assigned to drivers, producers of certain vehicles or fuels, or those responsible for road maintenance. In other cases, it is not easy to separate the effects, hence the responsibility is shared.

Vehicle pollution may be global, regional, or local. Noise is a local problem, whereas smog is local or possibly regional. Carbon dioxide ($CO_2$) emissions are global, because the negative effects of $CO_2$ on the environment are cumulative and affect the whole world independently of time or location. Furthermore, because the amount of $CO_2$ emitted is directly proportional to fossil fuel use, a relatively straightforward policy instrument such as a differentiated fuel tax could be economically efficient; the main problem in implementing such a policy is international coordination. In congested cities, the combination of congestion and local health, attributable to vehicle emissions, makes the design of policy instruments much more complex. The "first-best" instrument might be approximated by environmentally and geographically differentiated road pricing, but practical considerations of transaction and administrative costs might favor some combination of technology restrictions, zoning, fuel taxes, and fuel standards, for instance.

In Chapter 19, I describe the complex damage function for environmental damage caused by the transportation sector. Chapter 20 addresses environmental road pricing as one specific approach to paying for this damage. When it comes to regulating fuel efficiency (see Chapter 21), most European countries have sizeable fuel taxes, whereas the United States relies on regulation. A good deal of policy in transportation is regulation of fuel and vehicles: instruments for fuel quality and vehicle inspection and maintenance are discussed in Chapter 22. A particularly difficult challenge is to find suitable policy instruments in the megacities of low-income countries, where they are needed the most. Some lessons learned are presented in Chapter 23.

## Notes

1. More than half a million people are killed each year in road accidents. In low-income countries, the yearly fatalities are almost 1 per 100 vehicles—about 30 times more than in

high-income countries. In India, only 5% of traffic fatalities in recent years were in vehicles; the rest were pedestrians and cyclists. In Kenya, losses associated with road accidents have been estimated as roughly 1.3% of gross domestic product (World Bank 1996).

The relationship between accidents and mileage is complex. Depending on assumptions concerning speed, care, and other endogenous variables, the accident rate may be roughly proportional to traffic flow. It can then be argued that no external accident cost is due to increased mileage. However, such external effects may still be due to costly adaptations (in careful driving) by the road users (Johansson 1997c).

2. Industries are another source of varying importance. Also, in many locations, unpaved roads are a major source of particulate matter (PM), although typically these particulates are larger and less harmful, so that the vehicle share of PM <10 micrometers (particularly <2.5 micrometers) is important.

# CHAPTER 19

# Environmental Damage Caused by Transportation

*I*N A "FIRST-BEST" WORLD, it would be optimal to correct for an externality by imposing a correcting charge equal to the short-run marginal external cost or a Pigovian tax. The transportation sector shows how this simple idea can be complicated because the damage function is so complex. The global, regional, and local components of the damage, each has a different logic. In this chapter, I describe the damage function and discuss how it can be sufficiently simplified and stylized to serve as a starting point for the discussion of policy instruments.

## Vehicles

The main global damage associated with transports is the emission of carbon dioxide into the atmosphere and the consequent climate change. Climate change and climate policy have many complexities, notably, the coordination between countries, the inclusion of all the relevant climate gases, and the role of sinks. However, in the context of determining the damage caused by driving a mile in a vehicle, global damage is the least problematic because it depends on only the fossil carbon content of the total fuel consumed.

Regional problems such as acidification differ considerably with respect to the location of emissions because the current pollution pressure and the sensitivity of the ecosystems, soil, and underlying rock vary dramatically (see Chapter 24); within a region, the differences are typically smaller. However, the fuel and engine characteristics that give rise to various emissions differ considerably.

Local externalities include congestion (see Chapter 3), noise, and air pollution. The modeling of the damage from pollutants must account for several complicated technical, atmospheric, chemical, ecological, and health aspects. First, the vehicle emissions contain thousands of chemicals and may vary considerably depending on fuel and vehicle characteristics (described below). The pattern of atmospheric reactions and the transport of these pollutants involves many factors that depend crucially on the weather as well as the topography of the city. There-fore, translating emissions into ambient levels of pollution is complicated. The next level of complication is estimating the damages (to health, ecosystems, and capital) caused by these levels of pollution. The damage depends largely on pop-ulation density. The dose–response functions for human health are complicated by the long-run gestation of many of the induced medical conditions. Effects also vary with respect to timing (e.g., problems related to tropospheric ozone are highest during the summer, and noise is worst at night). Ultimately, these dam-ages then have to be valued, which introduces yet another series of difficulties. Translating mortality, morbidity, suffering, and days of work lost due to sickness into monetary values presents methodological as well as ethical problems. Various methods—stated preference, revealed preference, and simple loss of production values—have been used in the literature.

For the purposes of this discussion, it suffices to state that these damages vary dramatically with respect to the exact time and location of the emissions. The health costs depend on the number of people affected and are considerably mag-nified in a city center; location and timing can account for differences of several orders of magnitude. In addition, emissions exhibit a similar range of variation between vehicles. When these factors are multiplied, it becomes apparent that the cost of driving a dirty vehicle (with poor exhaust characteristics) one mile in a city during rush hour is different from that of driving a relatively clean vehicle one mile in the country. Environmental damages also depend on such factors as fuel choice, driving style, and weather conditions. When there is heterogeneity in damage and abatement costs, the risk of incurring large costs by inappropriate design of instruments is the highest (Chapter 12), and in the case of transporta-tion, both types of heterogeneity exist. Damages incurred by driving one mile vary strongly by location, suggesting the need for geographical (and temporal) differentiation.

Abatement costs also vary because they incorporate the costs of driving less often, scrapping cars, retrofitting vehicles, and similar actions—the costs of which vary enormously between vehicles. As a result, the potential gains from market-based instruments are large.[1]

## Location

Damages to health often make up a large percentage of local pollution costs. Exhaust emissions tend to disperse quickly, which makes the exact location of emissions a crucial factor in determining the extent of damage. Driving in an area where the density of exposed population is high results in higher environmental damage than driving in an area where few people are exposed to emissions.[2] This factor can easily vary by several orders of magnitude—hundreds or thousands of times.

Other geographical and climatologic conditions are equally important. Many of the worst affected urban areas have distinctive features, such as high altitude or being surrounded by mountains, such as Mexico City. Table 19-1 presents an estimate of the importance of geographical location for Gothenburg, Sweden (population roughly 0.5 million). The values build on conditions that include

Table 19-1. *Environmental Values and Geographical Location*

| Pollutant | Regional environmental effects (US$/kg) | Local environmental (health) effects (US$/kg) | | |
|---|---|---|---|---|
| | | Country | City average | City center |
| VOCs | 1.7 | 0 | 5 | 25 |
| NO$_x$ | 4.0 | 0 | 5 | 25 |
| PM | 0 | 18 | 90 | 450 |

*Notes:* VOCs = volatile organic compounds; NO$_x$ = nitrogen oxides; PM = particulate matter. Values were converted at 10 Swedish krona = US$1.

*Source:* Johansson and Sterner 1997.

population density, climate, fuel, vehicle stock characteristics, and willingness to pay for averting respiratory diseases—which are related to income levels and many other specific conditions. The exact data are not transferable to other contexts, but the values do reflect large differences in health values in the inner city, the suburbs, and the country. This variation is derived mainly from the differences in the number of exposed people and might be similar in similar towns. In large, densely populated cities, the differences may be greater.[3]

## Combining Vehicle Age and Location

The power of technical progress in the area of engine and exhaust technology is considerable. Table 19-2 presents one estimate of the effects of technical progress on vehicle emissions by model year. Because road-pricing systems and other policy instruments are designed to work for a relatively long time, it is important to consider expected future emissions. Again, the important point is not so much the values per se but their approximate rates of change.

The values on the left side of Table 19-2 show which emissions rates may reasonably be expected of new vehicles in each respective class and model year. From the 1988 cars to the 2000 cars, emissions of volatile organic compounds (VOCs), nitrogen oxides (NO$_x$), and particulate matter (PM) improved on the order of 5–10 times. In previous decades, similar progress (broadly speaking) took place; the advent of the catalytic converter alone resulted in reductions of one to two orders of magnitude for many pollutants. Similarly, future emissions could be much lower than the present ones.[4] Within vehicle model years, another difference is normally found between individual vehicles and the actual vehicle stock. At any one moment, there are many vehicle model years, so the overall difference in emissions characteristics (and thus environmental damage) is much greater than that shown in Table 19-2—probably three orders of magnitude (1,000:1) when comparing the dirtiest vehicles to cleanest vehicles in industrialized countries. The divergence between new and old cars can be expected to be even bigger in poor countries because of the tendency to keep cars that would be scrapped in richer countries (and even to import old cars from other countries).

The combined effect of vehicle model year and location of emissions can be assessed by multiplying the emissions factors (for VOCs, NO$_x$, and PM in Table

Table 19-2. *Estimated Emissions and Costs for Different Vehicles*

| Vehicle model year | Estimated average emissions | | | Local and regional environmental damages (US$/1,000 km) | | |
|---|---|---|---|---|---|---|
| | VOCs (g/km) | NO$_x$ (g/km) | PM (mg/km) | Country | City average | City center |
| Passenger cars, gasoline | | | | | | |
| 1988 | 2.50 | 1.53 | 37 | 12 | 35 | 127 |
| 2000 | 0.46 | 0.17 | 7 | 1.6 | 5 | 20 |
| 2010 | 0.08 | 0.04 | 1.2 | 0.3 | 1 | 4 |
| Buses, diesel | | | | | | |
| 1988 | 1.30 | 13.2 | 500 | 53 | 140 | 490 |
| 2000 | 0.40 | 7.3 | 150 | 30 | 74 | 250 |
| 2010 | 0.15 | 3.2 | 70 | 13 | 33 | 110 |

*Notes:* Values are corrected for cold starts, climate, driving cycle, and deterioration of emissions reduction systems over time. Future emissions factors are based on decisions within the European Union and foreseeable engine and fuel improvements. They are not forecasts but "assessments of the technologically and economically possible." The original tables have more details for more years and more categories of vehicles. The value for trucks, for instance, is similar to that for buses. Values were converted at 10 Swedish krona = US$1.

*Source:* Johansson and Sterner 1997. Emission factors are based on Ahlvik, Egebäck, and Westerholm 1996.

19-2) by the environmental values for different locations from Table 19-1. The resulting values are "local and regional environmental damages" in Table 19-2. The individual values are uncertain, but illustrative numbers are required to give the orders of magnitude to be able to discuss the likely allocative gains from environmental road pricing relative to their cost. As shown, the difference in estimated environmental damage between a 1988 vehicle in the city center and a new vehicle in the countryside is enormous, even for the average vehicles used here.[5]

## Engine Temperature and Other Factors

Many engine parameters are important for the rate of emissions. One is engine temperature, which has a large effect in cold climates, because combustion is much less complete before the engine and the catalytic converter have warmed up (Table 19-3). In cold climates, a significant share of total VOCs and carbon monoxide (CO) is typically emitted during the first kilometer driven, even of a fairly long trip.[6] Because most urban trips are fairly short (often averaging under 5 kilometers), cold start–related emissions typically make up more than half of total emissions for VOCs and CO (but not for NO$_x$). These emissions can be reduced by using cleaner fuels, preheating the engine or the catalytic converter, or using faster catalytic converters. Considering the importance of this factor, special attention may be warranted because it is unlikely to be picked up by other policy instruments that policymakers choose.

**Table 19-3.** *Effect of Temperature on Emissions for a Car with Catalytic Converter*

| Outdoor temperature (°C) | First kilometer driven | | Second kilometer driven | | Warm engine | |
|---|---|---|---|---|---|---|
| | VOCs (g/km) | CO (g/km) | VOCs (g/km) | CO (g/km) | VOCs (g/km) | CO (g/km) |
| 22 | 2.6 | 21.0 | 0.07 | 0.16 | 0.02 | 0.12 |
| −7 | 15.7 | 123.1 | 1.38 | 11.0 | 0.25 | 0.80 |

*Source:* Laurikko, Erlandsson, and Abrahamsson 1995.

Similarly, several other factors (e.g., weather, fuel, driving style, and congestion) have a strong influence on total environmental damage. The most notable weather factor is wind (maybe a factor of at least 3, according to Leksell and Löfgren 1995). In some locations, patterns of weather known as *thermal inversion* cause pollutants to accumulate over several days, reaching very high levels—cool air traps a warmer bubble of air over a city so that the air cannot disperse. Fuel quality (e.g., chemical composition, the use of reformulated fuels, and the addition of alcohols and ethers) can have substantial effects on health, which are strongly reinforced during periods of thermal inversion. One striking example of the importance of fuel composition is the use of lead additives in gasoline (see Chapter 22). For driving behavior, significant environmental improvements are possible (Rouwendal 1996), and good logistical planning in transportation companies could cut fuel use (and emissions) considerably.

Congestion causes a double effect. First, the time cost of a vehicle mile rises rapidly with increased congestion, because the addition of a vehicle to an already congested network increases travel time for many other passengers (see Chapter 3). Furthermore, this effect tends to interact with emissions: because the average speed is reduced to levels that are far below the optimal operating speed for vehicles, the rate of emissions per mile increases as well (Johansson 1997a).[7]

# Notes

1. Consider, for example, the cost of requiring all car owners in India to retrofit their cars with catalytic converters to curtail urban smog in Calcutta.

2. In fact, even the population characteristics may be important. Children, pregnant women, and people with certain ailments or illnesses are typically more sensitive to pollutants; thus, the damage would be higher near a maternity ward, hospital, or school than when driving elsewhere.

3. Many cities in the United States, Canada, and other countries have low population densities (i.e., 10 people/hectare compared with around 50–70 people/hectare in many European cities and more than 100 people/hectare in Tokyo, Japan). In U.S. cities, the environmental damage per mile driven is thus bound to be smaller. On the other hand, this kind of urban town planning has created the need for a personal car as well as the large average distances per vehicle that lead to high levels of total urban pollution in the United States.

4. However, no mechanism ensures that this potential technical progress occurs. Whether it does depends on the buyers' preferences, the price of cleaner technology, and the policy instruments used.

5. These values build on a high regional environmental damage for $NO_x$ that reflects the relative seriousness of acidification (and eutrophication) in Sweden (see Part Five). In areas where this problem is less serious and where cities are larger and more densely populated, the differences between rural and urban values may be even greater.

6. For VOCs and CO, emissions during the first kilometer driven are about half the total emissions of a 100-km trip. The pattern for particulates is probably similar. Fuel consumption and carbon dioxide emissions also increase for the first kilometer driven during cold weather, but not much. The engine typically does not produce higher levels of nitrogen oxides on cold starts; the reason for the emissions increase is that the catalytic converter does not work properly when cold (Holman, Wade, and Fergusson 1993). Relevant policy instruments are discussed in Chapter 27.

7. Johansson-Stenman quotes Krawack (1993), who states that a reduction from 40 km/h to 20 km/h doubles the emissions of carbon monoxide (CO) and volatile organic compounds (VOCs) for a passenger car equipped with a catalytic converter. Because VOCs and CO are 250% higher under congested conditions than during free-flowing traffic, congestion and air quality are directly correlated.

# CHAPTER 20

# Environmental Road Pricing

O NE SPECIFIC APPROACH to paying for the environmental damage caused by the transportation sector is environmental road pricing. Current interest in pricing traffic efficiently includes all externalities, health effects, regional environmental effects, global warming, noise, barrier effects, road damage, and accidents. European political interest in using pricing more efficiently within the transportation sector appears to be increasing, as reflected in a green paper on fair and efficient pricing in transportation (European Commission 1995). Whereas no country has yet been able to implement advanced environmentally differentiated road pricing, some sophisticated examples of road pricing, area licensing, and mileage taxes include advanced traffic-management schemes in Singapore, toll roads in Norway, and a road pricing scheme in Switzerland that uses the Global Positioning System (GPS). The main reason for this interest in road pricing—besides a general increase in environmental awareness—is presumably the fact that modern information technology has made various road-pricing systems more realistic options.

## Calculating Environmental Damage from Road Transportation

The cost that is relevant for a Pigovian tax is the sum of damages incurred (see Chapter 6). For a certain vehicle, this cost is not easy to calculate because many complex processes govern emissions and because of the causal chain from emissions to ambient pollution, exposure, and damages. To simplify the calculation, a stylized damage function is used:

$$D = D(e,g,t,w) \tag{20-1}$$

where $D$ is damage, $e$ is vehicle emissions, $g$ is location, $t$ is time of day, and $w$ is weather. Vehicle emissions are equal to distance ($m$) times emissions rate ($\xi$),

---

Parts of this chapter build on earlier work joint with Olof Johansson-Stenman.

---

## Box 20-1. A Feasible Scheme for Environmentally Differentiated Road Pricing

As an illustration, consider a simple multiplicative fee structure that approximates Function 20-1 as a variable tax ($T$) per mile:

$$T = v \times g \times t \times w \qquad\qquad (20\text{-}2)$$

In an intermediate system, vehicle characteristics ($v$) could be a number from 1 to 10, where 1 is the least-polluting vehicle (which might be electrical, for instance, but still causes wear and tear of roads, congestion, and thus some other externalities—indirectly at least) and 10 is the most-polluting vehicle. (Actual emissions rate could be measured with on-board diagnostics, but for simplicity, it is replaced here by $v$, which would be based on make, age, and periodic inspections or standardized tests.) The rating "$v$" might possibly be contingent on the use of a special fuel, because, for instance, diesel engines have different emissions characteristics when running on alcohol or on different kinds of conventional or reformulated diesel fuel.

Geographical location ($g$) might range from 1 (in the country) to, say, 10 or 100 in a city center. This variable might seem to be difficult to monitor for mobile sources but actually can be quite cheaply done by using Global Positioning Satellite (GPS) systems or even telephony or transponders located along the roads or at toll booths.

Time ($t$) might have only two values, such as 10 for rush hour and 1 for any other time, or it might be more directly related to a measure of current congestion. Similarly, for weather ($w$), a high value might be reserved for special conditions such as thermal inversion or smog episodes. This variable could be broadcast over the radio—making it possible for authorities to limit traffic (by raising the fees) on exceptionally bad smog days without actually banning traffic.

---

which is a function of vehicle characteristics ($v$), fuel characteristics ($f$), outside temperature ($t_o$), road conditions ($o$), and a vector of driving-related variables ($z$) that include speed, vehicle maintenance, acceleration patterns, and engine temperature [$\xi = \xi(v, f, t_o, o, z)$]. In practice, one of the main determinants is congestion, which determines average speed. A perfectly differentiated Pigovian tax like this is not possible, but some form of environmentally differentiated road pricing might be (see Box 20-1).

For an instrument such as the one in Box 20-1 to work, the driver must have knowledge of the current fee, and so the charges are assumed to be calculated and displayed on a device rather like a taxi meter. All the components for this kind of system are currently available at a cost that would not make a significant difference to the price of new cars. Many of these kinds of equipment would be cheap if produced in large quantities. (For instance, consider the cost of adding a compact disc (CD) player to a radio today—very small.)

In many industrialized countries, the auto industry is interested in introducing GPS or similar systems for other reasons—as a service that can offer drivers information about best routes and parking. GPS monitors are fairly inexpensive and becoming standard on many large and even medium-sized private boats. Many Swedish taxi companies have several years of experience with this kind of system. Together with CD-ROM–based maps, the taxis have a fully computerized system that gives verbal and visual instructions, enabling the driver to easily reach any destination, even in a new city. This system saves fuel, time, and money by emphasizing good logistics: minimizing search and route selection time (de Mattos and Willquist 1999). With this equipment, the car already has the hardware necessary for the type of charge system described in Box 20-1. All that is needed is a program to calculate the fees.

Potential problems include issues of privacy, tampering, and the period of "transition" before all vehicles have this equipment. In terms of privacy, monitoring the locations of all vehicles would be unacceptable (suggesting associations to a "big brother" society) but is not necessary for such a system to function; it is sufficient for each vehicle to know its own position. The issue of tampering with such devices must be addressed because the whole scheme hinges on the correct functioning of a GPS and software that incurs considerable cost to motorists. Monitoring could be difficult in countries or regions where inspection programs are inadequate. The severity of these issues will depend on the construction of the tariff system. The mechanism must be designed so that the compliant motorists who have clean vehicles and maintain them properly are rewarded.

The calculation in Equation 20-2 (in Box 20-1) is a simplification. It does not account for the environmental characteristics of different types of fuel, the disturbance caused by noise from traffic at night in residential areas, the environmental effects of speed (on emissions and external risks), and the particular risks of driving in sensitive areas. However, the potential advantage of incorporating such variables into the equation might not warrant the additional complexity of the instrument that would be required to do so.

Some of these variables are addressed separately as, for example, speed limits and no-traffic zones and as quality requirements and specifications for fuel. If possible, it is preferable to linearize the damage function and use one instrument for each subgoal. Unfortunately, it is not always possible when the damage function is truly nonlinear, as in Equation 20-2. If one instrument is used for the location factor and another for the vehicle, then the strong interaction between the two variables is lost. The effect will be restrictions that are much too tough for vehicles driven in the country and for urban vehicles that drive short distances, but much too lax for vehicles that are driven heavily in city centers (e.g., taxis, buses, and delivery vans). Practical policies will have to trade off the costs of complexity and the suboptimality of simpler fee systems.

## Simpler Pricing Schemes

It is not possible to build a single monolithic tariff system that relies on all vehicles having all the relevant equipment before the system can operate. No town in the world has yet implemented full-fledged environmental road pricing, and in fact, it may be too expensive to obtain all the potential cost savings because of the high costs of information, administration, and transactions.[1] Therefore, the instruments that are available in the transportation area—some of which are simplifications of a fully developed environmentally differentiated road pricing scheme—should be considered a useful benchmark for comparison. In this chapter, I present various schemes for road pricing, including area licenses, mileage taxes, and road tolls and also discuss differentiated vehicle taxes.[2]

Considerable health and environmental benefits can sometimes be gained with simple regulatory measures such as phasing out leaded gasoline, retiring the most-polluting vehicles, paving roads, rehabilitating public transport, managing traffic, or enhancing driver education. (For a discussion of such priorities in developing-world megacities, see Chapter 22.) One example of the system costs of sophisticated road pricing that probably exceeds potential savings is for cold starts. Obviously, a system would have to be complex to distinguish the first kilometer driven, when the catalytic converter is cold and emissions are high. This monitoring could be done with on-board instruments, but it would be costly. Technological fixes (e.g., the more widespread use of electric preheating of the motor and/or catalytic converter) probably can solve this problem more efficiently. Presumably, administrative ways of encouraging and mandating such solutions might be simpler and cheaper than building the variables into a road-pricing model.[3]

Another example of system costs exceeding the potential savings is global warming, where environmental damage does not depend on the time or location a vehicle is driven but on the original source of the vehicle's fuel. On-board systems for road pricing would have to be extremely sophisticated to differentiate between fossil methanol and biomethanol, so a simple tax on the fossil carbon content of the fuel is a superior instrument for this situation.

A third example of the cost–benefit trade-off is the protection of schools, hospitals, and residential areas from excessive levels of pollution and nighttime noise. Zoning and regulation are probably the most efficient approaches to this kind of problem.

### Area Pricing

One simple road-pricing scheme is the area pricing system, $T = f(g)$, or possibly a more sophisticated system that differentiates by time of day and vehicle characteristics, $f(g,t,v)$, which mainly targets the problem of congestion and pollution hot spots (for a general survey of theory and applications of road pricing, see Victoria Transport Institute 2001). Norway has road pricing in Trondheim, Oslo, and Bergen. In 1991, Trondheim—Norway's third-largest city, with a population of 140,000—implemented a "toll ring" around the city that is used by a large majority of drivers entering the city. Frequent drivers use an electronic card system, and infrequent drivers use coin machines. Rates are higher between 6:00 a.m. and 10:00 a.m. than during the rest of the day. As a result, traffic has declined by some 10% during rush hours while trips at other times and by bus have increased. Revenues are being used for road infrastructure, public transit, and pedestrian and bicycle facilities.

The best example of a modern area pricing system is in Singapore (see Box 20-2). The concept is spreading to cities in Europe and the United States. Recently, the mayor of London announced the first plan to introduce congestion fees in London, and reactions from the majority of business and transport organizations are positive (see Transport for London 2002).

## Box 20-2. Transport Management in Singapore

In Singapore, which has a population of more than 3 million people in an area of only 646 square kilometers, traffic and pollution are regulated with a combination of policies.

- Land zoning minimizes the need for transportation of both people and goods.
- The public transportation system is good, and parking fees are actively differentiated.
- Strict fuel standards include a phaseout of lead (with differential taxes) and reductions in sulfur.
- Vehicle standards are strict, and inspection and maintenance standards are enforced.
- Import duties for cars and registration fees are extremely high to discourage the owner-ship of personal vehicles. These fees are reduced when a new car replaces an old one.
- The Off-Peak Car Scheme reduces fees somewhat for vehicle use only on weekends (and some other nonpeak days). Off-peak vehicles are identified by red license plates.
- Under the Vehicle Quota System, potential car owners must purchase a vehicle entitle-ment at a monthly auction.
- Electronic Road Pricing charges motorists for road use according to vehicle, time of day, and target level of congestion. In 1998, this system was automated with the introduction of "smart cards," which allow motorists to pay electronically and avoid toll delays. A selection of the fees is shown below.

|  | Time | | | |
|---|---|---|---|---|
|  | 7:30–8 a.m. | 8–8:30 a.m. | 8:30–9 a.m. | 9–9:30 a.m. |
| *Restricted zone, Nicoll Highway* | | | | |
| Cars | 0.50 | 2.50 | 2.50 | 2.00 |
| Motorcycles | 0.25 | 1.25 | 1.25 | 1.00 |
| Buses etc. | 0.75 | 3.75 | 3.75 | 3.00 |
| *Restricted zone, remaining areas* | | | | |
| Cars | 0 | 2.00 | 2.50 | 2.00 |
| Motorcycles | 0 | 1.00 | 1.25 | 1.00 |
| Buses etc. | 0 | 3.00 | 3.75 | 3.00 |
| *Portsdown to Alexandra* | | | | |
| Cars | 0 | 0.50 | 1.50 | 0 |
| Motorcycles | 0 | 0.25 | 0.75 | 0 |
| Buses etc. | 0 | 0.75 | 2.25 | 0 |

*Notes:* Fees are in U.S. dollars. In the complete information, individual rates are specified for 10 time intervals and six categories of vehicles.

*Source:* Land Transport Authority 2002.

An advanced system is the satellite-based GPS system, which allows the charges to be related to distance (and varied between zones in the city). It is more efficient to charge vehicles per mile driven within the zone than just per entry or per unit of time. As usual, the system that is most expensive provides the largest potential efficiency gains. At the other end of the scale are simple systems that, for instance, require all vehicles in a certain jurisdiction (such as a town) to display a valid pass on the windshield. Although this blunt instrument does not account for the number of miles traveled, it can easily be differentiated by vehicle characteristics. In principle, such passes could reflect environmental characteristics, even if to date, the systems mainly distinguish broad categories (e.g., cars, buses, and trucks). In Singapore, vehicle passes also give access to public transport to emphasize that these modes of transportation are substitutes.

Finally, intermediate systems might consist of semiautomatic road tolls in a ring around a town. The choice between systems must be based on the trade-off between environmental effects and administrative costs. In such a case, the simplest and the most advanced systems may be superior to the intermediate ones.

## Mileage Taxes and Road Tolls

Two common road-pricing instruments are mileage taxes and road tolls. They typically have fees per mile $[T = f(m, v)]$ that may or may not be differentiated by various characteristics of the vehicle or other factors. Their simplicity keeps down the cost of monitoring and fee collection (two of the main arguments why roads should be public goods). In several locations worldwide, vehicles that travel certain toll roads and bridges may use passes that transmit a signal to a transponder. If a vehicle does not have a valid pass, its license plates may be photographed and a bill sent automatically to the vehicle's owner.[4] A new enthusiasm for privately financed toll roads is partly due to the decreasing cost of monitoring and fee collection. Mileage taxes are a national equivalent of the road toll, but they do not have any particular point of collection. They apply to all roads and are assessed on the basis of periodic readings of vehicle odometers.

The environmental effect varies depending on the type of road tolls. Tolls (on existing congested roads) can be used to reduce congestion and pollution in an area, but this role must be distinguished from the financing of new roads.[5] In the United States, congestion is a major cost in itself. Drivers in California's most congested urban areas have suffered an annual loss of more than US$14 billion in excess time and gasoline. Congestion costs the average driver in San Francisco, California, around US$1,000 in addition to substantial environmental damages.

Mileage taxes and road tolls can, at least in principle, be differentiated with respect to environmental aspects. Usually, road tolls are based on the size or weight of the vehicle, which is an important factor determining the wear and tear on roads as well as the level of emissions. Since 2001, German expressway user charges are differentiated not only by time but also by exhaust characteristics, making it cheaper for trucks that meet the Euro II standard to operate. This program is called Emissionsbezogenen Autobahnbenutzungsgebühr (Bundesamt für Güterverkehr 2002).

The mileage tax was used for diesel-fueled vehicles in Sweden for many years. The motivation for its implementation and later abolition illustrates some of the intricacies of environmental policymaking. The reason for the tax was to address the substitutability between diesel and fuel oil. Gasoline is heavily taxed in Sweden (see Chapter 21). In contrast, fuel oil (used mainly by industry, where international competition is at stake) was hardly taxed at all, and fuel oil and diesel are very similar in composition (in some cases, almost identical). Diesel could not be taxed for fear that people would use fuel oil instead. On the other hand, diesel vehicles had to be taxed in some way to prevent people from evading the gasoline tax by switching to diesel vehicles.

The solution was a mileage tax for diesel vehicles, differentiated by type of vehicle and by weight (for trucks only). The full environmental potential was thus not realized, but there was some environmental steering effect, because weight is related to emissions. For trucks leaving the country, the odometer was "stopped" and sealed so that they would not have to pay mileage taxes for driving in other countries. Anecdotal evidence indicates that some vehicle owners left the country to get the seal and then smuggled the vehicles back in to evade the mileage tax, but given the high level of vehicle control in Sweden, this kind of behavior must have been an exception to the rule. Nevertheless, various methods of evasion made the mileage tax more susceptible to cheating than gas taxes. The mileage tax was not directly compatible with E.U. regulations, and although there was a good deal of interest in the Swedish system within the European Union, Sweden dropped the mileage tax when it joined. (It was replaced with diesel taxes, again creating a need to organize a system of detailed controls for distinguishing fuel oil from diesel.)

Interestingly, the GPS is being used to register truck mileage in a new fee-collection system in Switzerland [$T = mf(v)$]. This fee is part of an agreement with the European Union to increase the permissible weight of trucks. Switzerland has a large share of transit traffic and a strong local opinion against them. The fee is applicable to all trucks that weigh more than 12 tons, on all Swiss roads, and is paid per ton, per kilometer.[6] They are differentiated by three emissions categories (see Table 20-1) but not by geographical zones within Switzerland.

*Differentiated Vehicle Taxes*

Environmentally differentiated vehicle taxes [$T = f(e)$ or $f(v)$] are an intermediate option between road pricing and the pure command-and-control strategy based on emissions standards. Taxes are usually levied on vehicle sales (especially new vehicles) as well as annual registration fees.

In Sweden, cars with catalytic converters were given tax credits as early as 1986, and starting in 1992, sales taxes and annual taxes for vehicles were differentiated according to three environmental classes that depend on emissions factors (one of which was the legal minimum). However, the European Union objected to the Swedish differentiation of sales taxes, and the system was abandoned. Currently, there are two standards for passenger cars, and the strongest standard (Environmental Class 1, from 1995) is exempted from annual tax on car ownership for

**Table 20-1.** *Fees in the Swiss Heavy Vehicle Fee System*

| Emissions category | Fee (cents/ton/km)[a] |
|---|---|
| Euro 0 | 2 |
| Euro I | 1.68 |
| Euro II/III | 1.42 |

[a]Measured as cents per ton of vehicle weight per kilometer driven (based on a conversion to 2001 U.S. dollars).

*Note:* For more information on the categories (European regulations for new heavy-duty diesel engines), see DieselNet 2002.

the first five years. Such classes may also be used for regulatory and other purposes (see Chapter 21).

Several arguments favor a system of taxes based on environmental classes for vehicles. First, it creates incentives for dynamic efficiency, that is, for car manufacturers to make cost-efficient improvements beyond current emissions standards. Although theoretically less efficient than perfect road pricing (because it does not specifically target vehicles with high mileage or vehicles that drive in cities), such a system provides incentives at a modest administrative cost.

Second, such a system may be part of a "second-best" strategy because of incomplete and asymmetric information. The authorities have limited information concerning the costs of tougher emissions standards, and environmental classes may be a way of obtaining such information. If the difference in tax between the environmental classes is sufficient to create an output response from manufacturers, then the additional cost is less than or equal to the tax difference. This kind of information is crucial for future policy decisions.

Third, the same environmental classification (extended with classes for older vehicles) can serve also for road-pricing purposes. Fourth, such a system would provide incentives to environmentally aware consumers and to companies with a "green" image to buy environmentally friendly (maybe green-labeled) vehicles.

Finally, a large share of new vehicles are bought as company cars. A few years later, they typically are sold and become private cars. The buyers of the used cars have much less influence on environmental characteristics than the original buyers. Policies that can affect the environmental image of the companies buying company cars thus have an important effect on the whole fleet. Most countries, at least in Europe, have different sales taxes for new cars; cars are frequently differentiated by weight and sometimes by emissions category. However, the differentiation is rarely strong enough to have a significant influence.

Environmental differentiation of the sales tax of new cars might be more effective than a corresponding differentiation of the annual taxes because the latter are implicitly discounted at a higher-than-market rate of interest. Because new cars are thought of as luxury items and bought by people with higher-than-average incomes, some people may find the high excise tax warranted. However, such taxes may be detrimental from a strictly environmental viewpoint. Old cars typically are more polluting; hence, to decrease air pollution, it is desirable to speed up the turnover rate of the vehicle fleet (as is consciously done in Japan).

### A Gradual Evolution of Road Pricing

One problem with advanced schemes for environmental road pricing is their complexity. Because the damage function is complicated, the optimal tax also would be complicated. Ideally, such systems should be introduced in a large area such as the European Union.[7] It would need modern equipment, such as GPS, which only some vehicles would have. However, including the older, more polluting vehicles that normally do not have any such fancy equipment is particularly important. The system also must be flexible enough to detect tampering efforts and to account for temporary visits by vehicles from outside the jurisdiction. Because real policies evolve within a given context and have to be acceptable to many stakeholders, the best way to introduce an advanced system of taxation with this kind of complex control technology may be to build on simpler systems first.

It is difficult to imagine how motorists, town planners, and the vehicle and oil industries could be forced to collaborate on a new system that requires considerable investments in new technology if they totally resisted the idea. Many people are opposed to the instruments that are most efficient. In one study, a few thousand randomly chosen citizens (half of whom had cars) were asked to rank various measures for reducing vehicle pollution. The most popular were a ban on private cars in city centers; technical requirements on motors or fuel, even to the extent of allowing only electric cars in city centers; road tolls; and speed limits. The least popular were higher taxes for larger cars, carbon taxes, and local taxes on mileage (Bennulf et al. 1998). The fact that the most efficient instruments are the least popular certainly poses a challenge to economists to be more pedagogical. It also reflects the importance of the distributional effects of taxes.

To introduce road pricing gradually, policymakers must allow for the parallel operation of different fee systems. It is simply not possible to introduce a completely new differentiated system of road pricing overnight. One introduction strategy is to make it attractive for some group of motorists to reveal themselves as less polluting by voluntarily joining a more advanced fee scheme. Because less-polluting vehicles are likely to have more sophisticated equipment, the new, more sophisticated fee systems must be made attractive to the drivers of those vehicles. Environmental road pricing would lead to less opposition if it were formulated not as new taxes on vehicles that pollute more, but as exemptions for vehicles that pollute less. This is one way of using "self-revealing" policy instruments (see Chapters 12 and 13).

For example, in areas where motorists already are charged high fees per day or per mile for city traffic, pay high parking fees, or are subject to other vehicle-related restrictions, fee reductions may be attractive to drivers who can show that their vehicles are cleaner and cause less environmental damage than many other vehicles. Following this line of reasoning, policymakers might exempt really clean (e.g., hybrid or electric) vehicles from such fees in cities that have schemes that otherwise charge per-day fees. Similarly, vehicles that are very clean (but not electric) might be (partly) exempted from mileage taxes, road tolls, parking fees, or other fees, and vehicles that use nonfossil fuels might have fuel taxes reduced to reflect the vehicle's environmental benefit.

Vehicles with certified odometers that can prove they drive only short distances could be eligible for reduced vehicle taxes or refunds on fuel taxes (to compensate for low-income people in rural areas, where the externalities of driving are small). Vehicles with GPS and certified odometers that could prove they drive only short distances in urban areas would have the possibility of paying for the miles they drive instead of paying an area-wide fee.

## Notes

1. As the price of this technology falls, the possibilities increase, but there is always a trade-off between allocative efficiency gains and system costs (for a derivation of such second-best regulations, see Verhoef, Nijkamp, and Rietveld 1995).

2. Tradable permit schemes would fulfill an analogous role. In such schemes, motorists would be required to acquire additional permits per mile for more-polluting cars and additional (or separate) permits for sensitive zones. Much of the discussion of the fee scheme also would apply to this kind of permit scheme.

3. Examples include making electric motor heaters mandatory in cold climates as well as providing the necessary electricity at parking meters and so forth. In the long term, new developments in technology may make such external electric preheating unnecessary.

4. This system is used in some locations in Norway. In some contexts, it might be considered an invasion of privacy.

5. Ideally, congestion fees should be charged on congested roads, not new roads. Optimal road pricing must have a substantial steering impact on the magnitude and composition of traffic. A system designed to raise revenues without affecting traffic (as is commonly proposed) wastes resources. In the United States, the road lobby seems to embrace the toll concept to continue building roads as public funds become less plentiful.

The number of new toll roads is increasing quickly in Chile, where they are used to finance road building. By 2000, 11 total toll roads yielded combined revenues of more than $100 mil-

## Supplemental Reading

### Fuel Economy and Transport Economics
Harrington 1997
Small 1992
Small and Kazimi 1995

### Pricing
De Borger et al. 1996
Hughes and Lvovsky 1999
Kågeson 1993
Kågeson and Dings 1999
Komanoff 1997
Maddison et al. 1996
Mayeres 1993
Newbery 1990
Verhoef 1994

### Congestion
Button and Verhoef 1997
Hau 1992
Johansson and Mattsson 1995
Lewis 1993
Morrison 1986
Transportation Research Board 1994
Vickrey 1963
Walters 1961

### Tolls
ETTM 2002
Gomez-Ibanez 1996
Hakim, Seidenstadt, and Bowman 1996
Ramamurti 1996
Viton 1995

lion, or 20% of the highway department's budget. Within a few years, more than 50 are expected. No fully automated tolls exist as yet, but such a toll is planned on the Costanera Norte (see Goldmann 2002).

6. This wise feature would not currently be possible within the European Union because the Directive on Road Tolls (1999/62/EG) allows only member countries to charge for highways. This regulation encourages diversion into secondary roads where the wear and tear caused by heavy vehicles would require expensive maintenance.

7. The effectiveness of road pricing (or of a system of environmental classes) may be scale-dependent. Some of the benefits (such as technology development) would be limited if implemented only in one country, whereas many of the costs (per vehicle) would be much larger. On the other hand, the problems (as well as the tax base) are highly local, and local politicians and motorists are likely to want the proceeds used locally, which would appear to favor some form of fiscal federalism.

# CHAPTER 21

# *Taxation or Regulation for Fuel Efficiency*

Τ HIS CHAPTER DEALS WITH FUEL TAXATION and its counterpart: regulation of fuel efficiency. A degree of fatalism sometimes creeps into discussions concerning environmental policymaking. Those who suggest drastic taxes are seen as unrealistic idealists. However, the mere range of fuel taxes that exists today (in the developed as well as the developing countries) is amazing. In some countries, fuels are still essentially subsidized, and in others, the level of taxation has raised the effective market price to three or four times the world market price. Although the main motive is not always environmental, this diversity should help illustrate the fact that the range of possible policies is somewhat wider than what everyday policy debates assume.

In this chapter, I not only discuss the importance of prices and taxes for consumption levels but also touch on the political economy of fuel taxes—what it is that makes fuel taxes so much easier in some countries than in others—and how effective the alternatives are in terms of fuel efficiency regulation.

## Fuel Taxation

In general, fuel taxes have not been imposed for environmental reasons, and in fact, they are not ideal instruments for addressing the complex set of environmental problems associated with road transportation. However, they are almost optimal for global climate issues, although the tax should be on fossil fuels only and not restricted to the transportation sector. Because fuel taxes have varied a great deal between countries and over time, good empirical evidence shows how well they work. The intent need not be environmental for the taxes to be environmentally effective.

Domestic fuel prices vary considerably, even among the developed market economies, primarily because of differences in tax rates.[1] Fuel prices in the

---

Parts of this chapter are based on joint work with Åsa Löfgren and Henrik Hammar.

United States are sometimes as low as one-quarter of those in several European countries. Prices in Canada and Australia also are lower than in Europe. Within Europe, differences in fuel taxes and prices between countries have been considerable, such as between Italy and Germany; during the 1990s, these differences were largely eliminated. Luxembourg, however, remains an extreme case. Its low tax rates attract customers from neighboring E.U. countries, and as a result, Luxembourg reaps big tax receipts (hardly a good example of European solidarity). In developing countries, variations over time are considerably larger, between both countries and petroleum products.

The oil-producing countries in the developing world tend to subsidize domestic consumption (Sterner 1989; Sterner and Belhaj 1989). Mexico, Nigeria, Saudi Arabia, and Venezuela have low fuel prices, whereas some of their importing neighbors tax fuels heavily, partly in an effort to reduce imports. In addition, many of these governments use heavy cross subsidies to favor different consumer groups; the favored fuel might be gasoline (to please the urban middle class), kerosene (rural poor), diesel for buses (urban poor) or tractors (farmers), or heavy fuel oil (industrialists). Gasoline, kerosene, diesel, and fuel oil are joint products from the same crude oil. The difference in taxation of the many distillates illustrates the politics of distributional effects (see Chapter 15). Complications arise with the pricing of close substitutes such as alcohol. The biggest difference is generally found in poor countries or countries with large, poor rural populations, such as Bolivia, Guatemala, Haiti, Jamaica, Mexico, Nicaragua, and Peru. In Jamaica, during the radical Manley regime, the price of kerosene was 20% that of gasoline. These policies could not continue in the long term because they led to large-scale adulteration of gasoline with kerosene.[2] One of the few countries in Latin America where no cross subsidization appears to take place is Chile, where market principles are engrained in policymaking.

Another rationale for these tax differentials is based on the notion that intermediate goods should not bear general revenue taxes; however, they should bear taxes intended to internalize externalities (see Chapter 14 and the introduction to Part Four). If it is administratively impossible to distinguish between fuel for commercial use and fuel for private use, then policymakers may believe that diesel is generally used for commercial transportation (e.g., buses, trucks, and tractors), whereas gasoline is used primarily for private vehicles and that a tax differential between the two may approximate the idea of not taxing intermediate goods.

Variation in product prices is even greater in Africa, where large-scale smuggling has taken place (Sterner and Belhaj 1989). Some countries of the Sahel subsidized different fuels in the 1980s to help low-income people and to reduce the risk of fuel wood collection in a fragile ecological environment, but the policies were not coordinated between countries. Senegal chose to subsidize one fuel— liquefied petroleum gas (LPG) in small 2.7-kilogram bottles, intended for household use—to avoid adulteration and industrial use. This policy caused considerable smuggling from Senegal to neighboring Gambia, where the same LPG cost more than six times more. At the same time, gasoline was being smuggled into Senegal from Guinea and other countries that had subsidized gasoline instead. Substantial "trade flows" (of contraband) in opposite directions were created between neighboring countries in response to differences in subsidies and tax policies. The price reforms that accompanied structural adjustment programs and other developments during the 1980s and 1990s have generally reduced or abolished subsidies. However, most African countries still try to shield their consumers from price and exchange rate variations through subsidies (Wright 1996).

Even in OECD countries, taxes differ considerably, so diesel is still much cheaper than gasoline in many countries. There may be several reasons for this. At one time, diesel was preferred over gasoline because it is generally more energy efficient and was believed to create less-toxic exhaust emissions; the role of particulate matter (PM) and the dangers of diesel to human health are fairly recent observations. Therefore, until catalytic converters reduced the emissions from gasoline engines, diesel fuel was favored. At the time, energy efficiency had a greater weight than health issues in policymaking, and diesel was more efficient (also an advantage from the viewpoint of global warming). Today, concerns over the health effects of PM have increased, and more attention probably will be paid to diesel.[3] Diesel fuel is used by taxis, buses, and tractors and thus affects politically sensitive interest groups. In addition, any diesel policy must address the substitutability of diesel and fuel oil (see Chapter 20).

### Effect of Fuel Taxes and Prices on Consumption

It is not uncommon for people to think that fuel taxes will have little or no effect (aside from the perverse effects such as smuggling) because they view transportation as being necessary and relatively inflexible. There is some truth to this belief in the very short run, but on the whole, it is a prime example of common sense leading the noneconomist astray. At the individual level, it might seem that people must make the same journey to and from their jobs, irrespective of the fuel price. At the social level, in the long view, many mechanisms of adaptation exist: people can buy more fuel-efficient cars, carpool, take public transportation, or move closer to their work; such adaptation explains a large fraction of the long-run fuel demand elasticity (Johansson and Schipper 1997). At the level of the economy, the direction of technical progress may be affected as more fuel-efficient cars are produced; even the architecture of towns and societies, public transportation, telecommunication systems, and cultural patterns of interaction may change.

Many of these effects are embodied in or complementary to other investments and become apparent over time, so the final result is complex and best studied by econometrics. A tax increase in one year most likely will not be reflected in a decline in consumption the same year or even the next year. During any given period of time, many conditions change in the economy. The results of policy changes occur in the long run and are difficult to observe because an observed trend must be compared with an unobserved counterfactual situation. A comparison of countries that have had stable but different policies for a long time can reveal some direct effects. For example, countries with cheap fuel (such as the United States and Canada) have much higher consumption rates than many of the European countries, where fuel is more expensive (see Figure 21-1). Similarly, in Latin America, cheap fuel in Mexico and Venezuela leads to higher consumption than in Argentina or Brazil (Rogat and Sterner 1998). In addition,

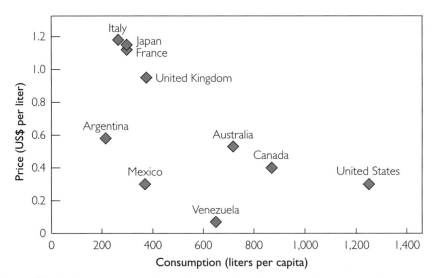

**Figure 21-1.** *Domestic Gasoline Consumption Levels and Prices in Some OECD and Latin American Countries*

many other factors that determine gasoline consumption cannot be captured in a single diagram but require econometric analysis.

In studies of gasoline demand, surveys show that the choice of model, data set, and estimator does influence results (see Box 21-1).[4] Generally, short-run elasticities are low, but long-run demand is responsive to income and fuel price (with elasticities not far from unity) (Table 21-1). The price elasticities for Latin America are on the whole lower than for OECD countries. Two explanations are likely: less flexibility to local prices (because cars generally are not tailored to Latin American prices but to prices in the United States or other car-producing countries) and the correlation between domestic prices and income in some oil-exporting countries (e.g., in Mexico and Venezuela, increased oil prices mean increased income and lower gas prices, whereas in Brazil and Argentina, they mean lowered income and higher gas prices). In addition, some factors are country-specific, such as the rapid decline in the use of gasoline due to the motor alcohol program based on domestic sugar production in Brazil.

Values for some African countries are in the same range as those for Latin America. In the United Arab Emirates, Qatar, Kuwait, Saudi Arabia, and Bahrain—countries with high incomes and cheap fuel—income elasticities typically are modest and price elasticities extremely low (Eltony 1996). This finding is important because the oil exporters account for an increasing share in fossil fuel consumption.[5]

*Political Economy of Fuel Taxation*

Although every country has different characteristics, gasoline demand is generally sensitive to price—and thus to taxation. This sensitivity is not great in the short run but is in the long run. Fuel taxation can be a powerful instrument to limit the

## Box 21-1. Models and Data Handling for Estimation of Fuel Demand Elasticities

To select policies, it is necessary to understand policy response, which is often summarized in elasticities. However, measuring elasticities is not a trivial task. Results vary substantially with the model chosen. The simplest model of fuel demand is the static model, $M_0$:

$$G_{it} = c + \alpha P_{it} + \beta Y_{it} + \mu_{it} \qquad (21\text{-}1)$$

where $G$ is gas consumption, $P$ is price, and $Y$ is income; indices $i$ and $t$ refer to countries and years, respectively; $c$, $\alpha$, and $\beta$ are parameters; and $\mu$ is an error term.

However, static models do not capture dynamic adaptation, at least not with time-series data. Demand for fuel is derived from transportation demand and has at least two components of adjustment: vehicle use and vehicle stock. Vehicles are long-lived, and adjustments take place over time. Furthermore, the complements and substitutes to vehicles include the entire transportation infrastructure (roads, town planning, etc.), which has an extremely long life.

The adjustment can be modeled either by including all the relevant variables (such as vehicles) or by using dynamic models. An early but widely used representation of dynamic behavior is the partial adjustment model, $M_1$:

$$G_{it} = c + \alpha P_{it} + \beta Y_{it} + \eta G_{it-1} + \mu_{it} \qquad (21\text{-}2)$$

This model assumes inertia in adaptation, leading to the inclusion of a lagged dependent variable, $G_{it-1}$. The lagged endogenous model assumes geometrically decreasing lags. Less restrictive models are available (e.g., the inverted "V" lag model, $G_{it} = c + \alpha P_{it} + \beta Y_{it} + \eta G_{it-1} + gP_{it-1} + dY_{it-1} + \mu_{it}$, and the polynomial distributed lag model, $G_{it} = c + \Sigma_\tau \alpha_\tau P_{it-\tau} + \Sigma_\tau \beta_\tau Y_{it-\tau} + \mu_{it}$, with appropriate restrictions on $\alpha_\tau$ and $\beta_\tau$ where $\tau$ is the number of years logged).

Among data strategies, one possibility is the *time-series approach*, in which each country is analyzed separately. It highlights country specifics but does not fully use the country-to-country information. Another strategy is *pure cross section*, which gives separate elasticities for each year. The most efficient method must be combined *cross section time-series* (Baltagi and Griffin 1983). The choice of model must take into account the character of data used (Pesaran and Smith 1995).

Additional approaches can be used to summarize all the information in one elasticity estimate: pooling estimators, for example, with country dummies (which is referred to as "within"); taking mean group estimates, which are averages of individual country (time-series) estimates or cross section estimates of averages over all the years of the data (referred to as "between"); or aggregating all the country data and performing a time-series analysis on the aggregate.

*Notes:* For the partial adjustment model, assume that equilibrium gas consumption ($G^*$) is $G^*_t = c + \alpha P_t + \beta Y_t + \mu_{t1}$ and partial adaptation $G_t - G_{t-1} = s(G^*_t - G_{t-1}) + \mu_{t2}$ (where $s$ is the rate of adaptation). For more information about the Koyck transformation, see any econometrics textbook (e.g., Gujarati 1988).

use of fuel but cannot be used without considering the availability and prices of substitutes such as public transportation. It is not necessarily easy to pursue an independent tax policy without considering the policies in neighboring countries. Finally, fuel taxes are not an optimal instrument for most transportation-related externalities. If they were levied on the fossil content of the fuel, they would be a correct instrument for global warming (see Box 21-2). They still would not be an efficient instrument to regulate local pollutants, however,

**Table 21-1.** *Elasticity Estimates in OECD Countries (1963–1985) and Latin American Countries (1960–1994)*

|  | Price elasticity | | Income elasticity | |
|---|---|---|---|---|
| Estimation technique | SR | LR | SR | LR |
| Pooled OLS | −0.12 | −1.39 | 0.05 | 0.58 |
|  | −0.10 | −0.50 | 0.21 | 1.17 |
| Pooled (fixed effects: "within") | −0.22 | −1.27 | 0.13 | 0.75 |
|  | −0.13 | −0.54 | 0.15 | 0.62 |
| Cross section ("between") |  | −1.19 |  | 1.09 |
|  |  | −0.79 |  | 1.24 |
| Mean group estimates | −0.25 | −0.85 | 0.37 | 1.15 |
|  | −0.17 | −0.58 | 0.23 | 0.69 |
| Aggregate time-series | −0.31 | −1.28 | 0.29 | 1.19 |
|  | −0.22 | −0.58 | 0.14 | 0.40 |

*Notes:* Top values in each box are for OECD countries; bottom values are for Latin America. Models used are $M_1$ except for "between," which uses $M_0$ (see Box 21-1). As a result, the pooled estimates may be slightly biased upward (Pesaran and Smith 1995). LR = long run; SR = short run (in this context, one year); OLS = ordinary least squares.

*Sources:* Sterner and Franzén 1995 (OECD); Rogat and Sterner 1998 (Latin America).

because they would do nothing to encourage technical abatement such as improvements in motor performance to reduce exhaust emissions.

The "solution" of Box 21-2 is conceptually simple but politically difficult. Why are fuel taxes not more frequently and uniformly used in OECD countries? Most countries declare an intention to reduce emissions, and most economists would agree that the most efficient way to do so is to increase taxes. Taxes are not designed to fulfill only one goal; in fact, they often are not "designed" at all, at least not in an abstract sense. Instead, they evolve through political processes in which various political stakeholders fight.

Most fuel demand studies assume that causality runs from taxes and prices to demand. Suppose that the causality runs the other way; the motorists (and voters) in countries with historically high levels of fuel consumption may be most adamantly opposed to tax increases, whereas those in countries with low consumption might find it easier to tolerate them. Gasoline taxation was examined in several OECD countries using Granger noncausality tests; the analysis confirmed that in addition to the demand relationship, the existence of a separate causal relation from quantities back to prices cannot be ruled out in many OECD countries (Hammar, Löfgren, and Sterner 2002).[6] This finding supports the notion that the political momentum built up in support of one institutional arrangement may influence the choice of policy instrument.

Models of vote maximization (e.g., Hettich and Winer 1988) assume that successful politicians will avoid overtaxing their voters. However, governments need money to operate, and something must be taxed—income, wealth, property, or consumption goods. Consumers may resist any tax, and fuel taxes more so if those

## Box 21-2. Fuel Taxes or Permits: A Politically Possible Instrument for Global Warming?

As part of a very crude policy experiment, assume that fuel demand is determined by $G = YP^{-0.8}$, where $Y$ is income, and $P$ is price (as suggested by the results in the section Effect of Fuel Taxes and Prices on Consumption) and that these elasticities are valid over a large interval. Assume also that a 50% increase in income is predicted and a 50% decrease in carbon emissions desired (a long-term goal for the whole economy suggested by the Intergovernmental Panel on Climate Change; these values are for the transportation sector, but similar policies would be applicable to other sectors).

   In this case, the suitable market price could be calculated from the formula $P = (G/Y)^{-1.25}$, and the price needed would be four times higher: $4 = (0.5/1.5)^{-1.25}$. In principle, such a price hike could be achieved through either taxes or tradable permits. To many people in countries with cheap fuel, it may seem impossible. However, many countries (primarily in Europe, but also in oil-importing countries in the developing world) already have shown that it is indeed possible

who bear the greatest burden have more political power than other groups. Fuel taxes also might be resisted less if they are seen as doing some environmental good or benefiting groups with limited influence (or groups with a high ability to pay).

The political attitudes toward mobility and environmental pollution may be decisive, and the costs and benefits depend on population density and other factors. An analysis of interstate differences in fuel tax within the United States presents the following observations (Goel and Nelson 1999):

1. *The presence of an oil industry leads to lower gas taxes.* The interpretation of this statement in terms of vested interests, employment, and lobbying seems straightforward. Similar mechanisms appear to apply to Mexico, Venezuela, and the countries of the Persian Gulf (but not to Norway or the United Kingdom), leading to disproportionately low taxes and high consumption levels.
2. *Higher highway tolls are associated with lower taxes.* The tolls are presumably seen as alternatives to fuel taxes for managing demand or financing highways.
3. *Higher population density appears to give higher taxes before 1981 and lower after.* Population density increases health costs while reducing the costs of building highways and providing public transportation. Low population density makes the automobile a necessity, whereas high population density reinforces the externalities. The question is why the relationship reversed after 1981. The cost of highway construction is suggested as the main factor, because this factor was strongest in the first period.
4. *Higher compliance with environmental standards means higher taxes.* This statement appears to be logical (although causality could go in either direction).
5. *Nominal rates tend to be adjusted to inflation, and higher real (pretax) prices of gasoline lead to lower taxes.* These statements suggest that politicians tend to seize the opportunity to raise taxes whenever it is relatively easy, for instance, when (pretax) gas prices decline.

The causality between political economy and fuel taxation could result from several scenarios. In high-consumption countries such as the United States, consumers own vehicles and property, and their lifestyle hinges on a high use of fuel. Thus, they stand to lose a lot of money from any fuel tax. Many businesses—car producers, gas stations, oil companies, and even amusement parks and shopping centers—have vested interests in keeping the cost of gas low. Employees of these businesses share this interest, to the extent that their jobs depend on the profits of their employers. U.S. oil companies, a notoriously powerful lobby, oppose fuel taxes. Therefore, the political representatives of all these parties have a lot of popularity to gain in making the case against fuel taxes. At the same time, the parties who have something to gain from higher taxes are either few or diffuse and unorganized. On the one hand, providers of alternative forms of transportation—conceivably, those employed by or with interests in public transportation, bicycles, and so forth—might gain from higher fuel taxes. On the other hand, in principle, the general public might gain from a better tax system and the resulting improvement in allocation within the economy, but this abstract concept is not likely to attract much support.

In a country like Italy, where little oil is produced, the balance of interests might be somewhat different. A reasonable strategy to keep oil imports down is to conserve energy by raising its price. The fact that this strategy happens to derail some of the resource rent from foreign producers to the national treasury can hardly be a problem. Most motorists have small vehicles and live close to their jobs. Fuel prices have been high for so long that motorists have adapted, so they do not have much to lose from even higher taxes. Furthermore, the owners, employees, and subcontractors of Fiat know that a high market share of its small, fuel-efficient automobiles depends on high fuel prices. The employees of the public transportation sector also benefit. The fact that income taxes have proven notoriously difficult to collect in Italy may be the real pragmatic reason behind the high gasoline taxes, which are considered an easy source of tax revenue. Thus, Italians who feel that the government needs its revenues—because they appreciate the services it provides or because they are government employees—may implicitly support the fuel taxes.[7]

The United Kingdom used to be a low-tax country by European standards. However, during the 1990s, fuel taxes rose 4–6% yearly. By 2000, the United Kingdom had the highest fuel taxes in the European Union, and protests began to mount. It had carried out a "green tax reform" without giving much publicity to the concept. The fact that taxes rose fairly quickly may explain the fairly strong protests to some extent. Protests led to the abandonment of the so-called fuel tax escalator, and currently the tax is supposed to rise only to compensate for inflation. Germany has a similarly programmed increase of 25 pfennig/year.

The tradition of earmarking fuel taxes also differs across countries and may influence the support for or opposition against such a tax. In the United States, most fuel taxes are earmarked for highway construction. This practice may appear odd to economists used to preaching the virtues of not earmarking but has general political support in the United States. In the United Kingdom, efforts to "hypothecate" the gasoline taxes are regularly vilified. In fact, a special road fund was created in 1909 by Lloyd George, then chancellor of the exchequer (minister of finance). This fund has been consistently raided by finance ministers who have considered the idea (i.e., that gasoline taxes could be hypothecated) absurd. In

France, a similar fund (*Fonds d'investissement routier*) was created in 1952. Still, the principle of the unified budget has dominated French political and economic thinking (Hayward 1983; Nivola and Crandall 1995), and the ministry of finance regularly diverted money away from the fund, which was fully abolished in 1981. In contrast, in the United States, when President Richard M. Nixon sought to impound some highway funds in 1972, the states challenged this action in a lawsuit and won.[8] Two years later, the U.S. Congress enacted legislation forbidding such impoundments.

At a general level, cost distribution is bound to be an important factor for the politics of industrial policy. In the United States, the absence of alternative means of transportation coupled with an uneven income distribution make fuel taxation particularly sensitive for low-income groups. In many European countries, public transportation offers reasonable alternatives and shelters low-income households somewhat from the cost of higher gas prices. In countries with a more even distribution of income, gas price would not be such a major concern. In poor countries, where automobiles are a luxury item, a fuel tax should be progressive. However, the affected elite often have such power that they can block the implementation of such taxes.

The argument is commonly made that if fuel (particularly diesel) prices rise, then all other prices in the economy will rise as a result of the increased cost of transportation. To an economist, this argument does not appear to be strong because the cost share of transportation services is typically small in the final price of most goods. Also, if fuel prices rise as a result of taxes, then these taxes may be assumed to replace other taxes (e.g., on labor, which also would have raised some prices). A large share of the resistance against fuel taxes often comes from professional truckers and farmers. For commercial vehicles, driving is not a consumable good but an intermediate good (i.e., service) and thus should not be subject to taxes intended to raise revenues (see Chapter 14). However, nonrevenue motives such as internalizing external costs are not affected (optimal road tolls under distortionary taxation are discussed in Calthrop and Proost 1998 and Calthrop, Proost, and Van Dender 2000). The actual taxation rate and price of the fuels do not necessarily have to differ at the pump. The differentiation can (at least in principle) be achieved by tax deductibility (for the revenue-raising taxes) in the accounts of the firms that use the fuel for commercial purposes. Such an approach may help to reduce the political tension surrounding fuel taxes.

Another important determinant of fuel consumption is population density. Few international studies of fuel demand include this variable, so it is difficult to show the connection empirically. It may be because population density is defined nationally, whereas the most important determinant is local population density. However, such a variable is difficult to construct because it is partly endogenous. Most U.S. cities have population densities much lower than those in Europe (in fact, lower than many "rural" areas in developing countries). With around 10 people/hectare, Detroit, Michigan, differs from London, Paris, and Berlin, which have densities of around 50–75 people/hectare, and Tokyo and Hong Kong, which have more than 100 and almost 300 people/hectare, respectively (Sida 1998; World Bank 1996). It is not surprising that fuel consumption in the dispersed U.S. cities, which often lack satisfactory public transportation, is four times as high as in typical European cities. Although a large share of the difference is due to habits and vehicle characteristics that would adapt to changed fuel prices within 5–10 years, another large share is due to differences in urban architecture, which would take considerably longer than 10 years to adapt.

## Regulations Instead of Price Mechanisms

### Emissions Standards

Despite all the discussion thus far about market-based instruments, the most usual policies for road transportation are still physical and regulatory: traffic management, speed controls, mandatory equipment (e.g., catalytic converters), fuel quality regulation, mandatory inspection and maintenance, and town planning (see Chapter 22). The policy instrument most commonly used to reduce vehicle pollution is emissions standards. In a first-best world without control costs, the standards would be superfluous and inefficient because emissions standards give excessively clean cars in the country and insufficiently clean ones in city centers.

In principle, an optimal set of road pricing charges would make it more expensive to drive in the city center, and an optimal mix of vehicles would result without implementing emissions standards. In reality, several factors make it unwise to abandon emissions standards. Because of the economies of scale in car production, the high costs of monitoring, and uncertainties about the actual effects of road pricing, keeping some form of conventional emissions standards seems prudent. However, with road pricing, the role and character of the compulsory emissions standards would presumably change.[9] Their importance for pushing technology ahead would diminish, and their role might be more of a safe minimum standard applicable to general conditions. The main driving forces for cleaner vehicles would be reductions in environmental road charges.

Another policy is zoning. Swedish municipalities are allowed to declare certain urban zones "sensitive" and ban certain kinds of vehicles within these zones. Typically, environmental vehicle classifications are used. Swedish authorities continue to lobby in favor of environmental classes, and these classes may be used by local government in selecting goods for public procurement, for instance (Naturvårdsverket 2002).

### CAFE in the United States

Surprisingly, the country generally thought of as the foremost supporter of free markets has relied almost exclusively on direct regulation of fuel efficiency instead of using fuel taxes. In the early 1970s, reducing fuel consumption was a universally accepted goal in the United States, not for environmental reasons but for reasons of national security. Hence, it would appear that the United States tried to choose effective policy instruments when it mandated the Corporate Average Fuel Economy (CAFE) standards in 1975 as the main instrument for influencing transportation fuel demand. The CAFE standards required all manufacturers of motor vehicles sold in the United States to meet a fleet-average fuel economy of

18 miles per gallon (mpg) by 1978. The standard was intended to increase the fleet-average fuel economy to 27.5 mpg by 1985, but by then, fuel prices had dropped and Ford and General Motors managed to lobby the U.S. Department of Transportation to lower the standard to 26.0 mpg temporarily; since 1990, it has been 27.5 mpg. Fleet-average fuel economy has improved by about 50% over 1978 cars, and in this sense it might seem that CAFE standards helped reduce fuel use. Presumably, they have, but not necessarily in an efficient way.

Several phenomena reflect the inefficiency of CAFE as a policy instrument. In a classic example of the market finding its way around physical regulations, mini-vans, light trucks, and sport utility vehicles (SUVs) have become increasingly popular, and their sales as personal vehicles have increased significantly. (Mini-vans, light trucks, and SUVs are not covered under the same stringent regulations as cars.) From 1980 to 1993, the number of passenger cars sold in the United States fell from 8.98 million to 8.52 million per year, whereas the number of minivans, light trucks, and SUVs sold more than doubled from 2.2 million to 5.0 million per year.

Whereas some researchers claim that the CAFE standards had little effect (Thorpe 1997), others disagree (Greene 1990). The level of penalty was relatively modest (US$5/vehicle for every 0.1 mpg shortfall).[10] Several European manufac-turers simply paid the fines, whereas domestic producers had access to other mechanisms. U.S. companies usually did not even pay the fairly modest fines. For several years, Ford and General Motors succeeded in their lobbying to roll back the standards. Another allowable strategy was to use "banked" credits from earlier overcompliance (i.e., scoring under the average in some earlier years) to compen-sate for penalties. Credits also could be gained through the sales of special alterna-tive-fuel vehicles, and the companies were able to avoid legislation by moving production abroad (or back to the United States) because the quotas for locally produced cars and "imported" cars (which included U.S. companies that manu-factured vehicles abroad) were different.[11]

According to other studies, the welfare cost of CAFE regulations was much higher than if a corresponding fuel tax had been used, and some studies even doubt that the standards had any aggregate fuel-saving effect at all (see Nivola and Crandall 1995 and studies cited therein). According to one study, the welfare cost (to consumers and producers) of the standards was on the order of US$0.60/gal-lon saved, compared with US$0.08/gallon saved if taxes had been used (Leone and Parkinson 1990). Although these figures are uncertain and the studies some-what contradictory, this type of standard probably would not be efficient in the allocative sense, and it is instructive to consider why the CAFE standards were chosen as a policy instrument. The simple answer is that raising taxes was politi-cally infeasible. Various administrations have unsuccessfully tried to raise gas taxes in the United States; for instance, the Clinton administration managed to approve only a small increase of US$0.043/gallon after proposing a significantly larger increase first.

The situation appears to be a clear case of the political power of lobbies, but why are they so successful? Models of political influence find that the strongest cases are when the benefits are enjoyed by a few and costs are spread widely. Although the allocative costs of CAFE regulations are spread widely, so are the benefits of low gas prices. However, effective lobby organizations seem to take care of organizing the motorists. Another explanation is that Americans attach little value to added government revenue. In that case, people may see the gas tax as a loss (much like the effect of rising international prices), particularly if they do not believe that other taxes will be reduced or that increased government spending will increase welfare. This aspect seems to be at least one important part of the explanation, because the past two decades have been a period of intense political opposition to "big government" and taxation in the United States. In one study, motorists in southern California were somewhat more amenable to paying for congestion fees if part of the fees were refunded to motorists (Harrington, Krupnick, and Alberini 2001).

Other reasons why CAFE standards were chosen are related to the political economy of gasoline taxation: Americans have become dependent on cars because of inadequate public transportation, the culture of mobility, and low population density—even in towns and cities, where "urban sprawl" is becoming more and more commonplace. In turn, population density is low and public transportation inadequate because of the large number of cars and the prevalence of low gas prices for almost a century. In addition to lobbying by industry and motorists, in several states (e.g., Wyoming and New Mexico), congressional representatives will fight fuel taxes for fear that their states will lose tourism revenues or that their constituents will be hurt by higher costs of commuting. The fact that the car is a means of mass transportation in the United States, more than anywhere else in the world, also suggests that income distribution concerns may weigh more heavily. In fact, one study finds that gas taxes are mildly regressive in the United States, although the shares of income are small and vary little except at the ends of the distribution (i.e., top and bottom decile) (Poterba 1991).

## Notes

1. The relationship between taxes and domestic prices is complex, and there are many reasons for the differences in domestic fuel prices between countries. Part of the price at the pump is the cost of service, refinery, fuel transportation, and so forth. These costs are determined by geography, scale, efficiency, cost levels, and company and income taxation in each country. Generally, however, they are a small part of the total price, and the main reason for variations in domestic prices is differences in excise tax. I focus primarily on gasoline prices because the data tend to be clearer for gasoline than for diesel, where the substitution for fuel oil causes some problems.

2. Adulteration and smuggling are not restricted to developing countries. In 1998, about 12.5 million liters of gasoline were adulterated with thinner to avoid taxes in Sweden.

3. Knowledge is always incomplete, and ironically, some signs indicate that the preference for gasoline over diesel may be reversed once more. New filters and other technologies reduce the particle emissions from diesels, and Otto engines (which use gasoline) reportedly produce more of the finest (nanometer-sized) particles, which may have worse effects on human health than larger particles.

4. Dahl and Sterner (1991a, 1991b) present a survey that stratifies results by model and data. In subsequent work, Sterner (1991), Sterner and Dahl (1992), Sterner and Franzén (1995), and Sterner, Dahl, and Franzén (1992) estimate different models for one consistent data set from

OECD for 1960–1985. Rogat and Sterner (1998) estimate the same models for some Latin American countries.

5. Gasoline consumption in Libya, for instance, is more than 10 times the African average. Five countries (Algeria, Egypt, Libya, Nigeria, and South Africa) account for one-third of Africa's population and just over one-half of the continent's gross domestic product but account for two-thirds of all petroleum product consumption and four-fifths of all gasoline use (author's unpublished calculations based on statistics from OECD, the International Energy Agency, and the United Nations).

6. In complicated systems, many causal relationships may overlap (e.g., prices influence demand, and demand, in turn, influences tax policies and thereby prices through political pressures). The Granger test of noncausality is one statistical method of differentiating these relationships.

7. Ironically, even the oil companies may benefit. In high-tax countries such as Norway, Sweden, Italy, and most notably Japan, the high taxes are combined with high pretax prices of gasoline. This combination seems odd (higher taxes should squeeze the margins of the oil companies) but may have a political explanation in the implicit acceptance of higher profits in exchange for high taxes. One can imagine the authorities desiring a high price level to promote conservation and "sharing" the rent with the gas companies.

8. *State Highway Commission of Missouri v. Volpe* 1973.

9. This is not the same as mandatory technology such as the catalytic converter. In principle, it is more efficient if companies are free to test new technologies intended to meet a specified goal.

10. This penalty applied only to sales of more than 10,000 vehicles/year in the United States. Even a dramatic shortfall of 5 mpg would warrant a fine of only US$250—less than 1% of the price of a new car, or the cost of the extra fuel for just one year of driving (assuming 25,000 miles/year and US$1/gallon).

11. Ford used this strategy at least twice, moving the production of rear axles to Mexico and then back again for "CAFE-related" reasons (Thorpe 1997).

# CHAPTER 22

# Fuel Quality, Vehicle Standards, and Urban Planning

CHAPTER 21 FOCUSED ON POLICIES for fuel efficiency and for carbon emissions. In this chapter, I turn to the policy instruments that are most frequently used to deal with regional and local environmental problems caused by transportation. These instruments commonly target fuel quality, vehicle standards (for new vehicles as well as the inspection and maintenance of older vehicles), traffic management, logistics, and urban planning.

## Fuel Quality and the Phaseout of Lead[1]

In addition to vehicle design and traffic management, fuel quality is an integral factor in formulating policy for the transportation sector. The use of lead in gasoline has become a symbol of the health effects caused by transportation. Originally, health effects were among the indirect reasons for phasing out leaded gasoline; the main reason was that catalytic converters become "poisoned" (i.e., they cease to operate) when contaminated by even small amounts of lead.

Catalytic converters were introduced for environmental reasons, but of a different kind. The original main concerns were smog-related emissions—nitrogen oxides ($NO_x$), volatile organic compounds (VOCs), and carbon monoxide (CO)—in sunny climates such as in California. However, as knowledge of the health effects of lead exposure increased, it became clear that exposure to lead can cause loss of intelligence, nervous disorders, hypertension, and even life-threatening consequences in cases of high exposure (CDC 1991; Hayes et al. 1994; Hu et al. 1996; Kim et al. 1996; Lovei and Levy 1997; Needleman et al. 1979; Schwartz 1994; WHO 1987). Children are at greatest risk, mainly because their nervous systems are developing. Malnutrition, particularly the lack of essential trace metals, appears to increase the negative effects of lead exposure, making it particularly dangerous for the poor.

*Technical Background on Lead Additives*

Lead (usually tetraethyl lead) used to be routinely added to gasoline in most of the world because it was a cheap source of increased octane and a good valve lubricant. Although leaded gasoline is still used in some developing countries, phaseouts continue worldwide.[2] Gasoline is not a homogeneous substance but a mixture that differs considerably across companies, countries, and even seasons. It is made up largely of alkanes (saturated hydrocarbon chains, straight or branched), alkenes (unsaturated), aromatics (carbon rings), and sometimes ethers, alcohols, and other compounds (Lefler 1985). The composition determines properties such as the flash point, evaporation temperature, partial vapor pressure, and octane number. These properties determine the gasoline's technical and environmental performance.

Refineries mix components to obtain the desired properties at minimum cost. Depending on how advanced a refinery is, it can more or less easily synthesize the necessary ingredients from various qualities of crude oil. If the base products do not meet the mark, then more expensive additives must be mixed in. As an octane booster, various combinations of aromatics, light hydrocarbons (e.g., butanes, alcohol, or ethers) can replace lead. Butanes are strongly volatile (i.e., tend to evaporate), which may cause environmental problems, and aromatics are highly undesirable because they are carcinogenic and contribute to the formation of ground-level ozone (i.e., smog). MTBE (methyl *tert*-butyl ether) and ETBE (ethyl *tert*-butyl ether) are preferred gasoline additives in many high-income countries. Alcohols sometimes play the same role and are popular in the United States and Brazil.[3]

The valve seats of older vehicles can be overly soft. Hard driving conditions with unleaded fuel can cause valve-seat recession, damage to the engine that is expensive to repair. This possibility appears to have been considerably exaggerated in support of leaded gasoline; valve-seat recession occurs only under extreme conditions. In addition, compounds less poisonous than lead (e.g., sodium) can replace the lubricating function of lead, and at a reasonable price—Powershield 8164 from Lubrizol costs only US$0.003/liter of gasoline at the recommended dosage (World Bank 1998a, 1998b). In fact, evidence indicates that lead reduction reduces overall maintenance costs because leaded gasoline typically contains halogen additives that build up deposits and corrode exhaust valves, spark plugs, and other engine parts (V.M. Thomas 1995).

The cost of phasing out leaded gasoline depends on market size and structure, refinery technology, type of crude oil, and other factors. Average cost is US$0.02 to US$0.03/liter of gasoline if starting from a high level of 0.6 grams of lead/liter or US$0.01 to US$0.02/liter if starting from the more common 0.15 grams/liter (World Bank 1998a, 1998b). The cost also changes over time, as technology changes, and may depend on the time horizon for the phaseout. For example, an individual refinery may refurbish machinery and extend some of its processes, change its crude oil mix, or buy various additives or components such as alcohols or ethers. At the national level, additional options include liberalizing imports, allowing foreign companies to set up refineries, tightening fuel specifications, and possibly implementing incentive-based policies. In some countries, the relevant authorities introduced several grades of gasoline; where only one grade is available, it must be a fairly high-octane gasoline designed to suit the most demanding vehicles, even though the majority of vehicles do not need such formulations. The availability of several grades can thus lower the total demand for octane boosters.

## Policy Instruments for Eliminating Lead from Gasoline

Before considering policy design, the "optimal" pollution level must be determined for each source. The desirability of simply eliminating lead exposure as much as possible is universally accepted in developed countries, but this issue might not be so clear in countries where people have very low incomes—particularly because they must carefully decide how to allocate scarce resources between various sources of lead poisoning.

Studies show that lead exposure results from lead-based paints, plastic additives, ceramic glazes, water pipes, solder on food cans, cables, ammunition, (car) batteries, and cosmetics in addition to leaded gasoline and direct deposition from lead smelters. Despite the extensive use of lead in gasoline, gasoline was not and is not a major source of lead in the environment; however, vehicles are the largest source of atmospheric lead in many cities and account for up to 90% of all lead emissions into the atmosphere (Brunekreef 1986). Because lead in this form is easily accessible, such emissions account for a large share of exposure and health effects for many people. In industrialized countries, many uses of lead are banned or heavily regulated.[4] In developing countries, water pipes, glazing, and other sources close to the food chain typically account for a large share of exposure and should be the top priority, although lead in fuel is also important.

No single policy instrument is optimal for phasing out leaded gasoline because the character of technology and markets vary considerably over time and between countries. The set of experiences currently available point to certain advantages and disadvantages of each policy instrument in different settings. The U.S. phaseout of leaded gasoline in the 1970s was different from a phaseout in El Salvador today. Knowledge of the alternatives, refinery technology, effects on motors, and so forth has increased. In addition, the situation in a large country with complex markets and many refineries that have individual interests is very different from that in a country that has only one state-run refinery. The approaches that have been used thus far include tradable lead permits, direct regulation, preferential tax treatment, eco-labeling, and lawsuits.

The United States was the first country to phase out leaded gasoline. Some of the background factors (e.g., smog in California and catalytic converters) have already been discussed in many previous chapters. In 1973, the U.S. Environmental Protection Agency (U.S. EPA) had already succeeded in drastically reducing the lead content of gasoline. In 1982, U.S. EPA wanted to phase out the remaining lead as quickly as possible, which turned out to be just over five years (starting in 1982).[5] Although the total benefits outweighed the costs considerably, refineries worried about meeting the fairly rapid pace of the phaseout showed serious resistance (Tietenberg 1994). Producing high-quality unleaded fuel requires appropriate technology, which can be expensive for some refineries, depending on their age and size. For a refinery that is in a process of expansion

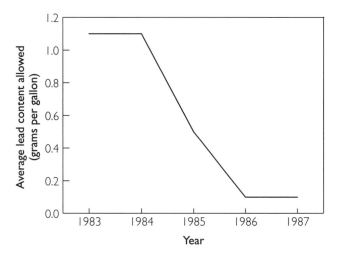

**Figure 22-1.** *Phaseout of Leaded Gasoline in the United States*

and modernization, the additional cost of complying with new regulations for removing lead may be almost insignificant. But for a refinery at a different stage in the investment cycle, the additional cost might be devastating unless it can be delayed to coincide with regular overhaul and maintenance. For some kinds of small refineries, the costs might be prohibitive if appropriate small-scale technology is unavailable. In the 1970s, the United States had many small refineries with small output but sizeable influence.

Almost any environmental regulation places some firms at a competitive disadvantage vis-à-vis their competitors and causes legitimate complaints by industry. U.S. EPA feared that lead phaseout legislation would be challenged in the courts and that companies would be able to delay it considerably. This expectation was far from hypothetical; up to 80% of U.S. EPA regulations are challenged in court (Davis and Mazurek 1998, 23). Therefore, U.S. EPA sought a mechanism that would allow individual plants some flexibility without delaying aggregate compliance. The mechanism chosen was a kind of tradable emissions permit that targets pollution intensity—in this case, the lead contents of gasoline; it was not intended to affect lead use by affecting gasoline demand itself (see Chapters 7 and 14).

Figure 22-1 illustrates the U.S. lead phaseout at the aggregate level. A credit was earned by a refinery that used less lead than the standard for that quarter and could then be sold to another refinery that used more lead than the standard. In 1984–1985, small refineries (<5,000 barrels/day) typically used at least 1.5 grams lead/gallon. The possibility of buying rights from the larger refineries apparently reduced the compliance costs significantly for the small refineries. Starting in 1985, U.S. EPA also granted refineries the right to "bank" credits for later use, and this system was used heavily. As a result, larger-than-required reductions in 1985 made it possible to exceed limits somewhat during 1986 and 1987, when the program was terminated. A significant amount of both banking (10.6 billion grams [Hahn and Hester 1989]) and trading took place, and the estimated savings were several hundred million U.S. dollars.

Although the U.S. lead phaseout scheme was successful in accomplishing its goal, it was not necessarily an optimal or efficient policy instrument. However, it did avoid deadlock in the courts and phased out leaded gasoline in a timely fashion. One unexpected by-product of the system was rent seekers—companies that were set up to blend fuels and then resell them to claim "lead rights." They were a nuisance to U.S. EPA and to the refineries that were "cheated" out of these rights. The details of this kind of program have to be worked out carefully; allocating rights in proportion to output—which is essentially what this program did—entails an implicit element of subsidy and risk of rent-seeking (see Chapter 14). These problems would have been avoided if the rules applied only to existing refineries and allowed U.S. EPA to vet new applicants before making them members of this phaseout scheme.

Other countries have not copied this scheme, even though most have phased out or are phasing out lead from gasoline, in many cases a decade or two after the United States. Scandinavia, Germany, and Austria phased out lead rapidly, and the E.U. Council of Ministers banned leaded fuel after 2000 (with some exemptions until 2005), but some European countries lag behind. Japan, like the United States, was early in introducing catalytic converters and phasing out lead in the late 1970s, apparently with no other policy instruments than the "advice" of the Japanese Ministry of International Trade and Industry (quoted in Nivola and Crandall 1995). In most countries, the main driving forces in addition to the concern for public health were catalytic converters. This fact tends to make the automobile industry an ally. Once a company has converted to producing cars with catalytic converters (which, in some cases, might have been a somewhat reluctant process), it becomes a strong advocate and wants to limit the use of lead, raise its price, and do anything that helps promote the new "clean" cars that it wants to sell. The pace of technological renewal varies between companies, and these differences shape different national policies for the phaseout of leaded gasoline.

During the 1980s, German delegates in the European Union were very interested in combating $NO_x$ emissions, thereby—and perhaps incidentally—promoting German auto industries, which were among the first in Europe to introduce catalytic converters. Some say that the French (whose industry was less advanced on catalytic converters) were less enthusiastic about combating $NO_x$ emissions but wanted to focus on reducing sulfur oxide ($SO_x$) emissions from fossil-fueled power plants, thereby helping the strong nuclear industry in France. Reducing sulfur emissions was bound to make fossil energy more expensive and thus help give nuclear power a competitive edge.

Most European countries chose a mixture of policy instruments, including fuel specifications and mandatory catalytic converters for new cars or, as in Sweden, subsidies for the addition of catalytic converters to older cars.[6] Such changes created a demand for unleaded gas, so the oil companies had to start supplying it. When catalytic converters became universal, drivers of new cars had no choice but to choose unleaded gasoline (the risk of inadvertent fueling with leaded gasoline was minimized by making the gas pump nozzles and gas tank access smaller for unleaded gasoline than for leaded). To speed up the transition, many countries then used subsidies or differential taxation on leaded and unleaded gasoline.

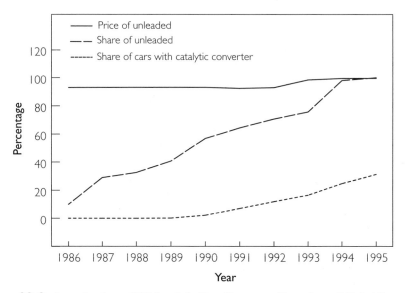

**Figure 22-2.** *Introduction of Unleaded Gasoline as a Function of Subsidies and Introduction of Catalytic Converters in Denmark*

*Notes:* Price of unleaded = percentage of the corresponding price for leaded gasoline. (Unleaded was roughly 7% cheaper from 1986 to 1992); share of unleaded = market share of unleaded gasoline as a percentage of total gasoline sold; share of cars with catalytic converter = percentage share of the total stock of cars that are equipped with catalytic converters.

For example, Sweden raised the tax differential on leaded gasoline drastically, to about US$0.08/liter. With a tax wedge much larger than the production cost differential, both retailers and consumers were anxious to switch. Many countries adopted similar tactics; see Figure 22-2 for the case of Denmark, where unleaded gas was about 7% cheaper, resulting in a 50% market share even before the cars with catalytic converters had any market share and 100% unleaded when these cars had only 25%. In this case, the subsidy appears to have had a large effect, but the effect varied between countries, depending on various factors.[7] The general conclusion is that the lead phaseout was quick because politicians used powerful policy instruments. The instruments used in Europe corresponded to a shadow price of US$200–500/kilogram of lead (Kågeson 1993).

*Phaseout of Lead in Developing Countries*[8]

Prejudice might suggest the phaseout would be slower in poorer countries, but many developing countries have already phased out lead completely: Bolivia, Colombia, El Salvador, and Guatemala. Overall, Latin America has decreased lead use by 85% over 10 years—roughly the same pace as Europe (UNDP/World Bank ESMAP 1996). Exceptions are the oil-rich countries (Peru, Venezuela, and the Trinidad Antilles, which is a special case), which should have the financial means and know-how to implement a lead phaseout. Ironically, the existence of a strong national oil company trying to resist imported technology could be setting these countries back.

Several eastern European countries have made good progress: Slovakia made catalysts mandatory in 1993 and phased out lead in five years. Poland reduced the maximum limit from 0.32 grams/liter to 0.15 grams/liter in 1993, made catalysts mandatory in 1995, and decided to phase out lead in 2000. Bulgaria, Croatia, Hungary, Latvia, Poland, Slovakia, Slovenia, Turkey, and Ukraine (about half of the 18 eastern European countries mentioned in DEPA 1998) are using economic incentives such as lead taxes to facilitate the transition. Most of the older vehicle stock appears to be designed for low-octane fuels, which makes the phaseout easier. The recent imports of (used) "western" cars that need higher octane make it important to phase out lead before demand increases.

The literature on phasing out lead in developing countries teaches several interesting lessons:

- Refineries are complex examples of joint production, making the design of policy instruments tricky. However, if there is a single local monopoly, regulation may be preferable.
- In small countries that have only one refinery, the phaseout can be relatively easy if lead-free fuel or additives can be imported. The experience of El Salvador (less than one year) and other Central American countries shows the advantages of a rapid phaseout, avoiding the cost of dual distribution systems. The introduction of an additional low-octane gasoline may capture a share of the market, allowing the oil companies to save their octane boosters for motorists who really need it, thereby reducing refinery investments.
- Pricing can be an important policy instrument.[9] If unleaded gasoline is more expensive than leaded gasoline, then a phaseout proceeds slowly. Vehicles with catalytic converters risk damage due to intentional or unintentional "misfueling." Policymakers often balk at the political cost of raising the price of leaded gasoline. However, in Jamaica, production cost was lowered by reducing the octane number, such that revenues could be increased without raising the price to consumers.
- In Mexico, a type of labeling was used: taxis and buses that operated on lead-free fuel were painted green instead of yellow so customers could choose.[10]
- In Azerbaijan, Kazakhstan, and Uzbekistan, gasoline is produced and sold in micro operations, and government control is so weak that lead content cannot be monitored (World Bank 1998a). In fact, lead content may be higher than legal limits. A tax on the lead content of fuels would not be feasible, but a tax (or ban) on imported lead additives might be.
- Brazil is a special case with its large alcohol program. Lead is completely phased out despite the country not having catalytic converters on even half the vehicle stock. Forty percent of vehicle fuel is ethanol, and in the rest, ethanol (which has a high octane value) is used as a blending agent.
- Auto and oil companies often support the lead phaseout (at least in the later phases, when they have already been forced to adapt to a ban on lead in domestic markets and find that they can reap the benefits of product standardization by requiring other countries to follow suit); the World Bank has organized

conferences and published reports supported by such companies.[11] The World Bank has provided some technical assistance, but lending to refineries has been a minor component. Its main function appears to be as a catalyst, using its credibility to encourage other institutions to finance the main investments.

## Policies for Fuel Quality in Sweden and Other Countries

One of the factors driving the need for the lead phaseout was the harmonization of fuel quality between neighboring countries. Although the health effects of lead are especially dangerous, numerous other additives, impurities, and components (e.g., polyaromatics, sulfur, oxygenates, and phosphorus) also must be controlled. Countries have different fuel requirements that hinder trade in refined products, increasing costs and decreasing opportunities for collaboration and learning. One might not think that fuel requirements would lend themselves to economic instruments, but some examples exist. For example, for some time, Sweden has used environmental classes for gasoline and diesel fuels, providing oil companies the opportunity to be proactive. The cooperative company OK tried green labeling and marketing to get larger market shares by being first to introduce environmentally improved fuels, such as unleaded gasoline. In the early 1990s, Sweden also used strong tax differentiation on fuel oils by environmental class. It encouraged the rapid introduction of cleaner diesel fuel by compensating the oil companies for increased production costs (see Figure 22-3). The oil companies even started to sell "environmental" heating oil, capturing 36% of that market in 1993. This result was hardly the intention of policymakers, because environmental advantages are limited outside the transport sector. The subsidy was subsequently removed for this category.

The availability of cleaner diesel had many positive secondary effects. First, it allowed Sweden and Finland to provide an early market for the introduction of particle filters for diesel-powered vehicles. It appears to have been a good example of technology forcing, because early on, there was some doubt as to how these filters would work, and successful demonstration was important for further international marketing. It also allowed environmental authorities to tighten emissions standards for diesel-powered vehicles. Additional secondary benefits were that the combination of cleaner diesel and particle filters made it possible to install better exhaust gas recirculation systems on some buses and trucks, greatly reducing $NO_x$ emissions. Still, various problems have surfaced, including the inability of the current tax law to distinguish between fossil-based diesel and diesel made from biological raw materials, which still has to pay a carbon tax. Another anomaly is that some clean fuels do not qualify for the lower Class 1 taxes because they fail to meet technical specifications such as cetane index or density, which is not central to the environmental issues.

A similar program for gasoline quality defines Environmental Classes 1, 2a, 2b, and 3. This program had somewhat similar effects to the programs for unleaded gasoline and environmental diesel. The introduction of tax rebates to offset increased production costs was used to phase out the more polluting fuel. In this case, a much smaller tax differential (<US$0.01/liter) was sufficient to cause a

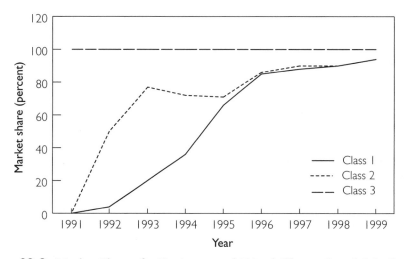

**Figure 22-3.** *Market Shares for Environmental Diesel Classes 1 and 2 in Sweden*

virtual takeover by the two improved qualities. Several dozen technical parameters are specified, including the maximum contents of olefins, certain alkanes, and aromatics as well as total and individual oxygenates, such as alcohols and ethers. Standards for phosphorus, sulfur, and lead are ambitious by international standards and have been tightened over time.[12]

These details have caused some conflicts in which smaller producers of new (sometimes biomass-based) fuels have found that the standards have been designed to suit the large fossil fuel–based companies. It is extremely important that the environmental classes (and all policy instruments) be designed from neutral, general principles related to actual environmental damage. Building them on the solutions or technologies that happen to be available at one point in time is tempting but risky and detrimental to long-run technical progress. The temptation lies in the relative ease. Regulators may otherwise be accused of creating rules that are impossible to abide by or taxes for categories that may never exist. If a concrete technology can be given as an example, regulators feel protected—but run the risk of creating regulations that are overly influenced by a particular brand of technology.

Some inconsistencies are found in the regulation of diesel fuel. Fuels with biological origin still have to pay carbon taxes, and the taxes on different environmental classes cannot claim to be based on estimates of damage. In fact, the relevant tax law is written in such a way that any fuel or fuel ingredient used in a gasoline-powered (i.e., Otto) vehicle is taxed the same as gasoline. This ruling even applies to water and, for instance, hydrogen produced from wind power. By reasonable judgement, some fuels produce lower emissions but still do not qualify for the best environmental class. Since Sweden joined the European Union, its ability to decide environmental classes on its own has been limited; Environmental Class 3 (the lowest) has been harmonized with the E.U. requirements.

With increasing globalization, all countries are exposed to increasing pressure to harmonize standards, regardless of whether they are members of a geographical trading block. This trend has both positive and negative aspects. On the one hand, it may limit countries that wish a faster pace of improvements, but on the other hand, it provides a powerful force for modernization and improvement in environmentally less advanced countries. In many countries, the World Bank and the United Nations (e.g., through their joint Energy Sector Management Assistance Program [ESMAP]) are active in promoting the harmonization of fuel specifications (UNDP/World Bank ESMAP 1998). Another important factor is the advent of powerful and relatively inexpensive technologies for detection of fuel adulteration. Thus, in 2000, even in Kenya, the diesel was being sampled in purity inspections that are able to detect even small quantities of kerosene (a popular fuel used for adulteration because it is a lot cheaper than diesel).

## Vehicle Standards, Efficiency, and Distributional Concerns

*Progress in Pollution Control for (New) Vehicles*

Aside from fuel, the main environmental policy options are traffic and vehicle control, and when it comes to the latter, most worldwide attention is paid to new vehicles. New vehicle control or standards date back to the late 1960s, and in the United States, federal emissions standards for new vehicles requiring catalytic converters (among other things) went into effect in the mid-1970s. However, progress in engine technology is much more than catalytic converters; it is a broad design issue that interacts heavily with driving patterns, road characteristics, fuel development, and other parameters. To understand engine technology, different engine types must first be distinguished: Otto, diesel, two-stroke, electric, and hybrid engines.[13] Similar (but distinct) processes of technological progress exist for gasoline and diesel engines (engine design, fuel injection control, turbocharging and exhaust gas recirculation, exhaust traps, oxidizers and catalysts, fuel modification, and so forth). However, because the importance lies not in these details per se but in their consequences for policymaking, I discuss only gasoline technology.

The Otto has spark ignition and is usually called the "gasoline engine" after its most common fuel. This label is somewhat confusing because Otto (and other) engines can run on other fuels, such as alcohols or ethers, and the development of fuel quality is intimately intertwined with engine development. An engine whose performance is optimized for one fuel may not have good emissions characteristics when run on another fuel. This variability illustrates the disadvantage inherent in dealing with fuels and vehicles separately (as compared with "first-best" instruments such as environmental road pricing). By switching fuel (or by adding extra equipment), some vehicles might actually lower their emissions as much as the difference between, say, Euro I and II or Euro II and III classifications. However, obtaining credit (or certification) for this kind of emissions reduction generally would be very difficult. Before controls were required, engine crankcases were vented directly to the air. In the United States, these emissions were eliminated by requiring crankcase vent ports starting in the early 1960s. Another major source of emissions from these cars was the evaporation from the carburetor and fuel tank, and these emissions now are fed back to the engine or stored in charcoal canister filters.

The efficiency of technology varies depending on fuel volatility. Exhaust emissions consist of hydrocarbon (HC), CO, $NO_x$, and many other compounds that depend on several technical parameters, such as fuel quality, fuel-to-air mixture, ignition timing, and combustion chamber configuration. These parameters have been modified in several ways, including through the use of electronic controls and on-board diagnostics. To meet regulations and further reduce some emissions components, three-way catalytic converters (which reduce the three pollutants $NO_x$, CO, and HC) were first introduced in the United States by Volvo in 1977. They require precise control of the fuel-to-air mixture and thus have indirectly led to the development of better fuel management.

The technologies mentioned in this section have been developed and introduced mainly through a process of tightening standards for emissions rates that were commonly formulated as maximum emissions per mile or equivalent. The United States has been a prominent driving force with the U.S. Clean Air Act and its various amendments, but there have also been important initiatives from Japan and some European countries. In 1990, California took the lead when the California Air Resources Board (CARB) adopted a visionary scheme of progressively tighter emissions requirements, and vehicle manufacturers were forced to sell minimum shares of vehicles fulfilling successively tighter environmental criteria (see Table 22-1). For instance, 25% of vehicles manufactured after 1997 had to be low-emission vehicles (LEVs).

Several details of the CARB program have been modified, but there is no doubt that it has been an important mechanism for encouraging new technological developments. The current legislation still aims for 10% zero-emission vehicles (ZEVs) for the period 2003–2008, after which the target will increase gradually to 16% in 2018. For the near term, part of this obligation can be fulfilled by credits earned from partial ZEVs (PZEVs) or super-ultra-low-emission vehicles (SULEVs). Early introduction of ZEVs (before 2003) is an added "flexibility mechanism" that gives additional credits. Whereas standards such as the LEV are regulations, they are not as inflexible as a design standard (such as mandatory catalytic converters) because they allow manufacturers to choose how to meet the standard (see Chapter 6). Despite dramatic improvements in fuel, engine design, and emissions reduction, the most promising future in this area probably belongs to other technologies, such as electric and hybrid vehicles.[14]

Many developing and recently industrialized economies, such as Brazil, Hong Kong, Malaysia, Mexico, Republic of Korea, Singapore, Taiwan,★ and Thailand, have adopted regulations for new vehicles (and for fuels) that are not far behind those adopted in the United States and Europe. The vehicle industry exerts considerable pressure because it prefers standardized requirements in all markets and because standardization gives large international vehicle manufacturers a competitive edge over small local producers. This pressure may give rise to tensions vis-à-vis local national interests. Although these developments are encouraging, from

---

★Throughout the remainder of this book, the economy of Taiwan, China, will be referred to as "Taiwan."

**Table 22-1.** *Overview of CARB Regulations, 1990*

| Vehicle emission category | Requirements (grams/mile) | Minimum percentage of vehicles |
|---|---|---|
| TLEV | HCs < 0.125 | 10% 1994–1996 |
| LEV | HCs < 0.075; $NO_x$ < 0.02 | 25% 1997; increasing by year |
| ULEV | HCs < 0.04 | 2% 2000, 15% 2003 |
| ZEV | 0 (locally) | 2% 1998, 10% 2003 |

*Notes:* TLEV = transitional low-emission vehicle; LEV = low-emission vehicle; ULEV = ultra-low-emission vehicle; ZEV = zero-emission vehicle; HC = hydrocarbon; $NO_x$ = nitrogen oxides.

an environmental standpoint at least, they do not have as much effect as in industrialized countries for several reasons: in developing countries, vehicles have a long lifetime and are poorly maintained, many used vehicles are imported, and a large fraction of the vehicle pool are two-stroke engines (e.g., motorcycles, mopeds, and three-wheeled tuk-tuks).

The two-stroke engines warrant special attention. Even though technological solutions such as computer-controlled fuel injection (in fact, many of the technologies used for four-stroke engines) could be applied to two-stroke engines, some problems related to lubrication remain because the lubricating oil is a source of dangerous emissions. Also, two-stroke engines typically do not have sophisticated technology; they are cheap and operate under rough conditions with little maintenance. Many other kinds of equipment with two-stroke engines also deserve attention, such as chain saws, machinery for road building and agriculture, and motorboats. Some countries have already banned certain high-polluting vehicle types (see Box 22-1). Another solution is to require the use of ultraclean fuel (e.g., alkylate petrol, which consists essentially of straight alkanes) for two-stroke engines, which have inherently poorer combustion properties.[15] Again, the complex interplay between fuel and motor regulation is highlighted. Sometimes cleaner fuels (with alkanes and/or alcohol) can serve as a substitute for cleaner combustion; sometimes they are complements (as with lead elimination, necessary for the operation of catalytic converters).

## Inspection and Maintenance for Used Vehicles

Despite technological and regulatory advances, new vehicle standards are not sufficient to achieve ambient goals (if vehicles deteriorate rapidly or owners keep old cars instead of scrapping them). In fact, Draconian standards on new cars combined with lax regulation on older vehicles would be a good example of new source bias (see Chapters 12 and 16) and could have a negative effect on the ambient environment if it led to a significant delay in the turnover rate of the vehicle stock. In Japan, inspection and maintenance requirements become extremely stringent when cars are more than three years old, so the vehicle stock turns over rapidly. The used cars are exported en masse to neighboring countries. This approach benefits at least some interests in Japan: the auto industry gets high sales figures, and the local environment is subjected to fewer and fewer emissions.

U.S. EPA has encouraged states to establish inspection and maintenance programs to maintain a minimum standard for vehicles on the road. Because a large

## Box 22-1. Polluting Vehicles Banned in Kathmandu

The government of Nepal has banned polluting vehicles from Kathmandu (Gajurel 1999). Because the Vikram Tempo (the diesel-fueled three-wheelers operated locally) is the most polluting vehicle, it has been targeted first. The program has widespread support in environmental circles, but many locals resist the ban, which eliminates the most affordable means of transportation in favor of Danish-made electric vehicles (Safa Tempo) that are more expensive to purchase and to operate. The program has widespread support, but many ordinary people demand alternative means of transportation.

share of pollution comes from a few vehicles, concentrating on the most polluting vehicles should improve ambient air quality efficiently. Although the potential gains of inspection and maintenance programs are large, the practical results to date are not impressive (Harrington and McConnell 1999). In 1990, the U.S. Congress established in the Clean Air Act more stringent requirements for enhanced inspection and maintenance programs, which started to be implemented in 1995. Particularly in the United States, overseeing such a program involves a principal–agent structure with several stages. U.S. EPA monitors the state programs and requires them to include various features in their enhanced inspection and maintenance programs. Compliance gives the states "credits" toward the fulfillment of their state implementation plans, which help determine whether the state (or parts of it) has designated "nonattainment areas." The states act as agents vis-à-vis the federal agency (U.S. EPA) and as principal vis-à-vis the local controlling agents down to the individual motorists.

The U.S. programs fail to deliver emissions reductions at reasonable costs for several reasons (Harrington and McConnell 1999):

- The assignment of liabilities to the individual motorist results in high transaction costs.
- Most of the resources in the program are used for emissions testing and little for actual repair. Failure rates at tests are low, suggesting that more selective testing might be beneficial.
- Motorists have many opportunities to evade required repairs. They can test vehicles numerous times, until they happen to "get lucky" and pass. This is referred to as Type I errors in testing.
- Many jurisdictions (outside large cities) do not require emissions testing as part of inspection and maintenance; hence, vehicles can be sold to someone or even easily (although illegally) registered in such an area. In Arizona, 22% of vehicles that fail the emissions test never come back for a retest (Ando, McConnell, and Harrington 2000). It is not clear where they go, but presumably, some go to areas that do not require a test, or the owners drive the vehicles illegally.
- Allegedly, some mechanics specialize in temporarily "fixing" vehicles to pass the emissions test with little or no permanent effect.
- The system includes numerous "waivers." A motorist who is over a certain age and drives few miles may be totally exempt, and a motorist who can prove

## Box 22-2. CUT-SMOG Smoking Vehicle Program

California's South Coast Air Quality Management District (AQMD), the smog-control agency for all or portions of Los Angeles, Orange, Riverside, and San Bernardino Counties, has regulated regional air emissions since 1987. The AQMD developed the Smoking Vehicle Program to reduce visible exhaust from vehicles in the area and established a complaint "hotline" to which people could voluntarily and anonymously report (by e-mail or by calling AQMD's toll-free number) vehicles emitting excessive amounts of exhaust.

The AQMD cannot impose fines, but it does send a letter to the owner of each reported vehicle. The letter advises the owner that the vehicle was reported and recommends having it repaired. The letter also advises the vehicle owner that excessively smoking vehicles can be fined $100 to $250 for a first-time offense.

By 2000, nearly 1 million people had volunteered information to the CUT-SMOG program. On average, more than 40% of the "offending" vehicle owners have returned compliance forms to the AQMD, claiming that they have had their vehicle inspected and repaired, when necessary.

*Source:* AQMD 1996.

having spent a certain amount of money (not much—maybe $50, $75, or even $15) on repairs may not be required to make further repairs, even if the car fails the emissions retest.

Waivers for cars driven short distances make sense because the damages are small. Income distribution is also important, because the oldest and dirtiest cars tend to be cheapest and are owned by the poor. More stringent requirements may hit vulnerable groups. However, if these vehicles have high mileage (and are driven mostly in inner cities), then the environmental damage may be considerable. In such cases, it does not make sense to help low-income people by absolving them from these important environmental standards.

Limitations in enforcement reflect negatively on program efficiency, and as a result, dirty cars are seldom found and are repaired properly (or retired) only rarely. In the worst of cases, the inspection and maintenance program degenerates into being a hassle for owners of new and fairly clean cars, for which the program provides little environmental benefit. U.S. experience shows that vehicle testing should generally be separated from repair. Combined "test and repair" facilities have incentives that are far from efficient. Also, mandating enhanced inspection and maintenance with a tougher mandatory minimum repair cost of more than $450 was difficult and led to considerable public resistance, ultimately causing U.S. EPA to back down. On the other hand, other mechanisms in society may compensate for the lack of formal monitoring. In California, using citizens' complaints in the search for the worst-offending vehicles appears to have been successful (see Box 22-2).

Most other countries have inspection and maintenance programs. In the Swedish program, all cars more than a couple of years old are inspected every year in an integrated environmental and safety inspection and maintenance program.[16] Sweden is not a society in which unregistered cars can be driven for very long; monitoring is rather strict, and cars with deficient safety or emissions tests will definitely be stopped. There are no cost-based or other "waivers" (except for antique cars or museum pieces). The Swedish program is thus likely to be effective in stopping dirty cars, although at a high cost (expenses of inspection, waiting, maintenance, repairs, lost years of use from scrapped vehicles, and so forth)— particularly for cars that get driven short distances in the country, where the potential damage to the environment is minimal. Concerns relating to income distribution may be less applicable in Sweden because income is not so unevenly distributed and because public transportation is readily available; the (relatively) poorer vehicle owners are not so dependent on old, polluting personal vehicles, and they can take public transportation if needed. The share of income spent on vehicles and fuel does not vary strongly with income in Sweden.

Although U.S. and Swedish experiences differ, they have some traits in common. Both have transaction costs due to the fact that liabilities are assigned to the individual motorist, who has an incentive to avoid the maintenance needed for better environmental results. Other possible assignments of liability or other instruments could help alleviate the principal–agent aspect of this monitoring problem (Harrington and McConnell 1999):

- The liability could be forced to a greater extent back to the vehicle manufacturers. This happens in other industries, such as the airlines, which lease engines from manufacturers, who are thus forced into taking maintenance into account. Such a tendency is observed for vehicles, too—with demands for extended emissions warranties covering more years and miles. This has led to much lower emissions at 50,000 miles for 1991 vehicles than for those made earlier. However, problems of moral hazard enter into extending an unlimited warranty: because the manufacturer has no control over fuel quality or everyday maintenance, the owner of a really old vehicle may have an incentive to damage the vehicle if someone else is liable for all repairs. Furthermore, long liabilities may produce practical difficulties (e.g., car manufacturers disappear from the market).

- Liability could be centralized, either through leasing (rather than purchasing) vehicles or through the separation of vehicle ownership from emissions liability, whereby the latter could be auctioned off (for whole classes of automobiles) to specialized companies that would receive payment from the government in exchange for taking care of, say, "all 1992 Fords." Performance would be monitored by remote sensing, which is not precise for individual cars but sufficiently exact for large numbers of randomly sampled cars. These specialized companies would have to determine how to entice individual owners to inspect and repair, and they would have a clear incentive to do so efficiently.

- With individual liability, the monitoring problem could be reduced by shifting the burden of proof through a deposit–refund type of mechanism, such as a high annual fee that could be reduced (refunded) on proof of low emissions values (possibly also on other grounds for exemption, such as the use of vehicles only on a farm or in the country). This approach changes the monitoring and repair incentives. Singapore uses a combination of high registration and annual taxes with rebates for certain categories (such as weekend cars) or for owners who replace old cars with new ones (see Chapter 20).

• The last of these alternatives might have the greatest relevance in the polluted megacities of developing countries, where the car stocks are commonly old. They urgently need efficiency but cannot afford inspections that do not lead to improvement, so better targeting is needed. Income distribution effects vary between countries, but in the poorer countries, private cars are luxury goods; hence, concerns should be limited. Still, among motorists, the poorer will almost inevitably have the most-emitting vehicles. This does not necessarily mean that those motorists lose out from a long-run policy of requiring maintenance; the rising costs of maintenance will presumably tend to be "capitalized" in a lower purchase price. This effect means that the former owners of the vehicle—including the first owner—share in the incidence of these expected costs. However, when new rules are made in periods of transition, those who have bought vehicles that they later find out are in need of emissions-related repairs—or, in the worst-case scenario, cannot be driven at all—definitely have a lot to lose.

## Urban Pollution in Developing-World Cities

Box 22-3 is one example of the rapidly increasing attention to urban environmental issues, including air pollution in the megacities of developing countries. In poor countries, there is always a heightened need for prioritization. The number of severe problems is great, and the resources to tackle them small. There is a lot to be said for integrated urban pollution management programs in such cities as Mexico City, Mexico; Santiago, Chile; Dhaka, Bangladesh; Cairo, Egypt; and São Paulo, Brazil. The experiences, particularly from Latin America, have shown that four broad strategies need to be pursued (Faiz, Gautam, and Burki 1995):

• improve vehicle technology through engine replacement, inspection and maintenance, and vehicle scrappage;
• use cleaner (reformulated or alternative) fuels that can be targeted (by price policies) at particular areas of application;
• control and manage traffic; and
• improve urban transportation infrastructure.

---

### Box 22-3. Banning Lead and Two-Stroke Engines in Bangladesh

"In Dhaka, Bangladesh, pollution is blamed for 15,000 deaths per year out of a population of 9 million and Dhaka's air during the November–January dry season contained one of the highest concentrations of lead in the world. About 160,000 motorized vehicles are plying the streets, about one-third of them having no fitness certificates and constantly emitting black smoke, according to the Bangladesh Road Transport Authority. In 1999 Bangladesh banned leaded fuels and pollution-belching two-stroke motors. The IDA, a soft-loan window of the World Bank, provided a $177 million credit line to finance a project to fight air pollution and improve traffic conditions."

*Source:* Temple 1999.

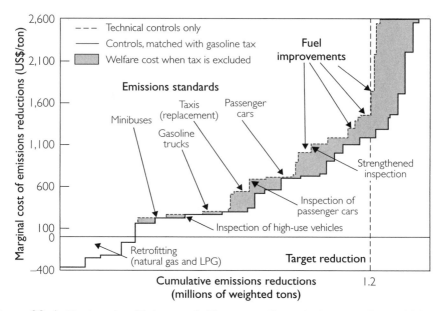

**Figure 22-4.** *Engineering Estimate of Abatement Costs in Mexico City, 1992 Estimates*

Source: Eskeland 1994.

In practice, the number of steps that can be taken to improve the air in a city like Mexico City is large (Eskeland 1994; Goddard 1997). By organizing these steps in ascending order of marginal cost, an engineering estimate of the marginal cost of abatement curve can be produced for the whole city (see Figure 22-4). At first, improvements include retrofitting trucks and minibuses with liquid petroleum gas and recovering gasoline vapors at gas stations and depots at low or even negative cost.[17] Then come inspection and maintenance for various kinds of vehicles—starting with high-use vehicles such as taxis or delivery vans, moving gradually to private passenger vehicles. Other measures include fuel improvements (tighter standards, removal of impurities, the inclusion of certain additives, and lowered vapor pressure), paving of roads, new vehicle standards, and so forth. Some improvements can be done to various degrees with increasing marginal costs (e.g., inspection and maintenance has a high cost-efficiency for professional vehicles, but not when extended to all vehicles). The most difficult but not necessarily least important are the most systemic improvements, such as transportation control and zoning measures.

Considerable savings are possible through a good combination of policy instruments, including a fuel tax and a combination of physical reduction measures; without the gasoline tax, an otherwise well-designed program of controls would be almost 25% more expensive (or US$110 million/year in Mexico City) (Eskeland and Devarajan 1996). This is a good example of a case in which two policy instruments may be combined in a "second-best" strategy (given that environmentally differentiated road pricing is unavailable). The fuel tax helps

**Table 22-2.** *Benefits and Costs for Several Policies in Manila, Philippines*

| Abatement measure | Avoided effects | Benefits (million US$) | Costs (million US$) | Approximate benefit/cost |
|---|---|---|---|---|
| Address gross polluters | 160 deaths, 4 million RSD | 16–20 | 0.08 | 20–25 |
| Clean vehicle standards | 895 deaths, 24 million RSD | 94–116 | 10–20 | 7 |
| Vehicle inspection and maintenance | 310 deaths, 8 million RSD | 30–40 | 5.5 | 6 |
| Improved diesel quality | 94 deaths, 2.5 million RSD | 10–12 | 10 | 1 |

*Note:* RSD = respiratory symptom days.

*Source:* Adapted from Shah and Nagpal 1997 (table ES2).

reduce overall demand, whereas the technical improvements reduce emissions per mile. In addition, the tax proceeds can be used to finance abatement. Generally, economists emphasize the inefficiency of such earmarked funds. In some institutional settings (where the environmental authorities would otherwise not be able to compete for any funds), this revenue recycling might provide additional arguments for pollution fees.

Somewhat similar results were found for Manila, Philippines, a city of 10 million where 80% of the population lives in areas where the national standards for total suspended particles (TSPs) are exceeded (Shah and Nagpal 1997). TSP levels are frequently more than five times the WHO guidelines. Some measures can be 20–25 times as cost-effective as others (Table 22-2). One key measure is overall traffic planning and management. Because calculations are inherently difficult, I simply present one good example and one bad example.

## Traffic Management through Rationing that Backfires

One policy widely considered an abysmal failure is the "Day without a car" ("*Hoy no circula*" or "*Día sin carro*") policy implemented in Mexico and in many major cities in other countries. It was designed as a rationing scheme by which each car was banned from driving one weekday. Several appealing features of this plan were touted:

- Compliance is easy to monitor; the license plate number indicates which day the vehicle is banned. (Anecdotal evidence suggests that some people might have had two plates, but monitoring was simple—partly because the way the police monitor can in turn be monitored and complemented by any casual observer.)
- The program is "fair" because it targets the rich and poor equally.
- Commuters who travel only by car supposedly will learn about the public transportation system available to them and perhaps realize its benefits.

In fact, none of these expected features turned out as planned, and instead, problems inherent in the program became apparent (Eskeland and Feyzioglu 1997a, 1997b):

- Households or drivers put off trips to other days (weekends) or carpooled. Although this result is part of the desired outcome, distance driven was not necessarily reduced. Households also acquired additional cars, many of which were older, more-polluting vehicles. From 1983 to 1989—before the regulation—Mexico City exported used cars to the rest of the country at a yearly rate of 74,000 used vehicles per year (typically replaced by newer ones because average income is higher in the capital). Immediately after the introduction of the program, however, the capital imported 85,000 used vehicles from the rest of the country.

- The regulation should not necessarily be seen just as a ban on one day a week but rather as the issue of (nontradable) driving permits for six days a week. The allocation mechanism of six permits per vehicle created a perverse incentive to acquire additional vehicles (an example of a perverse output effect, see Chapter 14). Overall, fuel demand, congestion, and pollution was increased by the policy (Eskeland and Feyzioglu 1997a, 1997b).

- There was nothing particularly "fair" about the policy. In fact, the wealthier households appeared to be the ones who had had the greatest ease in getting around the regulations by buying extra cars. No strong evidence indicates that getting drivers "acquainted" with the overcrowded and run-down public transportation system in Mexico City was a positive experience.

### Traffic Management through Foresight and Planning

One example of rare foresight in town planning is in Curitiba, the capital of Paraná, Brazil, where local authorities have long insisted on sustainable transportation and have consistently favored buses over heavier, more expensive rail (Meadows 1995; Prefeitura Municipal de Curitiba 2002; Rabinovitch 1992; Tlaiye and Biller 1994). While not a megacity, Curitiba is large and growing quickly: population was 0.5 million in 1965, 1.5 million in March 2002. What sets Curitiba apart is the long-run commitment to make Curitiba "livable" with ample green space, reasonable garbage handling, and sensible traffic management.

The original plan came out of a public competition in 1965. Authorities then set up the Curitiba Research and Urban Planning Institute. Coherent town planning and zoning limits developments in the center, which has large pedestrian zones and emphasizes cultural heritage. The first express busways were inaugurated in 1974. Instead of the usual mix of unstructured roads and expensive rail found in most world cities, Curitiba has followed an almost Napoleonic vision of planned road construction. The hierarchy of roads is based on the central linear axes with a three-way structure: at the center are exclusive express bus lanes, flanked by two local roads. On each side of the central road (a block away) are high-capacity one-way roads for traffic into and out of the city. These major arteries are complemented by a system of four levels of roads (labeled "structural," "priority," "collector," and "connector"), the smallest of which are the minor residential roads.

One of the factors tipping the balance in favor of express buses rather than rail systems is simply the capital costs. Curitiba's direct-route busways have a construction cost of less than $0.2 million/kilometer compared to $20 million to $100 million/kilometer for underground or aboveground rail systems. To be able to choose this alternative, city planners have to acquire the necessary land before the city gets large and congested—a necessary aspect of long-term planning.

Currently, Curitiba's public transportation system carries more than two-thirds of the population each day. It is operated by private bus companies that are selected by competitive bidding and work on tightly regulated concessions with the municipal government. The bus fares go to a municipal bus fund, and the companies are paid per kilometer served. Even though Curitiba has a high rate of car ownership (more than most major cities in Brazil), there is no serious congestion, and public transportation is estimated to save 25% of fuel consumption. Several other large cities are trying to copy the Curitiba concept, and some find it difficult precisely because they feel the infrastructure should have been put in place before the cities grew to their current size. However, this is no cause for delay. Most forecasts show continued dramatic urban growth in poor countries, and now is the time to plan for the megacities of the future.

## Notes

1. Parts of this section are based on joint work with Åsa Löfgren and Henrik Hammar.

2. *Octane number* is a complicated parameter, but it is roughly the ability to avoid "knocking" (premature ignition). This term is called *octane* because it was scaled so that a particular isomer of octane (*iso*-octane) was given a value of 100.

3. MTBE is also a health hazard, and MTBE leakages reportedly have contaminated some water sources. The use of alcohol leads to emissions of ketones and aldehydes, which may also cause health problems.

4. Leaded paints are banned in most countries. Lead batteries are subject to charges or deposit–refund schemes in some countries, including Sweden, where the military, in its environmental action plan, has decided to phase out the use of lead in bullets.

5. Actually, some lead was left for another few years, but the quantities were not significant.

6. In 1970, maximum lead content in Swedish gasoline was 0.70 grams/liter. It was lowered to 0.40 grams/liter in 1973 and then to 0.15 grams/liter in 1980. In 1987–1989, new emissions requirements made catalytic converters practically mandatory. Various incentives also were introduced to subsidize catalytic converters. By 1994, the share of leaded gasoline had fallen significantly, and it was banned; the oil companies had to switch to other metal lubricants, such as sodium.

7. An econometrical analysis of the phaseout of lead in the European Union shows the importance of tax differentials, income levels, and catalysts (Löfgren and Hammar 2000). With a model of market shares as a function of average price difference (and share of cars with soft valve seats), elasticities are significant and high (Lovei 1998). In an analysis of price differences within U.S. states, Borenstein (1993) found that a five-cent additional tax on leaded gasoline might have reduced the phaseout time in the United States by about two years.

8. This section builds on a number of World Bank reports and other papers (Lovei 1998; Onursal and Gautam 1997; Sayeg 1998; Shah and Nagpal 1997; Shah, Nagpal, and Brandon 1997; UNDP/World Bank ESMAP 1996, 1998; World Bank 1998a).

9. However, some countries (e.g., Panama and Mexico, at least during some years) let the oil companies charge more for unleaded gasoline because it costs more.

10. Anecdotal evidence seems hypothetical or even counterintuitive. Noticing that unleaded gas was more expensive, I asked a couple of Mexican taxi drivers if it was not a disadvantage to have a cab with a catalytic converter; they said they got more customers that way.

11. The World Bank's Clean Air Initiative in Latin American Cities conference (Dec. 2–4, 1998) is one example (Lovei 1998; Onursal and Gautam 1997; Sayeg 1998; Shah and Nagpal 1997; UNDP/World Bank ESMAP 1996, 1998; World Bank 1998a).

12. This situation means that there are two parallel avenues for improvement. Not only have the market shares for Classes 1 and 2 been increasing, but the requirements for each have become increasingly stringent over time. Currently, lead content is required to be <5 milligrams/liter of fuel; sulfur <50 and <100 ppm for Classes 1 and 2, respectively; and benzene <1%. Some technical standards (e.g., vapor pressure) also vary geographically and seasonally.

13. Much of the information in this section builds on Walsh 1994 and 2000.

14. In a speech to the *Automotive News* World Congress, William Clay Ford Jr. (great-grandson of Henry Ford and chairman of Ford Motor Co.) said, "Hybrids and fuel cells will finally end the 100-year reign of the internal combustion engine as the dominant source of power for personal transportation" (cited in Walsh 2000).

15. Such ultraclean gasolines are used for chain saws in Sweden (important because the user's nose is close to the exhaust) and have been suggested for outboard motors in sensitive coastal areas by Swedish marine ecologists (Östermark 1996).

16. Some countries only inspect and test for environmental vectors, but testing safety features seems logical, too, because faulty brakes can also cause external damage.

17. Economists are skeptical of "free lunches," but presumably, some attention cost is not counted that can explain the apparent paradox of negative cost abatement. In fact, economists are fairly wary of this kind of cost curve estimated solely on the basis of engineering data, but in some cases, they may still be the best models available.

# CHAPTER 23

# Lessons Learned: Transportation

THE CHAPTERS IN PART FOUR attempt to cover a lot of ground, from the abstract to the practical aspects of policy design. Part Four started with the damage function, which is complex and varies dramatically depending on vehicle, fuel, location, time of day, and weather. The strong interactions between these variables suggest that an integrated approach may be highly appropriate in order to save costs. However, the complexity makes the design of an all-encompassing "first-best" system (such as A in Table 23-1) appear impossible. Although developments in information systems may improve the situation, such information systems are so far used mainly for traffic management (K), as in Singapore.

Typically, most policies aim to reach subgoals regarding fuel quality, vehicle standards, fuel efficiency, and traffic management and to find suitable instruments for each. Fuel taxes (B) are commonly used. Although not always recognized as— or even intended to be—environmental policy instruments, they have a strong effect on fuel use and thereby can be effective in combating global warming. This effect is usually not a top priority in poor countries, but reductions in overall fuel use yield ancillary benefits, such as decreases in local emissions and other externalities. A fuel tax is generally not the best way of achieving all the environmental goals related to the transport sector, but it may be an important part of a package of policy instruments. However, a fuel tax may be regressive in countries such as the United States, where total reliance on the automobile is combined with a fairly unequal income distribution. In countries with more equal income distribution, the issue may be less significant; in many developing countries, fuel taxes may be progressive because poor people have little access to individual motorized transport. The social distribution of the costs of a fuel tax are important for its acceptance. One political aspect of fuel (and other) taxes is the development of more or less powerful lobbies in favor of or against each particular tax.

Fuel quality, vehicle standards, and urban planning are the three most important policy areas for local (urban) environmental management. The choice of policy instruments depends on the context. Urban planning (C) is a regulatory

**Table 23-1.** *A Policy Matrix for Some Environmental Aspects of Transportation*

| Policy | Fuel | Vehicle | Traffic | Integrated |
|---|---|---|---|---|
| Regulation | D | H | J | C |
| Bans | G | N | O | |
| Price (tax) | B | P | K | A |
| Information | F | Q | L | |
| Deposit–refund or two-tiered instruments | | I | | |
| Tradable emissions permits | E | | | |
| Public provision | | | M | |

*Notes:* A = environmentally differentiated road pricing ("first-best" system); B = fuel taxes (possibly differentiated); C = urban planning; D = fuel quality regulations; E = marketable permits (e.g., for fossil carbon); F = green labeling of fuels; G= bans on certain fuels or fuel components; H = regulations concerning vehicle emissions, for new vehicles or for inspection and maintenance; I = two-tiered systems that would, for instance, entice cleaner vehicles to "self-reveal" and opt for a certain option in a menu of policies; J = zoning; K = tariffs for public transport; L = green labeling of transportation services; M = physical structures, such as railways; N = retirement, removal, or repair of the most-polluting vehicles; O = restricting traffic in certain zones or at certain times of day or night; P = taxes (or subsidies) on vehicles; and Q = green labeling of vehicles.

tool and relies heavily on zoning (O) and infrastructure investments (M). Physical regulations and reforms of the pricing and tariff structure are important determinants of demand (e.g., the pricing of public transportation (K) and, more generally, the organization of markets for bus and other public transportation services).

Fuel quality is a complicated and technical area in which technical standards, bans, and norms are the most common basic instruments of regulation (D). However, experience has shown that many improvements—notably, the transition to unleaded gasoline—have been facilitated by the use of tradable permits (E) in the United States and by differential tax treatment (B) in many other countries. In some cases, information- or reputation-based policies such as green labeling or green marketing (F) or simply bans (G) were used. The choice of instrument appears to depend on the sophistication of the economy, among other things. In some diverse, large economies, the flexibility afforded by market instruments allows cost savings (and preempts political opposition). In smaller economies, the scale economies may tip the balance in favor of simpler instruments to avoid the extra costs of multiple standards and delivery systems. The regulation of new vehicles and particularly vehicle emissions technology (H) is traditionally one of the most central instruments. It also can be supplemented by information and labeling (Q) (see Chapters 21 and 22).

The interactions between fuel and vehicle technology can be intricate. The substitutes available as well as the political dynamics differ by situation (e.g., for diesel, alcohol, and gasoline), because the fuels are used and produced by different categories of business. Sometimes the fuel and vehicle industries try to shift all the blame onto each other (and onto vehicle owners) for problems that essentially have joint causation, raising issues of liability and moral hazard and making it important to have real monitoring and flexibility in the application of policy instruments. For instance, it is not inconceivable that a truck classified as Euro II might, with special ultraclean fuels, or with additional exhaust treatment such as particle filters, meet the emissions requirements that correspond to the next class (Euro III). On the other hand, with poor maintenance and ordinary or poor-quality fuel, its emissions might correspond to an inferior class. If policy instruments are solely based on the vehicle type or the fuel type, then the important interaction between the two will be missed.

Green marketing also has been used in the context of transportation services (L). The electric trams in Gothenburg, Sweden, have an environmental label partly because they run on "green" electricity. Similarly, buses and taxis in Mexico that converted to unleaded gasoline were painted green instead of the traditional yellow.

Because a disproportionately large share of all pollution comes from a few of the worst vehicles, targeted inspection and maintenance might be a powerful instrument (H), but the exact design of such a system is tricky. One question is the degree of homogeneity versus flexibility. Damages due to emissions depend strongly on location and vehicle type. Requiring inspection and maintenance only in cities would be an effective restriction except that it would provide an incentive to register vehicles in rural areas. Flexible systems make monitoring and enforcement considerably more complicated. Introducing a kind of market-based instrument such as a deposit–refund system (I) might alleviate the monitoring issue by placing the incentives for revelation or signaling with the individual motorist. The idea is to create incentives for the motorists who cause the least environmental damage to reveal themselves through voluntary reporting of some sort.

As shown by such positive examples as Curitiba (see Chapter 22) as well as the many negative counterexamples, conscious long-run planning is a crucial element for improving the urban environment. Such planning includes zoning, infrastructure, institutions, public transport pricing, and choice of technology (J, K, L). The fact that many of these instruments are long-run instruments should in no way be taken as an excuse for implementing them immediately. However, they must be tailored to future technologies, of which knowledge is incomplete. As in Mexico City, combinations of many policy instruments that include both fuel taxation (B) and regulations (H) may be significantly better than either instrument separately. In contrast, partial driving bans (such as "*Hoy no circula*") do not appear to be efficient.

Because technical change is so rapid, policymakers should be skeptical to invest in systems with large and expensive ground-based infrastructure. This applies both to physical structures (such as subways) (M) and infrastructure for monitoring (such as toll stations, if one considers that such systems may soon be out-competed by technically more advanced systems with less fixed infrastructure). Because the development of information technology in general and intelligent traffic systems in particular is rapid, it seems likely that the advantage of more sophisticated systems will increase over time. On the other hand, vehicles also will become less polluting over time as a result of technological progress, tending to somewhat reduce the benefits of this kind of instrument.

Although vehicles and roads have long useful lives and much emphasis must be placed on long-run planning, numerous vital policies must be pursued in the short run, too. Such policies include fuel reformulation (which includes abolishing lead and sulfur); retirement, removal, or repair of the most-polluting vehicles from sensitive areas (N); and removal of perverse subsidies such as tax breaks for older vehicles or subsidies for the manufacture of obsolete vehicle models (P).[1]

The transportation sector presents a range of environmental problems with considerable technical challenges. In some respects, it presents more sociopolitical difficulties than in, say, industry, because monitoring and enforcement are difficult and because the owners and polluters are not professionals, which introduces several political constraints. This area also has experienced remarkable technological progress. Consequently, the cost of additional emissions reductions may frequently be higher (at the margin) than in other sectors. Because mobility is so highly valued and because cars have status, it may be possible to forge ahead with environmental goals faster in the transportation sector than in other sectors. Doing so will require a continued refinement of policy instruments to ensure that the potential promises of new technology are harnessed.

## Note

1. At least as late as the mid-1990s, Mexico had vehicle tax exemptions for vehicles older than 10 years. Mexico and Brazil long protected their domestic production of vehicles such as the Volkswagen Beetle, which is hardly on the cutting edge of emissions technology.

# PART FIVE

# *Policy Instruments for Industrial Pollution*

THE SOURCES OF POLLUTION ASSOCIATED WITH road transportation (addressed in Part Four) are small, numerous, and mobile. These characteristics pose several challenges to policymaking, particularly in the area of monitoring.

In contrast, the sources of industrial pollution addressed in Part Five are typically large, except perhaps in the informal sector of developing countries. Because polluting firms are usually large and fixed in location, they commonly are prime targets for environmentalists, particularly in the initial phases of environmental awareness. Abatement may be easier to monitor in large industries than in road transportation, but obstacles still exist. Difficulties in monitoring and enforcement stem from the power and influence that industries wield through the economic resources at their disposal as well as the fact that so many people depend on industry for employment, business, and tax receipts (see Chapter 16).

When industries are international, regulations tend to "spill over" from one country to another. Part of the reason may be copying successful legislation, but considerable evidence indicates that multinational industries have a strategy. They initially resist legislation or standards in one country. However, after they have adapted, they proactively promulgate the same standards in other countries to achieve a competitive edge by avoiding the cost of producing for many specifications in different markets. This is one of the many factors that distinguish policymaking in developed countries (Chapter 24) from that in developing and transitional countries (Chapter 25).

# CHAPTER 24

# *Experience in*
# *Developed Countries*

$I$N THIS CHAPTER, I ADDRESS many topics chosen to illustrate various aspects of instrument design for the industrial sector:

- efficiency in the abatement of sulfur emissions,
- abatement of nitrogen emissions,
- green tax reform in Sweden and Germany,
- prohibition, taxation, and regulation of hazardous solvents,
- liability and Superfund,
- information provision and voluntary agreements (VAs) for toxic emissions in the United States,
- global ozone policy, and
- global climate policy.

First, I discuss the energy sector, in which many countries combine regulation with market-based policies to promote energy efficiency, affect the structure of energy supply, or specifically limit emissions (e.g., of sulfur oxides or nitrogen oxides). This discussion provides an opportunity to compare policies proposed for similar goals in different settings. One of the most important issues related to the energy sector is acid rain, and I compare the tax approach commonly used in northern Europe with the permits approach of the United States. I also look at green tax reform in Germany and Sweden.

The mainstays of environmental policy still are legislation, liability, and public spending; therefore, I have selected three illustrative case studies. In the first, I compare a ban (Sweden), a tax (Norway), and a stringent regulation (Germany) of trichloroethylene, a particularly noxious solvent. In the second, I describe Superfund (United States), a liability and public cleanup program that started in a hazardous waste site in New York state. Finally, I discuss information provision as an instrument for addressing toxic emissions in the U.S. Toxics Release Inventory program. To conclude, I present some issues related to main global environmental issues: ozone depletion and global warming.

## Abating Sulfur Emissions[1]

Acid rain was one of the first environmental problems to achieve global recognition and has come to symbolize environmental issues in many countries. It is hardly new; in fact, its effects were recognized almost 2,000 years ago, when copper smelting released large quantities of sulfur, killing vegetation in the Roman Empire.

When Sweden and Norway started to associate the death of lakes and fish with sulfur emissions in the 1960s, the connection was first rejected by Great Britain, Germany, and other sources of large-scale emissions. With the *Waldsterben* (death of forests) in Germany and better research, the connections are well understood today. They are somewhat complicated, because the sensitivity of the underlying rock and soil can vary so much from one country to another. Countries with old rock, like Scandinavia, can be several orders of magnitude more sensitive than countries that have a lot of calcareous rock with a high acid-buffering capacity, like England. The incentives to abate sulfur emissions depend on the perceived environmental damage and vary by country. The response also depends on abatement costs and political factors as well as the treaties that countries are bound by (in this case, mainly the Sulfur Protocol) and the efficiency of the policy instruments they select (e.g., Klaasen 1996).

Table 24-1 shows that several country groups can be distinguished by emissions and rate of emissions reduction. Perhaps not surprisingly, Austria, Norway, Sweden, and Switzerland—countries that are sensitive to acidification—have low per capita emissions and reduced sulfur emissions significantly in the 1980s and 1990s. The United States, Canada, and some southern European countries have larger emissions per capita and have made smaller reductions; Greece and Portugal have even increased sulfur emissions. Emissions in the United Kingdom used to be high but decreased rapidly during the 1990s (mainly as a result of switching from coal to gas). During the past decade, European collaboration has been successful, and many (western) European countries have made considerable efforts to abate. In the transitional economies, emissions have been significantly reduced but usually not as a result of specific sulfur policies; rather, the reductions are due to industries closing down or increased energy efficiency. Many but not all of these countries still have high emissions.

### European Policies: Regulation and Tax

Some countries have found "abatement" easy by switching energy sources from oil and coal to gas (United Kingdom) and hydropower and nuclear (Sweden and France), or because large industries were closed (e.g., when East Germany and West Germany merged). Also, the need for energy for heating and the structure of industry and transportation systems are relevant factors that vary considerably between countries. Distinguishing the effect of intentional policies from that of other changes in the economy (e.g., fuel switching, energy savings, and structural change) is tricky; judging the effectiveness or efficiency of the policy instruments used is even more difficult.

Most of the policymaking in this area is regulatory: performance standards (the sulfur content of fuels) or design standards (mandatory technologies for abatement).

Table 24-1. *Sulfur Emissions in Several Countries*

| Country | Emissions per capita (kg of SO$_x$), 1997 | Change (%), 1980–1997 |
|---|---|---|
| *Low sulfur emitters* | | |
| Switzerland | 4 | −78 |
| Austria | 7 | −86 |
| The Netherlands | 8 | −75 |
| Norway | 8 | −78 |
| Sweden | 8 | −86 |
| *Medium sulfur emitters* | | |
| France[a] | 16 | −73 |
| Germany | 18 | −80 |
| United Kingdom | 19 | −66 |
| Finland | 20 | −83 |
| Belgium | 22 | −74 |
| Denmark | 22 | −76 |
| Italy[a] | 23 | −64 |
| Portugal[a] | 37 | 40 |
| Ireland | 41 | −26 |
| Greece[a] | 49 | 36 |
| Spain[a] | 49 | −37 |
| United States[a] | 65 | −26 |
| Canada | 91 | −41 |
| *Transitional economies* | | |
| Russia | 17 | −66 |
| Lithuania | 19 | −75 |
| Ukraine | 23 | −71 |
| Poland | 56 | −47 |
| Hungary | 66 | −60 |
| Czech Republic | 70 | −69 |

[a]1995 or 1996 data.

For example, the European Council's 1975 resolution on energy and the environment was intended to reduce the sulfur content in light fuel oil (LFO) and regulate the use of heavy fuel oil (HFO) (Johnson and Corcelle 1995) (see Table 24-2).

In addition to regulating contents, several European countries use differentiated energy or fuel taxation to encourage a reduction in sulfur use; some also use permit trading (see Chapter 7; see also Klaasen 1995). Sweden, Norway, and Denmark charge a high tax on sulfur emissions (US$3,000, US$2,100, and US$1,300/ton, respectively), whereas taxes in Italy, France, Switzerland, Spain (a regional tax in Galicia), and Finland (only diesel) are all less than US$50/ton. Although all taxes should, in principle, have some effect, the effect of very low taxes is unlikely to be significant. Even in the high-tax countries, the effect is

**Table 24-2.** *Sulfur Content Limits in Liquid Fuels (Gas Oil) in European Member States*

| Article | Came into force | Limit (% by weight) | |
| --- | --- | --- | --- |
| | | Light fuel oil | Heavy fuel oil |
| 75/716/EEC | Before 1980 | 0.5 | 0.8 |
| 75/716/EEC | Oct. 1, 1980 | 0.3 | 0.5 |
| 87/219/EEC | Jan. 1, 1989 | 0.3 | |
| 93/12/EEC | Oct. 1, 1994 | 0.2 | |
| 93/12/EEC | Oct. 1, 1996 | 0.05 | |

*Source:* Ercmann 1996.

hard to separate from the effects of concurrent policies. In Sweden, the sulfur emissions tax was estimated to be responsible for 30% of the reduction between 1989 and 1995 (SEPA 1997).[2] The decline in oil use was the main factor behind decreasing sulfur emissions in Swedish manufacturing, and this decline depends on the general policy of high fuel taxation combined with a rapid expansion in nuclear power during the same period (Hammar and Löfgren 2001). However, the sulfur tax also appears to have been important in reducing the actual sulfur content of fuel oil below the legal limits. For instance, the limit for LFO was 0.2% in 1976, but actual sulfur content was lower. In 1991, the actual sulfur content was 0.08%, and in 1994 it was 0.058% when the legal limit was still 0.2%; this result must be mainly a result of the tax.

Poland deserves special mention in a discussion of sulfur taxation because among eastern European countries, Poland has most consistently applied tough economic instruments to reduce pollution and to finance the abatement of pollution; the fees are paid into funds (see Chapter 25). Poland's sulfur tax was between US$60 and US$100/ton between 1990 and 1996 (Klarer, McNicholas, and Knaus 1999)—although low by Scandinavian standards, it was higher than that of any other OECD countries. Considering the nation's low average income and high overall pollution levels, this tax is sizeable (comparable to sulfur permit prices in the United States). The emissions levels may not have declined as quickly in Poland as in other formerly planned economies, but to a large extent, the smaller amount of de-industrialization is responsible. The tax is believed to have been successful, through both its incentive effect and the funds for abatement investments that it generates.

## U.S. Policies: Tradable Permits

The other major sulfur policy experiment is the U.S. trading scheme. Title IV of the 1990 Clean Air Act Amendments (CAAA) established a market for tradable sulfur dioxide ($SO_2$) among electric utilities.[3] Title IV was the first large-scale, long-term environmental program in the United States to rely on tradable emissions permits. Although the trading of abatement credits had been allowed in numerous earlier programs, it typically had been an auxiliary mechanism, whereas this time, it was the primary mechanism. Earlier air pollution regulations in the United States (and many other countries) targeted individual sources and their emissions rates. In this program, the focus was on the aggregate emissions, which were capped (although only from the electricity sector).

The legislation was intended to reduce emissions by roughly half, that is, by 10 million tons per year. (Total U.S. emissions in 1990 were about 21 million tons, of which 17 million tons were emitted from the electricity sector.) To achieve this goal, permits equivalent to about 9 million tons were issued.[4] The reductions are targeted to the electricity sector, and all fossil-fueled plants are obliged to hold several permits that correspond to their emissions. At the end of the year, utilities must prove their compliance: the allowances in their accounts with the U.S. Environmental Protection Agency (U.S. EPA) Allowance Tracking System must equal or exceed annual $SO_2$ emissions. There is a 60-day grace period to purchase additional $SO_2$ allowances. Remaining allowances may be sold or banked for future use. Trading is unrestricted. Individual plants decide how much to abate and how many permits to buy, and anyone can buy permits—utilities, fuel suppliers (who may, for instance, bundle the sale of coal with that of permits), environmentalists (who might want to reduce aggregate pollution by holding but not using permits), and brokers speculating in rising permit prices.

When the CAAA sulfur reduction was being planned, the main avenue for reductions was thought to be scrubbing and other technological forms of abatement. Estimates of abatement costs varied considerably; values of US$250–350 for Phase 1 (from 1995) and US$500–700 for Phase 2 (starting in 2000) were calculated (Joskow, Schmalensee, and Bailey 1998). The early (1990) estimate from U.S. EPA was US$750, and pre-1989 industry estimates were US$1,500 for emissions permits, which was also the U.S. EPA price for direct sales (Burtraw 1998a, 1998b). Another upper limit is the penalty payable for failure to hold permits, which is US$2,000/ton. Yet other estimates are illustrated in Figure 24-1.

One of the reasons for choosing this design was that earlier policies had some undesirable features. They targeted the new sources more than the old ones (referred to as "anti-new-source bias") and required similar technologies or emissions reductions of all plants without properly accounting for the characteristics of different kinds of coal. Because abatement costs appeared to be considerably heterogeneous, the potential efficiency gains from a market-type policy instrument were expected to be large (see Chapter 12). As it happens, the main variation in cost has been between western producers of "cleaner" low-sulfur coal and eastern high-sulfur coal mines. CAAA required power plants in the west to include scrubbers on new plants that operated with fairly clean coal, whereas old eastern plants could continue, with fairly limited regulations, to burn high-sulfur coal. This policy was both unfair and inefficient.

The current (1990) program coincided with deregulation of the railways and provided considerable impetus to the transport and marketing of low-sulfur coal across the United States. Freight rates dropped dramatically, and the high-sulfur coal has been almost outcompeted by the cleaner coal from the western states. As a result, power stations have been able to meet a large share of their desired reductions cheaply by switching coal suppliers instead of investing in technological improvements such as scrubbing. The price of permits has been much lower than forecast (see Figure 24-1). This result is often taken as evidence that permit

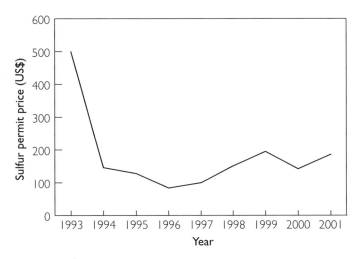

**Figure 24-1.** *Sulfur Permit Prices*

*Notes:* Value for 1993 is an approximate average for pretrading predictions, which ranged from $309 to $981.

*Sources:* U.S. EPA 2002a, 2002c.

trading is an efficient instrument, and although it is somewhat true, the rapid and large fall in permit price is a reflection of "technological development" or "substitution" that has been unusually successful. The brilliance of the policy instrument would be if this development were due uniquely to the trading, but at least part of this fuel switch might have occurred under other policy instruments. Several observers note that the reduction in permit price was largely a result of reduced freight rates, extra allocations of permits, and local state rules that provided an extra constraint in some cases, leading to large abatement investments (Ellerman and Montero 1998; Burtraw 1998a, 1998b; Joskow and Schmalensee 1998; Joskow, Schmalensee, and Bailey 1998). However, the cost savings stemmed from tradable permits rather than regulatory rules, which "could have made abatement more than twice as expensive" (Fullerton, McDermott, and Caulkins 1997). This example illustrates the importance of using general policy instruments that allow economy-wide processes to work instead of targeting narrow technological solutions (see Chapter 16).

Sulfur emissions exhibit heterogeneity in both abatement and damage costs. As discussed in Chapter 12, market-based instruments have considerable potential to reduce costs when abatement costs are heterogeneous. However, when damage itself is heterogeneous, market-based instruments are less suitable, and instruments such as zoning have an advantage. In the case of permit trading to reduce sulfur, the relevant risk is that health damages might increase if the patterns of emissions change unfavorably. More detailed models show that trading and banking led to sizeable shifts in the geographical (and temporal) distribution of emissions (in some states, effects may be as large as 20% compared with a baseline that was half the original levels [Burtraw and Mansur 1999]).

The overall effects of these shifts are beneficial, not detrimental, in terms of total health costs because emissions appear to be moved away from the most densely populated regions. This effect is partly due to regulatory pressure on abatement; allegedly, "downwind" states (such as New York) also pressure their utilities into not trading with "upwind" states (such as Pennsylvania and Ohio). Banking led to more rapid decreases in the 1990s (and thereby increased future emissions). The aggregate cost savings from trading and banking are estimated to be 13% of compliance costs in 1995 and 37% in 2005. Potential annual cost savings due to trading are US$800 million (1995 dollars) or 43% of compliance costs under an "enlightened" command-and-control program in which the regulator adopts a performance standard approach (Carlson et al. 2000). The savings would be even greater when compared with a less enlightened command-and-control program with uniform technology standards.

Permit allocation has received a great deal of attention in the literature. The expected value of the permits at the onset of the program was around US$5 billion/year, and thus lobbyists were prepared to fight for every detail in the allocation (Joskow and Schmalensee 1998). The main mechanism used for determining permit allocation in the United States has been somewhere between grandfathering and output (or rather input) allocation (see Chapter 7). Permits have been free and allocated based on historic values—not in proportion to historic emissions but past values of heat input.

An allocation based on emissions (traditional grandfathering) would have been somewhat more beneficial to the "dirtier" plants, whereas an allocation based on electricity produced would have favored the more thermally efficient plants. In actual policy, important issues that affect the decisions include the timing of allocations, the lifetime of the permits, the variables and data on which to base decisions, the choice of base year, and the categorization of plants that have been built since or were closed (part of) the base year, among others. One of the vital questions, for instance, is whether to have permanent allocations or to update them. Lobbyists from the affected firms and politicians from the concerned towns or states find all kinds of arguments in favor of one or the other principle or additional allocation rule. Presumably, the decisionmakers settled on permanent allocation to avoid repeated opportunities for manipulation. (Allocations are for 30 years but are automatically updated, making them effectively quite permanent.)

Because this legislation had to pass a Congress that was fairly skeptical to environmental legislation, it met with many pitfalls. For the proponents of the law, it was vital to secure support wherever it could be secured, even if it meant compromising somewhat on permit allocation rules. One might think that the political allocation process would severely skew the result. However, although the high-sulfur-producing states were able to impose considerable (abatement and environmental) costs on the rest of the country in the regulations of the 1970 Clean Air Act and its 1977 amendments, they generally were not successful in getting a disproportionate share of allowances in the more market-based revision of 1990 (Joskow and Schmalensee 1998).

In one positive summary of this policy experience, U.S. sulfur trading is portrayed as an efficient instrument along the lines of those illustrated in economics textbooks: the "environmental goals were achieved on time, without extensive litigation and at costs lower than projected" (Joskow and Schmalensee 1998). One point essential to this efficiency is ensuring that the permit market really develops; in this case, U.S. EPA held annual auctions for 2.8% of the permits that had been withheld (that would otherwise have been allocated to individual units). This sale was revenue-neutral, because the proceeds of the auctions were handed over to the plants that would otherwise have received these permits. The auctions helped start the process of price discovery by the participants in the market. They also helped allay the fear that the initial permit recipients would "hold on to them" and that there would thus be insufficient trading to satisfy the needs of new sources.

The sulfur-trading program is a big step ahead in mechanism design. Future programs may develop even more efficient mechanisms and will have to adapt to the characteristics of individual pollutants. It is difficult to say whether taxes are preferable to permits. As discussed in Chapter 14, taxes (and auctioned permits) have output and revenue-recycling effects that free permits do not have. Presumably, the biggest difference is in political acceptability, and the best-suited instrument may depend on the pollutant and the political setting. For the extent of sulfur reduction, the most crucial factor is probably not the choice of policy instrument but the effective tax level (or number of permits and thus permit price). Looking back at the unexpectedly low permit price, some observers consider that the number of permits was too high and the goal of 50% sulfur reduction much too modest (see Table 24-1). It may partly be a reflection of U.S. priorities; acid rain is not as high a priority as it is in Scandinavia. However, for the purposes of this discussion, it would be interesting if something in the mechanism choice biased the outcome. Grandfathering of permits is a fairly polluter-friendly policy. Other policies, such as taxation, probably would not have had a realistic chance of being implemented at this level of effective control. However, it is worth considering what a step-shaped function (such as that suggested in Requate and Unold 2001) could have achieved. For sulfur emissions, such a mechanism might have resulted in a smaller (free) allocation of permits combined with a menu of call options (or auctions) for additional permits (see Chapter 13).

# Reducing NO$_x$ Emissions from Combustion[5]

Nitrogen oxides (NO$_x$) are the other major precursor of acid rain. Its chemical effects are not the same as those of SO$_2$, but they are perhaps more comparable for these two pollutants than for many other pairs. Several technical differences are relevant for their abatement costs, monitoring, and the politics of instrument choice. In general, NO$_x$ emissions reduction has been less successful than that of SO$_2$, largely because of inherent technical difficulties. Most of the NO$_x$ is formed by the effect of high temperature on atmospheric nitrogen. Unlike sulfur, NO$_x$ does not come from an impurity in the fuel. As a result, NO$_x$ emissions cannot be easily predicted, and complicated, expensive monitoring equipment is required.

## Refunded Emissions Payments in Sweden

For geological reasons, Scandinavia is especially vulnerable to acidification; most of the area is on granite bedrock. Consequently, nitrogen oxides and sulfur oxides have been important environmental policy targets in Sweden and Norway. Even so, the $NO_x$ emissions reduction in Sweden between 1980 and 1995 was approximately 20%, whereas emissions of $SO_2$ were reduced by 80%. Part of this difference related to the complexity of $NO_x$ formation and monitoring, which has made it impossible to impose a broad-based tax on all $NO_x$. Instead, the Swedish Parliament decided in 1990 to introduce a charge of 40 SKr per kilogram of $NO_x$ emitted from all combustion plants that produce at least 50 gigawatt-hours (gWh) of useful energy per year. The charge came into effect in January 1992 and affected about 200 plants. The level of the $NO_x$ charge (roughly US$4,000/ton) was determined by a Swedish Environmental Protection Agency (SEPA) study in 1987.[6]

The design of the charge is unique. On the one hand, it is high—more than 200 times higher than the French charge and much higher than U.S. permit prices. On the other hand, its proceeds are returned to the polluting companies, relative to the amount of energy produced by the specific plant. This refunded emissions payment (REP) scheme (see Chapter 9) means that the polluting industry as a whole does not pay anything to society—presumably, this fact has made the charge politically feasible. The design mechanism was chosen partly because only large combustion plants are obliged to pay the charge. This decision was based in part on the high costs of metering, which (together with abatement costs) were considered unreasonable for smaller plants. If a tax were applied to only a subsection of an industry, then that subsection would be unfairly disadvantaged compared with other firms in the same industry. If the tax were applied only to the large plants, then companies such as Gothenburg Energy would have an incentive to set up several small combustion plants instead of one large one, and this response typically is not desirable (in any respect, including the emissions of $NO_x$ and other pollutants).

As the system has developed and because it appears to be effective, costs for abatement and metering have decreased, and the criterion for inclusion has been lowered twice. In 1996, plants producing at least 40 gWh of useful energy per year were included, and in 1997, the level was lowered to 25 gWh. In 1998, 400 plants were subject to the charge, compared with 200 plants in 1992.

## Tax Levels, Use of Tax Revenues, and Administrative Costs

In a REP scheme, the cleaner-than-average firms make a net gain (which may or may not be larger than abatement costs), whereas the rest have to make a net payment. Total revenue from the Swedish $NO_x$ charge (which is refunded to the plants) was the equivalent of between US$50 million and US$100 million between 1992 and 1998 (SEPA 1997; Naturvårdsverket 2002).

The charge is administered by SEPA, and administration costs have been kept low—to approximately 0.3% of revenues collected. Metering costs are estimated to be approximately 3% of total charges paid. $NO_x$ emissions are measured hourly by sophisticated SEPA-approved equipment. If a firm cannot measure its emis-

**Table 24-3.** *Summary Statistics Regarding Combustion Plants Subject to $NO_x$ Charge in Sweden, 1992–1998*

| Year | No. of plants | Energy produced (gWh) | $NO_x$ emissions (tons) | Emissions coefficient |
|------|---------------|----------------------|-------------------------|-----------------------|
| 1992 | 181 | 37,465 | 15,305 | 0.41 |
| 1993 | 189 | 41,158 | 13,333 | 0.32 |
| 1994 | 202 | 45,193 | 13,025 | 0.29 |
| 1995 | 210 | 46,627 | 12,517 | 0.27 |
| 1996 | 274 | 57,150 | 16,083 | 0.28 |
| 1997 | 371 | 54,911 | 15,107 | 0.28 |
| 1998 | 374 | 56,367 | 14,617 | 0.26 |

*Note: Emissions coefficient* is measured in kilograms of $NO_x$ per megawatt-hour of energy produced.

*Source:* SEPA 2000.

sions temporarily due to technical problems, a high emissions factor is assumed for that period; therefore, most firms have a very high rate of compliance. All plants are obliged to submit a form with information regarding their production and $NO_x$ emitted. SEPA audits the firms and randomly selects several firms each year for detailed review. The goal is to review each firm on average every fifth year. At the beginning of each calendar year, each plant submits a declaration of its emissions and energy produced for the previous year. Charges are due before October 1, and SEPA refunds the plants before December 1 of the same year. Each firm can have several production units monitored separately.

### Environmental Effectiveness, Efficiency, and Links to Other Instruments

The $NO_x$ charge is generally thought to have been rather successful. Firms adapted quickly to this economic policy instrument, and emissions reductions started immediately (see Table 24-3). The positive environmental effect was biggest in the beginning, but improvements continue, although at a decreasing rate. These reductions are somewhat obscured by the fact that new, smaller plants (producing between 25 and 50 gWh/year) with fairly high emissions factors are being brought into the system, whereas the average emissions for the original (larger) plants had fallen to 0.26 and 0.25 for 1997 and 1998, respectively.

In addition to the $NO_x$ charge, many plants are subject to local regulations. Hence, observers argue that the reduction in $NO_x$ emissions is due to both the charge and local standards. According to one study, approximately two-thirds of the reduction can be attributed to the charge and the rest to the local regulations (SEPA 2000). The method for arriving at this estimate is shaky, however, and almost all the plants seem to be operating below their legal requirements. As a result, the charge alone might have given almost the same effect without the standards, which makes it difficult to allocate shares between the two instruments.

The dual advantage of a REP scheme instead of an "ordinary" tax is that the incentive for reducing emissions is strong while the liable firms are not subjected to a heavy economic burden. The fact that the net payment for the companies is smaller than a tax (and 0 on average) has made this instrument acceptable to firms. In the United States, firms have considered grandfathered tradable permits preferable to taxes because they place a smaller burden on firms and are more neutral regarding competitiveness. The Swedish $NO_x$ example shows that it is possible to create a financially neutral tax as an alternative to tradable permits.

One potential drawback with refunding the charge is that the instrument thereby does not satisfy the more stringent interpretation of the polluter pays principle and therefore does not give an optimal resource allocation (i.e., there is no output or revenue-recycling effect). An ordinary environmental tax (or permit scheme) would increase total and average production cost, hence raising the market price of the final good. The changed price due to internalizing the environmental damage in the production cost gives signals to the economy that would, in the long run, induce structural changes. In an economic market model, these price responses are part of the optimal adjustment. They are not as strong with a refundable charge; the product price will rise because of abatement costs but not because of tax payments. From a general equilibrium viewpoint, this limited price response is unfortunate. However, for many participants in the economy, it may be seen as an advantage: production, employment, and so forth are not cut back; prices do not rise (as much as with a tax). In fact, $NO_x$ might be reduced not by decreasing overall consumption and production of energy but only through technological abatement. In such a case, the refunded charge would be the ideal mechanism.

Furthermore, refunding makes it feasible to set the charge at a relatively high level. A tax of the same magnitude would meet enormous resistance from the entire polluting industry, whereas the refunded charge is a benefit to the cleaner companies. It has the effect of splitting the industry lobby. Finally, the effect on prices depends on what it is measured against. Compared with doing nothing, price increases due to the cost of abatement. It is still a small increase that reflects abatement costs—roughly equivalent to the increase if (optimal) regulations had been used to achieve the same reduction. Compared with the use of taxes of the same magnitude, product price increases less.

### Policies in France, Spain, Italy, and the United States

In 1985, the French government started to apply the polluter pays principle to tax $SO_2$ emissions. This law was renewed and expanded several times to incorporate several other air pollutants, including $NO_x$. In 1998, the rates were increased to the equivalent of US\$40/ton of $NO_x$ or volatile organic compounds (VOCs) (and US\$25/ton of $SO_x$ or HCl). One assessment of the tax estimates that $NO_x$ emissions have been reduced by 6% since the mid-1990s (ADEME 1998). This reduction can hardly be due to the direct incentive effects of the tax, because the tax is low and based on estimated emissions rather than actual emissions. However, the proceeds of the tax are used partly to give grants for abatement technology and research, and the latter are assumed to have led to emissions reductions.

Italy has a tax on $NO_x$ of around US\$100/ton, which is higher than that of Spain and France but not Sweden. The tax applies to only large combustion plants (>50 megawatts/year) and to emissions that exceed the standards specified in the E.U. Large Combustion Plant Directive. It seems to suggest that it is applied only in special cases as a form of penalty. Similarly, the autonomous region of Galicia, Spain, taxes $NO_x$ at the equivalent of about US$30/ton (Bokobo Moiche 2000). No evaluations appear to be available yet.

In the United States, $NO_x$ is considered primarily not as a problem related to acid rain but as a precursor to ground-level ozone. Because the severity of environmental effects varies considerably, $NO_x$ is subject to several distinct abatement programs that differ from state to state. Title IV of the 1990 CAAA specifies a two-part strategy to reduce $NO_x$ emissions from coal-fired power plants: reduce annual $NO_x$ emissions in the United States by 400,000 tons per year between 1996 and 1999 (Phase 1) and by approximately 1.2 million tons per year beginning in 2000 (Phase 2). These reductions correspond to less than 5% of total emissions and, according to U.S. EPA, translate to abatement costs of a few hundred dollars per ton. Essentially a traditional regulatory program, the program does contain some provisions for optional trading. It allows utilities to "average" emissions over separate units but not to bank or trade with other utilities.

A couple of other U.S. programs also have $NO_x$ targets. One is RECLAIM (see Chapter 7), an ambitious program that aims for an 80% reduction of $NO_x$ (and $SO_x$) from 1994 to 2003 in the south coast region of California. It is not completely a cap-and-trade program but rather a conventional "bubble" trading program. Each licensed source has to reduce its overall emissions (from a whole facility), and if a firm overcomplies, it can trade the excess permits.

A regional U.S. emissions trading program known as $NO_x$ SIP Call targets only summertime $NO_x$ emissions in several eastern states. Its structure is more in line with the cap-and-trade sulfur emissions program, and its projected reductions are significant. In the SIP Call region (which, at the time of this writing, is scheduled to take effect in 2004 in 19 eastern states and the District of Columbia), the program would lead to annual reductions of 34%, from projected baseline levels of 3.51 million tons to a target of 2.33 million tons in 2007. In the summer season, the program is expected to reduce emissions by 62%, from 1.5 million tons to 0.56 million tons (U.S. EPA 1998, table 2-1).[7]

The allocation mechanisms, which differ from state to state, have been the subject of much debate. Many states use a form of output allocation instead of the usual grandfathering or fairly frequent updating of allocations. This makes this so-called Ozone Transport Commission (OTC) program more similar to the Swedish REP policy for $NO_x$ (and to the U.S. EPA Lead in Gasoline Phaseout program).

Table 24-4 summarizes the main differences between the instruments used in these four countries plus Sweden (in addition to local standards used in each country).[8] Perhaps the most striking difference is the level of the charge itself, which is highest in Sweden. It presumably was only politically possible because of the special construction of the policy with refunding. It is somewhat difficult to compare the results analytically because Sweden has a much higher degree of detailed monitoring, as is expected with such a high charge level.

A high charge is probably the only really effective way of achieving a sizeable reduction. In five or six years, the Swedish system verified the reduction of more than 30%, starting from a level that was low already because the country had been pursuing an aggressive policy on the precursors of acid rain for many years.

Table 24-4. *Comparison of Policy Instruments for $NO_x$ Phaseout*

| Criterion | Sweden | France | United States | Italy | Galicia, Spain |
|---|---|---|---|---|---|
| Charge level or permit price (US$/ton) | 5,000 | 40 | 200–500 | 100 | 30 |
| Pollutants covered | $NO_x$ only (taxes for $SO_x$) | $NO_x$, $SO_x$, HCl, VOCs | $NO_x$ | $NO_x$ | $NO_x$, $SO_x$ |
| Instrument | Refunded charge | Charge | Tradable permits | Fee for excess pollution | Tax |
| Use of funds or allocation of permits | Returned to companies | Abatement, research, etc. | To companies | | General (5% for environmental restoration) |

*Notes:* $NO_x$ = nitrogen oxides; $SO_x$ = sulfur oxides; HCl = hydrochloric acid; VOCs = volatile organic compounds.

Lower tax levels would have less of an effect, but using the revenues to subsidize abatement equipment (as in the French system) should also reduce emissions. However, in Sweden, continuous monitoring indicated that the mere installation of equipment generally did not have as great an effect on emissions as it should according to engineering projections. To realize the potential emissions reductions, companies had to experiment with trimming and operational adjustments, which can be done only if there is continuous-time monitoring and if plant operators can strive for incentives (such as permits or charges based on actual measured emissions).

## Green Tax Reform in Sweden and Germany

Green tax reform has become a popular issue in Europe.[9] The idea is to use environmental or resource taxes that are needed anyway to finance a reduction in other (more distortive) taxes. Although this idea has merit in some cases, it has been greatly oversold and very heavily criticized (see Chapter 14 on double dividends). Although environmental taxes are powerful instruments for solving environmental issues, they do not necessarily generate funds that are large enough to have a significant impact on the overall budget or on other taxes.

One area in which significant revenue might be generated is in the taxation of energy carriers or their associated environmental problems. Two countries that have taken some modest steps that could be referred to as "green tax reform" are Sweden and Germany, which have tried to introduce energy taxation as part of a more general tax reform. Much remains to be done, because the tax systems are far from satisfactory with respect to the environmental parameters involved. Many environmental aspects are complicated, and the exact environmental damage depends on many local details (see Chapter 4). Other problems are global (or regional), and thus national taxes are the wrong level of action. Reform meets considerable resistance in the form of legal issues and special interests of energy suppliers, customers, and trade (and other) organizations.

## Sweden

During the past half-century, Sweden has been one of the most ambitious countries in financing welfare through taxes. Income taxes had a fair amount of progressivity, which meant that the tax rate was much higher for higher incomes. Because the law set tax rates with respect to nominal income levels, the high inflation of the 1970s created a situation in which the marginal tax rates for ordinary income earners reached levels that had originally been intended for the very highest incomes (>80%). With the liberalization of labor and financial markets in the 1980s, the system became untenable. The tax reform of 1991 resulted in a considerable reduction in income taxes and thus provided a political opportunity for the government to increase energy taxation because it was a reasonably popular way of limiting the budget shortfall.

In principle, this increase affected all energy consumers—households, commerce, and industry. However, several exceptions date back to a law on tax reduction from 1974, the main purpose of which was to ensure that Swedish producers would not be unduly disadvantaged compared with international competitors who faced lower energy taxes. The main thrust of the law applied to certain heavily energy-intensive industries and to the horticultural sector. Energy-intensive firms with energy taxes of more than 1.7% of the total product value (or even of a single product line) could apply to the government for a tax cap of 1.7%; the horticultural sector had special rules and had to pay only 15% of the (energy and carbon) tax levels. The number of industrial companies receiving tax concessions according to this rule varied between 100 and 150 between 1978 and 1990. Toward the end of this period, the value of tax reductions was roughly 700 million SKr (somewhat short of US$100 million). Although this amount did not in itself constitute a large fraction of total energy taxes, it did play a significant role in several respects.

The tax concession could, in some cases, perhaps be a decisive contribution for an individual firm. This tax concession was granted not by the tax authorities for a whole branch but by the government for a specific company, for one year at a time. This arrangement was not only impractical, anachronistic, and expensive but also subject to suspicions of nepotism and corruption. In an international context, such tax concessions could easily be considered illicit subsidies and used as an argument by foreign competitors urging sanctions, quotas, or tariffs. Foreseeing the need for harmonization with European Community legislation, the government appointed a Committee on Industrial Energy Taxation (SOU 1991).

To formally avoid subsidies for some companies (for the sake of trade agreements), the entire system of energy taxes was changed. The ironic effect is that, as a means of abolishing "exceptions" to the energy tax, the *whole* energy tax was lowered considerably for all of industry (but not for households). The main argument was the risk of competitive disadvantage for Swedish energy-intensive

**Table 24-5.** *Energy and* $CO_2$ *Taxes for Selected Fuels, 1990, 1991, and 1993*

| | 1990 | 1991 | 1993 | | |
| --- | --- | --- | --- | --- | --- |
| Fuel | Total | Total | Energy | Carbon | Total |
| Oil (m³) | 1,078 | 1,260 | 0/5–540ᵃ | 230/920 | 230/1,460 |
| Coal (tons) | 460 | 850 | 0/230 | 200/800 | 200/1,030 |
| Natural gas (1,000 m³) | 350 | 710 | 0/175 | 170/680 | 170/855 |
| LPG (tons) | 210 | 855 | 0/105 | 240/960 | 240/1,065 |
| Electricity (mWh) | 70/92 | 50/72 | 0/35–85ᵃ | 0 | 0/35–85 |

*Notes:* When two values are given, the first applies to industrial users and the second to all other users. Value-added and excise taxes on sulfur, nuclear power, and gasoline are additional. Peat, waste, and biofuels pay no carbon or energy tax. All taxes are in SKr (1 SKr ≈ US$0.10).

ᵃEnergy taxes for diesel are differentiated by environmental quality: 5 for Class I, 290 for class II, and 540 for class III (see Chapter 22). Electricity taxes are also further differentiated.

industries. Presumably, this case illustrates the considerable lobbying power of industry in this area.[10]

The final tax reform as of January 1993 led to a total exemption of all energy tax for industry and 75% of the carbon dioxide ($CO_2$) tax for all industrial (and horticultural) activity. This cut was financed by additional energy tax increases for private consumers, particularly in the carbon tax (which increased from 250 to 320 SKr/ton of $CO_2$) and the general tax on electricity (an additional 13 SKr per megawatt-hour). Thus industry was to pay 80 SKr/ton of $CO_2$, which is more than in most other countries but much less than in 1991. These changes resulted in much higher taxes for private consumers (see Table 24-5). In 1997, the Green Tax Report, yet another parliamentary investigation, heavily criticized the lack of logic in energy taxes and suggested a system of tax levels that would be environmentally motivated, but it has not been implemented (SOU 1997; see also Chapter 22 on the taxation of clean fuels).

At present, when consumers and producers in Sweden (as in many other countries) purchase the same fuel oil or diesel, they pay completely different prices. This situation leads to considerable incentives for tax evasion by (diesel) consumers and the need for special arrangements such as color additives to distinguish one from the other. Anecdotal evidence indicates that some users developed allergies to the green color additive, providing additional incentive (or perhaps excuses) for using the (much cheaper) fuel oil!

## Germany[11]

Because there is some resistance to E.U. environmental legislation (particularly from the "cohesion countries" of southern Europe), the more environmentally proactive countries make unilateral decisions. This behavior raises issues of competitivity, particularly in small open economies. It also was evident in the German government SPD/Grünen (Social Democrat/Green) coalition proposal for ecological tax reform of November 1998, which had a long, detailed list of industries suggested for exemption from the new tax (Deutscher Bundestag 1998, 5). This list included several kinds of glass, cement, steel, aluminum, fertilizer, and other energy-intensive industries. These industries are the most affected—but they are also the ones that need to be affected (both in their choice of technology and their contribution to the overall output mix of the economy) if the policy is to have the intended effect.[12] Exceptions of this kind go strongly against the grain of current thinking in public policy, which tries to promote transparency and homogeneity of rules. In the final law approved in 1999, the exceptions were not included. Instead, the proposed tax increases were reduced heavily for all industries, which in fact pay only 20% of the new ecological energy taxes borne by others (e.g., households). The result is somewhat similar to the reduction in Swedish energy taxes described above.

One of the most contentious areas for the German tax was electricity taxation. The general tax on electricity is now around US$10/mWh. Originally, this tax was going to be differentiated by type of electricity production, to give preference to ecologically sustainable means of producing electricity. This proposal ran into problems related to trade deregulation within the European Union. Legislation does not allow taxes on imported electricity to be higher than the lowest of the applicable taxes within the country (see Box 24-1). This law makes it almost impossible to differentiate electricity taxes by source of production.[13] Because the European electricity net is becoming both physically and legally fully integrated, it appears to make any national environmental taxation of energy difficult. Taxing inputs such as coal or oil would have been an alternative at the national level, but technically this means that the direct national tax on electricity is zero. As a result, imported electricity also has to be free of tax, even if it is produced in a polluting way. If this pollution is transboundary, the importing country may be affected by the pollution but unable to do anything about it.

Another interesting case is the KohlePfennig, a direct cross subsidy to the coal industry abolished in 1997. Prior to 1997, a dedicated electricity fee had subsidized domestic coal production. (In addition, regulations required German power and steel producers to burn a minimum share of German coal.) The coal industry fee was abolished because the German constitution does not allow this

## Box 24-1. E.U. Trade Rules and Domestic Environmental Taxes

Finland tried to impose an environmental tax on imported electricity at the average tax rate on domestic electricity. This practice was stopped by the E.U. Court of Justice (Court of Justice 1998):

> The first paragraph of Article 95 of the EC Treaty precludes an excise duty which forms part of a national system of taxation on sources of energy from being levied on electricity of domestic origin at rates which vary according to its method of production while being levied on imported electricity, whatever its method of production, at a flat rate which, although lower than the highest rate applicable to electricity of domestic origin, leads, if only in certain cases, to higher taxation being imposed on imported electricity.

This ruling makes it impossible to have special low environmental taxes for ecologically benign options such as solar and wind power.

kind of earmarking. Taxes are to be for general purposes, and the Bundestag must be able to assign money as it pleases.[14] Instead, direct, overt subsidies are now granted to the coal industry each year.

## Prohibition Compared with Other Policies: Trichloroethylene[15]

### Background

Chlorinated (or halogenated) hydrocarbons are good solvents for degreasing and dry cleaning; they are chemically active and dissolve fats (including those that protect nerve cells). These same properties have made use of these chemicals controversial since environmental problems and health hazards were discovered. Many chlorinated or halogenated hydrocarbon compounds are hazardous or toxic directly or indirectly (e.g., they may form extremely toxic dioxins when burned)—for example, the effect of chlorofluorocarbons (CFCs) on the ozone layer. In this section, I focus on prohibition as a policy instrument in Sweden for one of these solvents, trichloroethylene (TCE), and briefly compare it to the policies followed in two other countries: in Norway, TCE use is taxed; and in Germany, TCE use is strictly regulated.

Degreasing is common in the metals industry. As soon as an object is processed, surfaces must be greased to prevent corrosion. The grease must then be removed for later steps such as welding, lacquering, assembly, and delivery. Health and environmental issues related to these chemicals are complicated, and priorities change with knowledge. This changing nature poses challenges for the design of policy instruments. For some time before damage to the ozone layer was discovered, several ozone-depleting substances (ODSs) such as the CFCs were introduced as substitutes for other kinds of chlorinated solvents because they were less hazardous to human health. After the phaseout of CFCs to protect the ozone layer, there was some substitution back to TCE, perchloroethylene (PER), and methylene chloride. Other alternatives to chlorinated solvents in metal degreasing are water–alkaline processes and low-aromatic mineral oils.

### Sweden

PER is chemically close to TCE and is used mainly in dry cleaning. The only regulation of PER in Sweden is a tough maximum exposure limit in the working environment of 10 parts per million. Although PER is not banned, its use has decreased by 85% from 1988 to 1995 mainly because of a transition in dry cleaning from open to closed systems. However, some open systems still exist, and, according to SEPA (1997), many are poorly maintained and have high emissions. Only 20% of the open machines have carbon filters, and significant low-cost reductions could be achieved (23–47 SKr/kilogram). Because many dry cleaners are located in apartment buildings, the risk of unintended exposure is relatively high. Paradoxically, the use of TCE is prohibited, but not PER.

In 1991, the Swedish Parliament passed a law prohibiting the professional use of TCE and methylene chloride, effective Jan. 1, 1996. The use of TCE in con-

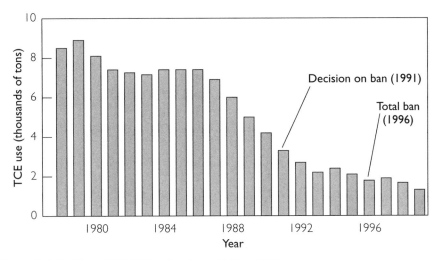

**Figure 24-2.** *Use of TCE in Sweden, 1978–1999*

*Source:* Naturvårdsverket (the Swedish Environmental Protection Agency).

sumer products was banned starting in 1993. A complete prohibition on all use appears to be a strong policy instrument but has not been wholly effective. The reason for this appears to be that the strength of the instrument is in some sense its weakness: the ban creates a strong opposition among users who find it particularly difficult to replace TCE, or simply disapprove of the timing or policy method. The ban also has been controversial because the evidence on the toxicity or environmental hazards of TCE is less clear-cut than for some other solvents. TCE was sometimes preferred because it was considered less hazardous than other alternatives, such as CFCs and highly aromatic petroleum products (e.g., in the early 1980s, the SKF factory in Gothenburg used 3,000–4,000 tons of TCE per year—half of Sweden's total consumption—as a replacement for CFCs and highly aromatic petroleum products).

Figure 24-2 shows the decline in TCE use in Sweden up to 1999, three years after its prohibition. By the time the ban was decided, consumption already had been reduced from almost 9,000 to around 3,000 tons/year. This reduction may partly have been an effect of the ban, because industry could have observed the process of legislation and decreased TCE use in anticipation of its prohibition. On the other hand, several other countries experienced similar decreases in use. During the late 1980s, large industries were faced with tight emissions standards, which led them to replace technologies and solvents, or to adopt closed systems and carbon filters.

SKF is an interesting case study not only because of the company's size and significant use of solvents but also because its precision demands are unusually exacting. When removing water droplets from ball bearing components, SKF requires the elimination of all water within 30 seconds; after that time, corrosion becomes unacceptable. If SKF can eliminate the use of TCE, then most other industries should find it possible. Reduction in TCE use at SKF was driven by two factors:

the trade union demands for the working environment and SEPA requirements for reduced emissions. SKF was ordered to reduce emissions from 250 to 15 tons/year and to quickly install active carbon filters. The installation alone led to a two-thirds reduction, to about 80 tons/year. Additional reductions required process changes. A new degreasing process was installed that uses water and low-aromatic oils; in addition, new packaging and storage routines were combined with the use of lighter oils instead of wax for the conservation of ball bearings. Finally, only small amounts of TCE were used, so the costs of maintaining the handling, storage, and filter facilities became disproportionately high. Also, SKF saw potential benefits in environmental image by being proactive (instead of reacting passively to SEPA instructions) and in standardizing processes across facilities. The company therefore decided to phase out all use of TCE in all SKF plants, in Sweden and abroad, even if local authorities did not require it. This decision is one example of the international spillover of environmental standards.

The Swedish ban on TCE also led to bitter opposition and protest; petitions were written, and several companies fought the legislation in the courts. I summarize only some of the more salient points here. In 1994, 39 firms wrote an open letter to the prime minister, saying that the prohibition was poorly motivated and should be withdrawn. They went so far as to say that their competitivity was threatened and that they would have to move abroad if the prohibition were to be enforced. In later interviews, several company representatives admitted that the points had been exaggerated. Most of the companies eventually did phase out TCE use, and none of them moved abroad. However, the tone of the letter reflected their strong resentment.

In the first year (1996), some 500 companies were given waivers; however, after that, only companies that could show they were making serious efforts to substitute for TCE were to be granted exemption. Exempted firms also were charged an exemption fee of 150 SKr/kilogram, which was not only an environmental tax but intended to remove any unfair disadvantage that a complying company would suffer vis-à-vis its noncomplying competitors. Hundreds of companies continued to apply for exemption, and when rejected, they appealed the decision. The Stockholm court reversed the decision by SEPA, and several additional rounds of appeals progressed to higher courts.

The Netherlands, the United Kingdom, and the E.U. Commission also started to criticize the Swedish ban. Sweden had no domestic production of TCE (it was all imported), and the ban was thus a barrier to trade in this commodity. The case of one company, Toolex, was referred by a Swedish court to the European Court of Justice (case C-473/98) to determine whether the Swedish prohibition was in accordance with the free movement of goods. Meanwhile, the Chemical Inspectorate (Kemikalieinspektionen) modified its rules for exemption, dropping the requirement of a phaseout plan and removing the fee, because the E.U. Commission considered it "out of proportion" to the environmental damage. The only requirement left for exemption was that the company actively seek other alternatives and avoid harmful exposition. Several interesting principles of European law underlie this case. On July 11, 2000, the court ruled that the Swedish prohibition is not counter to E.U. legislation on the free movement of goods. The reasoning behind the ruling was as follows:

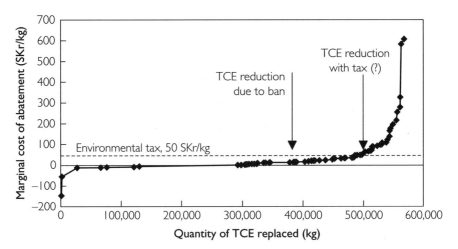

**Figure 24-3.** *Marginal Abatement Cost and Effects of Tax Compared with Ban on TCE in Sweden*

*Note:* SKr = Swedish krona (1 SKr ≈ US$0.10).

*Source:* Data from Slunge and Sterner 2001.

- The basis for the prohibition is concern for health and environment.
- The European Union has classified TCE as toxic and carcinogenic.
- Member countries have the right to stricter environmental legislation.[16]
- There is no basis to assume the prohibition was motivated by an attempt to stop trade.
- There are reasonable possibilities of obtaining a waiver.

Hence, one might conclude, with some exaggeration, that the prohibition was accepted because it was watered down by more generous standards for obtaining waivers.

To discuss alternative policy instruments, the structure of the abatement costs must be known. Based on applications for exemption as well as interviews, estimates for the entire TCE-using industry were compiled (Slunge and Sterner 2001), as illustrated in Figure 24-3. In most cases, TCE substitution is relatively cheap, but it may be expensive under special circumstances: if an individual firm lacks space (water-based equipment is typically larger), time, or finances (i.e., liquidity problems or difficulty borrowing money for this kind of investment). In addition, in small companies, all personnel may be "essential," and even one week lost on "side issues" such as degreasing can carry a significant cost that is hard to quantify but may be reflected in strong feelings about the ban. Requiring firms to comply quickly may also exacerbate costs considerably (see Chapter 12).

It is difficult to distinguish the firms that are bluffing from those that really face compliance difficulties. Companies that face tough environmental restrictions often protest that the requirements are too tough, but they sometimes find it easier than expected to make the change—and may even find that the new processes are cheaper (see Chapter 17 on the Porter hypothesis).

Is a ban a good policy instrument in this case? A traditional "*P*-vs.-*Q* reasoning" suggests that if the marginal cost of abatement is steep close to the zero limit (and the damage curve not so steep), then taxes (or other price-type instruments) would be better (Weitzman 1974). If technological progress is expected to be fast but unevenly distributed between specific application areas, then this preference for taxes would be strengthened by considerations of dynamic efficiency (see Chapters 12 and 13). When regulation is the instrument used, the company has a stronger incentive to overestimate abatement costs than when taxes are used (see Chapter 13).

Figure 24-3 shows that a large share of emissions have a low marginal cost of abatement. Some of the high estimates may be tactically exaggerated, but at least 90% of the abatement probably would have been achieved with an environmental charge of 50 SKr (about US$5) per kilogram. The reduction with the ban has yet to reach 90%. An increase to a higher charge would not have had much of an effect because the costs of abatement rise dramatically at the end.

### Norway

Since 2000, Norway has taxed TCE and PER at a rate of 50 Norwegian crowns or US$6/kg. It means raising the market price about five times and may be assumed to have a large effect, as suggested by my firm-level inquiries. Industry has not been enthusiastic but is still fairly complacent and appears to appreciate that the tax allows firms much more flexibility than a ban. This tax is so new that it is too early to evaluate its effects. However, data for 2000 indicate a drastic decline in TCE use of more than 80%. Even if there was some increase in 1999 (attributed to pretax hoarding), the fee still appears to have been effective.

The fee has partial deposit–refund characteristics in that half of it is refunded on delivery of TCE-containing sludge for proper treatment. This feature reduces net payments from the industry to the state, which reduces political resistance. It also specifically targets emissions rather than use, making closed systems attractive. Because the emissions cause health and environmental problems, this approach seems logical. The fee proceeds also could have been used to support research into alternative degreasing methods or even to provide loans on favorable terms for special cases. This alternative runs counter to the usual public economics arguments against earmarking, but these funds are both minor and transient. The interest for the treasury is bound to be minute, and the politically important aspect is to gain the support and understanding of—and preferably, partnership with—the industry concerned to avoid the kind of confrontation provoked by the Swedish ban.

### Germany versus Sweden and Norway

The Swedish TCE ban has not been fully effective. The legal battles were costly, and SEPA lost prestige vis-à-vis the complying companies that, in good faith, followed regulations only to find that their efforts were "unnecessary" and their

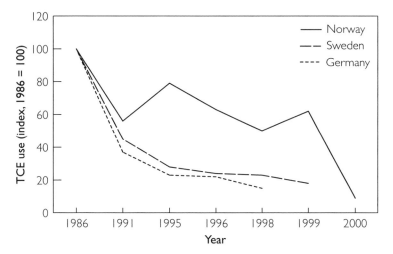

**Figure 24-4.** *Use of TCE in Norway, Sweden, and Germany*

competitors did not even have to pay environmental or compensatory fees. Prohibition should only be used in clear-cut cases (when health or environmental damage is specifically targeted or international opinion more coordinated).

Other policy alternatives worth considering are stricter emissions requirements, a deposit–refund scheme, and special environmental charges distinct from taxes. Germany has neither tax nor prohibition but tough technological requirements concerning emissions instead. This approach has led to a much more rapid decrease in TCE use than the European average, and even faster than in Sweden (see Figure 24-4).

SEPA probably would have been more successful with a fee, as in Norway, but perhaps applied over a somewhat broader range of chemicals. If the proceeds had been earmarked for environmental collaboration with the industry association, then the results might have been even more enhanced.

The conclusion must also be seen in the broader context of all chemical hazards. There is no way the legal or administrative systems could handle such complicated, expensive, and time-consuming conflicts concerning each individual chemical. Many thousands of industrial chemicals are as dangerous as (or more dangerous than) TCE. Traditionally, the regulation of chemicals has suffered from new source bias because the new chemicals have to pass complicated tests and administrative barriers to be registered, whereas older chemicals continue to be used by default. (Gasoline is one example; some of its ingredients, such as benzene, are extremely toxic and carcinogenic.)

In a current review of its chemical strategy, the European Union has started to discuss reform in the way chemicals are tested, which would mean that chemicals would be categorized by their level of hazard. For those that fall in the more dangerous (carcinogenic, mutagenic, or bioaccumulating) categories, the requirements for testing will be made much more stringent, and the responsibility will be placed squarely on the producers and importers.

## Liability and Superfund[17]

Another area involving the handling of hazardous materials is the dumping of highly toxic chemical wastes. One hazardous waste site that has become notorious is known as Love Canal, slightly upstream from Niagara Falls, in upstate New York. In the 1890s, the location was planned to be a bypass on the Niagara River; however, the canal was never finished, and the land eventually was sold for use as a municipal and chemical landfill. Between 1942 and 1953, companies deposited around 21,000 tons of toxic chemicals at the site.

The site was simply covered with dirt and clay in 1953 and sold to the city of Niagara Falls. In 1955, a school was built on the site to serve the surrounding neighborhoods; over time, houses also were built over the former chemical landfill. Eventually, the hidden pollution started to seep through: water became contaminated, and odors permeated the area. Studies indicate that numerous toxic chemicals had migrated into surrounding areas; the runoff from Love Canal drains into the Niagara River 2.8 miles upstream of the intake tunnels for Niagara Falls' water treatment plant, which serves about 77,000 people. Local residents suffered a statistically significant increase in pollution-related diseases and other medical complications (e.g., a 47% greater incidence of serious pregnancy-related problems than elsewhere).

In 1978, about 200 houses closest to the site were purchased by the state, and in 1979, another 700 residents were evacuated. Between 1977 and 1980, New York state and the federal government spent about US$45 million at the site: US$30 million on relocating residents and health testing, US$11 million on environmental studies, and US$4 million on a demonstration grant to build a leachate collection and treatment system. U.S. EPA eventually sued for US$125 million, and private lawsuits reached around US$14 billion. The area has been fenced off, the clay covered by an impermeable cap, and the school demolished. The Love Canal story attracted U.S. attention to the problem of "hidden" hazards from sites that had been used for waste dumping and led to the creation of the Superfund program.

The Comprehensive Environmental Response, Compensation, and Liability Act (CERCLA, commonly known as Superfund) was enacted by the U.S. Congress in December 1980. It authorized a tax on the petroleum and chemical industries, the funds to be used for the cleanup of hazardous substances that threaten public health or the environment. In the first five years, more than a billion dollars was collected to finance the cleanup of hazardous waste sites. The law was amended several times, and by 1995, US$11 billion had been collected. By the end of 2000, 757 Superfund sites were classified as "construction completion"—that is, all physical construction at the sites had been completed (U.S. EPA 2002c). More than 40,000 sites had been assessed, and more than 1,000 sites had final cleanup plans approved. U.S. EPA tries to recover the costs of cleanup from the responsible parties, and over the life of Superfund, U.S. EPA has reached settlements with polluters totaling more than $16 billion (as of 1999). To allow for this recovery, CERCLA stipulates strict and retroactive as well as joint and several liability (see also Chapter 10).

Tracing causality and responsibility in the case of health or ecosystem damages from hazardous waste is bound to be complicated. The technological processes and issues involved are complex, and the time periods are often long. The fact that the liability is strict and retroactive somewhat facilitates the possibility of suing for damages, although many of the companies responsible may no longer be in business. Joint and several liability means that any company involved in creating the damage can be sued for total damages. The company that owned a waste site may be gone, for instance, and the companies that transported its goods may be small, but it is still possible to sue a large chemical firm that used the site—for instance, one that has large assets or is sensitive about its reputation. The company sued then has the right to try to recover part of its costs from other responsible parties, but this situation is different from a damaged party having to recover costs from multiple parties.

Several lessons can be learned about policy instruments from Superfund and from other countries' experiences. First, illegal disposal of wastes ("midnight dumping") is not even a short-run solution. It may be done to avoid disposal costs, out of ignorance, or because current technology does not provide easy means of avoiding the creation or destruction of the wastes. Decisionmakers also may not realize how hazardous the wastes are or may be considered in the future.[18] This point is almost obvious when considering cases such as Love Canal in retrospect, but presumably, it also applies to decisions today, which may not always be as obvious. The most general lesson is perhaps the value of the precautionary principle and of not being too short-sighted. Which instruments to choose is a more difficult question to answer, but a few comments are tentatively appropriate.

- Regulation may be inadequate in hazardous waste policy, partly because of limited knowledge, which makes precaution important. Liability and the risk of lawsuits may (assuming functioning legal institutions) be potent policy tools because they partly internalize the risk implicit in incomplete knowledge. To do this, the liability must be strict and retroactive as well as joint and several. In countries with weak institutions, some agents (maybe including foreign firms) may strongly discount the risk of being apprehended. However, in the United States, litigation is common (see Box 24-2).
- Market instruments such as taxes, charges, and permits may be even more difficult to use than standards. Because hazardous pollution from chemical plants is often complex and poorly understood, society probably cannot correctly and sufficiently quantify the relevant parameters for these kinds of instruments before the effects become apparent. Furthermore, taxes are hard to change quickly. Responsible waste treatment and disposal must be economically reasonable for companies. There may be a role for small taxes or fees to create funds for research and treatment. (The initial funds for Superfund came from a tax on petroleum and chemical industries.)
- Ultimately, the public sector is responsible for providing public goods and thus some form of cleanup, particularly in cases of historic pollution when the polluter either is not held accountable or no longer exists. The public sector may try to avoid costs by requiring liability bonds or some form of deposits, or by trying to recover the costs ex post through a charge on an entire industry, as in the case of Superfund. The burden of costs has been shared differently in dif-

## Box 24-2. Suing for Health Effects

Thousands of cases of environmental litigation show the importance of environmental liability in the United States. In one case, the Pneumo Abex company recycled railroad equipment in a foundry in Portsmouth, Virginia (confidential personal communication, May 2, 1999). The metal contained lead, which was removed from the workplace to protect workers, but for decades, the company simply dumped the leaded filters on an unfenced field near the factory. The site overlooked low-income public housing built in the 1960s—without leaded paint, because the dangers of lead poisoning had already been well established. Lead-laced sludge eventually leaked from the field where children played.

More than a dozen of the children of the housing development had severe lead poisoning, and many more were affected to a lesser extent. Their families sued Pneumo Abex for negligence, alleging the lead caused disease, reduced their IQ, and thereby reduced future earnings. The company settled out of court, and although the amounts are not publicly available, it is not uncommon for a court to award as much as one or several hundred thousand dollars per victim for this type of suit.

The idea of suing firms seems to be spreading internationally. At the time of this writing, the British Geological Survey is being sued (in one of the largest suits ever) for failing to test for arsenic when contracted by the U.K. Overseas Development Administration to investigate the feasibility of drilling wells for drinking water in Bangladesh in 1992. This water is the cause of millions of cases of arsenic poisoning that now afflict the country's population (Clarke 2001; BGS 2001).

ferent countries. Canada's cleanup was unique in that the federal and provincial governments paid all the costs, whereas in Austria, VAs shared the costs between the state and responsible parties (Davis and Mazurek 1998). In the United States, Superfund cleanup has succeeded in cleaning up many sites but still has been criticized as deficient in environmental results and as benefiting mainly the lawyers who litigate the cases. However, joint and retroactive liability is bound to lead to fairly high transaction costs because of incentives to minimize one's share of the total costs.

## Information Provision and VAs on U.S. Toxic Emissions

In this section, I focus on two projects: one of information provision, and one of VAs. They are intimately connected because the VA to reduce chemicals builds directly on (mandatory) information provision. The Toxics Release Inventory (TRI) program was enacted by the U.S. Congress in 1986 as a part of the Environmental Protection and Community Right to Know Act (EPCRA, which is also related to Superfund). It is designed purely to provide information to the public about releases of toxic substances, most of which are not subject to standards or other regulation. U.S. EPA has modified the list of chemicals through a petition and review process; for instance, 286 chemicals were added in 1994, bringing the total to more than 640 individual chemicals.

The TRI states that firms with 10 or more full-time employees that also use 10,000 pounds or more of a listed chemical per year or import, process, or man-

**Table 24-6.** *Total Releases for All Firms in the Toxics Release Inventory*

| | Total releases (millions of pounds) | | | Reduction (%) |
|---|---|---|---|---|
| | 1988 | 1995 | 1998 | 1988–1998 |
| Number of facilities | 20,470 | 20,783 | 19,610 | −4.2 |
| Air emissions | 2,183 | 1,201 | 921 | −57.8 |
| Surface water | 165 | 37 | 45 | −72.9 |
| Underground injection | 162 | 143 | 115 | −29.3 |
| Total on-site releases | 2,968 | 1,688 | 1,427 | −51.9 |
| Total releases | 3,396 | 1,977 | 1,857 | −45.3 |

*Source:* U.S. EPA 2000b.

ufacture 25,000 pounds or more of a listed chemical per year must file an annual report on each of the chemicals in the plant. The reports focus on the quantity, type, and frequency of toxic releases as well as the medium into which the chemical is released. The information is available to the public and does not free firms of their usual obligations to report emissions to state and local authorities.

To complement and reinforce the TRI program, U.S. EPA initiated the 33/50 Program, a prime example of a VA (see Chapter 10). It set national goals for 17 prioritized toxic chemicals of a 33% reduction by 1992 and a 50% reduction by 1995 compared with 1988 levels. These reductions were to be achieved voluntarily, and compliance was to be measured by the TRI reports. The program emphasizes pollution prevention rather than end-of-pipe control. The initial invitation list of 555 companies with substantial chemical releases was subsequently expanded to include 5,000 companies. Some 1,300 corporations ultimately agreed to participate and collectively reduced their emissions by more than 50%, a total of 757 million pounds of pollutants, by 1994—one year ahead of schedule. The total releases of all the TRI firms were reduced by 42% in 1995 and 45% in 1998 (Table 24-6).

Several interesting conclusions can be drawn about the effectiveness of the TRI. The information disclosure itself appears to be the main driving force. The VAs may have speeded up the reduction somewhat. If the participating companies reduced emissions by 50% by 1994, then the nonparticipating companies must have had reductions just over 30%, but it is hard to separate the "effect" of the VA program because there is an obvious sample selection problem (i.e., the companies with the easiest abatement tended to be the ones that participated in the program). Another and maybe more worrying fact is that reductions seem to have leveled off after rapid decline for the first few years. This slowdown may indicate that this policy works best for low-cost abatement.

Another notable effect is the level of interest that nongovernmental and other organizations have developed for TRI data. Local environmental groups and media use it to pressure or report on local industries. Investors and ordinary citizens use it to plan the location of investments, including real estate. Organizations such as Scorecard (www.scorecard.org) facilitate these kinds of uses by processing data to make it more readily available in a user-relevant form. Scorecard provides detailed maps of the United States that can be zoomed in to the street level to give all the TRI data as well as other information (such as health effects) in a format that is relevant for local decisionmakers. Information disclosure greatly enhances market performance in the sense of incorporating or internalizing information on environmental effects (Tietenberg 1998). To be more specific, access to information operates product, labor, finance, and insurance markets as well as general neighborhood relations. As shown in Table 24-6, on-site emissions have decreased by more than 50%, and the emissions to surface water and air—particularly sensitive topics for public opinion—have been reduced more (73% and 58%, respectively).

Finally, information provision can serve as a good starting point for new regulation, perhaps in areas where pollution is particularly bad or the inhabitants for other reasons wish to push ahead to attain public health goals. California's Proposition 65, established by popular vote in November 1986, requires firms to notify all people who may be affected by TRI chemicals: in addition to informing U.S. EPA, the firm must actively inform workers, customers, and neighbors. Proposition 65 also gave these groups greater power to sue the companies if they were not notified.

## Global Policymaking: Protecting the Ozone Layer

International response to scientific evidence that the stratospheric ozone layer is being depleted by the emissions of anthropogenic ozone-depleting chemicals has been fairly rapid. It is a good illustration of the process of creating rights to the environment by international agreement (described in Chapters 5 and 10). First, in 1985, several nations agreed to form the Vienna Convention for the Protection of the Ozone Layer. The Montreal Protocol on Substances that Deplete the Ozone Layer, which builds on the Vienna Convention, was agreed upon later that year and proved to be far tougher than many observers would have expected only a few months earlier. It came into force, on time, on Jan. 1, 1989. By then, 29 countries and the European Economic Community—representing approximately 82% of world consumption—had ratified it. More than 170 countries (including about 130 developing nations) are now parties to the convention and the protocol.

The negotiations were diplomatically tricky because the distribution of gains and costs is uneven despite the global nature of the problem. Stratospheric ozone (as opposed to tropospheric or ground-level ozone) plays an important positive role by absorbing most of the biologically damaging ultraviolet sunlight (UV-B), stopping it from reaching the Earth's surface. Many experimental studies of plants and animals have shown the harmful effects to humans of excessive exposure to ultraviolet radiation. Ozone depletion in the stratosphere leads to more ultraviolet radiation.[19] The effect is unevenly distributed: hardly any effect at the equator and large effects towards the poles (areas in which mainly high-income people live). The effect of moderate radiation will primarily be an increase in skin cancer for people with fair skin and is thus a medical problem mainly for rich, fair people who live long enough to worry about cancer. For people with darker skin in countries where large percentages of children still die of diarrhea, health issues such as clean water and refrigeration (to keep food and medicines fresh) are much more important. If the benefits are unevenly distributed to the advantage of the rich, it is unfortunately the opposite with some of the costs, because the poor people are forced to pay more for refrigeration and air conditioning. Thus, in essence, the treaty asks poor people to pay more for a benefit that would largely accrue to richer countries.

The agreement had to take into account these differences between countries, and the 1990 London Amendment to the protocol contains clauses that allow the developing countries certain delays and create mechanisms for the transfer of technology and funds (the Global Environment Facility) to cover the special circumstances of several groups of countries, especially developing countries. The protocol is also flexible; it allows for adjustments in case the scientific evidence changes, without having to completely renegotiate the agreement. Instead, new amendments are added, for instance, relating to new chemicals with a weaker ozone-depleting effect.

After the Montreal Protocol, most countries adopted individual plans for the phaseout of ODSs. The main policy in the major producing countries such as the United States was a fairly brisk progressive prohibition of both production and consumption of ODSs. According to Section 604 of Title VI in the U.S. Clean Air Act, "it shall be unlawful for any person to produce any class I substance in an annual quantity greater than the relevant percentage specified ...." The text goes on to give specific percentages for each year and chemical (e.g., for carbon tetrachloride, 100% in 1991, 90% in 1992, 80% in 1993, 70% in 1994, 15% in 1995–1999, and 0% thereafter). The most prominent ozone-depleting chlorinated solvents were regulated in turn, depending on their ozone-depleting potential and areas of use. To create some flexibility and to facilitate the phaseout, permits could be traded but not banked. Because permit allocation was based on grandfathering and provided an advantage to some firms, policymakers imposed taxes as a complement to limit windfall profits (see Tietenberg 1995). These taxes were fairly important and collected almost US$3 billion in the first five years. They appear to have been an important factor in decreasing the use of ODSs.[20]

The main policy instruments used in ODS-importing countries such as Sweden were import restrictions. Between 1988 and 1994, the use of ODSs in Sweden decreased by 93%; thus, the Swedish strategy was fairly successful—also fairly easy, because there was no domestic production.

## Global Climate Change: Domestic Policies and New Technology[21]

One of the largest international environmental issues at present is global climate change. Because the effects of the targeted pollutants are truly global, policymaking must be international, and one important aspect is the structure of international negotiations. (The theory behind these negotiations is beyond the focus of this book but is mentioned briefly in Chapter 17.)

In practice, the history of international climate negotiations is long and complex. In 1992, the United Nations Framework Convention on Climate Change

(UNFCCC) established broad goals for climate protection and sustainable development and created the milieu for international negotiation on more specific goals and measures. In December 1997, the Kyoto Protocol (to the UNFCCC) was negotiated. Through the Kyoto Protocol, the so-called Annex 1 countries (mainly industrialized countries, members of the OECD, plus the transitional economies of central and eastern Europe and the former Soviet Union) agreed to various percentage reductions (averaging around 5%) in national emissions relative to a 1990 baseline by 2008–2012. In practice, because of ongoing growth in energy use from 1990 to the 2008–2012 commitment period, the required reductions are much larger than 5% (for the United States, they are estimated to be more than one-third).

Climate negotiations were deadlocked after the sixth Conference of the Parties to the UNFCCC, held in November 2000 in the Hague, the Netherlands. In early 2001, the United States rejected the Kyoto Protocol. Experience thus far illustrates how difficult it is for the world community to take even small steps toward reducing greenhouse gas emissions. Especially telling is that the Kyoto Protocol is itself only a small step compared with what would be necessary in the long run to stabilize atmospheric concentrations of greenhouse gases (i.e., reduce net emissions by more than half). The reasons for the difficulty are that climate change is uncertain and complicated (ecologically and technologically), abatement (and damages) will be costly, and the distribution of the cost burden is a contentious issue.

International agreements require policy instruments for their implementation. The Kyoto Protocol is an innovative environmental agreement from an economic perspective in its explicit incorporation of various forms of international emissions trading, flexibility in the gases to be controlled, and the use of carbon-absorbing "sinks" such as forests. Nevertheless, one element of discussion (and disagreement) before and after the negotiations has concerned the selection and design of policy instruments at the national and international levels; the main incentive-based alternatives are taxes and permits.[22] Other contentious points have included how to address sinks within the protocol (i.e., how to quantify carbon uptake and whether to allocate corresponding permits) as well as the more general issues of permit allocation and obligations between countries. In July 2001, a compromise was reached concerning implementation of the Kyoto Protocol. Almost 180 countries signed, with the notable exception of the United States, which voiced strong reservations about both the targets and the policy mechanisms.

Within the framework of the Kyoto Protocol, negotiations to date are based on percentage reductions from 1990 emissions levels by industrialized countries; this approach is analogous to the use of grandfathering at the national level. "Prior appropriation" is applied as the main concept of property rights in this particular global environment area: large emissions in the past are taken to create rights to large emissions in the future. Many other allocation principles could be used, for example, per capita entitlements or a composite of formulas that account for factors such as climate (extra credits to cold countries), economic level (extra credits to developing countries), population (extra credits to sparsely populated countries), or energy resources (extra credits to countries without hydropower). Unfortunately, criticisms of allocation approaches are often formulated as attacks on the idea of permit trading as such, whereas the larger issue of contention is the allocation of permits. Without trading, there is no chance of even approximately equalizing marginal costs of abatement, and numerous opportunities for efficient trades would be forgone.

In fact, the distribution of current and future burdens appears to be the main underlying problem. The United States has expressed concern that China, India, and other major developing countries have no obligations under the Kyoto Protocol. Other critics consider that countries of the former Soviet Union (mainly the Russian Federation and Ukraine) are unduly favored because they have received generous permit allocations (often referred to as "hot air") compared with current emissions because a large share of their heavy industry has collapsed since 1990, the baseline year for the emissions. For their part, developing countries have made clear their opposition to the prior appropriation concept implicit in the Kyoto Protocol emissions targets for industrialized countries and have argued for per capita allocations.

Permits are not the only possible mechanism. An international tax, in principle, is another alternative. Such a tax would have the advantage of directly raising the international price for fossil fuels, creating predictable and equal incentives for abatement everywhere. However, it also would create tremendously large rents, which would make resistance to the tax quite high. (These rents exist with quotas or permits, too, but they are not made immediately visible.) Countries are not willing to hand over such a large amount of income—or the political power that goes with such taxation—to any international or supranational body. In the United States, there would be resistance to the tax per se, and in Europe, it would be difficult for governments to displace the existing national energy or carbon taxes. Taxes might be automatically refunded to various states by some mechanism, but this option will lead to long negotiations concerning the refunding mechanism.[23] One also could conceive of, in principle, internationally coordinated taxes that are levied nationally. They would circumvent the problems related to the international transfer of rents.

Several arguments favor a price-type instrument. The accumulation of warming gases in the atmosphere is best described as a stock pollutant (see Chapter 4), and in the absence of strong nonlinearities or threshold effects, the environmental damage function would be expected to be fairly flat.[24] In this case, an application of the Weitzman result would favor a tax-type instrument (see Chapter 13). Considering the vagaries of inflation and exchange rates, together with the fact that the tax would have to be updated frequently and thus passed through national assemblies in all the countries of the world, such an instrument would not necessarily be easier to manage than quotas.

Therefore, assume that negotiation on the basis of quantitative national emissions targets continues. These targets could, like the Kyoto Protocol targets, be percentage reductions from historic levels, per capita levels, or some formula that combines various principles. Such an international treaty would have definite consequences for the choice of the national policy instruments for fulfilling the agreed quotas. These national policies would not necessarily have to be the same. In principle, the same choice between tax- or quantity-based instruments exists at the national level. However, if the goal of national policy is the fulfillment of a certain quantitative target that has been negotiated, then quantitative instruments will also have an advantage at the national level. With a tax instrument, the state has to estimate and adjust the taxes in such a way as to fulfill a given target. Admittedly, it can presumably buy or sell excess international credits (or use other flexibility mechanisms, such as banking) if the target is missed. However, if a country uses permits domestically, it will automatically fulfill its obligations at the same time as it allows local agents to seek the cheapest combination between abatement and purchase of permits—not only locally but on the global market.

In countries that already have high levels of energy taxation, these taxes probably would not be relinquished. Instead, various combinations of instruments are likely (e.g., existing taxes combined with permits; new hybrid instruments, such as increased but partially refunded taxes; or permits that require partial payment—one share of permits might be grandfathered and another share auctioned). Permit systems also give rise to tremendous rents, but the allocation of these rents depends critically on the permit allocation. Given the magnitude of the task and the size of the rents involved with tax and permit approaches, there is considerable scope for instrument development to try to meet various goals concerning efficiency, equity, and political feasibility.

U.S. experience shows how efficiency and cost savings really hinge on using trading mechanisms. Some parties, particularly Europeans, have wanted to limit trading and require a certain percentage of domestic abatement. This position may be partly due to a lack of understanding of the benefits of trading. It may also, to some extent, reflect a feeling that the size of the U.S. quota share is unfair. If this were the case, then it would be better to argue for a larger U.S. reduction than to argue against the trading per se. It is also noteworthy that in late 2001, the European Commission announced a major step forward in creating the architecture for trading among large (point-source) emitters throughout the European Union.

Given the difficulties incurred in taking the first step toward global agreement, one needs to consider the whole issue afresh. Because climate change is a result of stock pollutants that accumulate fairly slowly, there is still time to take action. Several studies have shown that long-term atmospheric targets for greenhouse gases can be met in many ways over time. More gradual approaches appear to offer substantial cost advantages in terms of slowing the obsolescence of capital and allowing for investment in ever-improving new technologies. On the other hand, more aggressive action early on will stimulate technological development, and it may be important to reserve the option for more stringent long-term greenhouse gas control as information about the potential risks is refined (for views on these issues, see Toman, Morgenstern, and Anderson 1999 and Azar and Schneider 2001).

The current round of Kyoto Protocol negotiations concerns the first commitment period (2008–2012), but they should be looking forward to the second and third commitment periods. To be cynical, one might say that the exact emissions of any particular country in 2008 or 2009 are not crucial. What is crucial is to build partnerships, mechanisms, institutions, and trust for future periods. All countries need to gain experience and understanding to face the challenges of more stringent emissions reductions in the future.

One of the complications of this area is that a good understanding of both the natural science and social science issues is needed. Natural scientists tend to look for only the physical causes and technological solutions (e.g., new sources of energy, new sinks, or enhanced energy efficiency) without considering what policy instruments are needed for these solutions to be adopted or the associated costs. On the other hand, economists tend to be ignorant of the scientific and technological issues. Two vital questions face policymakers on this issue:

• Can substantial technological progress be expected to help solve the climate change problem?
• If so, what are the likely characteristics and determinants of this progress (so policy instruments can be designed to encourage it)?

For the foreseeable future, policies probably will be weak, politically determined compromises that combine several instruments and principles. Even if these policies are far from ideal, they provide the necessary time for learning and building consensus on the issues as well as incentives for the long-run process of technological innovation.

## Some Technological Options

As usual, opinion diverges widely on the future of technology for application to the effects of climate change. Some optimists believe in the massive availability of renewable energy and energy-saving potential (Johansson et al. 1992). Others, such as the Club of Rome, are much more pessimistic. Economics is nowadays generally considered to be optimistic in contrast to the famous "dismal science" reputation it sported a few hundred years ago. Presumably, the transition is the effect of watching the effects of technological progress over the centuries.

However, many economists in the area of global warming appear to believe that $CO_2$ is the one pollutant that cannot be abated easily or cheaply.[25] Noncarbon substitutes (e.g., nuclear power and renewable energy sources) exist, but they are perceived to be problematic, limited, or expensive. Energy efficiency is thus one of the main alternatives, and considerable opportunities for increases in energy efficiency exist. Still, economic growth will increase the need for energy, and the right to emit carbon will be in high demand.

I do not pretend to understand the whole range of technologies possible or to be able to judge their possible future contributions. However, reality probably will be midway between the optimistic and pessimistic positions. Technological progress will not be so easy that concerns about efficiency and policy instruments become obsolete; neither will it be so difficult that prices increase astronomically and saving energy becomes so Draconian as to stifle the economy. Instead, there is considerable scope for technological progress, but it is difficult to forecast exactly where the progress will be most successful. Internationally recognized policy instruments must be put in place that will encourage the research, development, and adoption of such technologies, but considerable care is needed in the design of those instruments.

Even if only $CO_2$ emissions are discussed, a wide range of options is available.[26] Options include counteracting the effect of emissions, reducing emissions, or adapting to climate change. Among the options to counteract their effect are various ways to increase the rate of removal of carbon from the atmosphere through enhanced fixation in forests or in the sea. "Fertilizing" the sea (with iron—apparently a limiting factor—among other substances) to increase carbon fixation has been suggested and might be effective but would risk serious ecological side effects.[27] Planting forests (both afforestation and reforestation) is a conventional option that has now even been included as an eligible type of project for the Clean Development Mechanism (CDM) in the Kyoto Protocol. The potential to sequester or release carbon from forestry, agriculture, and the ecosystems in general is large. However, multiple problems concern additionality and credible baselines, and for these reasons, "avoided deforestation" is not currently eligible for CDM projects. This distinction is logical but may cause problematic and perverse incentives to clear-cut forests (with no effect on national targets or obligations), thereby creating the opportunity for future "reforestation" projects.

Options for reducing carbon emissions include increasing energy efficiency and switching to less-polluting fuel, and the variety and progress of various renewable options should not be underestimated. The combinations of new solar or biological techniques with progress in energy carriers and conversion may provide some positive surprises.

Yet another possibility has received somewhat less attention: $CO_2$ separation, sequestration, and storage. To begin, $CO_2$ may be separated (by adsorption, membrane separation, amine scrubbers, or various other methods) after normal combustion of the fuel. Then, the fuel (coal, oil, natural gas, or biomass) may be converted into hydrogen and $CO_2$ through reaction with water. The $CO_2$ then could be easily sequestered (i.e., captured and withdrawn). Most of the energy content would be with the hydrogen molecules and could be converted into heat or electricity with a high efficiency (e.g., by using fuel cells). Chemical-looping combustion is another (but still experimental) process in which the main combustion of the fuel is achieved not through oxygen in air but through fine metal oxide particles (which are reduced to metal and then oxidized again in a separate reactor). The exhaust stream thus contains only $CO_2$ and water vapor, which can be condensed (see Golmen 1999).

Separated $CO_2$ then must be stored, which is not completely infeasible. In fact, there are currently at least two serious options: ocean disposal and underground disposal. Ocean disposal is feasible because the ocean is already a sink for human-made emissions of $CO_2$ (roughly one-third is naturally absorbed this way). If $CO_2$ is injected in liquid form at more than 3,000 meters deep, it would be denser than water and sink to the floor of the ocean, where it would at least partially form solid hydrates. The costs and ecological consequences of this option are not yet known.

For underground disposal, more is known, because many areas have sedimentary and porous rocks, such as those that contain petroleum and gas basins. European aquifers and petroleum reservoirs that could be used for such storage are estimated to be sufficient to store all $CO_2$ emissions from E.U. power plants for several hundred years (Lindeberg 1999). Large-scale $CO_2$ injections are already used in some U.S. oil fields as a method to enhance oil recovery. Economically, this practice provides a double benefit.

Presently, the separation of carbon, rather than its disposal, is expensive. Disposal costs are estimated to be US$4–8/ton of carbon, but the costs of separation are of the order of US$100/ton of carbon. Carbon sequestering and solar hydrogen are predicted to be the two major energy technologies by around 2050, and together, they have the potential to allow for a sharp reduction from present emissions levels of 6 billion tons of carbon per year to less than 1 billion tons of carbon per year before the turn of the next century (Lyngfelt and Azar 1999).

*Policy Instruments To Encourage New Technology*

The technological details presented in this chapter are just a sample of the future possibilities. For instance, plants for deep sea injection of $CO_2$ might be combined with plants that produce power from the difference in temperature at different levels of the sea (known as ocean thermal energy conversion). The important question to address now is what kind of policy instrument will best encourage the development and adoption of new technologies.

Numerous aspects determine the incentives for technological innovation and for the spread and adoption of technology, particularly in markets that are not characterized by perfect competition and perfect information and for technologies that require investments in large, indivisible, and fairly irreversible technologies. There may for instance also be effects on the incentives for exit and entry and on the incentives for truthful reporting (see Chapters 12–14). Table 24-7 is one attempt at summarizing the comparative properties of some policy instruments under particular assumptions. Some policies (e.g., command and control) have no positive effects, whereas others (e.g., taxes) provide strong incentives for both innovating and adopting firms. Yet other instruments (e.g., tradable permits) have distinct effects depending on how they are allocated. Auctioned permits have the strongest positive effect on innovation but do not provide such strong incentives for other firms to copy or adopt (because the permit price falls as a side effect of the innovation).

A carbon tax or permit scheme (preferably auctioned) would, if correctly designed, encourage the use of nonfossil energy sources as well as $CO_2$ retention and storage. However, to do this, the instrument must target carbon emissions. If the tax or permit targets carbon purchased, then firms must be allowed to gain credits or tax rebates (almost like a deposit–refund system) for carbon sequestered and stored. Since the creation of a Swedish carbon tax in 1990, the annual use of biomass (particularly forest residues) in district heating has increased rapidly, by about 10 petajoules per year (Kåberger 1997). However, with the current formulation of this tax, there is no incentive for biomass in transport fuels.

In Norway, the carbon tax has caused large-scale storage of $CO_2$ in industrial aquifers. On the Sleipner Vest gas field, $CO_2$-rich natural gas is being stripped of its $CO_2$, which is injected into a saline aquifer at about 1,000 meters deep. Technologically, this process is not unique; it is similar to enhanced oil recovery in the United States. However, the uniqueness of the Norwegian process is that it is being done purely for environmental reasons and as a result of the carbon tax. Roughly 1 million tons of $CO_2$—a nontrivial 3% of Norway's total $CO_2$ emissions—are captured and injected.

**Table 24-7.** *Incentives for Innovation Created by Environmental Policies*

| Policy | Direct gains to innovating firm | Potential rents from adoption |
|---|---|---|
| Command and control | None | None |
| Best available technology | –<br>New standard raises overall compliance costs. | +++<br>Tighter standard increases incentive to adopt. |
| Performance standards | ++<br>Limited to existing abatement costs. | ++<br>Limited to existing abatement costs. |
| Emissions tax | +++<br>Lowers abatement costs and taxed emissions. | +++<br>Lowers abatement costs and taxed emissions. |
| Auctioned emissions permits | +++++<br>Lowers abatement costs and costs of all permits purchased. | +<br>Buying permits becomes cheaper alternative. |
| Grandfathered emissions permits | ++<br>Lowers abatement costs. | +<br>Buying permits is cheaper alternative. |
| Tradable performance standards/output-allocated permits | ++<br>Initial abatement costs are higher but lowers output subsidy. | ++<br>Initial abatement costs are higher, but permits become cheaper. |

*Source:* Fischer 2000a, 2000b.

An ideal "first-best" instrument for climate policy would discourage all activities that inappropriately alter the balance in the atmosphere and reward all activities that redress the balance, including research. Such research that led to a major contribution to solving the problem of climate change would be very profitable, because it would avoid the costs of permits or taxes for firms. However, the world community has been unable to commit to strong instruments, and property rights to research are notoriously difficult to protect. Thus, incentives for research may fail to be appropriate.

Because incentives for innovation are so important, policymakers might have to consider subsidizing relevant research. Subsidies have disadvantages, and they are not included in Table 24-7. However, if the world community is unable to come up with permit or tax schemes that provide credible incentives, then subsidies might be considered on a larger scale to support long-term basic science, which has public good characteristics. Such subsidies would need to be carefully targeted (informed by both technical and social information) and harmonized with other instruments in international negotiations. If, for instance, future carbon sequestration from the atmosphere (e.g., through forest plantations) or enhancement of marine sequestration is important, then special instruments must be designed to encourage their development, because current schemes focus mainly on emissions.

# Notes

1. This subsection and the next benefited from helpful comments from Dallas Burtraw of Resources for the Future.

2. Much of the reduction in sulfur use in Sweden predates the tax. Emissions were more than 900 kilotons in 1970, 500 kilotons in 1980, and 136 kilotons in 1990. Only the last (but perhaps most difficult) reduction, to 66 kilotons in 1999, was (partly) attributable to the tax. The Swedish tax applies to sulfur emitted but is levied on fuels on the basis of their sulfur content; a rebate is provided for sulfur removed by filters and so forth.

3. The emissions rights were called "allowances" but sometimes are called "permits" or "rights." They were not property rights in the formal sense because, in principle, the U.S. Congress could reduce or even abolish them without the owners being able to raise compensatory claims (Joskow and Schmalensee 1998).

4. The figures apply to Phase 2. The initial phase (Phase 1, 1995–1999) was a transition that focused on reducing emissions at a few large firms. The program also includes separate targets of 2 million tons of $NO_x$ reduction.

5. This section is based mainly on Höglund 2000, Sterner and Höglund 2000, and SEPA 1997.

6. The sulfur oxides tax and the nitrogen oxides charge are of similar magnitude. The sulfur tax is US$3/kilogram of sulfur (the same as US$1.50/kilogram of $SO_2$), whereas the $NO_x$ tax is US$4/kg of $NO_2$ (US$2.60/kg of NO). The two most common oxides, $SO_2$ and NO, happen to have roughly the same acidifying effect per unit of weight (1,000 grams of NO gives $1,000/30 \approx 33$ moles each of NO and $H^+$) because the molecular mass of NO is 30. Similarly, $SO_2$ gives $1,000/64 \approx 16$ moles of $SO_2$, but $1,000/32 \approx 31$ moles (twice as many) of $H^+$ because there are two hydrogen molecules in sulfuric acid. However, $HNO_3$ may be denitrified in nitrogen-deficient ecosystems, reducing its acidifying effect. On the other hand, $NO_x$ also leads to eutrophication and the formation of ground-level ozone (which is the main concern in California, but not in Sweden).

7. The reductions pertain to U.S. EPA's original program, which covered 22 states plus the District of Columbia. The U.S. EPA baseline includes only Phase 1 controls in the Ozone Transport Commission (OTC). Although states may develop their own strategies and rules for reducing $NO_x$ emissions, U.S. EPA has strongly encouraged participation in regional trading. The U.S. EPA program for the SIP Call region will eventually subsume the smaller OTC program in the northeast.

8. Many countries use other instruments, such as standards based on (implicit) shadow tax rates. The Dutch ministry informally recommends a value of US$4–5/ton to local authorities for rule setting. If abatement technologies are less expensive, then the local government should insist on their adoption. The Dutch government has discussed a *Kostenverevening* program in which the higher-than-average polluters would pay those below average, which is akin to a refunded emissions program or a permit program (Kroon 2000).

9. Many discussion groups on green tax reform abound; one is organized by the Wuppertal Institute (http://www.wupperinst.org/).

# Supplemental Reading

### Policy Instruments
Carlson et al. 2000
Cason 1995
Kolstad 2000a

### Climate Change Risks and Policies
Bohm 1999, 2000
Climate Strategies 2002
Fischer and Toman 1998
Fischer, Kerr, and Toman 1998
Grubb, Hourcade, and Oberthür 2001
Müller, Michaelowa, and Vrolijk 2001
RFF 2002
Toman 2001
UNFCCC 2002

10. The Committee on Industrial Energy Taxation (CIET) analysis suggested that a reduction in energy taxes would lead to a long-run rise in output and employment by a few percent, but this outcome is uncertain. It seems unlikely that even the high 1991 energy taxes could be so damaging. Even among energy-intensive companies, most pay relatively small percentages in energy tax compared with costs for labor, capital, and raw materials. Furthermore, other variables in the economy determine the competitiveness of industry. Within a couple of years of publication of the CIET report, the value of the U.S. dollar (in local currency) changed by more than 50% and the real rate of interest by much more. In this context, the effects of a redesign of energy taxes should be minor. The lobbying power of these companies is considerable (see also Chapter 7 on the special Climate Change Levy Agreement for the energy-intensive companies in the U.K. trading schemes).

11. Thanks to Michael Kohlhaas (Deutsches Institut für Wirtschaftsforschung, Berlin, Germany) for interesting comments. For more information on German green taxes, see BMF 2002.

12. On the other hand, energy-intensive industries also are most likely to relocate to countries with cheaper energy costs. Thus, the "carbon leakage" argument can be used to defend temporary exemptions until other countries get similar taxes.

13. The German law allows exemption from electricity taxes for electricity produced sustainably if the electricity is used by the producer or supplied in an isolated net with exclusively "sustainable, i.e., tax-free, electricity," but this exemption is rarely applicable in practice.

14. Compare the U.K., French, and U.S. views of earmarking in Chapters 8 and 21. In Germany, where the production cost of German coal is 240 DM/ton and that of (cleaner) imported coal is 70 DM/ton, the principle served to abolish a subsidy that was perhaps detrimental to the environment. Under other circumstances, this principle may prevent the use of efficient environmental policy packages such as taxes on vehicle use combined with subsidies for public transport. Furthermore, because of strict emissions laws in Germany, the only really negative effects in this case were local environmental effects of mining.

15. This section is based on Slunge and Sterner 2001.

16. The fact that regulations and laws governing the use of TCE were not coordinated across E.U. countries was an important factor in favor of allowing Sweden's regulations. In areas of law where there is explicit harmonization (e.g., pesticides or vehicles), it would be harder to accept different national legislation.

17. This section draws heavily on Davis and Mazurek 1998; Probst and Beierle 1999; and information from the Internal Revenue Service website (http://www.irs.gov/prod/tax_stats/excise.html) and the Superfund website (http://www.epa.gov/oerrpage/superfund/sites/npl/nar180.htm).

18. Not only information but also preferences and location of activities change over time. Thus, what is considered a slight nuisance today may later become a real health hazard. Once-useless wilderness may become a residential area, a town, or a major tourist attraction (e.g., Niagara Falls).

19. Over Antarctica, up to 60% of the total stratospheric ozone is depleted during the spring (September to November). Scientific evidence shows that synthetic chemicals are responsible for the observed depletion. The ozone-depleting substances are hydrocarbons combined with one or more halogen atoms (e.g., chlorine, fluorine, or bromine) and often are called *halocarbons*. One important group are the chlorofluorocarbons (CFCs). CFCs were first produced in 1928 and became popular because they were usually less hazardous than the chemicals they replaced. This is partly due to their stability—the attribute that lets them survive until they manage to reach the stratosphere. CFCs include carbon tetrachloride and methyl chloroform, which are used in refrigeration, air conditioning, foam blowing, and cleaning of electronic components and as solvents. Today, they are replaced by HCFCs (hydrochlorofluorocarbons), which are only partly halogenated and thus are less hazardous to the ozone

layer (although they are not harmless, either). Another important group is the halons, used in fire extinguishers.

20. The United States started to tax selected ozone-depleting substances in 1990. The tax was determined by multiplying a base rate per pound of ozone-depleting substance by an ozone-depletion factor. Initially set at US$1.37/pound, the base tax gradually increased to US$5.35/pound in 1995 and then by US$0.45/pound per year. The ozone-depletion factors are based on each chemical's relative damage and were originally set in the Montreal Protocol. For example, methyl chloroform has a factor of 0.1, whereas halon-1301 has a factor of 10.0, so in 1995, the tax on methyl chloroform was US$0.53/pound and that on halon-1301 was US$53.50/pound.

21. This section has benefited from considerable discussion with Christian Azar and Mike Toman. The debate on global warming is extensive (although the actual implementation of policies is limited), and for this reason, the subject is given limited space in this book. However, I discuss several relevant instruments, such as carbon taxes and permits, fuel taxes (Chapter 21, especially Box 21-2; this chapter; and Chapter 25), and relevant forestry policies (Chapter 30).

22. Because seriously reducing emissions is a major environmental undertaking, costs will be appreciable. "Softer" instruments such as voluntary agreements (VAs) and informational policies will not be sufficient but may be valuable as auxiliary mechanisms. In Denmark, VAs were used to reduce carbon emissions during the late 1990s. The relative and partial success of this program does not mean that larger and more costly reductions could be achieved without monetary incentives. Command-and-control measures also are unsuitable in light of the potential costs and the resulting importance of policy cost-effectiveness.

23. Idealists see these large rents as a benefit and argue that they should be used to combat poverty and promote sustainable development in the poorest countries. Unfortunately, opinion is far from being in agreement, and such suggested uses only make taxes even less acceptable to the countries that do not want to do much about global warming anyway.

24. However, there may indeed be strong nonlinearities or threshold effects. The climate system may be less stable than commonly believed, partly because of changes in the ocean currents (Rahmstorf 2000).

25. Even in 1987, the U.N. Commission on Environment and Development concluded that the only way to reduce carbon emissions was to reduce the use of fossil fuels, despite the fact that scientists had already demonstrated various methods for "engineered" carbon sequestering.

26. Some other greenhouse gases are methane, chlorofluorocarbons, and $NO_2$. The principles of radiative forcing have been known for more than a hundred years (Arrhenius 1896), but the dynamics in the atmosphere and for instance the behavior of water vapor (the most important greenhouse gas) is complex and still not fully understood. Policies for the other gases are important but beyond the scope of this discussion.

27. Offsetting the effects of warming by dusting the atmosphere with particles that have good reflective properties has been suggested as an option but appears to raise serious ecological concerns.

# CHAPTER 25

# Experience in Developing Countries

$I$N DEVELOPING COUNTRIES, THE FIRST PRIORITY in policymaking is to provide employment and income opportunities. Pollution is typically thought to grow as income grows—at least at the lower end of the Kuznets curve (see Chapter 2). In many developing countries as well as in the more industrialized of the formerly planned economies in central and eastern Europe (CEE), pollution from industries creates the majority of emissions.[1] In the richer countries, environmental protection agencies typically targeted industrial pollution first because industries are relatively easier to control (because they are large, immobile, visible, and have at least some economic resources at their disposal). Most of the environmental protection agencies of poorer countries do likewise and start by regulating industries. One difference is that these agencies typically are more aware of the clean technologies that are available than the OECD countries were when they started their process of cleanup.[2]

It is sometimes asserted that developing countries are not interested in pollution control (e.g., World Bank 2000), but this assertion is little more than prejudice. In fact, many developing countries put considerable emphasis on pollution control, and although the stringency of control increases directly with rising income, this process starts at low income levels (Dasgupta and Mäler 1995). Most of the decline in pollution intensity comes at the low rather than the middle income level (Hettige et al. 1996). However, experience also shows considerable individual variation from one country to another depending on culture, politics, climate, and many other factors.

In this chapter, I illustrate the range of opportunities that exist by looking at some successful environmental policy instruments. Variations in instrument design are closely tied to variation in economic policy in general, and politics differ widely in developing countries—from a strong dedication to market principles to belief in import substitution and infant-industry protection, not to speak of the remaining partly planned economies. Many countries continue to subsidize energy, and Mexico and Brazil long protected domestic production of the

Volkswagen Beetle, which was produced for much longer in Latin America than elsewhere even though it was a heavily polluting vehicle.

Given the diversity of policymaking, it is difficult to generalize about the experiences of developing countries, but one fact does appear to apply broadly: physical regulation may be a starting point for environmental policymaking, but it is commonly insufficient where institutions are weak. Regulation is complicated, because emissions vary and are difficult to measure (even for point sources). If the environmental protection agency has little staff and other resources, the monitoring equipment is prohibitively expensive, and staff are underqualified and underpaid, then the risks of deceptive lobbying or even corruption are particularly high. Given the urge to industrialize and to attract investments, policies that build on tough enforcement are not attractive.

Experience shows that abatement costs in developing countries typically are low. Sometimes they are lower than managers think—especially if firms are given some suitable advice and sufficient time for flexible adaptation; mild economic or informational policies can be effective. In a situation where drastic fees or prohibitions would lead to antagonistic or deceptive responses and might be fought in court, small fees coupled with assistance and information may lead to a process in which firms voluntarily engage in substantial abatement efforts. Instruments that build mainly on reputation, such as labeling schemes, appear to have similar effects. In many cases, the best policy is a combination of instruments. This chapter addresses several topics:

- environmental funds in some formerly planned economies of eastern Europe;
- environmental fees and funds in China;
- environmental charges in Rio Negro, Colombia;
- voluntary agreements in emissions control by Mexican brickmakers;
- differentiated electricity tariffs in Mexico and Zambia;
- information provision in Indonesia (PROPER); and
- zoning and self-regulation of industrial estates in India.

The first three sections are dedicated to the environmental charges and funds in formerly planned economies. These schemes are not simple Pigovian taxes; rather, they are combined with other instruments, and much attention is placed on the use of the tax proceeds. The fourth section focuses on voluntary agreements and other policies in the informal manufacturing industry. The fifth section focuses on the design of tariffs in developing countries. The two last examples are intended to emphasize the importance of building appropriate institutions and of choosing policy instruments that are adapted to the resources at the disposal of a local environmental protection agency.

## Environmental Funds and Other Instruments: CEE Countries

During the past couple of decades, CEE countries have experienced high levels of industrial pollution and poverty (Anderson and Zylicz 1996; Zylicz 1994; Kallaste 1994). These countries (particularly the more industrialized ones, such as Poland and the Czech Republic) have the technical capacity to solve their problems but lack the appropriate institutions and incentives. The environmental issues are significant enough to have repercussions back to the dynamics of economic growth. They represent not only costs and threats to health and ecosystems but also a significant deterrent to much-needed foreign investments. In responses to a 1992 survey of the world's thousand largest corporations, executives ranked environmental liability as a prime obstacle to investments in Poland (OECD 1997, 162). They pointed to two problems of particular relevance: the levels of contamination and the lack of clear legislation concerning property rights and liability (see also Chapter 5 and Bluffstone and Panayotou 2000).

High initial pollution levels had several consequences for the choice of policy instrument. They indicated that damage costs were high, and (disregarding the difficulties of obtaining exact values) if Pigovian taxes had been set at these levels, then they would have been prohibitively high at the moment when firms would be struggling for survival and the countries struggling with considerable unemployment and social unrest (see Markandya 1997). The taxes would also come due when the firms would be trying to find funds to finance restructuring and abatement investments (see Chapter 8). The high level of emissions and the existing stock of environmental problems (hazardous waste sites and similar issues) also meant that environmental policymakers needed not only money for abatement investments but also considerable physical resources for cleanup. Understanding these factors is crucial to understanding the importance of environmental funds in the CEE countries. However, environmental investments are not always necessary for cleanup. Many of the environmental improvements, such as emissions reductions, were simply due to the introduction of world market prices for energy and resources (Bluffstone and Larson 1997).

Despite tight economic restrictions, the CEE countries managed to fund a high share of environmental investments (typically more than OECD countries did). How did they manage to acquire the necessary resources? One explanation is that the funds managed to attract international capital in the form of debt-for-nature swaps, Global Environment Facility funds, and environment-related funding from the World Bank, donor agencies, and so forth (see Chapter 17). These funds may have been psychologically and tactically important, but numerically, the international capital flows were insignificant—less than 10% of total environmental expenditures in key CEE countries (Zamparutti 1999, 20).

Another important source of finance for the funds—and thus a key instrument of environmental policy in CEE countries—was environmental fees or taxes. One prerequisite for their acceptance was that they be fairly low (and in all likelihood well below the marginal damage cost). Another important factor was that these taxes (and the use of earmarked funds) were not unknown within the older planning system. Several countries, such as Russia and Estonia, also had been using environmental taxes for some time (Gofman 1998; Gornaja et al. 1997; Kallaste 1994). It is perhaps surprising that some of the most carefully (at least superficially) designed environmental taxes were built within the frameworks of planned economies. Individual emission taxes were levied on emissions depending on potential damage, but taxes also were differentiated depending on size and sensitivity of the polluted area, according to

$$T = \Sigma_i \, e_i T_i g_i \tag{25-1}$$

where $e_i$ is the quantity of emissions of type $i$, $T_i$ is the tax rate for individual chemicals (often worked out as a function of earlier standards, which were taken as a reflection of toxicity—e.g., in Estonia in the 1980s, the tax rate for sulfur dioxide ($SO_2$) was 1 ruble/ton, whereas that for lead was 408 rubles/ton and that for benzopyrene was 57,735 rubles/ton), and $g_i$ reflects the sensitivity of the region (higher for urban or recreational areas than for rural areas).

These taxes had little incentive effect under the planning period because of soft budget constraints; firms did not depend on the market for their income but rather on a ministry, and the payment to managers did not depend on profits. If the "costs" (including environmental taxes) of an enterprise went up, then so did the financial allocations from the ministry to that firm. In fact, Estonian managers liked the environmental taxes for purely patriotic reasons: the money that went to the environmental fund stayed in the country, whereas "surplus" was siphoned off to Moscow (Kallaste 1994).

When this structure was put to work within the context of a market economy, however, the fees had an important effect as a source of revenue first, and considerable evidence indicates that they also had some incentive effect (Bluffstone 1999; Clark and Cole 1998; Zinnes 1997). It is usually stated that the tax is set below "the" abatement cost—as if there were such a unique concept—but the marginal cost of abatement varies, and typically some abatement would be profitable, even at low tax levels. To achieve more abatement and particularly to achieve the optimal level of abatement, a higher or optimal fee level is needed.

Environmental taxes will not be successful if they are based on self-reporting and the regulatory agencies lack the resources to monitor and control (Bluffstone and Larson 1997). Another important practical point is that tax laws take time and effort to change because they have to pass both bureaucratic bodies and parliament. In an inflationary economy, they quickly decline, even if they were originally set at a somewhat more realistic level (Chapter 12). Thus, it is important for countries to have some form of indexing or updating of the levels. Yet only some countries have this for their environmental taxes.

In eastern Europe, environmental policy instruments appear to have worked best, and environmental funds have attracted the most money in Poland. Poland has many specialized and regional environmental funds as well as one large national fund (Zamparutti 1999). The latter collected and disbursed roughly the equivalent of US$300 million to US$500 million per year during the 1990s, whereas the Russian Federal Fund collected US$10 million to US$20 million per year and less than US$2 million per year was collected in Ukraine. Even smaller Polish funds were larger than many of the funds in the New Independent States (NIS). One such fund, EcoFund, was financed by debt-relief agreements (debt-for-nature swaps) with the Paris Club of western creditors. It has addressed transboundary pollution specifically as well as several other environmental issues that were difficult to finance with other mechanisms. Through particularly careful attention to high benefit–cost ratios and professional evaluation of projects, it has avoided the pitfalls of inefficiency that are rampant in earmarked fund programs.

Hungary, the Czech Republic, Slovakia, Slovenia, and Estonia also had sizeable funds with respect to their population. Most of the CEE countries have some experience with environmental taxes or fees, and in fact, some have many environmental taxes that are not unlike those of western European countries. Taxes on gaseous emissions tend to be levied on sulfur and nitrogen oxides, carbon, particulate matter, and lead (differentiated fuel taxes). Other transportation-related fees include tolls, mainly on new roads. Several countries have water extraction charges as well as effluent or water pollution charges and fines (often integrated into water tariffs). Several countries charge waste management fees, and a few have penalties or fees for noise, fertilizers, and pesticides as well as special charges for natural resource use (land, forestry, mining, and so forth). Poland, the Czech Republic, Slovakia, and the Baltic states often use two-tiered fees based on pollution permits that carry a certain charge, and excessive emissions are generally charged a surcharge or penalty fee.[3]

The NIS countries, however—even the poorer CEE states, such as Romania and Albania—have generally found it difficult to attract foreign capital and to generate domestic tax revenues in this way. In fact, economic decline in many of these countries has been so harsh and brought such deep recession and social problems that environmental issues have typically been pushed far down on the political agenda. The only significant environmental improvements in these countries are of the somewhat perverse kind that arise due to the collapse of sizeable portions of the heavy industry. For some of the CEE and particularly the NIS countries, the most urgent task may still be to strengthen the capacity of the environmental protection agencies and other responsible authorities. Without supervision or technical assistance, wise judgement in formulating directives and regulations, and monitoring and enforcement, there is no hope of much progress in the area of environmental or natural resource management. No environmental instruments can change this basic fact, but if low fees are the only way of financing the buildup of capacity and institutions, then it may be an important first step. To take additional steps, the environmental authority must learn to prioritize issues and to be flexible (without being lax or corrupt).

## Environmental Fees and Funds: China[4]

China is not only the most populous country on the planet but also one of the poorest and most polluted. In this kind of a setting, economic allocation choices really become harsh. Although people would benefit from environmental cleanup, scarce funds also could be used to provide health care and education, improve housing, and ensure a clean water supply. Atmospheric pollution in Beijing is so bad that removing 100 tons of $SO_2$ per year (out of a total 300,000–400,000 tons) reduces mortality by one statistical life (Xu et al. 1994). The cost of this reduction—of saving one life—is thus about US$300 (Dasgupta and Wheeler 1997). Even in a poor country like China, it would appear to be the kind of public investment that should be high up on a ranking of public investments in terms of its social rate of return.

China's system of pollution charges is an interesting and important example of the application of market-based policy instruments in a developing (i.e., formerly planned economy, now transitional) country. In 1979, an environmental law was passed that stated, "in cases where the discharge of pollutants exceeds the limit set by the state, a compensation fee shall be charged according to the quantities and concentration of the pollutants released" (Article 18 of China's Environmental Protection Law). Almost immediately, a few municipalities began to enforce the regulation, and in 1982, the state council called for a nationwide implementation. Today, most of the country is covered, and several hundreds of thousands of factories are monitored and potentially subject to this fee. Already in 1994, more than 19 billion yuan (more than US$2 billion) had been collected from environmental levies (NEPA 1994).

Chinese environmental fees are not textbook examples of Pigovian taxation, but few such examples exist, and many other features of the Chinese economy deviate from those in the economics textbooks, too. One distinctive feature, for instance, is that fees are paid only for discharges that exceed a certain level. The fees thus resemble a form of noncompliance fine. However, the idea of a natural resource or user fee appears to be spreading, and some of the latest water fees and the charge for $SO_2$ emissions are straight Pigovian charges. Such fees can be interpreted as a limitation in the implicit property rights of the polluters (see Chapter 5).

Another distinctive feature of Chinese environmental fees is that they are used first to finance funds that are used by industry for abatement (roughly 70–80% of the fees) and then for central administrative control costs. After administrative costs, the rest of the funds are returned to the enterprises to be used only for environmentally beneficial uses. Thus, in a general sense, the fees also are a form of refunded emissions payment (see Chapter 9 and Chapter 24). Usually, the funds are to be used for abatement investments but may be used for current costs of environmental management. In some regions, the fund allocation is based directly on the amounts paid by the individual companies, whereas in others, it is supposedly a neutral selection of projects based on the propositions of the same firms. It is probably safe to assume that even in the latter case, firms that make large payments to the fund have some leverage in financing their own abatement projects. Thus, the firms may be less averse to paying such fees than other kinds of taxes that go only to the treasury.

Finally, the fees are relatively low and considered to be less than the marginal cost of abatement necessary to meet the Chinese emissions standards; hence, some observers question whether the fee is at all effective (NEPA 1992, 1994; Shibli and Markandya 1995; Qu 1991). However, researchers who studied the rate of industrial emissions per unit of output for chemical oxygen demand (COD) and the effective water pollution charge for 29 Chinese provinces and urban regions between 1987 and 1993 claim that the environmental fee has had positive effects (Wang and Wheeler 1996). Supposedly, the rate is the same across the whole country—a blunt regulation that might be considered "fair"—but it is not necessarily efficient, because damage and abatement costs vary considerably across the rather heterogeneous areas of this vast country and the pollutants are not perfectly mixed. For example, it would not be desirable to have the same

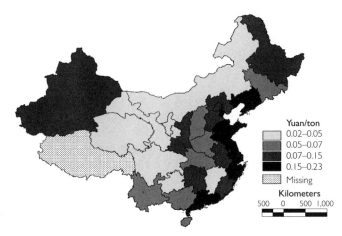

**Figure 25-1.** *China's Effective Pollution Levies, 1993*

*Note:* Levy is per unit of overstandard wastewater discharge.

*Source:* World Bank 2000.

standards (or charge level) in sparsely populated areas and in major cities. In practice, however, differences in monitoring and enforcement at the factory level can cause the effective fee to vary widely between provinces because some charges are only partially enforced or even waived. In this case, apparent inefficiency may improve the functioning of the overall system.

Figure 25-1 shows the significant variation in the effective levy rate between provinces. This pattern of variation is not random; effective levies are much higher in the densely populated (i.e., urbanized and industrialized) provinces of the country, particularly in the eastern coastal regions. These regions are not only more densely populated but also tend to have higher per capita incomes, which would explain why both objective damage measures and political pressures in favor of abatement would be higher. The degree of industrialization also would lead to higher pollution levels, which, under the most plausible assumptions concerning the damage function, is a separate, additional factor that raises the marginal value of damages. The charge also has varied over time, with large increases since 1987. From 1987 to 1993, COD intensities fell at a median rate of 50% in all the provinces, and total COD discharges declined at a median rate of 22%.

Both supply and demand were estimated for "the environment" in a study by Wang and Wheeler (1996). The demand for "a cleaner environment" depends on population and pollution densities as well as on education, income, and share of industrial activity in economic output. The estimated elasticity of abatement with respect to the environmental fee is large and significant (almost exactly −1). Also, results indicate that larger plants pollute less and that state ownership has little independent influence on pollution intensity, which may perhaps be surprising.

Using citizen complaints as a measure, demand for environmental quality has been explained by three factors: the level of education of the affected citizens, the income level of those individuals, and the level of visible pollution (Dasgupta and Wheeler 1997).[5] Industrial demand for the environment tends to depend on technical production intensities (e.g., wastewater intensity), production volume, and legal standards.

Management of the pollution charge may have been effective in China. Pollution levels, income, and population density appear to explain the demand for abatement. Education and bargaining power are important, too. In summary, preliminary analyses suggest that the central Chinese authorities may be wise in letting each province trade off the costs and benefits of effective implementation levels. However, despite the results presented here, many researchers doubt whether the Chinese fees are actually operative and effective. Much about the system remains to be understood.

## Environmental Charges: Rio Negro, Colombia[6]

Colombia provides an interesting example of an environmental fee that is working despite what at least some outsiders might think of as a difficult policy environment. The Colombian economy has been growing quickly over the past few decades. From 1985 to 1995, GDP grew by an average of 4.6%, and much of this growth was resource intensive (i.e., mining or agriculture). Water and air pollution also have been increasing; the major cities of Baranquilla, Bogotá, Bucaramanga, and Cali are heavily polluted. Most of the wastewater is untreated, and diarrhea and enteritis are the leading causes of death of children under five years old. Respiratory diseases are escalating rapidly in industrialized areas.

Law 99 passed by the National Congress in 1993 created the Ministry of Environment and local environmental protection agencies (*Corporaciones Autónomas Regionales*) and decreed that environmental policies in Colombia should account for environmental damage costs. It stipulates the use of economic instruments (pollution fees) to prevent, mitigate, and repair environmental damages as well as to finance the local agencies. The fees initially target effluents based only on two measures of pollution: biological oxygen demand (BOD) and total suspended solids. However, the program is more than a charge, because it also requires all key stakeholders (industries and municipalities) to actively negotiate pollution reduction targets over a five-year period. To guide the implementation of abatement measures, the Ministry of Environment has set a minimum national pollution fee. Actual fees vary between regions. The regional factor starts at the national level (a factor of 1) and is increased incrementally (to 1.5, 2, 2.5, ... times the national fee) every six months until the targets are met.

The program was first launched in the ecologically sensitive Rio Negro watershed area near Medellín, where the regional environmental protection agency, *Corporación Autonoma Regional del Rionegro-Nare* (CORNARE), was particularly interested in developing these new tools of environmental policy and had good working relationships with the key stakeholders. Within six months of implementing the charges, pollution from industrial sources in Rio Negro dropped drastically (by 28%, as measured by BOD). This reduction was a good start toward the 50% target agreed on for the first five years. However, there was little further progress during the subsequent two years, and some degree of conflict arose among the polluters. Parties in industry claimed the escalating fees (toward

**Table 25-1.** *Allocation of Revenues Collected from Colombian Environmental Charges, 1999*

| Revenue allocated to | Allocation (%) |
| --- | --- |
| Waste treatment plants | 50 |
| Clean technology investments | 30 |
| Research | 10 |
| Administration | 5 |
| Education | 5 |

*Source:* CORNARE 2001.

meeting the overall target) were unfair and that industrial sources had reduced effluents while the municipalities had done nothing. Meanwhile, the municipalities refused to pay their fees, alleging they had no money.

Currently, CORNARE is helping the municipalities to build sewage treatment plants. The focus will later expand to include nonpoint-source polluters (households and farms), which make the design and enforcement of policies more difficult. Nonetheless, tackling industry was presumably a necessary first step—without which it would have been pointless to approach other sectors. The charges collected from industry have provided funds not only for environmental investments in the industries but also for staffing the environmental agencies. This element of long-run capacity building is hoped to strengthen the institutions so that they will be better equipped to manage other sectors later.

One of the main areas of discussion and contention is the use of the revenues (see Table 25-1). Corporate acceptance of the charges hinges on the recycling of a significant share of the funds back to industry for abatement (see Chapters 9 and 15), whereas CORNARE wants to invest in municipal capacity and wastewater treatment. The allocation of charges in 1999 is shown in Table 25-1. Industry gets at least 30% of the revenues back in some form and perhaps a share of the 10% for research, too. In an analysis of industry's response to the system, the fees had had a significantly positive effect on firms' environmental investments and BOD reductions (Coronado 2001).

The charges are now gradually being spread throughout the 33 regional environmental agencies in Colombia; as of mid-1999, more than one-half of the agencies had a plan in place, and two agencies were in the process of implementation. Ideas for expanding the system include introducing charges for air pollutants, solid waste, and highly polluting segments of the agricultural sector (e.g., large-scale plantation crops such as banana and coffee).

Several features are considered to have been key to this relatively successful start in CORNARE:

- a high level of knowledge and commitment on the part of local staff;
- solid support at higher levels, particularly the ministerial level;
- good technical information and data on pollution;
- a good working relationship with the polluting industry; and
- a relatively small region, which may facilitate monitoring.

One of the most important changes in Colombia appears to have been the change in attitudes, evident in the two following statements by Vice Minister of Environment Fabio Arjona Hincapie (World Bank 2002):

> The idea that developing countries cannot use economic instruments ... negates the possibility, the assumption that competitive forces exist in developing nations.

> The experience we had with command-and-control mechanisms in 23 years, we have had little pollution control results and really high costs of pollution reduction. What we are trying to create is a fundamental change in the signals that we give to the industrial community.

## Voluntary Participation in Emissions Control: Mexico[7]

Small enterprises in the informal sector are an important component of industry in many developing countries; they provide valuable employment and production but also contribute to pollution. They typically are difficult to regulate because they operate in the "gray zone" of the economy, where reporting requirements simply do not apply. The use of conventional regulation, zoning, and incentive instruments (e.g., taxes) is difficult, and in some cases, the regulatory agency faces almost the same kinds of dilemmas as with nonpoint-source pollution.

Brick kilns are a common source of pollution in rural and urban centers throughout developing countries. In these contexts, brickmaking is relatively simple: clay and sand are mixed, pressed into molds, dried in the sun, then fired in primitive kilns. All work is done by hand. Although these businesses are not inherently polluting, they present a hazard to surrounding neighborhoods because poverty and competition drive entrepreneurs to use the cheapest fuel available: scrap, such as polluting refuse—plastic waste, car tires, or wood (with poisonous paint and other chemicals)—and used motor or other oil. The kilns also lack suitable filters and thus produce considerable emissions of black smoke.

The most common regulatory strategy, a ban on the use of dirty fuels, has had little success because it is difficult to enforce. The kiln operators have neither the financial resources nor the motivation to change, and those who do will see their profits undercut by competitors who do not (and thus have lower costs). Competition between brickmakers is fierce, and small differences in costs and prices determine their survival.

On average, kilns studied in Ciudad Juárez, Mexico, had a capacity of 10,000 bricks, employed 6 workers, and generated profits of US$100 per month (Blackman and Bannister 1998). The majority of the workers had only a few years of schooling (25% were illiterate) and lived in simple houses without sanitary facilities next to the smoke-belching kilns. Roughly 20,000 brick kilns of this type exist in Mexico, and large Mexican towns have several hundred. These firms are particularly interesting because they appear to be fairly representative of the informal sector, which is a major source of income for economies in Latin America and other countries in the developing world.

Pollution control in Ciudad Juárez did not rely on conventional regulation from superior authorities at all; it was essentially a grassroots initiative by a non-governmental organization (NGO) but had the crucial support and encouragement of the environmental protection agencies. The location of the city (on the border between the United States and Mexico) also made it a focal point of attention in the North American Free Trade Agreement (NAFTA) debate. As a seriously polluted area, it was an obvious target for protectionists and environmentalists alike, who could assert that free trade would lead to loss of jobs as well as increased pollution as industry relocated to places such as Ciudad Juárez.

Pollution from brick kilns is an issue of urban planning and zoning. Originally, many of the kilns were located outside or on the outskirts of Ciudad Juárez, but as a result of rapid urban sprawl, they are now well within the borders of the city, and thus their smoke is a health hazard to not only more than 2,000 workers but also many local residents. The brickmakers are typically organized in *El Partido Revolucionario Institucional* (the revolutionary party that was in power in Mexico throughout most of the twentieth century) or other organizations with strong ties to political parties. These organizations provided a valuable channel through which to communicate with the brickmakers.

Despite their economic importance, brickmakers have faced increasing pressure to reduce pollution during recent decades as an "environmental conscience" has been growing rapidly in Mexico. Efforts led to an initiative to persuade brickmakers to convert to clean and heavily subsidized propane. The initiative was led by the municipal council but later was taken over by NGOs better suited for the political task of communicating with the brickmakers' organizations. By 1993, 40–70% of the brickmakers had been persuaded to switch to the cleaner fuel. However, this considerable success was due to the confluence of several factors: subsidized propane from PEMEX (Petróleos Mexicanos, the Mexican state oil company), grants and subsidies for conversion, and a great deal of direct support and media attention during the midterm elections at the same time as the NAFTA debate became particularly focused on the issue of environmentally motivated capital flight. The media and political importance of the issues facilitated the acquisition of grants for propane burners, courses, and knowledge transfer. The NGOs even helped establish among the brickmakers a floor price for bricks, which was supposed to guarantee profitability to all brickmakers, even if they used more expensive fuel. (This agreement and an attempt at organizing boycotts against the producers using dirty fuels soon broke down under market pressure.) Local authorities also played a role by starting to monitor and enforce the regulations against the burning of debris that they had failed to enforce previously.

Political pressure to allow the use of "clean" waste products such as uncontaminated sawdust led to a weakening of the controls against the burning of debris. Then, the Mexican authorities faced an interesting dilemma in pricing petroleum-based fuels. The fuels are all close substitutes in various combinations, but some are environmentally detrimental in one situation and beneficial in another. If gasoline and diesel should be taxed because of negative external effects, then fuel oil and kerosene should be taxed, because they are close substitutes for diesel or gasoline in some uses. And propane should be taxed because it is sometimes a substitute for kerosene.[8] Propane was subsidized in Ciudad Juárez to encourage the use of clean-burning fuel (i.e., to discourage the burning of scrap tires and other waste). In Mexico, petroleum products have traditionally been cheap as part of a strategy for spreading the rents of the national oil heritage and, supposedly, also to speed industrialization and help people with lower incomes (for a review and critique of the petroleum product pricing policies in Latin America, see Sterner 1989). During the 1990s, Mexico was influenced by the World Bank, NAFTA, and other forces—both market and ideological—to abandon this form of state intervention in product pricing in favor of a more market-oriented approach. The energy-sector subsidies in Mexico have been an important cause of increased energy intensity, waste, and emissions. The abolition of such subsidies is, in general, good news for the environment. However, it had a negative effect in Ciudad Juárez because the vast majority of brick kilns abandoned propane as soon as its price increased. The price increase was sizeable—in 1992, propane and sawdust were almost the same price, and in mid-1995, propane was almost three times more expensive than sawdust.

At least some improvements in Ciudad Juárez appear to be permanent. Although most brick kilns have reverted from propane to sawdust, they have not reverted to the most polluting fuels, such as motor oils and tires. Some important conclusions can be drawn (see also Blackman and Bannister 1998):

- A naive "first-best" policy would be to tax the waste fuels, but such an approach might not be possible in a large informal industry with cutthroat competition and difficulty in monitoring.
- More likely to succeed, at least with rising income in the longer run, are policies to ban certain fuels or to use zoning for certain activities.
- In the short run, private-sector collaborative initiatives can work, at least with high-level public support.
- Subsidies that would otherwise be rejected might be a useful instrument if carefully targeted.
- The involvement of local grassroots organizations is crucial.
- Fancy abatement technology may be less appropriate than simple low-cost measures.
- In volatile developing economies, regulation of the informal sector by market-based (or any other) instruments is difficult.

## Differentiated Electricity Tariffs: Mexico and Zambia

The cost of providing infrastructural services such as water or electricity depends crucially on the physical supply and the number and consumption level of individual consumers. The density of consumers in urban areas makes the necessary investment per capita for their services much smaller than in rural areas. The cost of supplying a certain customer does not consist of only the (long-run) marginal cost for bulk electricity (or water) but also the costs for local distribution and associated costs of metering, monitoring, administration, and so forth. These costs are all essentially fixed, which means that low-volume consumers are much

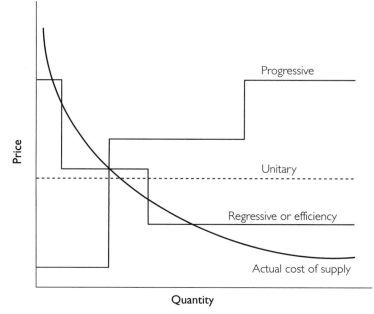

**Figure 25-2.** *Different Kinds of Tariffs*

more expensive to supply than high-volume consumers. The supply curve thus slopes downward for the relevant interval.

In setting tariffs, several factors must be considered together. Conceptually, tariffs can be described as regressive, progressive, or unitary depending on whether they rise, fall, or remain constant with rising consumption (see Figure 25-2). In recognition of the fixed costs of providing a service, one can argue that high-volume consumers should pay less; the tariff system then follows the supply curve, which promotes efficiency because it reflects actual economic costs for the supplier. I call this kind of charge an *efficiency tariff*. Each customer pays the costs incurred by the supplier, and thus the supplier has reasonable incentives to provide service to all customers. On the other hand, this tariff may be seen as regressive.

At the other end of the scale, the progressive tariff essentially takes income distribution or ethical (political) considerations as its starting point. It provides services (e.g., electricity or water) to poorer (lower-volume) consumers at subsidized rates and charges richer (higher-volume) consumers increasingly higher rates. This kind of charge may appear to be the "greenest" tariff construction and most appropriate in reducing wasteful consumption: the minimum (necessary) consumption is cheap, but the more wasteful uses are discouraged by a high price. However, when analyzed carefully, the issues are complex, and the progressive tariff is not necessarily attractive—for neither the poor nor the environment.

The unitary tariff is in a sense "progressive" in that it provides subsidies to low-volume consumers and taxes high-volume consumers relative to the actual costs of supply. The unitary tariff also is simpler to implement than other kinds of charges because cheaper meters and administrative routines can be used and there is less incentive for cheating.

It is a misunderstanding that progressive tariffs are needed for resource conservation. Conservation requires a correct value of the resource rent. For individual consumers, this value does not rise. As society uses more of a certain resource, the rent may rise, but this increase should then be reflected in rising tariffs for everyone over time—not necessarily a progressive tariff schedule for each consumer. Problems arise when the appropriate rent level is so high that poor people cannot afford their small but vital consumption.

The progressive tariff also makes the poor customers unprofitable for the supplying company. In principle, cross subsidies might be possible in some contexts (e.g., the company might make its profits from the rich, high-volume consumers and subsidize the poor, low-volume consumers) but unlikely on a large scale. Instead, the poor consumers end up receiving second-rate services—or no service at all. For instance, when an accident cuts off the electricity supply to a poor community, the electricity company has a perverse incentive to not restore power; the company will lose money as soon as the problem is fixed. As a result, power outages are more or less permanent in some areas (Köre and Widepalm 1993). This is not a good incentive structure. In fact, the general state of public goods supplied to poor communities in many developing countries illustrates this point all too well. The water is often of poor or uncertain quality, and supply may be unpredictable; electricity outages are frequent and bus services inadequate. Poor people want public services to be cheap, but their primary concern is that the services be functional. If tariffs are such that the supply companies or transportation organizations do not manage to break even, then the people will not get the services they need.

Progressive tariffs may even disfavor the many families who share an apartment, house, or room with only one service contract, because the total consumption levels of all these families is too high to be charged the low-volume tariff. On the other hand, some high-volume customers are able to cheat the system by acquiring several contracts in the names of different family members, and if each of these separate contracts has sufficiently low consumption, then they all will be charged the minimum tariff.

With efficiency tariffs, the requirement is clear: a mechanism to distinguish high-volume consumers from low-volume consumers. Usually, this distinction is accomplished by the use of a two-part tariff, so that the low energy price comes with a fixed meter or ampere charge that is related to the maximum capacity installed. This fixed charge is prohibitively high for low-volume customers, who are thus forced to reveal themselves or their type and thereby be charged the higher energy tariff (but lower fixed costs). (This simple example illustrates the revelation mechanism discussed in Chapter 13.)

Actual tariffs for water and electricity in various countries vary considerably from the progressive to unitary or efficiency models. For many years, Mexico has favored progressive tariffs, so large industries and high-volume consumers pay much more for electricity than low-volume consumers do. Table 25-2 shows a selection of electricity tariffs for Mexico in 1993, after some of the heaviest subsidies had been removed but considerable cross subsidies remained. If the price of high-tension power for large industries was 0.13 pesos/kWh, then the cost of supplying residential electricity was probably much higher. Yet the low-volume

Table 25-2. *Some Electricity Tariffs in Mexico, 1993*

| Tariff level | Electricity consumption | Rate (N$/kWh) |
|---|---|---|
| 1: Residential | 0–25 kWh | 0.06 |
| | 25–50 kWh | 0.09 |
| | 50–75 kWh | 0.11 |
| | 75–100 kWh | 0.13 |
| | 100–200 kWh | 0.15 |
| | >200 kWh | 0.47 |
| 2: Small commercial/industry (<25 kW) | 0–50 kWh | 0.32 |
| | 50–100 kWh | 0.41 |
| | >100 kWh | 0.45 |
| 3: Large commercial/industry (>25 kW) | | 0.22 |
| 9: Irrigation | 0–5,000 kWh | 0.10 |
| | 5–10 mWh | 0.12 |
| H-M: High tension (>1,000 kW) | | 0.13 |

*Notes:* N$ = nuevo peso; kW = kilowatt; kWh = kilowatt-hour; mWh = megawatt-hour.
*Source:* Comisión Federal de Electricidad 1994.

consumers paid less than half this rate for small quantities of low-tension power (0.06 pesos/kWh), while medium-volume consumers paid somewhat more and small industries paid more than five times as much. Irrigation received large subsidies, whereas commercial and small-industry tariffs were hardly subsidized. One result of this policy is that many low-income Mexican households still lack electricity because supplying them entailed large losses to the electric company for many years. Since the late 1990s, these cross subsidies have been removed or at least reduced, and low-tension tariffs have higher energy charges than high-tension tariffs.

In Zambia, where the mining industry creates a large-scale demand for electricity, millions of rural and even urban inhabitants live without electricity and depend on biomass fuel for household energy.[9] The country has considerable hydroelectric potential, but most of it is still untapped, partly for lack of funds in Zesco, the state electric company. For many years, Zesco had a monopoly on electrical service but was not allowed to charge prices that covered long-run marginal costs to most of its customers. Before 1994, at least 19 electricity tariffs were used, 12 of them for unmetered use by households, and all low (Mulenga 1999). For instance, low-volume household customers could choose connections with load restrictors of 1 to 15 amperes. A small household might have a 2-ampere connection (Tariff LC) and would have been charged a fixed fee of 15.7 kwacha per month in 1992. In contrast, under Tariff E1, intended for somewhat higher-income households (and higher-volume consumers), the energy charge was 17.35 kwacha/kWh. Supposing the latter had been a true reflection of cost, then the households with the unmetered 2-ampere connection would be paying for less than 1 kWh/month—the equivalent of using one lamp for 10–15 hours. The unmetered customers were getting a really good deal on electricity service, but most people who wanted a connection could not get one.

**Table 25-3.** *Electricity Tariffs and Long-Run Marginal Cost of Electric Supply by Tariff, Zambia, January 1998*

| Consumer category | Tariff type | Price charged Capacity (US$/kV-A) | Price charged Energy (US$/kWh) | Estimated cost (LRMC) Capacity (US$/kV-A) | Estimated cost (LRMC) Energy (US$/kWh) |
|---|---|---|---|---|---|
| Residential | R2 | | 0.033 | | 0.08 |
| Commercial | C1 | | 0.044 | | 0.15 |
| Commercial | C2 | | 0.059 | | 0.09 |
| Industrial | MD1 | 6.25 | 0.035 | 12.50 | 0.04 |
| Industrial | MD4 | 8.33 | 0.016 | 11.66 | 0.04 |
| Social services | H1 | | 0.042 | | 0.09 |

*Notes:* LRMC = long-run marginal cost; kV-A = kilovolt-amperes; kWh = kilowatt-hour. The LRMC does not include the cost of metering, billing, and marketing.

*Source:* Adapted from Mulenga 1999.

A somewhat more pragmatic policy has been followed for large industrial customers, which can choose tariffs that have lower marginal costs and higher fixed or capacity charges. However, the tariffs are updated infrequently, and with high inflation, their real value tends to fall below the long-run marginal cost (see Table 25-3). All the tariffs are still so low that none covers marginal cost. Furthermore, some tariffs carry an implicit subsidy that is larger than others. The commercial Tariff C2 caters to large customers and thus has a low marginal cost, but its price is higher (US$0.0586/kWh) than the commercial Tariff C1 (US$0.0436/kWh) for small consumers. This kind of tariff setting breeds corruption because the subsidized good inevitably is in great demand, and those in charge of meeting demand probably can grant favors to certain people. Moral character or culture does not necessarily determine the extent of corruption in a society; the determining factor is more likely to be technical matters that could easily be resolved. Political economy is very important here: there is a risk that tariff setting is influenced by the demands of the "insiders" (who already have electric connections) with insufficient regard for the interests of the outsiders, who will not receive electricity service unless it profits the supplier.

# Information Provision and Institutional Capacity: Indonesia[10]

Indonesia epitomizes many environmentalists' worries about economic growth. Economic growth has been spectacular—exceeding 10% for long periods—but it has been based on a drastic exploitation of natural resources such as oil and timber. Estimates of the country's real (net) growth (corrected by subtracting for degradation of natural capital) are much smaller, but still impressive (e.g., Repetto et al. 1992).

The country is also known for an authoritarian government in which the family and associates of President Suharto (1967–1998) wielded considerable power and influence not only in politics but in business and administration as well. Policymaking and implementation were far from transparent. Many of the owners of polluting industries were powerful, and the local environmental protection agency, BAPEDAL, was a relatively small and weak organization. The agency would be asking for trouble if it poised poorly paid, ill-prepared inspectors and regulators against the most well-connected and powerful business leaders in the country—who, if need be, could employ the best lawyers. The risks of ineffective monitoring, weak enforcement, and even corruption were obvious, at least if the program were to seriously threaten the interests of the ruling elite.

BAPEDAL eschewed the idea of regulations (which, in principle, already had been tried with no effect for many years) because it believed it had no chance of success against industry in trying to enforce closures, fines, or other penalties. The agency also was wary of a system of environmental charges, in part because they create a situation of negotiation between the polluter and individual officers of the agency, which might bring incentives for bribery. Instead, they chose a rating or labeling system, the Program for Pollution Control Evaluation and Rating (PROPER).

What is distinctive about PROPER is the care and attention put into all details of the design of the system. Fundamentally, it is a system for emissions reporting, evaluation, and control of emissions reports and ratings as well as assistance and advice to the firms. Each industry is graded, and ratings are based on several parameters in the reports. A good deal of effort was spent on designing the parameters. The purpose was to have an appropriate set of parameters: not so many as to make reporting burdensome and alienate the firms, but not so few that the reports would be easy to falsify. By experience, BAPEDAL officials know that stoppages, for example, cause the excess release of certain, although not necessarily all, pollutants. Such stoppages are related to other parameters that the firms report: production, electricity use, raw material consumption, water, and so forth. Therefore, the PROPER team collected many indicators and developed small but sophisticated programs for checking the correlation among the various data collected.

Engineering studies and programs have given the PROPER inspectors a valuable tool: knowledge. These inspectors must be well prepared for inspections and discussions with the firms in order to gain the firms' respect. They also must be perceived as helpful, impartial, and influential. In detecting a release or a problem, they must not be perceived as being "out to get" the company. Confidence, respect, and communication are vital. Many problems found during inspections can be resolved by the firms, and in many cases, the PROPER team has the expertise to help. Inspectors acquire and improve their expertise because firms respect them and share technical knowledge with them.

In addition, the inspectors must be seen as incorruptible. Even one incidence of corruption can ruin the reputation of this kind of system. The process of collecting, evaluating, and rating emissions data is highly automated in a formalized computer program, and a staff of fewer than a dozen well-trained people runs the whole scheme. The rating and evaluation procedure is also transparent, so firms know exactly what they have to do to improve their grades and other agencies, competitors, the press, and the public can verify and control not only the companies but also the work of the BAPEDAL regulators. Some neighboring commu-

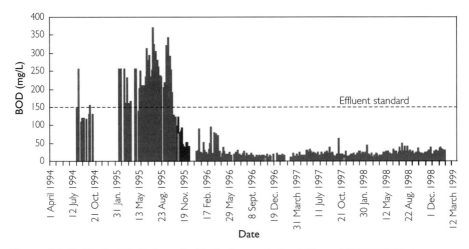

**Figure 25-3.** *Daily Variations in Pollution and the Effect of PROPER in Indonesia, after 1995*

*Note:* Effluent standard is marked at 150 mg/L.

nities affected by heavy pollution have challenged ratings (and thus a firm's emissions data), and in one case, a community succeeded in changing the rating of a firm that had cheated to receive a Green rating (when it should have been Black).

To understand at least partly the complexity of just monitoring, consider Figure 25-3, which shows data for one measure of effluent from one Indonesian factory. The data for this particular industry show a clear response to the PROPER program, but such diagrams may be difficult to interpret. Note in particular the daily variation. This kind of variation is fairly typical, and there may be a corresponding variation by the hour or minute, making actual measurement quite an art. If sampling is the chosen tool, then what mechanisms must be in place to guarantee that the sampling is representative? How can the system ensure that the company is not releasing massive amounts of emissions while the measurement system is inactive? Furthermore, Figure 25-3 shows only one measure (BOD concentration), but other measures of pollution include more detailed counts of certain emissions. Regulations must determine the percentage of observations that are reasonable under a compliance standard. If only a few measurements are taken, then the risk is high that the measurements are unrepresentative. The greater the number of observations, the smaller that risk—but then decision algorithms must be created to distinguish between allowable anomalies and systematic violation. If too much power of judgement were left to individual inspectors, then the transparency would diminish and the risk of corruption (or of being suspected of corruption) would surface.

Labels were carefully chosen to be simple, clear, and understandable to citizens who have little or no prior knowledge or understanding of technical and pollution issues. For simplicity, a color-coding scheme was chosen, but for coherence and credibility, the colors have precise meanings. The color categories are described in Table 25-4. The entire system builds on earlier regulation. Firms

Table 25-4. *Color Labels Used in PROPER*

| Rating | Technical requirements |
| --- | --- |
| Gold | World-class clean technology<br>Waste-minimization and pollution-prevention efforts |
| Green | Above legally required standards for environmental protection<br>Good maintenance and environmental work |
| Blue | At legally required standards for environmental protection |
| Red | Below legally required standards for environmental protection |
| Black | Serious environmental damage<br>No pollution-control effort |

that simply comply with environmental regulations are Blue. Firms that are pro-active and significantly exceed the environmental standards can be awarded a Green status, and the world-class Gold status has yet to be achieved by any firm in Indonesia. Firms that fail to meet minimum standards are Red or, in cases of significant environmental damage and no effort at abatement, Black.

Information disclosure has been handled carefully. The PROPER team started by evaluating 187 large and medium-sized polluting plants in 1995. Initial ratings indicated that two-thirds of the companies were noncompliant and would receive Red or Black ratings, showing how poorly regulation—which formally had been in place for a long time—had worked. The PROPER team then took two cru-cial steps. First, it went to considerable trouble to verify data, vetting the ratings at many levels, holding discussions with representatives of industry and NGOs, and having the ratings approved by not only the minister of environment but also the president. The knowledge that the president had seen and approved the rat-ings was crucial in giving them credibility and legitimacy. Second, the team went to great lengths not to antagonize industry, because it was crucial that the PROPER team not be perceived as "anti-business." They therefore started by informing all the industries and giving those with poor ratings six months to improve before public disclosure. Already, this first step had considerable effect. Then, they disclosed the names of only the handful of firms with really good rat-ings, which were congratulated in a highly publicized event with Vice President Tri Sutrisno.

In December 1995, full disclosure of all the ratings started, industry by indus-try, to receive maximum press coverage during a fairly extended time period. The changes in rankings from June 1995 to December 1995 and December 1996 are shown in Table 25-5. Response was immediate during the first six months of the program—even before disclosure—particularly among Black- and Red-rated plants. The incentive to avoid the disgrace of a bad grade appeared to be the strongest factor in increasing firms' ratings. All but one of the Black ratings improved, and 24% moved up from Red to Blue. The most important sign of progress was that in the first 18 months, overall effluents from the first 187 firms inspected decreased by 43%.

Another aspect of BAPEDAL's PROPER strategy was the gradual inclusion of more firms. Concentrating its efforts on the most polluting industries first, the

**Table 25-5.** *Change in Firms' Ratings Due to PROPER Disclosure in Indonesia*

| Rating | June 1995 | Dec. 1995 | Dec. 1996 | Change (%) |
|---|---|---|---|---|
| Gold | 0 | 0 | 0 | 0 |
| Green | 5 | 4 | 5 | 0 |
| Blue | 61 | 72 | 94 | +54 |
| Red | 115 | 108 | 87 | −24 |
| Black | 6 | 3 | 1 | −83 |
| Total firms | 187 | 187 | 187 | |

*Source:* Data from BAPEDAL (Indonesia Environmental Impact Management Agency) and World Bank 2000.

agency could target approximately 20% of the pollution coming from just 1% of the firms. Moving on to 10% of the plants would cover roughly 90% of pollution, and this effective strategy concentrates inspection efforts on firms where the environmental effect per control dollar will be highest.

Several possible mechanisms appear to be essential to the success of the PROPER scheme (see Chapters 3, 6, and 13). The most relevant are the following:

- Media and reputation: Firm managers feel a strong sense of civil responsibility. Indonesia has a strong culture of shame, so managers generally try to avoid bad publicity.
- Consumer reaction: All other things equal, a firm's reputation may influence consumers' purchasing decisions. This response is likely only in cases when the number of producers gives consumers an array of choices. One might think that this mechanism works only when the industries produce consumer products, but in fact, reputation also sometimes appears to be important for firms that produce inputs for other firms.
- Worker reaction: Employees, like managers, must take pride in their workplace to be motivated to do a good job. In extreme cases (high levels of pollution), employees may hesitate to work in a particular plant or demand a higher salary to do so. This factor presupposes that laborers have some choice in employment, so it applies mainly in an economy that offers reasonable levels of employment.
- Investor reaction: Like consumers and workers, investors may make investment decisions based on ethical, environmental, or other sensitivities. Even if they do not, they may believe that regulators or communities will react to environmental problems and thus may worry about the hidden liability risks of noncompliant firms.
- Community reaction: Local, regional, and national communities may react to pollution. Public disclosure is perhaps the instrument that makes this kind of reaction most likely, because it does not depend on the action of a few environmental protection agency officials. It can be influenced by NGOs and citizens who have access to information and can thus monitor the actions of not only the firms but also the regulators.

It is too early to judge the relative importance of all these factors in the success of the PROPER program. Preliminary studies indicate that the companies most sensitive to reputation pressure are multinational; publicly owned firms are also sensitive to this pressure. Perhaps the most important mechanism that encourages success is the simple provision of correct, reliable technical and environmental information to managers (Afsah, Blackman, and Ratunanda 2000). In this respect, PROPER is fulfilling the role of an environmental audit. Firms that apply for international licensing such as ISO 14000 also are keen to use information in this way.

PROPER appears to have an advantage for the kind of quick, broad-based cleanup effort that constitutes the first stage of environmental management. To deal with the diversity of pollutants and processes, the tools for monitoring, licensing, and information must encourage good management practices. In these contexts, charges and tradable permits are not easily applicable. In some of the formerly planned economies in eastern Europe, long and detailed lists of chemicals with specific charges or fees were attempted. This scheme did not work well and is likely to be untenable, partly because of the costs of properly supervising such a complicated procedure. In contrast, the cost of the PROPER scheme was modest (approximately US$100,000) for the first 18 months. Almost 200 plants were rated, so the average cost was approximately US$500 per plant, which presumably could be financed through license fees. Given the pollution reduction of 40%, the PROPER program was cost-efficient. Unfortunately, the severe economic crisis in Asia more or less killed the program before it became fully institutionalized, and little of it remains.

Several other countries have adopted similar programs. Most advanced is the Philippine EcoWatch program, which has many similar features, including the color-coding system. Under EcoWatch, in the initial evaluation of 52 plants in the Manila area, 48 were found to be noncompliant (i.e., Red or Black ratings). This failure of conventional regulation was the main motivation for the labeling scheme. Within just over a year, the number of plants rated Blue increased from 8% to 58% in 1998.

Mexico, Papua New Guinea, and Colombia also have begun public disclosure systems, and a handful more countries have announced similar intentions or interests. In Colombia, the program will complement the charges discussed above. In India, an NGO has started the Green Rating Project to monitor the environmental performance of large Indian firms (Kathuria and Haripriya 2000).

## Two-Tier Pollution Regulation: India[11]

The Golden Corridor through the Indian state of Gujarat is a collection of pollution hot spots. Gujarat, located in west India on the border with Pakistan, has become a "toxic haven," as reflected by the relocation of the hazardous ship-breaking industry from the industrialized world to Asia. More than one-half of the world's ships are dismantled in one Gujarat town alone (Alang). This hazardous task involves exposure to large quantities of asbestos, arsenic, residual oils, paints, and other toxins. The breakup is not done in shipyards; the vessels are simply driven up onto the beach, stripped, and cut up with hand tools (such as torches) by migratory workers who lack protective clothing and accessories. One of the worst sources of pollution in this area is a group of chemical industries located in the Ankleshwar Industrial Estate (AIE), not far from Alang.

During the past 30 years or so, many environmental laws have been passed in India. These laws provide for the use of command-and-control measures, economic instruments, and institutions that facilitate participation in environmental management. The Environmental Protection Act empowered the government through the Ministry of Environment and Forests to create central and state pollution control boards intended to prevent, control, and abate pollution. In Gujarat, the highest environmental authority is the Gujarat Pollution Control Board (GPCB). It can demand information from any industry, issue fines, and even order closure. The most important legislative provision that has made polluters more conscious of and susceptible to legal action is public interest litigation (see Chapter 11). At least on paper, Indian authorities have a wide array of instruments at their disposal, including formal standards, varied frequency of inspection, subsidies, taxes, and fees as well as information disclosure, the imposition of polluter liability, and zoning.

### Industrial Estates in India

A large share of organized (polluting) industries in India are located in industrial estates, which are zoning tools for industrial dispersal and the industrialization of "backward" areas. This emphasis on zoning potentially could include a strong focus on adapting the nature of industry to the carrying capacity of the ecosystems, or at least some consideration of infrastructure such as water supply and waste disposal. At present, nearly 1,000 industrial estates exist in India.

Industrial estates may play an important role in the regulatory process. They have some form of democratic or cooperative structure, and pollution from within their boundaries can be considered nonpoint-source pollution to some extent. For regulators and residents outside the estates, it is difficult and expensive (although not necessarily impossible) to monitor individual plants, but aggregate pollution from an industrial estate is noticeable and attracts attention that may lead to bad publicity, complaints, and even plant closures. A two-tiered structure of monitoring may arise because more detailed monitoring is cheaper and easier for insiders (i.e., in the estate). In addition, industrial estates have some incentive to regulate and discipline their own members because all the member industries in an industrial estate can be hurt if the estate's reputation is tarnished. It is therefore relevant to refer to the literature on common property resources and to the theory of nonpoint-source regulation (see Chapters 4 and 13).

Although pollution control boards are formally mandated to monitor individual plants in an estate, given the enormity of the task and the costs involved, they may monitor only the aggregate emissions from the estate if the estate takes the initiative to monitor its members. The intermediary position of the estate is tricky, however. Its only instruments include the provision of infrastructure (partly subsidized by central authorities) and voluntary systems of fees and sanctions. It holds no formal enforcement power. In the following subsections, I discuss the efficiency of these instruments.

### Policymaking and Mediation: AIE

The AIE is one of the largest chemical estates in India. It covers an area of 1,605 hectares and houses more than 400 chemical plants. These units manufacture more than 25% of Gujarat's (5% of India's) output of pharmaceuticals, chemicals, pesticides, dyes, and dye intermediaries. The Ankleshwar Industries Association (AIA) has estimated that its members generate between 250 million and 270 million liters of liquid waste daily and approximately 50,000 tons of solid waste annually. As a direct consequence, villages on the periphery of the AIE must bear heavy costs in the form of water (and air) pollution. The major local water body, Amlakhadi Creek, is completely void of biological life. Neighboring villages have suffered from the poor quality of the groundwater, which allegedly has contributed to the death of some cattle.

During the past few years, the AIE and other estates have started to take some preliminary steps toward cleanup—perhaps in part because of easier access to public information, public interest litigation, and highly publicized tragedies such as that in Bhopal, which seem to have made polluters more conscious of the risk of legal action. The estate also is situated along the Bombay–Delhi railway route, highly visible to passersby.

The AIE has used several instruments to coerce its members, including provision of information, direct regulation (of emission standards), effluent fees, and fines. These efforts were partly related to the creation of a common effluent treatment plant (CETP) for small-scale industries and a central landfill for sludge and solid waste. These facilities were subsidized by the government but operate on the principle of polluter payment (i.e., each unit pays according to pollution load and quantity of waste generated). An additional and presumably important policy instrument for AIE is publicity. In 1989, AIA created the Ankleshwar Environment Preservation Society (AEPS) to take a more proactive stance. The two aims of the society are to assist industries in disposing of wastes and liquid effluents and in controlling air pollution, and to set up a laboratory for testing samples of air and liquid effluents. The society has purchased monitoring equipment, offered educational programs, and planted trees. It was one of the first organizations in the country to obtain tax concessions for pollution-control projects.

One of the activities of the AEPS is the monitoring of effluents and emissions from chemical industries in the AIE. If the pH, COD, and other values of samples are found to be in excess of GPCB standards, then the AIA environment committee calls the concerned units for discussions. After a couple of notices, the AIA levies a financial penalty or other sanction. These penalties are carefully graduated and increase proportionally with the frequency and severity of the offense, pollution, or method of discharge (some forms of illegal discharge outside the designated area are particularly reprehensible). For example, the penalty structure for the illegal pumping of acidic effluents into drains is listed in Table 25-6; similar disposal outside the drainage system carries even stiffer penalties.

**Table 25-6.** *Penalties Imposed by the Ankleshwar Industrial Association for Illegal Pumping of Acidic Effluents into Drains at the Ankleshwar Industrial Estate in Gujarat, India, 1998*

| | Penalty (rupees) | |
| --- | --- | --- |
| Offense | Small enterprises | Medium and large enterprises |
| First | 2,500 | 10,000 |
| Second | 5,000 | 20,000 |
| Third | 10,000 | 40,000 |
| Fourth | Report to the Gujarat Pollution Control Board | |

Other infractions against COD or other rules have different but similarly strict penalties.

After repeated infractions by a unit (or when penalties are not paid), the AIA's last resort is to report that unit to the GPCB. The existence of this threat is a decisive mechanism. The AIA will not overreport to the GPCB, because overreporting is a sign of failure on its part. The AIA also hesitates to report because reports of one unit may lead to criticism of the entire industrial estate. Additionally, inspection of one unit by GPCB may trigger inspection of other units in the estate. All of these reasons could explain why by early 1999, the AIA had reported only three perennial defaulters to GPCB. The way that GPCB reacts also will be important for the success of the AIA "voluntary" fines in future monitoring and detection situations.

From 1996 to 1999, several units were subjected to mild penalties for violating GPCB standards. Penalties levied and recovered from the polluters are listed in Table 25-7. Although the amount of penalty levied is increasing nominally, the rate of collection over time appears to be decreasing. This trend indicates an apparent erosion of AIE authority. However, the number of observations is too few to be certain, and the incomplete figures for 1998 are particularly uncertain.

The decrease in the percentage of collected fees and levied penalties is attributable to the fact that AIA and AEPS are voluntary organizations and not authorized by law to collect fines, so defaulting units may not feel obliged to pay. Compliance hinges on voluntary participation, but all industries should realize that it is in their common interest to achieve minimum abatement to protect their collective reputation. However, this situation is a classic "Prisoner's Dilemma," and the rapid decrease in voluntary compliance with the fines may be one example of a move from an initial collaborative equilibrium to a Nash equilibrium (see Chapter 10). After a few firms succeed in getting their fines waived or realize that nothing really bad happens to the 20% of firms that did not pay in the first year, other firms quickly lose their incentive to collaborate.

Rules and institutions must be carefully designed to avoid this kind of problem (Ostrom 1990). People who collaborate to manage common property resources tend to be on a relatively equal footing. They commonly are also linked by history and culture, which creates trust, traditions, and norms. However, among a collection of industries, the disparities in size and power may be enormous, and the cultural factor is definitely weaker than in a fishing or farming village.

**Table 25-7.** *Fines Levied and Collected by Ankleshwar Environment Preservation Society/Ankleshwar Industrial Association at the Ankleshwar Industrial Estate in Gujarat, India, 1996–1999*

| Year (1) | No. of fines levied (2) | No. of fines collected (3) | Collection rate (% of no. of fines) (4) | Total amount of fines levied[a] (5) | Amount collected[a] (6) | Collection rate (% of amount collected) (7) |
|---|---|---|---|---|---|---|
| 1996–1997 | 150 | 36 | 24 | 6.9 | 5.5 | 80 |
| 1997–1998 | 196 | 38 | 19 | 9.8 | 6.0 | 61 |
| 1998–1999[b] | 186 | 37 | 20 | 15.1 | 2.7 | 18 |
| Total | 532 | 111 | 21 | 31.7 | 14.2 | 44.6 |

[a]Amounts in hundreds of thousands of rupees (also called *lakhs*).
[b]Data incomplete; to December 1998 only.

One of the most important institutions for collective self-governance is a system for monitoring, enforcement, and punishment, not unlike the AIA's graduated sanctions. Other important criteria concern congruence among duties, obligations, and democratic influence. One might suspect that disparity in size and influence between the large and small plants would be a major difficulty, and anecdotal evidence from Gujarat suggests that many companies suspected that firms that had executives on the boards of the AIA or AEPS would receive more lenient treatment or have their fines waived. However, the collection rate for the total fine amounts (Column 7) is much higher than the collection rate of the number of fines (Column 4) (see Table 25-7). These results indicate that the AIA failed to collect the smaller fines, not the larger ones. The main reason probably is that small firms simply have no cash, which suggests that these data do not support an undue influence by the larger firms.

Fines levied and collected serve a dual purpose because they are used to finance recurring costs as well as part of the capital expenses of the AEPS testing laboratory. A decreasing collection rate has a direct bearing on future monitoring. Records of the penalty letters issued for COD violations show that in the first 6 months of 1998, 64 plants were given a total of 202 penalty letters amounting to total fines of 6 million rupees, twice the actual penalty collected by the AIA in the previous three years. Surprisingly, the owners of three faulting units are members of the executive committee and the environment committee.

### Provision of Infrastructure: Effluent Treatment, Landfill, and Pipeline

A crucial component in addition to the monitoring and enforcement system is the provision of local public goods in the form of infrastructure for abatement, treatment, transport, and disposal of wastes. This task is intimately linked with the monitoring task. The estate has set up an effluent treatment plant (the CETP) and a landfill, and a project to build a pipeline to the sea is under way. These projects warrant some comment.

In principle, the CETP provides primary, secondary, and tertiary treatment and was built exclusively for the small-scale industries, which were believed to have neither the necessary finance nor the know-how to undertake such treatment on their own. The state provided a 25% subsidy for construction costs and subsidized land, but the CETP is run as a private enterprise according to the polluter pays principle. The sludge from the CETP is to be deposited at the waste disposal site with some solid waste—also for a fee. The disposal site was built according to modern specifications to avoid leakage and also runs on a commercial basis (the only subsidy was half the cost of the land needed).[12]

A couple of serious flaws are evident. First, at a technical level, the CETP concept is not credible, because the pooling of various chemicals creates an unknown mixture that is difficult to treat properly. If many toxins are present, then microbes will also be killed off, making it impossible to properly treat even the biodegradable fraction—waste that would have been fairly simple to treat in isolation. Tests carried out by Greenpeace show that the treated effluent and the sludge contain the same toxins, casting considerable doubt on the efficiency of the treatment.

Also, serious organizational setbacks have plagued the AIE. Many small firms appear not to be members of the CETP and are suspected of using open drains or other illegal means of waste disposal. Small firms complain about high charges and the inflexibility of the contract. The charges are fixed, irrespective of size, making the fee astronomical for some and insignificant for others. Small firms also complain that dumping fees are not flexible with respect to waste load (which varies with production). Finally, they complain that competitors in other states do not face the same environmental regulation and thus can undercut their prices. In addition, several large firms use the CETP, even though it was intended to be used only by small firms. Apparently, either their own treatment plants are insufficient, or they save costs by using the CETP. This practice may increase the operating costs of the CETP and, ultimately, the fees that the small-scale industries are paying, which aggravates the controversy surrounding this issue.

It is also interesting to note the reasons some chemical firms pointed to for joining the CETP—that is, the indirect benefits of being an AIE member. Membership entitles the unit to a "zero discharge" certificate, which reduces monitoring pressure. Several units specifically said that membership meant fewer visits from GPCB officers. Under the amended Water Act of 1988, the GPCB wields a lot of power; it can strip electricity and water supplies to offending units. Anecdotal evidence indicates that some firms feel obliged to treat the GPCB officers well to avoid any risks, and thus they highly value any reduction in the frequency of their visits.

The Amlakhadi Creek, which is extremely polluted, is 14 kilometers long and empties into the Narmada River. The creek carries effluents from the AIE as well as Panoli and Jhagadia Industrial Estates. It was reinforced in 1995–1996 at an estimated cost of 6 million rupees (which was recovered from the industries) to reduce the spread of effluents to groundwater. However, the condition of the creek remains unsatisfactory, and the AIE environment committee has suggested constructing a pipeline to the deep sea for "safe disposal of the effluents." This project has some support from the World Bank, and although the dumping of toxic wastes into the sea is tragic and might even violate international conventions, it would be a local improvement.

Although insufficient, several serious efforts are made to limit pollution in Indian industrial estates, including the provision of infrastructure and the use of various policy instruments. The instruments include the provision of information, direct regulation (pollution standards), fees, fines, and subsidies at various levels. The voluntary efforts of the AIE have led to monitoring of the polluting industries, the collection of some self-imposed penalties, the construction of the CETP and the waste site, and the prospect of building a pipeline to the sea. Monitoring and enforcement appear to be difficult, and the effluent treatment and disposal systems have serious flaws. However, central authorities have grossly insufficient resources, and the AIE may have the potential to play the role of an intermediate principal in a two-tier regulation game. However, experience to date suggests that playing this role is not easy.

## Lessons Learned

Three waves of environmental policymaking in developing or transitional countries have been described: command-and-control regulation, economic incentives, and provision of information (Tietenberg 1998). To this excellent and concise characterization I add a fourth category: the construction of institutions for the allocation of rights that are fundamental for any market mechanism. In practice, many specific institutions are needed, for example, bodies that make rules for liability, courts, and organizations for monitoring and enforcement as well as the news media and agencies that deal with economic and environmental regulation at local, national, and supranational levels.

These three or four categories of policymaking do not necessarily have to be repeated in the same order as they occur in the industrialized countries. They are mutually dependent and must be built up in a fine balance that is unique to each particular case. Experience shows that several instruments can work in different contexts. Environmental taxes have played an important role in some of the cases discussed, such as in China and Colombia; permit trading has had some success in pollution abatement in Chile (see Chapter 7) and in other countries' fishing policies. Information provision and other soft instruments (e.g., voluntary agreements or two-tier regulation) have played important roles in Indonesia, Mexico, and India.

One might think that institutions must be in place before other policy instruments can function, and considerable logic and experience supports such a notion. However, some issues are so pressing that they need immediate regulation, even before the institutions have been established. Building institutions requires time and money; building a functioning hazardous waste program takes at least 10–15 years, even when financial resources are relatively abundant and there is a culture of law abidance (Probst and Beierle 1999). For this reason, moderate (and earmarked) fees may be a suitable part of an early instrument package. Some issues are formally under the jurisdiction of national authorities, but in practice, multinational companies are keen on standardization. Their subsidiaries are sensitive to the pressures on foreign markets, and in such cases, information provision can be an extremely efficient tool. Even small fees and the obligation to provide correct information require some form of regulation and a capacity for monitoring and enforcement, so at least some institutional capacity must be established early. Instruments and institutions then develop in a process of continuous interaction.

## Supplemental Reading

Bečvář and Kokine 1998
Bluffstone and Larson 1997
Klarer, McNicholas, and Knaus 1999
Markandya 1997
Tarhoaca and Zinnes 1996
Zhang 1998

It is impossible to generalize about distinct areas of the environment and natural resources, or about the groups of countries that are labeled "developing." Some countries have an almost unbroken and cumulative historic development over many hundreds or even thousands of years that gives them special characteristics in such areas as rights and institutions that hinge on trust, culture, and norms. Other countries have a recent, conflictive, or chaotic history with great cultural diversity (e.g., created through colonialism). Some areas of environmental management are themselves dependent on complicated technological and ecological science (e.g., waste from chemical or nuclear facilities), whereas others are characterized by known technology but fraught with problems such as asymmetric information, as with the joint management of many natural resources.

Efficiency, however important, is not the sole criterion for policy selection. Considerable attention must be paid to distributional, informational, and political criteria as well as to the process of policy implementation itself. Particularly in developing countries, where monitoring and access to technology and credits are particularly difficult, it is crucial to consider policies that avoid antagonism and encourage cooperation and involvement.

## Notes

1. Chinese environmental authorities estimate that more than 70% of all pollution in China is industrial. However, in many middle-income countries such as Mexico and Brazil, motor vehicles and other sources create the majority of the emissions (World Bank 2000).

2. Finding descriptive labels for groups of countries is difficult. Words such as "rich," "industrialized," "OECD," or "western" carry similar but imprecise connotations, especially when countries change categories. Poland, for instance, is poor by U.S. standards and was more so 20 years ago. However, the country even then was heavily industrialized—in fact, industrialization is the reason for many of Poland's environmental problems. Although Poland could have been contrasted with OECD countries in the past, it is hardly appropriate since 1996, when the country joined the OECD (with Hungary, Korea, and Mexico). I refer to the generally wealthier group of "developing" countries (Poland, the Baltic States, the former Yugoslavian states, and so forth) as "central and eastern Europe" and the former Soviet Union (i.e., Russia, Belarus, and Ukraine) as the New Independent States. On the other hand, references to OECD countries typically mean the richer, more industrialized economies of western Europe, the United States, and Japan.

3. Two-tiered taxes are mentioned in Chapters 7 and 15. Parts Four and Six contain more examples from the countries of central and eastern Europe and the New Independent States on issues such as leaded fuel, water, and forestry.

4. This section builds heavily on Afsah, Laplante, and Wheeler 1999, Wang and Wheeler 1996, and Wang and Wheeler 1999. It has greatly benefited from comments and information from Lee Travers, Hua Wang, David Wheeler, and Jian Xie.

5. *Visible pollution* is a measurement of particles and dust. The effect of invisible air pollutants (sulfur dioxide) and water (chemical oxygen demand) was much smaller.

6. Much of the information in this section is from the World Bank New Ideas in Pollution Regulation (NIPR) website (World Bank 2002). Thanks to David Wheeler of the World Bank, Thomas Black Arbelaez of MINAMBIENTE, and Jorge Garcia and others of the Universidad de los Andes (Bogotá, Colombia) for valuable comments and insights. As with many other schemes described in this book, social unrest is a constant threat to this kind of instrument (as to any other efforts at building a sustainable society).

7. The section builds on Blackman and Bannister 1998. Thanks to Alan Blackman for comments. See also Dasgupta, Hettige, and Wheeler 2000.

8. In some regions, kerosene has been subsidized to halt deforestation, particularly in arid, periurban areas. These policies have been unsuccessful in reducing fuel wood consumption (Mercer and Soussan 1992).

9. Wood accounts for more than 70% of energy requirements (Mulenga 1999). Although the township electrification program has connected many urban residents at a zero hook-up cost, many get disconnected because they cannot afford even the low rates.

10. Thanks to Shakeb Afsah for valuable comments and information for this section. Other sources for this section are World Bank 2000; Afsah, Laplante, and Makarim 1996; Pargal and Wheeler 1996; and Afsah and Vincent 1997.

11. This section is based on Kathuria and Sterner 2002 (see also Murty, James, and Misra 1999; Kuik et al. 1997; Greenpeace 1999).

12. This landfill was the first site in India where a public consultation process was carried out as per World Bank/Ministry of Environment and Forests guidelines. It was due to a previous setback in 1996, when an attempt to set up a landfill near an adjoining village was stopped by stiff local resistance and a public interest litigation was filed against the landfill.

## PART SIX

# *Policy Instruments for the Management of Natural Resources and Ecosystems*

$M$ANAGEMENT OF NATURAL RESOURCES AND ECO-systems—the "green issues"—is crucial for the world and economically vital in most developing countries. Ecosystem functions are both numerous and technologically and ecologically complex. The intricacy of an ecosystem is often mimicked by the social institutions designed to manage it. Risk management, uncertainty and information asymmetry, the risk of free riding, and distributional concerns must be addressed in the face of a partly stochastic outcome from nature. I do not pretend to cover all categories of natural resources in the next several chapters but have chosen a representative few that are ecologically significant, economically important, and illustrative of some theoretical issues of interest.

The public sector has a natural role in providing (or at least monitoring) water supply because this vital "good" has many special characteristics: merit or public good, externalities, and "lumpiness" of productive investments. At the same time, managing water is much more than supply technology; pricing and tariff structure are crucial issues, as discussed in Chapter 26. Waste management is seen as a "brown" environmental issue rather than a "green" natural resources issue. Nevertheless, waste management requires natural resources in the form of ecosystem capacity to recycle or assimilate waste (see Chapter 27).

Policy instruments for large industrialized as well as small coastal fisheries are discussed in Chapter 28. Agriculture is addressed in Chapter 29, particularly regarding nonpoint-source pollution, property rights, incentives, and risk management. In Chapter 30, I discuss forestry, particularly the allocation of land between agriculture and forests. Understanding this allocation is essential to any discussion of policy instruments (e.g., title deeds, tenure security, common property management, logging fees, concession policies, stumpage fees, subsidies, and tariffs). Finally, I explore ecosystem functions and their preservation through game reserves and ecotourism in Chapter 31.

# CHAPTER 26

# *Water*

THE CHARACTERISTICS OF WATER as a resource are unique in production and consumption. Water is essential for several distinct needs (drinking, cleaning, and irrigation), and the determinants of its demand are different for each use.

For drinking water, quality is of paramount importance; contaminated water is a major source of disease and death in developing countries. Because poor people have no alternative but to drink available water (often without further treatment) and because quality is not easily visible, the need for water monitoring and public involvement is considerable. Water quality is to some extent a merit good because some diseases caused by poor-quality water (e.g., cholera) are contagious, and their costs affect all of society. For equity reasons, there is a tendency to argue for cheap water for the same reasons as with electricity tariffs discussed in Chapter 25, but the sale of water below cost leads to a long-run lack of funds for system expansion and maintenance. Therefore, tariffs must be carefully designed (see Howe 1996 on equity versus efficiency).

Water is an essential input in consumption and production, perhaps the one that is the most commonly overlooked and taken for granted. Depending on local scarcity, water can be worthless or priceless. Water policy is a fruitful area for environmental economists because demand and supply depend crucially on so many technical and ecological parameters that good work must build on collaboration with local experts. There is no point in asking for the average production cost or individual valuation of a liter of water, because its value depends on the exact circumstances, timing, location, purity, and other characteristics.

Policymakers in France, Germany, and the Netherlands were among the first to use an ecosystem approach to water resource policy; water management does not follow the usual arbitrary political or historical borders like countries but is based on catchment basins, which are often self-governing. The use of some form of pricing to cover scarcity, supply cost, and the costs caused by effluents is fairly common in Europe.

Even though water economics has been extensively researched, many people still fail to grasp the pervasiveness of water scarcity; western experts have been described as having "water blindness" (Falkenmark 1999). In temperate countries, water usually is abundant; few people realize that water can be physically scarce, even in tropical climates that receive high rainfall, simply because of the high rate of evapotranspiration. Water may thus need much more careful management in a tropical country that receives 1,000 millimeters of rain per year than in a temperate one that receives 500 millimeters of rain per year.

The geographical distribution of water also contributes to the complexity of water policy. Ancient civilizations were created along many of the world's major rivers, but colonialism created national boundaries that follow a completely different logic. The number of potentially serious international conflicts over access to water has been increasing. Some nations are almost entirely dependent on "imported" water (such as Egypt, from the Nile River) because they receive little rainfall.

Water typically has been considered a free good. In many cases, ignorance about protection of the water cycle, misuse, and lack of legal and management structures are the main causes of water scarcity or stress. Although water is scarce in some countries because of low rainfall, many of the current water shortages are due to the lack of adequate policy, particularly with respect to pricing (e.g., Worldwatch Institute 1993). In Costa Rica and Laos, for example, rain is abundant but water is scarce because of poor infrastructure management, a lack of financial resources, and misuse.

Water economics has many similarities with common property resources (CPRs) and public services or goods such as urban public transportation, electricity service, municipal waste management, and security. Traditionally, some parties have made a powerful argument for the provision of these services free or at low, heavily subsidized rates. In addition to the public good argument, considerations of equity or income distribution have been used, for example, "because this service is so vital to life, we must see to it that even the poor can afford it." This argument is morally compelling and politically attractive, but everyone who has traveled on a bus in Tanzania, drunk the water in Mexico City, or wondered when the electricity will be switched on in a rural Kenyan village understands the trouble with it: the municipal or private entities that provide the service receive practically no revenue, and the service ends up being inadequate. In the end, the poor individuals—not the rich—are most directly affected. The rich have their own cars, mineral water (or other drinks), and so forth, whereas the poor are forced to make do with the publicly provided services.

In many of the formerly planned economies, natural resources, water, and energy were provided almost free of charge. As a result, there was no culture of economizing. Sometimes, the only way to regulate heat in one apartment block in the winter would be to open the windows (because there was no way of turning the heating system off) while apartments in other areas would be without heat as a result of severe rationing and fuel scarcity. Water demand was similarly mismanaged; some disastrous attempts at increasing water supply entailed diverting rivers. Polish authorities found that simply metering water use reduced consumption by about one-third (Zamparutti 1999, 147).

Although the principles of market economics and charging consumers according to cost are important, full-scale privatization of water supply is not necessarily always beneficial or even acceptable. In many places around the world, former municipal water supply systems are currently being privatized, which often means that they are sold to large multinational companies. For such arrangements to be socially acceptable, they need to be governed by very careful legislation and contracting that details the kinds of service expected. In some cases, supply systems are sold without such regulation, and the results may be detrimental to welfare.

## Water Management and Tariffication

For various reasons, the market mechanism cannot be expected to work perfectly for water management, and public policy must intervene in some form. When services are used primarily by the poor, it is important to get the valuation right. If consumers are willing to pay more for service that is reliable, then the welfare gains may be considerable. If not (and if the public good arguments are not sufficiently strong), then improving the services could be a waste of public resources.

From an economist's viewpoint, one of the main problems of organizing systems for water supply is that the fixed costs for water systems are typically high, whereas the marginal costs are low. Also, substantial economies of scale in several parts of the water supply industry create a fairly flat or even downward-sloping supply curve, which may result in a marginal cost below the average cost—and thus low or negative net profits. Water supply may be a "natural monopoly" that provides arguments for public involvement. Investments in water systems are typically large and indivisible, and there is an important distinction between short- and long-run marginal costs (see Figure 26-1). In the short run, supply (and demand) may be inelastic at Point A in Figure 26-1, with low prices because all the fixed costs of the system are "sunk costs" and the marginal costs small. When demand grows and exceeds the supply capacity of the current system, expensive new investments may be necessary to provide the necessary supply. The long-run marginal cost may thus be far above the corresponding short-run curve, and the long-run equilibrium would be at Point B in Figure 26-1, with a higher price level.

For the purposes of analysis, aggregate supply curves should be distinguished from individual supply curves. The cost of providing water (and other similar) services to an individual consumer often entails large fixed costs for distribution, connection, metering, administration, and control. Together with economies of scale in water production, these differences make a careful distinction between water management in rural and urban areas necessary. The high density of consumers in an urban area makes the per capita investment for the improvement of water services much smaller than in a rural area. With efficient management and an appropriate tariff structure, urban systems may even be run profitably. One example is the water enterprise in Santiago, Chile (Dixon and Howe 1993).

Water tariffication can incorporate elements of taxation or subsidies for different consumer groups in different areas and thus be an income-redistribution instrument. Profits that originate in economies of scale can be invested to improve water supply in poorer (rural) regions, where economies of scale are not

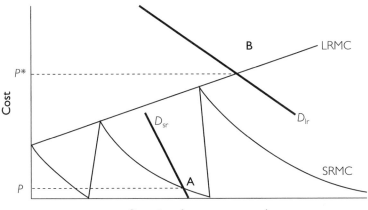

**Figure 26-1.** *Short- and Long-Run Costs*

*Notes:* $P$ and $P*$ = short-run and long-run equilibrium price, respectively; $D_{sr}$ and $D_{lr}$ = short-run and long-run demand, respectively; $A$ and $B$ = short-run and long-run equilibrium, respectively; SRMC and LRMC = short-run and long-run marginal cost, respectively.

attainable. Because of the supplier's fixed costs, small, rural customers are more expensive to supply than large, urban ones, and this point should be considered when constructing tariffs. Municipal water has several interesting features: The marginal supply costs are generally small in the short run but rise in the long run. Distributional concerns are important, and clean water is a merit good, the quality of which is difficult to monitor. Finally, administrative and distribution costs are sizeable, making efficiency important. Thus, tariff structure and organizational structure of the supply companies have important repercussions for welfare in this area.

The concepts of unitary, progressive, and regressive (efficiency) tariffs discussed in Chapter 25 for electricity also apply to water. Actual tariffs for water can follow any of these basic structures as well as more complicated "compromises" in which various combinations are used. In a dry country such as Morocco, the price of water varies by a factor of five between areas. China, perhaps for ideological reasons, applied a system of unitary water tariffs across the country; this approach is one of the factors that led to a concentration of heavy industry in the north, where water is less plentiful. Today, those areas are experiencing costly water shortages (Wang and Wheeler 1999). In the United States, tariffs have traditionally reflected the cost structure well with a decreasing block structure (i.e., a step-shaped structure like the efficiency tariff). This approach has been criticized for benefiting the large consumers and leading to waste, and lately, many regions have adopted increasing block rates instead. The increases are moderate, and rate-setting is greatly complicated by the existence of prior rights (see Chapter 4; see also Brill, Hochman, and Zilberman 1997 on efficient water pricing schemes for water agencies when members have property). Some interesting experiences from the Middle East, South America, and Africa as well as the case in which monitoring is not possible are presented later.

Despite the emphasis on tariffs in this chapter, mere tariff revision—however useful—may be insufficient in some situations. South Africa, for example, faces dire water shortages in a context where demand from different sectors (household, agriculture, and industry) could generate a good deal of conflict. Water policies must combine many elements, from supply enhancement to pricing, from information campaigns and the elimination of foreign tree species with high water requirements to the sale of permits for the right to plant trees (because of their water consumption).

## Tariff Structures in Some Middle Eastern Economies

Decisionmakers in many countries appear to be convinced that because the aggregate supply curve for water slopes upward, so should the individual tariff, but this is not necessarily so—at least not in terms of allocative efficiency. If a country faces (increasing) scarcity of water, then the tariff levels should be appropriately high (and maybe rising over time) but not necessarily rising over the span of consumption rates. The aquifer cannot discriminate between a small consumer consuming his first gallon and a large consumer consuming an additional gallon, and thus, there is no technical efficiency argument for progressivity.

Given the importance of water for health and welfare, it may be appropriate in some contexts to have a subsidized lifeline rate. This rate specifically applies to poor consumers and is logically tied to the volume of water consumed for personal use (i.e., washing, drinking, and cooking)—on the order of 10–20 liters/day. Sometimes the lifeline argument is taken too far, as in Saudi Arabia, where tariffs were set to almost zero (<US$0.05/m$^3$) for a monthly consumption of up to 100 cubic meters, which translates to use of more than 3,000 liters/day. Presumably, most families are on the first step of the tariff, which in practice removes the progressivity (see Figure 26-2). Even so, in many cases, the poor consumers who are supposed to be the main beneficiaries of the lifeline and increasing block tariffs do not have permanent, individual household connections and end up consuming poor-quality water collected locally or water purchased from water vendors at extremely high rates.

Geographical location and season are often the main determinants of water scarcity and thus the marginal cost of providing water. Reflecting these variations in tariffs can be politically difficult. However, not doing so can decrease overall allocative efficiency. In Kuwait and the United Arab Emirates, (unitary) tariffs are high because actual costs have been allowed to influence tariffs to a large extent. This gives some perspective on the subsidies implicit in the tariffs in Saudi Arabia. Even so, water experts at the World Bank indicate that Saudi Arabia appears to be modifying the "lifeline" rate by reducing the maximum monthly water consumption at the extremely low tariff.

## Water Tariffs in Chile

In recent decades, Chile has pursued decidedly neoliberal policies. Chilean authorities have set out to privatize (as much as possible) previously public water-

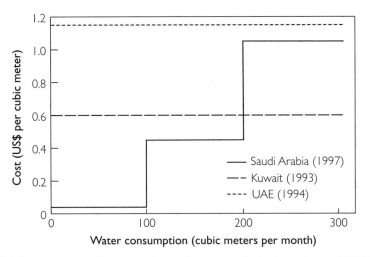

**Figure 26-2.** *Water Tariffs in Saudi Arabia, United Arab Emirates (UAE), and Kuwait*

management services. The intellectual defense is that the role of the state should be to regulate services, not provide them. The process of privatization was somewhat stalled by the return of democratic government in the 1990s, and currently, only a small percentage of urban water and sanitation are provided by truly private enterprises. However, they commonly are provided by public corporations that operate as independent legal bodies, clearly separate from the regulatory agencies. The fact that the state must have a regulatory role does not appear to have been disputed. Natural monopolies (sizeable fixed costs creating barriers to entry), external effects (pollution and use of scarce common pool resources), and asymmetries of information between producers and consumers are the major reasons behind the need for regulation.

To promote privatization, laws have been passed to require tariff construction to diminish the effects of information asymmetry and to make transparent the externalities that are being internalized by the tariffs. The tariffs must be based on calculations for "model enterprises" that allow for cost recovery and profitability (7% return) for efficient firms and must avoid cross subsidies between user categories. The recommended tariff structures include fixed charges for fixed costs (such as metering), capacity charges (depending on seasonal costs), consumption charges, and charges for waste treatment. All of these charges vary by region (and season, where applicable), depending on the actual levels and costs of water supply, infrastructure, treatment, transport, sewage, and so forth.

When the new tariff structures were applied beginning in 1990, the rate increase was particularly dramatic for the low-volume consumers, who previously had the most heavily subsidized rates (more than 450% for users of <20 m³/month in Region III). To soften the blow, the reform was implemented slowly over four years. In addition, a system of subsidies still exists for consumers who are unable to pay their water bills. To achieve transparency in public affairs, this subsidy is not automatic in the form of a reduced "lifeline" tariff but a direct subsidy. To receive this subsidy, consumers must submit a written application to the local municipality explaining their incapacity to pay and meet several conditions. One condition for eligibility is that a water bill for 20 m$^3$ must be more than 5% of the household's income.

In 1995, more than 15% of customers received subsidies, and this arrangement might be considered both efficient and sensitive to the needs of poor individuals while preserving transparency in the public use of money. However, the system incurs costs of administration (for the public agencies and for the poor consumers) and presumably may be somewhat humiliating for those who are forced to apply for the subsidy.

## Water Management, Laws, and Pricing in Southern Africa

Southern Africa experiences considerable water stress. Large areas are arid or semi-arid. Evaporation is high, and rainfall is moderate and erratically distributed over regions and time. These characteristics make it important to organize the storage, transport, and general management of water. Requirements for these functions are radically different in Botswana and South Africa. Whereas Botswana is mainly a rural, pastoralist society, South Africa is a highly diverse and complex society with many competing users.

### South Africa[1]

Despite the strong position of landowners in South Africa, riparian rights do not enjoy a strong legal standing (see Chapter 5). According to the 1998 National Water Act, all water rights are owned by the state; however, ownership and user rights are distinct. The legislation also recognizes some historic private user rights. In some respects, the catchment management associations and water users associations strengthen such private user rights through extended stakeholder participation in the allocation of user rights. However, the law states explicitly that water rights do not stem from the ownership of adjacent land (contrary to the riparian principle).

Decisionmakers have made a distinction between water resource management, which is regulated by the Water Act, and the provision of water, which is regulated by the Water Services Act. The Water Act creates new rights based on the principles of equity, sustainability, and efficiency. Under the equity principle, the act recognizes basic human water needs as a right—in practice, each person is entitled to 25 liters/day, irrespective of income. Poor consumers pay nothing up to this level. Under the sustainability principle, the act recognizes the need for an ecological reserve, sometimes referred to as the "instream flow requirement," that is not available for any economic use. Because the ownership of water is ultimately public and because ecological needs are a first priority, restrictions may be placed on land use and other activities that have effects on water availability or quality. One example is the restriction on planting trees (due to their demand for water) (see Box 26-1).

## Box 26-1. Prize-Winning Water Policies

Professor Kader Asmal, former minister of water affairs and forestry in South Africa and chairman of the World Commission on Dams, was awarded the 2000 Stockholm Water Prize for his extraordinary efforts in the field of water management in South Africa.

When Asmal was appointed to the ministry in 1994, more than 16 million South Africans did not have reasonable access to safe drinking water. In 1995, he created Working for Water, a water conservation program intent on removing invasive alien plant species that rob South Africa of up to 7% of its mean annual runoff, overtake its most productive lands, and threaten its biological diversity. He encouraged major water reforms in the 1998 National Water Act and the Community Water Supply and Sanitation Program, which focuses on providing access to the basic levels of service required to ensure health for all South Africans. More than 7 million people have already benefited, directly or indirectly, from these innovative programs.

One of South Africa's water management policies is aimed at controlling the planting of trees. Any commercial planting requires a license from the Department of Water Affairs and Forestry for what is called a "stream flow reduction activity." The licensing process also involves the Department of Agriculture and the Department of Environmental Affairs in a holistic approach to water management.

*Sources:* All Africa 2000; SIWI 2000.

Because of the rights created to provide for basic human water needs, the state is obliged to organize water supply, particularly in impoverished regions. The geophysical conditions of the country make this task expensive; they put considerable strain on already insufficient public funds and thus necessitate the involvement of the private sector through privatization or other arrangements. Although the Water Act promotes the role of the state as custodian of water related to basic human needs and ecological needs, the Water Services Act of 1997 strives for efficiency and makes provisions for privatization and competition at various levels along the water supply chain. Public–private partnerships are a new element of water management in South Africa (Conradie et al. 2001).

### Botswana

In Botswana, the role of water pricing in the overall management of water resources is limited in the informal and rural sectors, where a culture of consultation appears to make information and persuasion the best instruments (Arntzen 1995). During the long and severe drought of the 1980s, the water supply of southeastern Botswana was threatened, and educational campaigns proved effective in reducing water demand. This effect was not temporary; demand has decreased permanently. However, the pricing still needs revision in the rural sectors, because local councils charge farmers a flat fee for access to boreholes regardless of the number of cattle, which implicitly results in a decreasing price per liter and thus creates a perverse subsidy to overstock cattle.

In urban sectors (towns and large villages), water pricing is based partly on cost recovery and includes elements of progressive pricing (see Table 26-1) as well as

Table 26-1. *Water Prices in Gaborone, Botswana, 1992 and 1993*

| Consumption level (m³/month) | Rate ($US/month) | |
|---|---|---|
| | 1982 | 1993 |
| First 10 | 0.21 | 0.22 |
| Next 5 | 0.32 | 0.65 |
| Next 10 | 0.32 | 0.84 |
| Next 5 | 0.32 | 0.84 |
| Next 10 | 0.42 | 1.15 |
| Next 10 | 0.42 | 1.15 |
| >50 | 0.37 | 1.15 |

*Note:* Constant 1990 prices were converted to U.S. dollars.

*Source:* Adapted from Arntzen 1995.

geographically differentiated pricing (cheaper in the north, where water is less scarce). At the same time, state subsidies reduce the average cost. These subsidies should be either eliminated or better targeted to the people who need them most (or funneled into research on water-saving technologies) (Arntzen 1995). Other practices also should be discouraged, such as employers paying their employees' water bills (e.g., through taxes on such implicit benefits).

## Pricing Water When Metering Is Not Possible[2]

Pricing water for agricultural use is often extremely difficult because of riparian rights or practical reasons. It is nonetheless desirable when water is scarce. Irrigation water is commonly free or charged an acreage-based fee that might fund the waterworks or meet some notion of fairness. However, by definition, such a pricing scheme cannot have any incentive effect on economizing water. Sophisticated meters may increase the feasibility of metering. In some cases, technological progress can reduce monitoring costs significantly, changing the conditions on which the choice of instruments hinges. (For example, time-of-day meters for electricity and on-board or curbside monitoring of air pollutants from cars were unimagined just decades ago.) The sociocultural context also is crucial. In a closely knit society, monitoring by neighbors may be a viable alternative to monitoring by more distant central authorities or companies. In this case, CPR schemes may have comparative advantages. If compulsory metering is not possible, then it is impossible to have a direct incentive effect on water consumption.

A couple of options may still be considered. The first option is partial or voluntary metering. Requiring a category of users to install meters may be possible if they are distinguishable as particularly intensive (or wasteful) water consumers or if other special characteristics (such as location) apply. Second, offering some water users the option of metering may be successful if the alternative is a fairly drastic fixed charge. The third option is policy instruments, which may target the acquisition of certain types of capital (water-saving or water-wasting) equipment.

Finally, authorities may target the choice of crop and cultivation methods, which are more readily observable (although only indirectly related to water use). They would then design the charge so as to use the "revelation principle" (see Chapter 13), allowing water authorities some control over irrigation use even without metering. Designing such tariffs is mathematically rather complicated. However, the approach used is of great general interest because it could be applied to pollution charges or other natural resources fees. I illustrate the method for one concrete example, irrigation water.

Assume that the principal knows all farmers have a productivity type $\beta$ (depending on slope, crops, soil quality, and maybe skills) and that $\beta$ has a rectangular distribution over $\{0,1\}$ such that the farmers' harvest ($q$) will be a function of water use ($w$), as in

$$q = (1 + \beta)w^{0.5} \tag{26-1}$$

Assume that the farmer has private costs of irrigation and pumping ($c$) per unit of water and the price received for each unit of harvest is $P$. The farmer will then maximize private profits ($\pi$) as

$$\pi = Pq - cw \tag{26-2}$$

The water, however, has an external scarcity-related value, $e$. Social welfare considerations would thus be attained by maximizing the social profit function ($\pi*$) as

$$\pi* = Pq - (c + e)w \tag{26-3}$$

Inserting Equation 26-1 into Equation 26-3 and differentiating gives

$$\pi*' = 0.5P(1 + \beta)w^{-0.5} - (c + e) = 0 \tag{26-4}$$

Maximizing gives

$$w* = \frac{P^2(1+\beta)^2}{4(c+e)^2} \tag{26-5}$$

$$q* = \frac{P(1+\beta)^2}{2(c+e)} \tag{26-6}$$

where $q*$ and $w*$ are the socially optimal values for production and water use, respectively, and are distinct from unregulated market outcome ($q_m$ and $w_m$, respectively). Market values $q_m$ and $w_m$ are calculated by using Equations 26-5 and 26-6 with $e = 0$. With full information, the traditional Pigovian instrument would have been a tax of $e$ per unit of water. Tradable water rights with a price of $e$ would have worked equally well.

If monitoring is reasonably inexpensive, then the best policy is to require meters. However, if distributional effects and political factors block any reform that attempts to make water meters compulsory, then it would be in the interest of the entire farming community to show that meters do not work. A decentralized policy with somewhat better chances of successful implementation might be the imposition of a fixed fee on each user for the total scarcity cost combined with an exemption for users who adopt water meters (i.e., metered users would pay for actual water consumption). Large-volume consumers typically prefer a fixed charge, whereas low-volume consumers typically prefer monitoring. In general, when compulsory metering is dominated by partial adoption of meters by a subset of agents, a policy of decentralized monitoring can improve social welfare (e.g., Millock, Sunding, and Zilberman 2002). The size of the fixed fee is crucial. A low fee will lead only a small fraction of users to choose to monitor (thus, only that fraction will select optimal water use and production levels). However, a high fee will invite political resistance and perhaps questions concerning the feasibility of the policy. Imposing a high fee to encourage the use of water meters relies on two strong assumptions: the regulator's threat of a fixed charge is credible, and there are no transaction costs from taxation.

If these issues cannot be resolved or if monitoring is physically difficult (as it will be close to water bodies), regulating water use in a welfare-maximizing manner would seem to be an impossible task. However, possibilities exist for regulators who have information that is correlated to water use. Assume, for instance, that the regulator can monitor output perfectly and is the sole buyer of the crop (e.g., through an agricultural marketing board). The regulator could then offer a menu of two–part contracts with different harvest levels and water fees designed so that all farmers reveal their type ($\beta$). The contracts $[T(\beta), Pq(\beta)]$ would consist of a fixed water charge (or entry fee into the scheme), $T(\beta)$, and an offer to buy an output (harvest) that depends on each farmer's type, $q(\beta)$, at a price $P$. (In excess of this level, there would be no payment).

The revelation mechanism hinges on the existence of two components in the contract. Farmers might be tempted to declare a low $\beta$ to get a low water charge but then would have the right to produce and sell only a small harvest $q(\beta)$. The calculation of the fee scheme $T(\beta)$ that satisfies incentive compatibility Conditions 26-8 and 26-9 is given in Box 26-2. If the private cost of pumping water $c$ = 1, harvest price $P$ = 5, and the external scarcity rent of water $e$ = 0.5, then $T(\beta) = (25/9)(2 + 2\beta + \beta^2)$ (see Equation 26-10). Using this scheme, the payoffs can be calculated for farmers of different true types ($\beta_t$) if they claim to be other alleged types ($\beta_a$).

If a farmer reports $\beta_a > \beta_t$, then the marginal tax rate is higher than marginal gross profits (before tax). If a farmer reports $\beta_a < \beta_t$, then the marginal tax rate is lower, but then his or her allowable harvest is lower, which has a stronger effect on profits (see Table 26-2). For a farmer of type 1, the socially desirable production level (according to Condition 26-6) would be 20/3, or 6.7 units. The regulator will offer a contract intended for this type. A farmer of type 1 who agrees to this contract will be allowed to sell 6.7 units, giving a gross revenue of 33.3 and a gross profit of 22.1; the farmer will pay the fee of 13.9 and have a net profit of 8.2. If instead the farmer claims $\beta_a$ = 0.5, the water fee would be 9.0 instead of 13.9, but the farmer would have the right to sell only 3.7; therefore, net profit would fall from 8.2 to 6.2. Other types can be verified similarly.

The presence of transaction costs in raising tax revenue complicates the issue further, introducing informational rents and making the attainment of a "first-best" solution impossible (Smith and Tsur 1997). If transaction costs are too

**Table 26-2.** *Truthful Revelation of Types Applied to Water Contracts*

| True type | Alleged type | | |
| --- | --- | --- | --- |
| | $\beta_a = 0$ | $\beta_a = 0.5$ | $\beta_a = 1$ |
| Contracted quantity | 1.7 | 3.7 | 6.7 |
| Revenue | 8.3 | 18.7 | 33.3 |
| Tax $[25/9(2 + 2\beta + \beta^2)]$ | 5.5 | 9.0 | 13.9 |
| *Gross profit (net profit)* | | | |
| $\beta_t = 0$ | 5.5 (0) | 4.7 (–4.3) | –11.1 (–25) |
| $\beta_t = 0.5$ | 7.1 (1.6) | 12.2 (3.2) | 11.1 (–2.8) |
| $\beta_t = 1.0$ | 7.6 (2.1) | 15.2 (6.2) | 22.1 (8.2) |

*Notes:* $\beta_t$ = true type; $\beta_a$ = alleged type. Numbers refer to the example in the text. Net profit = gross profit minus tax.

large, then society may be better off without regulation. Regulation, and particularly regulation with transaction costs, generally appears to decrease the desirable level of production from farmers with low-quality type. This effect may raise distributional concerns.

This model is sensitive to changes in specification. Suppose, for instance, that there is some uncertainty about the exact shape of the function (see Equation 26-1); it might instead be $q = (1 + \beta)w^{0.6}$ or some other small deviation. Changing the elasticity estimate might be in the interest of low-output farmers (on poorer soils) who want to prove that they are not inefficient. Other changes in the formula would benefit other categories. If considerable amounts of resources depend on such numbers, then lobbying or rent-seeking is expected.

Notice also that the contracts designed do not provide incentives to economize on water after the contract is signed. The whole effect on water is indirect, working through the optimal output levels. Therefore, interactions with other policies might be expected. Many policies already target output levels: export promotion, agricultural support, poverty relief, crop insurance, income tax, and property tax, to name a few. When the policy proposed is implemented in addition to other such policies, the net effect may be blurred and even unpredictable. In the example above, this effect is partly indicated by the rising transaction costs. An increasing divergence between private costs ($c$) and social costs ($c + e$) also would tend to exacerbate the problem. In several cases, $c$ may in fact be close to 0, making the role of $e$ relatively more important.

The model presented in this chapter is highly stylized. That is the function of models—otherwise, cases cannot be analyzed. In this case, I assume that the regulators have no information about individual water use but full knowledge about all technologies as well as full visibility and control of crop output. In real life, crops generally depend on stochastic factors, adding moral hazard. It is not difficult to imagine that different crops have different monitoring costs. Subsistence farming would require a good deal more monitoring than would export crops such as coffee or cotton, which are sold to centralized agents, marketing boards, or factories. If some farmers were growing cotton or subsistence crops and the poorer marginal

## Box 26-2. Design of Self-Revealing Tariffs

If the range of technologies (Equation 26-1) is known, then contracts can be designed such that each farmer pays a fee ($T_\beta$) for the right to sell harvest ($q_\beta$). With increasing farm quality ($\beta$), both $q_\beta$ and $T_\beta$ will rise. The regulator sets $q_\beta$ equal to the level that would be optimal if $\beta$ were truthfully reported [i.e., $q^*(\beta)$]. The issue then becomes how to design the fees. The regulator knows the technologies and costs and can thus simulate what each farmer of type $\beta$ would do given his or her true type. The regulator must choose the fee so as to make each farmer pick the contract corresponding to his or her type. Note that profits are gross profits minus the fee, $T(\beta)$, and can thus be written as

$$T(\beta) = Pq(\beta) - cw[q(\beta)] - \pi(\beta) \tag{26-7}$$

where $P$ = harvest price, $c$ = private cost of pumping water, $w$ = water use, and $\pi$ = profit.

Consider a farmer of true type ($\beta_t$) contemplating the false reporting of alleged type ($\beta_a$). Because the allowed quantity of production and tax rate depend on the type reported, the farmer will report $\beta_a$ to maximize profits. The first-order condition for the farmer is

$$\frac{\partial \pi}{\partial \beta_a} = P \frac{\partial q(\beta_a)}{\partial \beta_a} - c\left(\frac{\partial w}{\partial q}\right)\left(\frac{\partial q}{\partial \beta_a}\right) - \frac{\partial T(\beta_a)}{\partial \beta_a} = 0 \tag{26-8}$$

The tax schedule $T(\beta_a)$ must be made to (at least) match the increased revenue due to false reporting. For the planner, the marginal social profit is

$$\frac{\partial \pi}{\partial \beta_t} = -c\left(\frac{\partial w}{\partial \beta_t}\right) \tag{26-9}$$

If $\beta_a = \beta_t = \beta$, then $\partial \pi / \partial \beta = P\partial q\ (\beta)/\partial \beta - c(\partial w/\partial q)(\partial q/\partial \beta) - c(\partial w/\partial \beta) - \partial T(\beta)/\partial \beta = -c(\partial w/\partial \beta)$ because $P\partial q(\beta)/\partial \beta - c(\partial w/\partial q)(\partial q/\partial \beta) - \partial T(\beta)/\partial \beta = 0$ from Condition 26-8. This is the envelope condition for truth-telling and is an empirical counterpart to Conditions 13-3 and 13-4 in Chapter 13.

The design of the optimal tariff can be illustrated using a numerical example in which $c = 1$, $P = 5$, and the external scarcity rent of water ($e$) = 0.5. It gives optimal values for harvest and water cost of $q^* = (5/3)(1 + \beta)^2$ and $w^* = (25/9)(1 + \beta)^2$, respectively. For the farmer who cheats, however, the values for harvest and water cost are $q(\beta_a) = (5/3)(1 + \beta_a)^2$ and $w(\beta_a, \beta_t) = [q(\beta_a)]^2/(1 + \beta_t)^2 = (25/9)(1 + \beta_a)^4/(1 + \beta_t)^2$. Equation 26-9 gives

$$\frac{\partial \pi}{\partial \beta_t} = 2c\frac{[q(\beta_a)]^2}{(1+\beta_t)^3} = \left(\frac{50}{9}\right)\frac{(1+\beta_a)^4}{(1+\beta_t)^3} = \left(\frac{50}{9}\right)(1+\beta)$$

when $\beta_a = \beta_t = \beta$. By assuming a uniform distribution of $\beta \in [0,1]$, integrate over $\beta$ to calculate the profit ($\pi$):

$$\pi = \int_0^\beta \left(\frac{50}{9}\right)(1+\beta) = \left(\frac{25}{9}\right)(2\beta + \beta^2)$$

Hence, from Equation 26-7, the optimal tax rate is

$$T(\beta) = \left(\frac{25}{3}\right)(1+\beta)^2 - \left(\frac{25}{9}\right)(1+\beta)^2 - \left(\frac{25}{9}\right)(2\beta + \beta^2) = \left(\frac{25}{9}\right)(2 + 2\beta + \beta^2) \tag{26-10}$$

continued on next page

## Box 26-2. Design of Self-Revealing Tariffs (continued)

To verify that this tax forces the farmer to tell the truth, calculate the marginal profitability of lying about $\beta$ from Equation 26-8:

$$\frac{\partial \pi}{\partial \beta_a} = \left[ 5 - \frac{2q(\beta_a)}{(1+\beta_t)^2} \right] \left( \frac{10}{3} \right) (1+\beta_a) - \left( \frac{25}{9} \right) (2 + 2\beta_a)$$

$$= \left[ 3 - \frac{2(1+\beta_a)^2}{(1+\beta_t)^2} \right] \left( \frac{50}{9} \right) (1+\beta_a) - \left( \frac{50}{9} \right) (1+\beta_a) = \left( \frac{100}{9} \right) (1+\beta_a) \left[ 1 - \frac{(1+\beta_a)^2}{(1+\beta_t)^2} \right]$$

Setting $\partial \pi / \partial \beta_a = 0$ as a maximum condition gives $(1 + \beta_a)^2/(1 + \beta_t)^2 = 1$, and thus $\beta_a = \beta_t$. This result shows that profit is (now) maximized by truthful reporting of $\beta_t$.

farmers had difficulty growing the export crop without large quantities of scarce water, then a simplified two-price mechanism might build on discrimination between crops and farm types. By requiring signed contracts ex ante with farmers for cotton and specifying an "entry fee" (as a proxy for the unobservable use of water) for such contracts, farmers with superior conditions for the relevant crop would self-select without interfering with the production of the subsistence crop. Water monitoring is often expensive, and in these cases, separating equilibrium may be interesting where the authorities charge for water based on acreage and crop. By providing farmers incentives for growing less water-intensive crops, the policies will have at least some effect on total water use.

One more category of situation should be mentioned. Other kinds of capital equipment may be needed for production that provide new possibilities to policymakers. The production function might be more complicated than Equation 26-1, and complementarities between capital and water may give valuable information. The French water agencies design contracts to make firms extend their water treatment more than motivated by the current tax, which is fairly low (A. Thomas 1995; see also Chapter 13). To set the correct abatement requirements for each firm, regulators need information about each firm's individual abatement costs, which is effectively acquired by offering firms a subsidy for this pollution control equipment.

## CPR Management of Water

CPR management systems for collective property rights can be seen as policy instruments through which users of natural resources at the local level manage common pool resources under difficult circumstances related to the stochasticity of natural conditions (see Chapters 5, 10, and 13). Irrigation schemes usually are used to illustrate CPRs.

In the Spanish towns of Alicante, Murcia, Orihuela, and Valencia, social organizations have evolved for regulating irrigation for the small-scale farms called *huertas* (Ostrom 1990). Rainfall is low and erratic in this warm, dry region, and the water level in rivers is variable. Thus, the success of agriculture is wholly dependent on irrigation. Over many hundreds of years, farmer organizations have evolved in response to the stochastic and somewhat harsh environment. Written records go back about 600 years, but the many Islamic influences in the procedures indicate that irrigation associations were in place several hundred years earlier, under (and presumably before) Muslim rule.

**Supplemental Reading**

Houston and Sun 1999
Sanford and Stroud 2000
Wood, Handley, and Kidd 1999

One of the characteristic features of the *huerta* irrigation organizations is the *Tribunal de los Aguas* (Water Court). In Valencia, for many centuries it met every Thursday outside the Apostles' Door of the cathedral. The *Tribunal* never has been an official part of the Spanish judicial court system, but its jurisdiction over the management of the irrigation canals is nonetheless respected. The governing officials of the *Tribunal* are democratically elected by the farmers. The officials hire guards and have the power to allocate water. The main allocation mechanism for irrigation water is a physical system of rotation. Each farmer is allocated water in turn and can take as much as wanted (except in periods of water shortage, during which the chief official or syndic can order additional rationing). A farmer who misses a turn by failing to open the gate when the water arrives must wait an entire rotation for another turn.

Under these circumstances, the farmers automatically monitor each other as they wait for their turns; however, the temptation to take water out of turn may be extremely high. The main tasks of the court are to solve conflicts concerning the incorrect taking of water and sometimes allegations of incorrect management by the syndic or his or her employees. Carefully graduated fines appear to play an important role in maintaining the integrity of the institution and avoiding the temptation to free ride. This longstanding practice was recorded in the fine books as far back as 1443, when more than 400 fines were collected in one year.

Details of *huerta* irrigation management in Alicante, Murcia, and Orihuela differ in numerous details from that of Valencia, showing the importance of adaptation to local conditions, both ecological and social. Only through this kind of adaptation are the motivation, credibility, authority, and flexibility of the systems maintained (see Chapter 10). Other irrigation schemes, such as the Zanjera irrigation communities of the Philippines (Ostrom 1990), have some elements in common with the *huerta* systems, perhaps due to the transfer of some experience in combination with Spanish colonization. However, the ecological and technical conditions are distinct.

Water scarcity is not such a prevalent problem in the Zanjera irrigation communities. Instead, the rivers are prone to excessive water flow and flooding, which destroy dams and other simple irrigation constructions built by the shareholder farmers. Farmers therefore spend 40–50 days of labor per year on heavy, risky repair and maintenance work. The temptation to shirk collective repair work and thus free ride on the efforts of others is perhaps the greatest challenge to be overcome in such a system—more important than the allocation of water in times of drought. The institutions are accordingly focused on the issue of collective labor supply for maintenance.

# Notes

1. This text is based mainly on Conradie et al. 2001 and Department of Water Affairs and Forestry 1996. Thanks to Martine Visser for valuable comments.

2. This section is based on Smith and Tsur 1997 (see also Millock and Salanié 1997; A. Thomas 1995). Thanks to Katrin Millock for good discussions on this topic. This section contains more mathematical detail than other sections to show how a self-revealing tax can be constructed; the most technical part is presented in Box 26-2.

# CHAPTER 27

# *Waste*

*F*OR MANY POLICYMAKERS, waste management is the epitome of environmental management. The dumping of solid wastes and the flow of untreated effluents can disrupt sensitive ecosystems, leading to the deterioration of water quality and thus poor human health. The handling of hazardous waste is especially worrying, given the developed countries' unfortunate trend toward dumping their wastes in developing countries. Related issues include the health problems of people who work in (and, in poor countries, even live on) waste sites.

Much effort is devoted to waste management, but sometimes the policies being pursued appear to be more symbolic than effective. In many countries, a large amount of political effort is devoted to promoting recycling. Germany is a good example, where extensive legislation requires the recycling of various categories of waste. It seems however that not enough effort has been devoted to the issues of what should be recycled, why, and which policy instruments should be used. The price mechanisms on the markets for used goods must be understood in order to discuss recycling. Some products have a high price, so the market "recycles" them automatically. Consider gold: have you ever heard of piles of gold (or other precious metal) waste? If a good becomes scarce, its price rises, creating incentives for reuse and recycling (see Chapter 4). Typically, this incentive does not apply to goods that can be produced industrially, including many minerals that are industrially excavated and processed. For many goods, the normal process of technological progress results in a decreasing price, at least measured relative to average income. A common description of a prospering economy is that income grows when measured at constant prices. Another way of describing it would be to keep income constant and say that the (income-deflated) prices are falling.

Some observers misinterpret this development as being primarily a change in attitude. They lament the fact that people no longer spend time taking care of rags and scrap metal, blaming it on a more wasteful lifestyle. This perception is somewhat incorrect because the primary reason for not recycling automatically is that waste products lack value relative to current salaries. By contrast, in a poor country such as India, a large fraction of what is thrown away by consumers is recycled.

A major problem with recycling in high-income countries is the lack of demand for recycled goods; without demand, this market mechanism does not work. People wonder what the use is of collecting used products if they are just dumped into a landfill. Almost inevitably, a country with rising income will produce more waste over time, and a government that insists on recycling may have to not only promote recycling but also establish and maintain a market for recycled goods. Recycling should not be based on a preconception that resources are becoming scarce and that all resources have some inherent value that the market does not understand.

One good reason for collecting at least certain categories of waste, even if they have no inherent value, is to avoid polluting the ecosystems where the waste is dumped. However, when too much focus is put on the categories of material that are not hazardous (e.g., glass, scrap metal, and paper), insufficient attention is paid to the hazardous components of the waste stream. Costs for monitoring hazardous waste should be an important component of the analysis; in some cases, waste management fees encourage destructive illegal disposal, so self-enforcing mechanisms like deposit–refund schemes should be considered (see Chapter 9; also see Fullerton and Kinnaman 1995; Jenkins 1993).

In this chapter, I discuss some policies for waste management in different countries, starting with variable waste charges and deposit–refund systems in Sweden and the United States. Next, I move to some developing-country issues, then the use of eco-labeling on dishwashing and laundry detergents (which are not waste in the usual sense but affect effluents and water quality). In the final section, I describe the use of a kind of tradable permit for packaging waste in the United Kingdom.

## Economic Incentives in Waste Management

Differentiated waste fees and other incentives for waste minimization, recycling, and increased composting have been used, particularly in the United States and in some of the more densely populated areas of Europe, such as the Netherlands. Results vary considerably depending on the exact details of the program as well as on local demographic factors (Fullerton and Kinnaman 1994; Repetto et al. 1992).

### United States

Municipalities in the United States, as in most other countries, have traditionally levied fixed collection fees for household waste, or even included them in general (property) taxes. This approach is inefficient because the effective marginal price of waste disposal is zero, whereas the marginal collection and disposal cost is positive. However, communities are increasingly charging for waste collection based on weight or volume of waste. Such programs have been implemented in more than 4,000 communities (roughly 10% of the U.S. population), and in Minnesota, some form of differentiated disposal fees is mandatory. Programs take different forms: prepaid garbage bags, prepaid stickers to be attached to bags or other waste for collection, collection fees based on the number or size of cans, weight-based systems, or two-part tariffs (a fixed minimum service at a base rate, then marginal charges for excess waste). Two-part tariffs, a relatively recent development, are popular because they are easy to implement, provide a stable source of revenue for collection services, and are perceived as offering a fair previously specified level of service at a fixed cost to most customers.

The effect of these programs has been satisfactory. Among 114 cities with variable rates and 845 communities with traditional fixed rates, the estimated reduction in waste disposal after beginning to charge US$1 for every 32-gallon bag is 44% (Fullerton and Kinnaman 1995). Other studies have found reductions of 10–75%, depending on circumstances (see U.S. EPA 2001 for a survey). Typical results have been much higher in cities where recycling programs are run simultaneously. Decreases in waste collection may be misleading if they lead to an increase in illegal waste disposal, which includes dumping, backyard burning, and placing waste in other peoples' containers (including public ones) (see Chapter 9).

## Sweden

In Sweden, several municipalities are now charging for waste on a per-kilogram or per-bag basis. In a municipality in Varberg (southwest Sweden), a weight-based billing system was introduced in 1994 for household waste, charging 1 SKr/kilogram (about US$0.26/pound) of waste. At the same time, recycling centers were set up, and a "green shopping" campaign was launched. This combined program led to a 35% reduction in waste collected within a couple of years (Sterner and Bartelings 1999).

In addition to survey data for the Varberg households, waste disposal was measured at the household level in Tvååker, a residential area (Sterner and Bartelings 1999). The most important explanatory variables for a household's waste were whether kitchen (and garden) waste were composted, living area (square meters of domicile), age of the residents, and the residents' attitudes about the difficulty of recycling. Economic incentives, although important, were not the only driving force behind the observed reduction in municipal waste: having the proper infrastructure to facilitate recycling in place motivated people more than just savings on their trash bill.

The use of deposit–refund systems has been increasing somewhat for conventional deposit items (e.g., bottles and cans) as well as for such items as scrap cars, where the intention is simply to avoid dumping. Deposit–refund systems are increasingly being used as an economic incentive and an information signal. However, paper is widely collected without any deposit scheme. It is therefore valuable to understand the role of deposits compared with comfort (ease of deposition), information, opinion, values, and habits. All these factors are important: the recycling of glass soft drink bottles is an ingrained habit, so recovery rates have been 98% for several decades (Sterner 1999; see also Chapter 21). For aluminum soft drink cans (same contents, sold in the same shops), recovery rates were initially much lower. However, with aggressive marketing and increased refunds, recovery rates rose from about 60% to 90% in the decade after 1984. Similarly, recovery rates for wine and liquor bottles increased from 40% to 70% over a 25-year period before the recycling program of the state liquor monopoly was closed down in the late 1990s. Price elasticities for all these bottles and cans were estimated to be around −0.2. However, the main result when the deposit increases seems to be an "attention" effect. When the deposit decreases (in real terms, through inflation), recovery rates do not appear to decline. On the other hand, voluntary paper and glass recycling programs have been successful without any financial incentive; estimated recovery is around 70% (compared with 20% a couple of decades ago). The considerable increase in the number of deposits for these materials indicates the importance of information, social norms, and ease of recycling.

For hazardous waste, proper household disposal has become much easier. Large recycling centers have been up in Gothenburg that accept whatever (hazardous and other) waste the public brings. Furthermore, smaller recycling stations are set up to receive oil, solvents, and car batteries. They generally are located at (and run by) gas stations or small marinas. Passive deposits can be set up for the battery collection, for instance, in the streets (close to collection points for glass and paper) or in appropriate shops. Collection vans collect the waste from these deposits and from the recycling stations. The same collection vans also were used for some time for neighborhood service. According to a posted schedule, they visited every neighborhood for hazardous waste pickup.

The passive deposit boxes and the small environmental deposits at gas stations have been particularly effective because they fit into people's everyday routine and thus entail little extra cost. The van was not cost-efficient because few people used it. Apparently, most people prefer a greater distance at a time of their own choosing to a short distance (or none at all) that requires attention to a schedule.

## Waste Management in Developing Countries

Waste management in developing countries is subject to the same general principles as in developed countries, but several important conditions are different. One of these is the abundance of labor with low "opportunity cost," that is, destitute people who have no other means of employment and thus are prepared to work in some way with the resource that waste represents. Another common characteristic is poor provision of public services (e.g., municipal waste management, water, and electricity). Consequently, in many cities in developing countries, waste is picked through several times after being placed curbside: first by people looking for food or saleable items, then—typically with some delay—by municipal sweepers or waste collectors.

Many other factors make each location unique. In some areas, urban agriculture allows for the recycling of organic matter. In other areas, uncollected garbage presents a significant health risk because of factors such as heat, humidity, insects, and animals. Waste sites themselves are often inhabited by thousands of people who sort through the waste under miserable working conditions. Waste pickers typically sell what they collect to middlemen, who then sell the items to recycling industries. The recycling industry is typically well developed and may be economically successful, particularly compared with those in richer countries (where the opportunity cost of labor is much higher and thus the value of recycled items lower) (UNDP/World Bank ESMAP 1998).

Because of these differences, the issues and the potential instruments for waste management are different from those in developed countries. Illegal disposal of small amounts of household waste (e.g., on the curbside) is not a major issue; in fact, it is the norm in many places. Waste collection fees may not be the best general instrument (except in more affluent districts). One important issue is the organization of the labor that is already dedicated to waste management or recycling. It is a natural target of aid programs. Some have emphasized the role of health services, whereas others emphasize the role of micro enterprises that can use the recyclable items. If scavengers were properly organized, their activities could have a great impact on the economy, and waste management could be considerably improved. The integrated solid waste management approach must be adapted to local social and technical conditions in developing countries (Beukering et al. 1999). Governments must integrate the contributions of existing informal waste collectors, recognizing that their contribution is high. Collaboration with all stakeholders is crucial to integrating the formal and informal recycling agents in Kathmandu, Nepal, where more than 15,000 people make their living scavenging (Bhattarai 2000).

Two topics related to developing countries deserve attention. One is waste production through tourism, discussed in the next section. The other is the dumping of hazardous waste or the siting of hazardous industries in developing countries (which may sometimes be almost the same). In principle, this practice is regulated under the Basel Convention, but many unscrupulous firms export hazardous waste under various false classifications (e.g., "fuel," "fertilizer," or "raw materials") to companies in countries that do not have the capacity (or the inclination) to monitor and inspect the materials (see Chapter 25 concerning hazardous wastes and industries in India). All trade is supposed to be registered under the Basel Convention, but the statistics are unsatisfactory. Import statistics for 1998 for non-OECD trade report almost 20 times the quantity that the corresponding export statistics show. Information concerning the treatment of waste (recycling, disposal, or other) are often missing.

Within the convention, work is in progress to formulate agreements on liability and compensation rules as well as to ban the most dangerous of the hazardous pollutants.[1] Information is a powerful tool in this context: through the Internet, any journalist—even in a poor country—can access information about hazardous wastes, then contact the press in the country of origin. In the future, this kind of monitoring may turn out to be at least as powerful as (and complementary to) formal legal methods.

## Tourism and Waste Management in the Caribbean[2]

Ecotourism is touted as a benign resource use for sensitive ecosystems. However, the effects of tourism are not always benign; charter boats and most other ships regularly dump wastes, effluents, and other emissions into the very coastal ecosystems that support them. Many of the small "paradise islands" are extremely sensitive—particularly coral islands, which are easily damaged by nutrients, temperature variation, and physical damage. Like many other popular sites, the Caribbean Sea is heavily affected but until recently has had difficulty protecting itself. (Several industrialized countries have started to demand that ships pay for waste disposal at every port of call, but this policy is novel to developing countries.)

At the same time, the management of locally generated waste is a severe problem. In 1995, the World Bank, the European Investment Bank, the Caribbean Development Bank, and the Global Environment Facility (GEF) joined forces to halt marine deterioration through a solid waste management project in the Organization of Eastern Caribbean States (OECS). The organizers of the waste management project recognize that

- the ecosystem is both sensitive to waste and valuable (as a source of tourism revenue and for its life support functions);
- the environmental problem has both local and international causes (making GEF finance an option; see Chapter 17);
- the generation and dumping of wastes is not always observable, making a Pigovian tax difficult to implement, but some source of local funding is crucial to make the project sustainable (the structure of fees was a fairly contentious issue); and
- it is difficult to locate suitable land for a landfill site on a small island.

According to the initial appraisal report, the waste management conditions in the OECS—particularly Grenada, the most heavily polluted of the islands—were bad (World Bank 1995). Waste collection service served only half the population and was severely underequipped. Tipping sites were poorly managed, with severe leakage (to aquifers and to the sea); underground fires; and significant odor, rodent, insect, and other health-related problems. In addition, tourists and tourist ships dumped their own waste. The funding agencies were keen to ensure the sustainability of the projects by insisting on a system of fee collection for both domestic and tourist wastes.

To agree to provide finance (US$50.5 million in a grant and loan package, of which US$8 million for Grenada), funding agencies required that the countries improve waste management collection, waste disposal sites, and local waste management and organization, which included implementing fee-collection systems. An appropriate tipping fee for domestic wastes would be US$20/ton, whereas transport and disposal of maritime wastes could be charged US$40/ton plus US$35/ton for haulage (World Bank 1995). Furthermore, an environmental fee of US$1.50 per visitor would be charged. These charges, together with a continued contribution from the public sector and the GEF grant, would be sufficient to pay for waste management services. Notably, the charges were discussed as methods of "cost recovery"; they were not designed primarily to minimize waste (monitoring was considered too difficult).

The construction of a system of fees and other components was carefully balanced to motivate all the parties and legitimize the project. This effort was necessary because there was still a good deal of resistance. Local low-income inhabitants who were used to disposing of trash informally, at no cost, initially were reluctant to pay; gradually, they were convinced of the importance of the program because they saw aesthetic and health improvements and because they perceived that they were not the only ones paying. For the sake of administrative convenience, the fee for waste management was linked to the electricity bill (and related to electricity consumption—far from ideal, but still a workable arrangement).

The largest conflict, which almost stopped the whole project, was with the cruise ship industry. The operators complained bitterly over the "arbitrary" US$1.50 environmental fee. In fact, their resistance was surprisingly ferocious, given that this sum was a relatively small share of the cost per passenger, per day. However, the industry lobbied each country, threatening to boycott the countries that imposed the tax and move their business to islands that did not. The World Bank organized meetings between the cruise ship associations and the governments of the region, but no agreement was reached until the countries imposed the fee unilaterally and simultaneously on June 1, 1998. In addition to these fees, Grenada also assesses a 2% waste fee on the importation of vehicles and household appliances as well as US$0.25 per beverage container. Although these instruments provide much needed finance for waste management, they are not really designed to minimize waste. A partial refund for returning used appliances and so forth has been discussed but not implemented.

The other major area of difficulty encountered by the project was the selection of a suitable disposal site on Grenada. The site initially selected had several good features but turned out to harbor Grenada doves—an almost-extinct species, of which fewer than 100 remain, only on the island of Grenada. This shy bird, which only dedicated ornithologists appear to have noticed, presented a serious threat to the project. Numerous delays and negotiations led to the creation of a dove sanctuary complete with guards, fences, a tourist center—and entrance fees.

By 2000, 95% (rather than 50%) of the local waste was being collected, new safe landfills were in operation, locals and tourists alike were being charged to recover the costs of waste management, and the dove population even showed signs of growth. Whether the cruise ships are only paying the fees or also engaging in any form of source separation and treatment (or dumping in international waters) is not known. The use of fees per ton might encourage both sensible and irresponsible treatment, and such fees might need to be restructured if they lead to perverse behavior. The project also needs to pass the test of sustainability when the donors pull out. According to World Bank project manager Usamah Dabbagh, the project would need an additional year or two of support to guarantee a smooth future (Peck 1999).

This example illustrates the importance for policymaking of a coordinated approach with several components that reinforce each other: collection service; disposal site; fee collection; and cost sharing among locals, tourists, and development agencies. Perhaps the most interesting response was from the cruise ship industry, which was not expected to object to a cleaner environment or to a fee so small that passengers would hardly notice it. The industry is either irrational or, more likely, worried about the precedent set for cooperation or "collusion," as they see it, among the island states off which it operates. This risk of collusion in other areas may be a by-product of environmental programs and instruments.

## Eco-Labeling of Soaps and Detergents

One important waste stream bypasses municipal collection: effluents that drain into sewage. Soaps and detergents—as well as the dirt that they are intended to remove—are one kind of such effluent waste.[3] Because of the nonpoint-source characteristics of these effluents, it is difficult to imagine any billing or even monitoring system that would detect the amount and characteristics of the waste disposed of through a household's drains. It is therefore natural for policymakers to choose an instrument that applies directly to the product. Standards, rules, liability, taxation, and several other policy instruments could be applied to dishwashing and laundry detergents. The large number of complex ingredients make taxation difficult, and thus the main alternatives are perhaps some form of direct control or voluntary instruments such as information disclosure and labeling. Labeling is also a strategy that may be used in several other waste-related issues, including the environmental regulation of trade.

Laundry detergents have three characteristics that make labeling particularly appropriate: the ecological and technical criteria are complex; the main hazards lie in the product, not the production process; and the product is bought mainly by households (rather than by industries as an intermediary good). The Swedish Society for Nature Conservation has its own eco-labeling scheme, Bra Miljöval ("Good Environmental Choice"), in competition with the more official Nordic eco-labels (see Chapters 10 and 12). In terms of market share, both labels have been successful. The market shares of eco-labeled shampoo, laundry detergent, dishwashing liquid, and sanitary cleaners have risen from 0 in 1990 to 90% in 1997 (50% for Bra Miljöval alone), and the eco-labels appear to enjoy a sound level of acceptance and credibility. If the eco-labeled products really are less environmentally damaging, then this policy will have significantly reduced the toxicity of effluents, decreasing the impact of an important category of household waste.

To become environmentally certified, companies must fulfill a set of criteria and pay a fee. The criteria focus on the ingredients or production process. Proponents of the eco-labeling scheme claim that the companies have been forced to decrease or eliminate the use of various harmful ingredients (such as phosphates) from laundry powders and that this change has resulted in considerable benefits for the environment. This claim was somewhat supported by a study that indicated that phosphates were reduced by more than 50% and perborates by 80% between 1990 and 1995 (Nilsson 1998). However, the criteria contain fairly lengthy lists of ingredients that no laypeople can compare, and the ingredients that are reduced have substitutes, so the use of other ingredients—notably, percarbonates, enzymes, and new kinds of tensides—is increasing. Several other studies report 45% reductions in those components that are the least biodegradable (Rosander 1998; Svenska Naturskyddsföreningen 1999; Wilske 1999) (see Table 27-1).

One of the criticisms against the voluntary Type 1 eco-labeling schemes is that they fail to take a life-cycle perspective of the product and fail to take into account the performance and user of the product (see Chapter 10). For instance, making a laundry detergent more environmentally friendly by reducing certain components is good from one perspective. However, if the new formula performs worse or consumers use more of the "green" product (or use it with much

**Table 27-1.** *Reduction of Poorly Biodegradable Ingredients in Swedish Detergents, 1988–1996*

| Ingredient | Function | Reduction (%) |
|---|---|---|
| Ethylene diamine tetraacetic acid | Complexing agent | 100 |
| Phosphonate | Complexing agent | 50 |
| Pigments | Coloring | 100 |
| Fluorescent whitening agents | Optical brighteners | 100 |
| Silicones | Defoaming agent | 15 |
| Perfume | Fragrance | 0 |

hotter water), then the ultimate environmental impact could be much larger than from using the detergent with the old formula. The notion that the eco-labeled laundry detergents are less effective (leading to higher consumption) appears to be incorrect, however, because total demand for laundry detergents in Sweden has fallen by more than 10% from 1989 to 1996—that is, during the period when eco-labeling was introduced, breaking a steady 100% increase from 1958 to 1989.

Another criticism against eco-labeling is that the criteria tend to be static and tend to be renewed almost automatically, without revision, which tends to "lock" the technology at one moment in time and make it impossible to benefit from technical progress (for the good of the environment) (Solyom and Lindfors 1998). Diverging opinions and even conflict concerning the ultimate usefulness of eco-labels is probably inescapable because the natural science, ecological, and technical details are complicated. If not, eco-labeling would not be needed. However, the complexity provides ample room for finding evidence in favor of one position or another that may be motivated by ideology or economic interest. Governments and environmental groups can be suspected of a desire to interfere and control with little interest in profitability or freedom of trade. They (at least the environmental groups) also may become dependent on the revenue generated and therefore may act tactically. In fact, governments and environmental groups risk becoming corrupt and viewing the market share of eco-labeled products (as well as the revenues they produce) as a primary objective, perhaps leading to lower standards and a lack of sensitivity to new research on environmental issues that might question the criteria used for eco-labeling. Although no evidence indicates that this is the case, it certainly should be kept in mind as a risk.

Some categories of industry might not want to have independent eco-labeling and therefore fight the principle with profitability arguments. (Industry is divided on this point; some categories of industry argue that the independence of the certifiers gives the certificate credibility and thus value.) An interesting question is why eco-labeling has been successful in the Nordic countries and hardly anywhere else, with a few niche exceptions.[4] Is environmental consciousness more developed in the Nordic countries than elsewhere? Are ecosystems (or consumers) more sensitive? Are other means of regulation not as available? Or are other means of regulation more available and thus a more credible threat if companies do not comply? Many individual studies would be required to answer these questions fully, and they have yet to be carried out.

At present, I can only offer some clues. One is the oligopolistic structure of retail business in Sweden: three major retail chains, one of which is a consumer cooperative. The environmental organizations appear to have managed to make eco-labeling acceptable and even attractive. The Good Environmental Choice label is a collaboration between the Swedish Society for Nature Conservation and the major retail chains, which is perhaps a practical reason for its relatively rapid success. The existence of a large cooperative may have been important. In soap products, mainstream manufacturers initially tried to ignore eco-labeling in Sweden, but this ignorance led to the emergence and rapid rise in popularity (as measured by market share) of a few small independent producers of supposedly more "ecological" laundry detergents. Because the label had turned out to be so popular with consumers, the large producers were forced to follow suit. However, the same firms continue to resist labels in other markets.

The battle in the rest of Europe has been different from that in Sweden. The large detergent producers are organized in the Soap and Detergent Industry's European Association, whereas the small green producers (such as Ecover) are organized in the European Association of Environmental Detergent Manufacturers with only 2–4% of the market (Nadaï 1999). The result of negotiation between these two organizations was E.U. detergent criteria that were so lax that they excluded only a few of the worst detergents on the market. Thus, the paradoxical result was that companies didn't use the label—because it had almost no information in it. The commission had hoped for 10% selectivity (i.e., only 10% of products would be eligible immediately), whereas the AIS wanted 95%—an exclusion of only 5%! This example illustrates the risk inherent in eco-labeling led by industry interests.

## Tradable Packaging Waste Recovery Notes[5]

In developed countries, waste management was originally a freely provided public service. The need to harness economic incentives in this area has led to a good deal of experimentation with deposit–refund system (DRS) and waste management charges. Several countries also have introduced landfill taxes and other policies.

The United Kingdom has a new policy that appears to be the latest and most sophisticated in the sense of harnessing the efficiency of the market in the quest for materials recycling. The background to this policy instrument is the E.U. Packaging and Packaging Waste Directive (94/62), which was transposed into U.K. law by the so-called Packaging Regulations, which set annual targets for each year up to 2001. It required the United Kingdom to meet targets for the recovery and recycling of packaging waste (set as a percentage of total packaging) by the end of 2001: 50% for total packaging with a minimum 25% recycling and an individual minimum 15% recycling for each material (e.g., plastic, steel, or paper).

The United Kingdom introduced the Producer Responsibility Obligations Regulations in 1997 (amended in 1999) to meet these targets. The regulations were modeled on the preceding producer responsibility laws in other countries, such as Germany, but the instruments have evolved considerably. They cover four sectors of industry and give each a percentage of responsibility. For a can of beer, the company that manufactured the steel might have a 6% obligation, the company that made the can a 9% obligation, the brewery a 37% obligation, and the shop that sold the beer a 48% obligation. Actual figures can vary by year, type of material, and other factors, giving decision-

## Supplemental Reading

Beukering 2001
Palmer, Sigman, and Walls 1997
Watabe 1992

makers flexibility. This instrument also distributes the burden of recycling. It is much more flexible than a mandatory DRS, which places the entire burden on the retail industry and the consumers.

All the parties involved in bringing a product to market have to meet their individual obligations. They do not, themselves, have to recycle or collect any waste. However, they do have to hold packaging waste recovery notes (PRNs) amounting to their obligation. PRNs are simply evidence that someone (known as the reprocessor) has carried out recycling or recovery. There is nothing to stop a reprocessor or a store from using deposits to encourage recycling. If a DRS is successful in collecting the cans or other packaging, then the organization that runs it can cover not only its own needs of PRNs but maybe collect more and thus have PRNs to sell on the market. In some cases, deposits may be the most effective way of collecting waste for recycling, but in others, special collection receptacles or some other method (such as separation from the waste stream) may be more cost-efficient.

The system chosen thus includes DRS as a possibility but also opens up for other more efficient methods that decisionmakers may not even have imagined. Several features of this system—encouraging technical and organizational progress, cost sharing, and including all the involved parties—are desirable attributes in any policy instrument.

## Notes

1. The Ad Hoc Working Group of Legal and Technical Experts To Consider and Develop a Draft Protocol on Liability and Compensation for Damage Resulting from Transboundary Movements of Hazardous Wastes and Their Disposal met for a tenth session in Geneva, Switzerland, during August and September 1999. The aim was to set international rules regarding liability and compensation for damage resulting from the transboundary movements of hazardous wastes and their disposal (see UNEP 1999).

2. Thanks to Phil Hazelton, John Dixon, and Julia Peck at the World Bank for information on this project.

3. There is some substitutability between conventional household waste and sewage. In the United States, garbage disposals (which chop up kitchen waste and feed it into the sewer) are common. They effectively convert solid waste into liquid effluent.

4. Organically grown foods have filled such a niche in many countries. "Green" electricity and "dolphin-free" tuna are two of the few U.S. examples. Indonesia's PROPER is one of the few developing-country examples (see Chapter 25).

5. Thanks to Frank Convery for information about and discussion on this instrument.

# CHAPTER 28

# *Fisheries*

$F$ISH ARE AN IMPORTANT SOURCE of high-quality food, particularly for the poor, for whom small quantities often provide the only protein in a diet based on a staple food such as rice. Fishing is an important source of employment for many coastal people.[1] The commercial value of fish presumably far outstrips the value of any other wild game that humans hunt. (Fishing also represents a considerable recreational value, which is not discussed in detail in this chapter). Finally, the stock of fish is a vital indicator of the state of the environment; fish live in the water that covers most of Earth's surface and also is the final deposit for a large share of industrial and agricultural pollution.

Cultivation of fish has increased significantly, and some people foresee a future in which wild fish will be displaced by aquaculture, as has occurred with terrestrial ecosystems. Although this possibility remains, it would pose some ecological and social problems. World production of fish and crustaceans reached 117 million tons in 1998, a decrease of 4.3% below the 1997 level. The landing of wild-harvested fish had decreased by 8% from 1997, but aquaculture had increased from 29 million to 31 million metric tons. Aquaculture already accounts for more than one-quarter of fish production and an even larger share of the fish directly used for human consumption.

According to the Food and Agriculture Organization (FAO), fish farming as a complement to agriculture is an old tradition that ought to be modernized, copied, and expanded. The oldest known references are from China (4,000 years ago), and China still takes the lead internationally in fish farming. Since 1984, world production has risen from 10 million tons to more than 30 million tons. The dominant producers in Asia (mainly China, India, the Philippines, and Indonesia) account for almost 90% of this production. Integrated farming systems could provide not only nutritional advantages but also income advantages by reducing risk (through income diversification), thus increasing harvests and profits. They also could provide associated environmental benefits—such as the bio-

Thanks to Sarah Gardner for research assistance and to Håkan Eggert for valuable comments.

logical control of mosquitoes, other disease vectors, and agricultural pests without the use of insecticides or pesticides—and complementary advantages because the ponds can be sources of both fertilizer and irrigation (FAO 2000).

To understand fisheries, several major kinds of fishes and fisheries must be distinguished as well as the complex ecosystems that they are a part of. On the one hand, *demersal* fish, which feed on continental shelf and ocean bottoms, are particularly abundant in shallow waters and typically are a main source of local subsistence fishing in developing countries. (The same is partly true of freshwater fish, aquaculture, and some nonfish species such as squid and crustaceans.) On the other hand, *pelagic* fish such as tuna, which live in deep water, are caught mainly by large commercial fishing fleets. There is a good deal of overlap—and conflict—between the various kinds of fishing.

Many of the major fish stocks appear to be harvested at (or beyond) their maximum sustainable yield. Total fish catch from 1970 to 2000 is illustrated in Figure 28-1. Some 10% of this catch is from inland waters; fish cultivation is not included. Prior to the late 1980s, catches of many of the economically important species increased rapidly for several decades. However, for most fisheries (including many of the most important commercial ones), these increases have been followed by stagnation or decline. In general, stagnation has not been due to less fishing effort. On the contrary, fishing fleets have grown in size, number, and technology. The largest trawling ships are enormous industries with nets so large that several jumbo jets would fit inside, and at the same time, these ships have all the best sonar, Global Positioning System (GPS), and other technology to find the schools. The industry rarely identifies an appropriate level and sticks to it. Time and again, overfishing—often despite clear warnings from fishery biologists—has resulted in crashes in fish stocks and numerous conflicts between fishing fleets. In some cases, these crises have led to postcrisis management schemes that may allow the fish stocks to recover enough to be fished sustainably. In 1999 and 2000, preliminary data show a slight increase, again discontinuing the dip in 1998. However, these world fishing data have recently been severely questioned by experts who argue that they are seriously misleading for two reasons: temporary effects related to El Niño and, more important, erroneous data from China, a leading fishing nation. The true picture is a decline by about 10% despite increased effort during the past 15 years (Watson and Pauly 2001).

In fisheries, "the market" generally fails without regulation for the simple reason that the basic requisite of resource ownership is not present (see Chapter 4). Thus, a policy to avoid overfishing is needed. However, the market failure usually is supplemented by policy failure, because the policies in place have the opposite effect. Instead of restricting effort, helping prevent overuse, and allowing stocks to recover, many policymakers simply "help the fishermen" as if fishing were any standard industry. Thus, credits and subsidies are provided for faster and bigger boats, longer and more finely meshed nets, and new fish-finding equipment. All these technologies lower the cost of fishing but expand capacity beyond the optimal—sometimes even beyond the sustainable—level, so they make overfishing worse. According to one estimate, the world fish harvest was worth $70 billion but cost $92 billion to catch (FAO 1983). Although these figures have been ques-

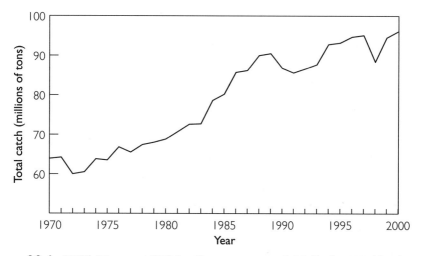

**Figure 28-1.** *Wild Harvest of Fish, Crustaceans, and Mollusks Worldwide, 1970–2000*

*Source:* FAO 2002.

tioned, they show that subsidies are—or at least were—an important element in lowering production costs and expanding fisheries. In more recent OECD estimates, subsidies appear to be lower; OECD subsidies in 1997 were calculated at US$6.3 billion, which is still sizeable. The costs of fishery management in Iceland, Norway, and Newfoundland (Canada) are considerably different, indicating differences in efficiency as well as the distribution of costs between fishermen and society. In Iceland, where the economy strongly depends on fishery surplus, management costs are kept low (3% of gross landings value), whereas in Norway, they fell from 13% to 8%, and Newfoundland was by far the highest with 15–25% (Arnason, Hannesson, and Schrank 2000).

Subsidies that lower the cost of fishing are bad policy if the main problem is overfishing as a result of open access. However, several policies appear to work, including partial property rights (such as the 200-mile exclusive zone), fishing licenses, restrictions (e.g., on gear or seasons), and fishing quotas. The most direct policy instrument is to set catch limits in the form of a total allowable catch based on maximum or optimum sustainable yields; however, they must be enforced, for example, through individual transferable quotas (ITQs). Other possible instruments would be taxes and information or voluntary instruments such as labeling schemes.

Restricting access is a popular instrument because it targets "outside" fishermen—from other countries, regions, or villages. This mechanism is fundamental at the local level of common property resource (CPR) management and also is the main purpose of the enclosure of the seas implicit in the extended fisheries jurisdiction (see Chapter 5), under which 90% of the world's fish catch is based on resources that lie in nationally controlled waters (Clark 1990). It will not necessarily lead to sustainable management but can be seen as a prerequisite to other instruments at the national or local level, such as CPR management or ITQs.

When the catch is too difficult to monitor accurately, technology might appear to be easier to regulate. The prohibition of destructive technologies such as dynamite and cyanide is a positive advancement. However, many other technologies are banned simply because they are too effective, even though they are not strictly harmful.[2] Fishermen who use simple, small-scale technology often perceive the more "advanced" equipment as unfair and lobby against it. Technology restrictions are common and, in some cases, effective in protecting stocks— but at the cost of depleting the rent that should be protected. A better approach would be to use the new technology but reduce effort and catches to an optimal and sustainable level (see Chapter 4). Regulating technology would mean monitoring each and every detail of fishing capacity (Brown 2000). If the number of boats is limited, then owners acquire bigger boats. When regulators fix the maximum length of boats, boats become rounder and more powerful. In areas with different length restrictions, boats have detachable "auxiliary" bows.

Restricting fishing seasons, another indirect method used when catch is difficult to monitor, also has several disadvantages. Resource depletion is not the only problem fisheries face—although it tends to be a dominant problem. Other important issues relate to occupational safety and to the value of the fish, which is strongly related to its freshness. Certain restrictions (particularly fishing seasons but also total physical quotas) tend to encourage derby-style fishing during the short period the fishery is "open." Some halibut fishermen adapted to a two-day season by investing in three identical electronic systems, where two were backups. This extreme example shows how restricted fishing seasons can lead to what is referred to as "overcapitalization" or "capital stuffing" (Homans and Wilen 1997). The extra capacity may be idle part of the year or migrate to other areas, contributing to overfishing elsewhere. Short seasons also lead to lower commercial value of catch, because it has to be sold frozen most of the year. Finally, it leads to fishing even in bad weather, which increases safety risks (NRC 1999).

A moral hazard issue can result from the link between fisheries management and safety (Bergland and Pedersen 1997). The authority (the principal) chooses public supply of effort on safety, then the fishing operation vessels (the agents) decide on levels of private inputs. Public engagement in safety issues is necessary to supply public goods such as communication and navigation systems used by all vessels. The moral hazard problem arises because the supply of public services might encourage individual fishermen to behave in a way that increases risk.

A tax would be an excellent solution if the focus were only on the allocative effects, but a tax is politically hopeless because the government would then remove the rents from the fishermen it purports to protect (see Chapter 4). Refunding taxes would not be an option because the level of fishing, not the technology of fishing, is being regulated (see Chapter 9). Refunded taxes will not have the "output effect" of a tax (see Chapter 9 and Chapter 14). If refunding were tied to output, then refunded permits would not be a reasonable way to reduce overfishing. However, a tax on catches, the revenue from which would be used for community development (e.g., infrastructure, aquaculture, tourism, or other public goods), might be an option.[3]

The appropriate instruments would reduce catches to a sustainable level while preferably allowing for some flexibility in this level as stock assessments vary. They would allow for the use of efficient fishing methods, leave the rents with the fishermen (unlike taxes), and preferably generate funds to compensate the fishermen who are forced to retire as a result of the program. Furthermore, credible monitoring is needed, and unacceptable social outcomes (such as massive regional unemployment) must be avoided.

In this chapter, I focus on two instruments: CPR management and ITQs. The former appears to be more appropriate for small, coastal fisheries and the latter for large commercial fisheries; however, this does not mean that policies such as fishing licenses and technology requirements are always inappropriate.

## Management of Small-Scale Subsistence Fisheries

Fisheries in most developing countries predominantly involve subsistence fishing, in which individual, often part-time, fishermen land their catch in small multipurpose boats. Such fisheries are difficult to manage centrally because of the highly scattered nature of landings and transport difficulties and the difficulty of enforcing regulations in remote areas (Baines 1982; Nietschmann 1984). Traditionally, these fisheries have been managed locally, not centrally, and they provide many good examples of CPR management regimes (see Chapter 10). Mechanisms for excluding nonparticipants are important for CPRs, and access to sea resources may be limited by three means: restricting information (e.g., fish location and suitable methods of fishing), actively defending fishing sites (by sabotaging rival gear), and formal exclusion (Durrenberger and Palsson 1987).

### Community-Based Systems

In traditional management systems, it is common to find the entire community involved in managing fishing resources. This kind of system has been documented in the Indo-Pacific region, the Caribbean, Latin America, and Africa as well as in many industrialized countries (Ruddle, Hviding, and Johannes 1992). In Japan, feudal rulers historically possessed fishing rights, which they gave to villages in exchange for taxes (see Chapter 5). Today, local fishing cooperatives control many of these fishing grounds (Akimichi 1984).

In the Pacific Islands, customary marine tenure systems are common. Membership in kinship groups includes user rights for both land and sea resources. Membership may be based on ancestry or obtained through marriage or residence, depending on the region, and complex rules sometimes govern user rights (Hviding and Baines 1994). Fishing rights may correspond to specified areas inside a barrier reef or another important fishing ground. Within these areas, outsiders must obtain permission to fish. Coastal residents may exchange fishing rights for farming rights with groups that live inland. Coast-dwelling clans may negotiate if a certain area is lacking in some species or experiences inclement fishing conditions. Generally, outsiders will be allowed to fish for subsistence but not for commercial purposes. Access to items that have only commercial value (e.g., pearls, coral) usually is severely restricted, enforced through social pressure, and violators are publicly disgraced. Compensation may also be exacted for illegal fishing. Resource owners may contact the local police to prosecute illegal fishermen, and habitual poaching or trespassing may result in social excommunication.

Traditional management centers on control of effort by controlling methods, location, and timing. Fishing gear and net size are commonly limited, explosives are almost always prohibited, and the use of poisonous plants and underwater spear guns is typically restricted. Certain species are not to be fished—such as eels, turtles, and crocodiles—or it may be taboo to fish a certain species at a certain time, such as juvenile fish in shallow waters. Seasonal closures on some or all fishing grounds are designed to allow stocks to recover. Certain areas may always be closed to fishing. Fishing sites are often rotated following cycles of the moon, tides, or wind, thus preventing overfishing in any one site. Rotation is a good method to distribute heterogeneous grounds fairly and also makes it easy for fishermen to monitor each other and enforce regulations at little extra cost as part of their own fishing activities (see a description of the Alanya fishery in Turkey in Ostrom 1990). Generally, the rules concerning allowable species and timing adapt to changing ecological conditions and to the needs of the community. The managers and fishermen have extremely detailed knowledge of the ecology of their systems; in one area, more than 60 fishing methods, 40 reef features, and 400 fish types are specified as well as the migration paths and spawning behavior of various fish.

This traditional kind of fisheries management is sometimes known as territorial use rights fishing (TURF). In the Caribbean, one of the few documented cases of TURF is in Jamaica, where fishing management within a reef area resembles that in the Pacific Islands. Social control of access to certain territories restricts fishing. Enforcement takes the form of interference with fishing traps by the controlling group. This kind of social management normally would not be expected to be efficient in the sense of maximizing rents, but it performs better than open access in this case; Jamaica had higher fishing yields than the average Caribbean fishery (Berkes and Shaw 1986). TURF management also may provide incentives for marine farming as well as marine parks and ecotourism: marine aquaculture has been introduced into TURF areas in numerous islands (Schug 1996), native groups in Panama and Papua New Guinea have requested that their traditional territories be made into national parks (Nietschmann 1984), and parts of Mafia Island (off Tanzania) are becoming a marine park (Andersson 1995; see Chapter 31 for more examples).

## Entry of Commercial Fishermen

In many countries, an influx of commercial fishermen into traditional fisheries has created hybrid management systems in which the CPR institutions tend to break down (Cordell 1984; Ruddle 1988). Often commercial and subsistence fishermen directly compete for resources. Commercial fishing can be a positive force, bringing new technology and higher prices; however, in the absence of clear property rights, commercial fishing has destroyed some fisheries in North Yemen, East Java, Thailand, and India (Thomson 1980). In Malaysia, powerful boats and gear used by commercial fishermen are destructive to traditional fishermen (Vincent et al. 1997). Commercial fishing creates pollution, degrades habitat, encourages overexploitation, and interrupts the fish food chain. In addition, heads of clans who sell resource rights to commercial fishermen and do not distribute the proceeds to everyone in the group have created social inequality in local communities.

Traditional systems may break down in response to many changes, for example, the effects of money transactions on previously non–profit-oriented fisheries, the loss of authority of traditional rulers, and the enforcement of new laws by colonists (Johannes 1978). Overfishing after commercializing previously subsistence fisheries has been reported for salmon in British Columbia (Rogers 1979) and for turtles in Nicaragua (Nietschmann 1972). Population growth is thought to play a role in the degradation of traditional management systems in Nicaragua (Nietschmann 1972) and in Micronesia (Johannes 1978). Growing populations in Eskimo and First Nation communities in northern Canada increases pressure on natural resources; government-sponsored resettlement of a scattered and nomadic population in the James Bay Cree fishery had similar effects. Resources need to be appropriated either through granting monopoly rights or quotas, in the case of commercial fisheries, or by giving small-scale fisheries control over their fishing grounds (Berkes 1985).

Opening up export markets and allowing the entry of foreign processors may be destabilizing to CPR management systems. Effort and capitalization increase, and fishermen borrow money to invest in larger crafts and more expensive gear. Even fishermen who realize that the fish stock is being depleted may continue to fish because they need to pay off their debts. Once the traditional systems have broken down, they are difficult to rebuild, because the traditional leaders have lost authority and because fishermen have become accustomed to having free access and resist attempts to limit their effort (Johannes 1978). In the island of Ouvea, one of the Loyalty Islands off the coast of New Caledonia, traditional authorities maintained their control even after the introduction of modern technology; when the sites were overfished, the tribe reverted back to its traditional conservation practices (Nietschmann 1984). Several traditional fishing communities have written constitutions to preserve and make explicit allowances for traditional fisheries management.

On Zanzibar Island, Tanzania, fishing effort has exceeded sustainable levels for several years (Mkenda 2001). Its fishing villages appear to have had a fairly effective CPR management originally, but various factors led to a rapid increase in the number of fishermen and vessels: the decline of the agricultural economy, subsequent unemployment, and massive entry into subsistence fishing as well as the entry of large commercial vessels. Some local decisionmakers appeared to believe that catches could be increased infinitely by increased effort.

## Effect of New Technology

Technology imports that are specifically aimed at increasing the standard of living of native fishermen may be counterproductive. If fish stocks are near their maximum sustainable yield as a result of population pressure, then technology improvements may further erode the stock and create income inequalities, because certain fishermen can afford new gear and others cannot. The least efficient fishermen may be driven out of the market as fishermen with better gear deplete the scarce supply. In San Miguel Bay, Philippines, the entry of trawlers in 1980–1982 caused a redistribution of income. In an effort to increase the dwindling catches of small-scale fishermen, the government subsidized motorized boats and nylon netting, exacerbating the overfishing problem. Meanwhile, the overall numbers of fishermen continued to increase, despite declining catches, due to poor employment opportunities elsewhere (Baines 1982). The Malaysian government supported traditional fishermen in the 1960s and 1970s in their competition with larger vessels by heavily subsidizing the traditional sector. When overexploitation resulted, the government implemented a restricted-entry licensing program for inshore waters in the 1980s and cut back its subsidy program. However, because large fishing firms had more political power than the traditional sector did, violations were seldom punished (Vincent et al. 1997).

Sakthi, a traditional fishing village in Kerala, India, was the site of a fishery development project in the 1950s. Some beneficial aspects of modernization reported include increased specialization and productivity, risk reduction as catch variance is lowered, and reduced effort. Motorized boats allow offshore fishing, breakwaters allow for fishing in rough weather, and storage facilities smooth price fluctuations. However, the negative effects include social stratification and loss of social cohesion. Owners of new mechanized boats were sometimes unable to pay for maintenance and repairs and lost ownership. Around the same time, the lucrative export market for crustaceans was emerging. Large investors such as Union Carbide, Tata, and Kelvinator entered the crustacean export business, and total catch increased dramatically. Soon, fish prices increased beyond the means of local residents. Previous small boat owners became crew members on the trawlers, and the line between crew member and boat owner became increasingly difficult to cross. The traditional sector decreased steadily in size; there were one-ninth as many traditional boats in 1978 as in 1953. The mechanized fishing sector experienced an inflow of workers from other industries, and land prices near fishing locations increased. More recently, signs of prawn overfishing have been apparent, in which case both boat crews and traditional fishermen will suffer (Platteau 1984).

Sometimes the asymmetries of information and the difficulty or cost of monitoring may lead fishermen to adopt input regulation. Fishermen have considerable difficulty on their own building up the institutions to limit harvest and thus protect stock, and as a result, they may resort to input control (Scott 1993). In addition to information asymmetries, the difficulty of dealing with heterogeneity, entry, and distributional issues are key factors. Ecological reasoning—such as the importance of age and size for recruitment—argues in favor of marine reserve areas or zoning as a prime instrument (Pauly 1997; see also Chapters 4 and 31). If ecosystems are very complex, it may be preferable to leave one zone fairly untouched as a source of larvae and other ecosystem "services" as well as a gene bank and a form of insurance in case the management of commercial areas is unsuccessful.

# ITQs in Fishery Management

## Background

Local community management may be effective for small-scale fisheries where the user group is homogeneous; for larger-scale operations, it is generally more efficient to use instruments such as transferable quotas (Berkes 1986). Security of property rights is vital, and whereas the extended fisheries jurisdiction provides some preconditions, it does not establish property rights at the appropriate level of the fishing agent. Establishing exclusive rights at the national level gives the government the possibility of applying other appropriate and effective policy instruments, such as ITQs (also called IFQs, where the "F" stands for "fishing"). Apart from pollution control, the only main area in which tradable quotas have been successful is large commercial fisheries. Fisheries appear to have several special characteristics that make tradable permits particularly effective: the high value of the resource, the mobility of the resource (and thus the difficulty of creating ordinary "private property rights" based on territory), and the strong negative externalities exercised by one fisherman vis-à-vis the others (Christy 1973). ITQs have become accepted and are, with some cautious reservations, endorsed by the National Research Council's Committee To Review Individual Fishing Quotas (NRC 1999). Additional characteristics of ITQ programs are listed in Table 28-1, which summarizes some information on how programs dealt with the tricky questions of allocation of rights and monitoring, excessive fishing effort, overcapitalization, efficiency, administrative and enforcement issues, and issues such as safety and fish quality.

Traditionally, fishing in the open seas (as opposed to coastal waters) was unrestricted. For obvious reasons, CPR management was difficult to organize. During recent decades, more and more conflicts have arisen between fishermen from different nations, and more and more regulation has been enacted to protect threatened or overexploited stocks. Earlier regulations—usually based on catch, exclusion, or technology—are being replaced by or supplemented with ITQs in several fisheries.

## Implementation: Iceland

Iceland is a particularly interesting example because it is a prominent fishing nation; fishing is such a dominant industry that policymakers take it seriously (its direct share in gross domestic product is more than 15%). In the 1960s, serious conflicts (the so-called fishing wars) broke out between Iceland and the United Kingdom that were an important factor in the formation of the Law of the Sea. Icelandic fisheries also have experienced many of the problems typically found in other countries' fisheries. The value of fishing capital in Icelandic fisheries increased by more than 1,200% from 1945 to 1983 while real catch value increased by only 300%, indicating that overcapitalization (and overexploitation in general) was a serious problem. Effort restrictions, moratoria, and overall quotas did not prevent fish stocks from being reduced to a drastically low level, at which point the fishing industry began to cooperate with management plans. In

Table 28-1. *Characteristics of Some ITQ Fisheries*

| Country: Area | Year | Species | ITQ allocation | Cost | Basis | Property rights[a] | Enforcement | Payment | Economic results |
|---|---|---|---|---|---|---|---|---|---|
| New Zealand | 1986 | 32 species | F | Initially free | H | Perpetual, full rights (except rock lobster) ($R_{max}$) | Auditing | Violation is a criminal offence | Output 0 Employment 0 Fish quality + Quota price + |
| Iceland | 1979 1984 1990 | Herring Demersal All | V | Free | H/C/E | Annual revocable vessel quota ($R_{conc}$, $R_{empl}$) | Auditing | Costs paid by industry | Catches +/0 Effort – Catch quality + Profits + |
| Canada | 1991 | Halibut | V | Free | 70% H 30% C | ($R_{conc}$) | Catch monitoring | Penalties: forfeiture of quotas; inform-ers get shares | Catch 0 Prices + Employment – Concentration restricted |
| Chile | 1992 | Red shrimp and cod | F | Auction | | ($R_{max}$); valid 10 years; annual auction of 10% | Weak; self-monitoring | Graduated fines | |
| Norway | 1973 | Herring, mackerel, blue whiting, capelin | V | Free | C | Restricted transfer subject to approval by Ministry of Fish | | | Rents increased |

| | Year | Species | Allocation | Price | ITQ basis | Property rights/tradability | Catch monitoring | Cost/enforcement | Outcomes |
|---|---|---|---|---|---|---|---|---|---|
| Australia | 1984 | Southern bluefin tuna | F | Free | 75% H 25% C | Freely tradable | Catch monitoring | Cost paid by industry (44%) | Catch/effort + Rents + Capital used − |
| United States | 1992 | Wreckfish | V | Free | 50% H 50% E | ($R_{conc}$); perpetual but not full property right | Monitoring by "catch coupon," landings inspection | Violation leads to fines | Overcapitalization − Efficiency + Effort − Price stabilized |
| East Coast | 1990 | Surf clam and ocean quahog | V | Free | 80% H 20% C | Quotas full property | Shoreside surveillance, cage tags, logbooks | | Efficiency + Catch/vessel, rent + Excess capacity − |
| Alaska | 1995 | Halibut and sablefish | V | Free | H | ($R_{conc}$) | Accounting and auditing (debit cards, fish tickets) | | Economic outcomes uncertain |
| Florida | 1992 | Spiny lobster | F | Fixed price | H | ($R_{conc}$) | Monitoring of trap tags; inadequate | | Number of traps − Landings stable Value of permit + |

*Notes:* ITQ allocation is to fishermen (F) or to vessels (V), and ITQ basis is historic (H), capacity (C), or equal (E).

[a]"Property rights" generally does not mean full property rights, but freely tradable except for retrictions as noted. Restrictions to free tradability or to property rights are based on share of fish or permit holdings ($R_{max}$), concentration ($R_{conc}$), or employment effects ($R_{empl}$).

*Sources:* Arnason 1996; Batkin 1996; Casey et al. 1995; Flaaten, Heen, and Salvanes 1995; Gauvin, Ward, and Burgess 1994; Geen and Nayar 1988; NRC 1999; Wang 1995.

1979, Iceland was the first country to implement an ITQ system, initially for herring. The program was expanded several times to include more species until 1990, when a system that encompassed all species in Iceland was created.

ITQs in Iceland were superimposed on the earlier management system; thus, a policy of limited access based on nontransferable licenses still exists. Initially, in the capelin (a small smeltlike sea fish) fishery, 33% of quota shares were allocated on the basis of vessel capacity and the remainder were divided equally; in the herring and shrimp fisheries, shares were divided equally. The Ministry of Fisheries sets the total allowable catch (TAC). TACs for certain species (such as cod) have been set above the biologically recommended limit allegedly "because of the dependence of the Icelandic economy on fishing." Quota trading has been strong: since 1986, 20–30% of quotas have been exchanged annually; in 1993–1994, 44% and 96% of quotas were traded for cod and saithe (the Atlantic pollack), respectively.

Fisheries management has been successful in restoring stocks of herring but not cod (because of the excessive TACs). ITQs appear to have increased the wasteful practices of disposing of immature fish and discarding less valuable fish when better fish are caught (highgrading) in Iceland's multispecies fisheries. Although the quota system was originally presented as a temporary emergency measure, it has caused a major restructuring of the fishing industry as well as massive regional unemployment. The industry has become highly concentrated, leading to some fierce distributional conflicts: the largest firms own 75% of the total quota in demersal fish. National strikes protesting the ITQ system offer several criticisms:

• The initial allocation of quotas to vessel owners has made vessel owners wealthy while crew members are disenfranchised.[4]
• The method of introducing the quota system was misleading.
• Most quotas are held by the largest firms.
• The profit-oriented exchange of fishing rights is considered morally wrong.
• The ownership of fish resources as private property is objectionable.
• Tenancy relations have emerged in fishing.
• Quota transactions are opaque.
• Massive regional unemployment and community dissolution have resulted.

The Ministry of Fisheries collects fees to cover the cost of monitoring and enforcement. The upper bound set for these fees is 0.2% of the estimated total catch value. Quotas are issued subject to a small annual charge for enforcement costs. Economic benefits vary by fish species. For herring, catches have tripled since 1977, and effort has declined 20% over the same period. In the capelin industry, catches have been constant while the number of vessels has decreased 40%. In the demersal fisheries, which account for more than 75% of the total value of the Icelandic catch, the annual growth of fishing capital has slowed while productivity has increased. Profit as a percentage of gross revenue was −5.8% in 1980 and 4.0% in 1995. In 1990, realized rents were more than 50% of the estimated maximum attainable rents. However, a government buyback program in 1994 suggested that the ITQ program was not fully successful.

## Implementation: New Zealand, Chile, and North America

The experiences of other countries vary but overall corroborate many of the Icelandic experiences. ITQs are generally a proportion of the TAC; in New Zealand and Iceland, shares were originally specified in terms of fish tonnage but currently represent shares that allow for easier regulation in response to variations in stock. Allocation is usually to individual fishermen (e.g., in New Zealand and Chile and for spiny lobsters in Florida) or vessel owners (e.g., in Iceland, Norway, and some U.S. fisheries and for halibut in Canada). Allocation to individual fishermen is more suited to and encourages small-scale operation and might perhaps in some contexts better satisfy distributional concerns.

The basis for allocation defines the character of each program. Some programs have a grandfathering approach whereby rights are based on historical catch. Other systems base rights on vessel size or capacity; commonly at the inception of programs, small boats are not included in the ITQ system and may catch as much as they like, although eventually (especially if there is a large noncommercial subsistence fishing sector), small boats must be regulated. Shares are sometimes allocated equally among all participating boats. Systems often use more than one of the above criteria for determining rights, but shares are almost always allocated free of charge. One exception is Chile, where rights are auctioned within the ITQ system; however, only two minor fisheries are involved in the Chilean ITQ system. In many countries, auctioned permits, like taxes, would generate considerable resistance. However, partial auctioning or allocating permits with some intermediate price might be an acceptable option.

New Zealand has (at least partially) defused the potential conflict between traditional and commercial fishermen. Commercial fishermen harvest their share of the TAC after tribal and recreational fishermen have fished. When the TAC is increased, commercial fishermen must purchase additional quota shares, whereas tribal and recreational fishermen have free access (subject to conservation laws), so social dislocation has been minimal. Another example where the dislocation was minimal is in the U.S. wreckfish fishery, which introduced ITQs without much economic dislocation, mainly because the fishery is relatively new: quotas allocated on the basis of catch history did not have a significant redistributive effect. In the U.S. halibut and sablefish fisheries, part of the TAC is allocated to six nonprofit communities as part of a community development program. Concerns over community development were one of the impetuses in introducing ITQs. The program does not apply to subsistence, treaty, or sport fishermen. Still, some dissatisfaction exists as to the initial quota allocation, which excluded crew members and processors

The nature of the ITQ varies from being a full legal property right (as in several cases in New Zealand) to being "perpetual" but not a full property right (as in the U.S. wreckfish fishery) to being nominally the property of the government (as in Iceland). However, the high cost at which permanent quota shares are traded in Iceland suggests that quotas are considered personal property that, although not private, is not likely to be revoked by the government.

In most cases, shares are permanent or of indefinite duration, and this certainty is considered an important aspect; in at least one instance, authorities have distributed permits and then been forced to revoke them (a rock lobster fishery in New Zealand [Breen and Kendrick 1997]). The exception is again Chile, where quotas are valid for 10 years. Quotas are sometimes subject to restrictions on trading, usually to protect share ownership or local employment in certain communities. Such is the case in the Canadian halibut fishery, where the ITQ program was initially a two-year experimental program that did not allow trading. In 1993, limited trading was allowed; currently, vessels cannot trade more than twice their initial allocation of shares and can lease shares only annually. Transfer restrictions are intended to limit consolidation of the fleet.

*Enforcement and Economics*

Enforcement in some areas began with a "game warden" approach with monitoring at sea, but in some places, this approach has been partly replaced with less expensive on-land audits. Violations are penalized with fines or sometimes forfeiture of quotas (as in Canada). In some cases monitoring is financed by the fishing industry. For instance, in the U.S. halibut and sablefish fisheries, fishermen are required to enter catch information at landings on debit cards. Halibut fishermen are also required to record landings on fish tickets, which are checked against buyers' records. The Magnuson–Stevens Act requires a cost recovery program: up to 3% of the value of landings can be recovered for direct management costs, but currently, costs are not being recovered.

Positive economic results of ITQ programs are evident to some extent in every program: increased efficiency in terms of effort per catch and catch per vessel, increased rents, increased quality of catch, reduced capitalization, and increased fish and quota prices. Consolidation of the fleet may inevitably exacerbate unemployment.

In the case of the Canadian halibut fishery, the background was bleak: seasons for commercial fishermen before the ITQ program had been increasingly shortened, from two months in 1982 to less than one week 1990. The resulting derby-style fishing had had a highly deleterious effect on fish quality and mortality. The economic effects of the ITQ system on the industry have been mainly a result of the switch from a frozen product to a fresh product as the season has been extended; prices have increased 55%. In addition, landings are currently being timed to reflect market prices and opportunity costs of other kinds of fishing activities. Also, productivity has increased with reduced crews. Designed and supported almost entirely by industry, the ITQ program is perceived to have increased retirement security, conservation, and record keeping and to have reduced waste, bycatch, and destructive loss of gear. However, entry has become difficult for new fishermen. The number of quota holders in halibut declined by 24% between 1995 and 1997; for sablefish, the number declined 18%.

In the U.S. surf clam and quahog fisheries, increased efficiency is reflected in increased landings per vessel, decreased fishing costs, improved earnings, and a lessening of excess harvesting capacity. Fisheries have accrued large capital savings. For original vessel owners in 1992, estimated surf clam resource rent was US$11.4 million under the ITQ program. Fishing hours per vessel have increased, indicating that the use of capital has increased. From 1990 to 1992, catch per vessel increased 96%. Industry restructuring has been rapid; within two years of starting the ITQ program, the size of the fleet had shrunk from 128 to 59 vessels. To secure supply, an emphasis on vessel ownership has shifted to ownership of quota shares.

In assessing the role of ITQs, concentration of shares remains as one of the main sources of contention in the Icelandic program.[5] The Icelandic Parliament has appointed a special Natural Resource Committee to consider various aspects of the allocation of these rights, including the possibility of taxing them. Taxation may ultimately complement ITQs and thus improve perceived fairness in Iceland.

A much more exciting possibility, however, is that taxation could entirely replace ITQs and be more efficient. In a model with uncertainty in the assessment of stock and recruitment, the most essential variable is the size of the surviving stock, which depends on the unknown (and stochastic) recruitment of new fishes. The inherent flaw of quantity regulation is that the size of the TAC and the quotas must be determined before the regulator knows these variables and thus the appropriate catch. The correct tax may be easier to know if, as commonly assumed, the cost of fishing varies with the density of fish. If the target stock is a certain minimum level, then all the tax has to do is remove the profitability of any fishing effort that would bring the stock below that level. Although such a tax scheme sounds indirect and complicated, a model with inherent ecological uncertainty that clearly favors taxes over ITQs from the viewpoint of efficiency has been developed (Weitzman 2002).

## Conclusions

As the contours of the last wild frontier that the oceans represent begin to become apparent, issues of management, efficiency, and equity surface, and resource economists have important roles to play. The role played by economists in renewable resources policy over the past quarter-century, particularly for fisheries, may be both applauded and criticized (Wilen 2000). The insight that ecosystems are capital and must be analyzed as such is fundamental. Other major achievements include the incorporation of open access and property rights into models that helped make ITQs an important and successful instrument. However, other policy-relevant issues such as property rights, monitoring, enforcement, and bycatches deserve more emphasis. Similarly, the creation of appropriate institutions, property rights, and policy instruments to address diverse ecological and social conditions with appropriate incentives for monitoring and reporting, fair distribution of benefits, and flexibility in the face of changing circumstances and complex ecology remains a great intellectual challenge to environmental and resource economists.

Successful ITQ systems illustrate the importance of property rights and restricted access. They can maintain sustainable stock levels as long as TACs are not set too high. ITQ programs typically result in less capital, more fish, higher quality and prices, and thus recovery of economic rents. The quotas allow catch to be reduced to optimal (or at least sustainable) levels. If the quotas are shares of a TAC, then they provide short-run flexibility to match stock variations. In the United States, the North Pacific Halibut Act of 1982 gave decisionmaking authority to Pacific Fishery Management Councils, which now set local annual limits for commercial catches.

ITQs foster responsibility and stewardship in the fishing industry, including increased research and cooperation with regulations, but disposal of bycatch is still a problem. New Zealand has added features to provide incentives to fishermen not to discard fish: the government will buy the bycatch at a set price, or fishermen may purchase quotas for the bycatch. Even with such options, the discarding of excess catch and low-priced fish appears to be a problem common to all ITQ systems (Copes 1986). It has not been a major problem in Iceland, but in tropical multispecies fisheries, it might be a considerable obstacle to ITQ implementation (Arnason 1996), as would the cost of monitoring.

Implementing an ITQ program risks concentrating wealth and creating concomitant social tension. Various allocation mechanisms can be designed to help meet social, cultural, and distributional objectives such as protecting regional fishing cultures. Some programs have sought to limit concentration by restricting trade. This approach also limits the efficiency gains, but distributional and regional concerns may make some restriction desirable. The U.S. National Research Council (NRC) recommends ITQs but says they are no panacea for all fisheries (NRC 1999). Instead, regional councils should decide whether to use them on a fishery-by-fishery basis. NRC recognizes the need for certainty but also cautions beneficiaries against viewing the ITQs as compensable rights; the government must retain ultimate stewardship of the resource and may, for instance, decide to limit fishing in certain areas to certain methods, and such restrictions should in no way lead to compensation. Finally, NRC emphasizes the need for fairness in addition to efficiency as well as requirements (such as "skipper-on-board") for limiting the accumulation of shares among too few parties.

Neither ITQ systems nor CPR management represents the optimal or definitive solution to all fishery management problems; fishery managers and communities have both instruments at their disposal, and a great deal of fine tuning is needed to adapt either instrument to specific socioeconomic and ecological conditions. CPR management is likely to be important in developing countries with dramatic diversity and large coastal populations dependent on coastal fisheries, whereas ITQ systems are more applicable to large, highly developed industries in fisheries with less ecological diversity. In the future, more hybrid instruments may combine ITQs with taxes or fees, and more attention may be paid to the allocation and duration of quotas. In addition, a more general approach to management may focus on ecosystem and recreational values.

## Supplemental Reading

Andersen 1976, 1979
Berkes 1977
Cordell 1978
Davis 1983
Hannesson 1998
Katsonias 1984
McGuire 1983
Watson 1982

# Notes

1. The impacts are large: several tens of thousands of fishermen were made unemployed by the decline of cod fishing in Canada during the 1990s. However, most of the shortfall in income has been replaced by landings of shellfish (Arnason, Hannesson, and Schrank 2000).

2. There might be many intermediate examples where effective fishing methods entail some increased risk to fish populations, such as fine-meshed nets, which catch a high proportion of juveniles. However, some technology is simply effective, such as equipment to localize or lure fish to one location.

3. Taxing has been successful when the proceeds were used to "buy out" fishermen from the industry, thereby further reducing excess capacity and keeping the funds within the group of fishermen.

4. The grandfathering of ITQs was based on catches from 1983 to 1988. This principle for the allocation of property rights has been challenged in the Supreme Court of Iceland, where the case was still pending in 1999.

5. In other countries, there has also been considerable concentration, even though it may not have engendered so much conflict. In the Australian program for southern bluefin tuna, the Western Australia and New South Wales fleets sold most of their permits to Southern Australia. The total number of vessels decreased by more than 70% from 1984 to 1988, but a good deal of the exit is considered to have been inevitable because stocks were declining.

# CHAPTER 29

# *Agriculture*

$A$GRICULTURAL ECONOMICS IS A VAST FIELD, and even the interface between agriculture and environmental issues is too large to cover in just one chapter. I have selected a few issues to highlight, primarily because of their importance and because they illustrate interesting applications of principles that are crucial for the design of environmental policy instruments.[1] Nonpoint-source pollution (NPSP; see Chapter 13) is a pervasive feature of agriculture. The economics of poor households that depend on natural resources such as small-scale agriculture or forestry typically are very different from the simple neoclassical textbook case of a profit-maximizing firm. The case is not that they do not care to maximize profit, but missing markets and information, a lack of reliable titles, and constraints in various markets (labor, credit, insurance, and input and output markets) routinely lead to "corner solutions" in which the simple first-order conditions of maximization do not hold.

The inherent riskiness or stochasticity of all natural resources–based activities is an important factor in explaining the "nonneoclassical" feature of NPSP. Share-cropping (see Chapters 3 and 13) has been labeled "inefficient" because the tenant receives only a share of marginal benefit but has to bear the whole marginal cost (Marshall 1920). Thus, cost and benefit are not equated, as would be expected in simple standard optimization models. This result is explained by stochastic outcomes, together with disparities in income and risk adversity as well as difficulties with monitoring and the ensuing moral hazard.[2]

Other important factors are property rights, returns to scale, incentives, and efficiency. Many economists have believed that the poor farmers who typically are sharecroppers are too small to survive in agriculture (or forestry) and thus are doomed to become landless laborers or possibly commercial farmers (e.g., Sadoulet, Fukui, and de Janvry 1994). One consequence might be that the problems associated with sharecroppers are transitional; therefore policy should focus on large-scale farms (either through policies that favor private concentration or through state or cooperative farms, depending on one's ideology). However, little definitive evidence supports this hypothesis (Binswanger and Holden 1998). Advantages to scale appear to be limited in most forms of agriculture (Berry and Cline 1979; Prosterman and Riedinger 1987; Rosenzweig and Binswanger 1994), and the incentive advantages of sharecropper agriculture may in some cases outweigh scale effects; however, excessively small, fragmented farms are inefficient. The prevalence of moral hazard gives an advantage to family labor over hired labor. Advantages to scale depend on geophysical and ecological conditions, among other things. For example, agricultural equipment might provide considerable advantages to scale in the plains but probably not in a hilly area.

Advantages to scale in the processing (as well as marketing and transport) of certain agricultural goods such as tea, sugar, palm oil, rubber, and coffee often have led to large-scale plantations of these crops. Close ties between harvesting and processing are natural but should not be inevitable. In fact, contractual arrangements between processors and small-scale farmers appear to be replacing the large plantations for sugar cane in Thailand and India and for oil palm in Indonesia (Binswanger, Deininger, and Feder 1995). The processing industry itself is often extremely contaminating despite the fact that abatement may be technically easy.

Small-scale farming, which is so common in the world, may be the best way of managing agriculture and related resources. However, some problems such farmers experience may lead them to unsustainable behavior. One of these is tenure insecurity, which, together with other factors, may lead to lack of interest in soil conservation. Another such problem area is the nexus between high discount rates and credit rationing. Both issues are discussed later in this chapter.

There is an important distinction between irrigated and rain-fed agriculture. Irrigated agriculture is associated with fertilizers, pesticides, high-yielding varieties, and usually a high degree of mechanization. Water supply is a major issue (see Chapter 26), as are nutrient runoff, water quality, nutrient depletion, and the risk of salinization in some semi-arid areas. Reliance on a few high-yielding crops may decrease biodiversity and increase exposure to risk. Intensive pesticide use is associated with risks of damage to ecosystems and to the health of farm workers and their families. However, pesticide application may be a rational response to risk in an environment where institutions for insurance and savings are absent.[3] Rain-fed agriculture, on the other hand, is associated with complex relationships among poverty, discount rates, hillside deforestation, soil erosion, and conservation as well as downstream effects, property rights, population pressure, and tenure.

The provision of infrastructure such as roads is a powerful instrument to affect land use. New roads may open up areas to agricultural exploitation (and the cutting of virgin forest). They generally facilitate the enforcement of property rights and thereby extend frontier access, pushing the managed system (e.g., farmland, grazing land, plantations) farther into the unmanaged area of natural forest. Roads also can have a large impact on the productivity of agriculture and other activities. They are generally high on the list of priorities of poor rural people. Increased accessibility to markets could conceivably more than compensate for any negative effects by increasing incentives for organized and sustainable agriculture, silviculture, and land management in general. Extensive and intensive road networks must be differentiated. Whereas an extensive network of roads erodes incentives for sustainable forestry and agriculture (by providing new cheap land to be mined), an intensive network of farm-to-market roads makes small-scale agriculture viable (Schneider 1995). Agricultural incentives can have profound impacts on forests depending on the stage of development. In the first stage, they have a direct effect on the natural forest by encouraging the clearing of land. In later stages, they may lead to incentives for managed silviculture.

In the Amazon, the kinds of policy mentioned here, the classical policies (e.g., agricultural extension and the provision of related services), and the role of government structures interact in a complicated way (Schneider 1995). Four attractive policy recommendations can promote growth as well as resource conservation:

- eliminate perverse subsidies that artificially benefit deforestation,
- disseminate (e.g., through agricultural extension) technologies that increase the productivity of existing land,
- enable indigenous communities to manage their lands efficiently, and
- provide infrastructure (rural roads, small-scale irrigation, and so forth) that provides "safety nets" for poor communities (Deininger and Minten 1996).

The broader issues of land use allocation between agriculture and forestry are addressed in Chapter 30.

Labeling as a policy instrument has the potential to benefit or harm developing country producers in the future. It seems reasonable to presume that many producers in developing countries might benefit from ecologically sensitive production (partly because it is labor intensive). Several initiatives already exist, such as shade-grown coffee and organically grown fruit. However, the costs of certification may be too high for small-scale farmers in this context (Gobbi 2000). Harm might result from labels that are designed to suit northern consumers (or producers) and come to act as trade barriers; Mexico makes this claim in the case for dolphin-free tuna (see Chapter 27).

## Managing Agricultural Runoff

Agricultural runoff generally has a nonpoint-source character. Various externalities (e.g., soil, nutrients, pesticides, odor, genetically modified organisms, and pests) spread from one farm to another and to various downstream (or downwind) recipients, but the source (individual farmer or field) is typically difficult to pinpoint. Thus, designing policy instruments may be difficult; however, it need not be the case. For hazardous pesticides, for example, a tax, ban, or other regulation stands a good chance of being effective (unless there is smuggling).

For agricultural practices such as manure application, soil terracing, and the use of inputs that are not classified as hazardous, designing the proper instruments is more difficult but still important. The application of large quantities of nutrients in intensive farming can be detrimental to the water bodies into which those nutrients ultimately flow. For instance, water bodies can become *eutrophic*— depleted of dissolved oxygen—a state that results in considerable loss in biodiversity and aesthetic as well as recreational and productive values.[4] It is caused not only by synthetic fertilizers but also by natural manures, which cannot be taxed, and agricultural practices such as draining wetlands and straightening the water-courses of streams and rivers. These processes remove the natural opportunities for denitrification and other retention of nutrients. The restoration of wetlands can be more cost-efficient than the building of sewage treatment plants when it comes to nutrient removal to stop eutrophication (Andréasson-Gren 1991).

Many freshwater lakes and even some coastal and marine areas in Europe are severely eutrophic. In Africa, Lake Victoria—which drains a catchment area in which 30 million people live—is severely affected by problems ranging from overfishing to the invasion of foreign fish and plant species (Nile perch and water hyacinth, respectively). The lake receives enormous amounts of nutrients from various sources: soil erosion caused by deforestation and hillside agriculture, agriculture close to the lake and on former wetlands that were drained, and poorly managed agricultural industries. Sugar plants, coffee mills, and other processing plants discharge large amounts of nutrients. The total surplus of nutrients in the lake is a vital factor behind the massive takeover of the water hyacinth, which in some years during the 1990s was so great as to stop not only fishing canoes but large ships and ferries. For long periods, the hyacinth mats were so dense that ships could not enter port and power stations had to be closed. The hyacinths also contributed to the spread of malaria and other diseases by providing habitat for mosquitoes and other disease vectors as well as providing habitat for snakes and other dangerous animals.

The studies on NPSP discussed in Chapter 13 assume uncoordinated selfish behavior in a Cournot–Nash equilibrium. This assumption does not account for the possibility of interaction (nonmarket competition, rivalry, or cooperation) among farmers. Still, neighbors do collaborate and have information about each other, and this information needs to be brought into the analysis. Sociological and political science literature show that effective monitoring is often possible by peers even when it is not possible by outside authorities. This finding has led to the development of schemes in which farmers are encouraged to collaborate to protect the environment in a watershed. The tax rate may be lower in the case of full cooperation, and the polluters thus have an incentive to cooperate (see Millock and Salanié 1997 for a model of an ambient tax with cooperation in abatement; see Shortle, Horan, and Abler 1998 for an overview of research issues in this field). However, they also have an incentive to free ride, and thus the group must have peer monitoring as well as a credible punishment strategy that can sufficiently prevent the polluters from deviating from the cooperative group. The national environmental protection agency can choose a two-tier system in which farming communities choose between tough conventional regulation (e.g., high taxes on pesticides) and this kind of collaboration. The result is to be monitored at a location such as a watershed, and if the collaboration fails, the environmental protection agency reverts to the conventional regulation as its "threat point."[5]

If farmers can be assumed to observe each other's efforts to reduce pollution, then peer monitoring might minimize the information costs of regulation in applying an ambient tax to control agricultural NPSP (Byström and Bromley 1998). The incentive scheme, based on nonindividual contracts between farmers and the regulating authority, can achieve the cost-effective outcome even if farmers do not have homogeneous risk preferences (i.e., there is adverse selection).

Also, because farmers are allowed to trade pollution abatement efforts, neither individual contracts with each farmer nor full information on the part of the regulatory agency is necessary to achieve cost-effective abatement, which will substantially lower the information costs of regulation. This result is more efficient than the (conditional) subsidies scheme to overpay the agents due to imperfect agricultural monitoring. The only weakness is that the system of collective penalties might have weak legal support because criminal law in general requires that liability be established at an individual level.

Common property resource (CPR) management is another relevant mechanism in this context because it has the potential to overcome the incentives for free riding and the costs of monitoring by using the cohesion of culture together with specially designed rules of management in a way that can be (at least partly) modeled in terms of repeated games. Peer monitoring is often superior to monitoring by agents of an outside authority (Ostrom 1990). Because peers have more relevant information and more legitimacy than outside monitors, monitoring is cheaper, more accurate, and perceived as being more fair. CPR management appears to be particularly useful for the marginal resources in relatively poor contexts in which yields are so low as to make the cost of fencing appear high or so variable as to make them inappropriate. (See additional examples and discussion in Chapters 26 and 31.)

One of the critical environmental issues in agricultural policy is how to compensate farms that provide additional ecosystem services. In many industrialized countries, the agricultural sector is subsidized, and farms may receive subsidies that are explicitly tied to agricultural practices that reduce runoff (e.g., leaving strips along waterways, planting certain crops, keeping herd sizes relative to pasture, letting land lie fallow, and timing and zoning fertilizer application). In poor countries, agricultural subsidies are typically less elaborate, and in general, the agricultural sector is taxed heavily (not necessarily directly but through the setting of relative prices). Compensation for ecosystem services is fairly new in developing countries and, in principle, can be funded by a share of ecotourism; hunting or fishing revenues; contracts with bioprospecting companies; increased revenue from eco-labeled products; or direct payments from parks, public authorities, or other downstream users (e.g., Pagiola 2000). Costa Rica has one of the most elaborate systems of direct payment for environmental services (Chomitz 2000; see Box 29-1 for some examples). Many of the sources of compensation for ecosystem services are likely to target forest practices, but some agricultural activities also may be included.

## Property Rights, Population Growth, and Soil Erosion[6]

Given the pervasive nature of hunger and famine, developing countries naturally focus interest on food security (Sen 1981, 1999). Meanwhile, observers worry about the long-run sustainability of agriculture and other natural resources–based activities in these countries (e.g., Scherr 2000; Templeton and Scherr 1997). Outside interests frequently offer well-intended advice concerning various aspects of agricultural management, soil conservation, water management, fuel

## Box 29-1. Paying for Ecosystem Services

The city of New York started an extensive, long-term process in 1989 to protect its water sources and their catchment areas. The program included land purchases, environmental projects, sewage treatment, and environmental agreements with landowners. Hundreds of millions of dollars were invested in land purchases alone, but these investments were much cheaper than the cost of water purification that otherwise would have been necessary. As a side benefit, New York has acquired an area for hiking and recreation.

Similarly, municipalities in Quito and Cuenca, Ecuador, pay to protect their water resources. Some downstream municipalities have agreed to make fiscal contributions to the Bosque El Imposible National Park in El Salvador in return for improved water quality.

*Sources:* City of New York 2002; World Bank 2002; Pagiola 2000.

---

wood supply and conservation, breeding strategies, the provision of seedlings, and other specific activities and technologies.

One problem of this approach is that it focuses on partial subgoals instead of the main goal of improving farmer welfare. Similarly, in other sectors, the strategy of focusing on one particular issue at a time—birth control, vaccinations, literacy, electrification, and so forth—has often been unsatisfactory. Much of the concrete advice may be valuable, but many projects and policies have failed, sometimes because they went against the interests of the supposed beneficiaries. For instance, the colonial administrations in many countries ordered the erection of physical soil conservation structures such as stone gabions and bench terraces. Because these administrations had a reputation of caring relatively little for the welfare of current (local) farmers, it was perhaps puzzling that they apparently cared so much for the long-run welfare of their descendants. The manner in which these soil conservation structures were ordered was nonparticipatory, to borrow a modern term; their construction was simply the result of forced labor, and their presence was strongly resented. After former colonies gained independence, the inhabitants abandoned or destroyed many of these structures, illustrating what the local farmers thought of them.

Even voluntary assistance programs often confused means and goals, so that the aim of projects (and the remuneration mechanisms) came to be formulated in terms of "kilometers of terraces" rather than "improved agricultural productivity." Modern approaches focus on enhancing productivity and participation. The primary goal now is simply to help farmers achieve higher sustainable income, although other goals include reducing runoff and downstream effects such as the formation of gullies that wash away other farmers' soil or destroy roads and other infrastructure. Maintaining the productive capacity of the soil is still important but now is typically achieved by biological (and less labor-intensive) methods, such as planting special types of grass, trees, or shrubs or simply leaving agricultural waste along contour lines to facilitate the trapping of water and nutrients and, consequently, the natural development of terraces.

Investment in land conservation depends on several factors (Dasgupta and Mäler 1995; Lutz, Pagiola, and Reiche 1994; Shiferaw and Holden 1999). One of the most contentious and serious issues is the effect of population pressure.

The pessimist Malthusian view is that increasing population will lead to falling marginal productivity, overuse, and resource degradation that will further undermine human health and deepen poverty (Cleaver and Schreiber 1994). Policy responses reflecting this view emphasize population control, environmental education, and controls on access to resources. The optimistic Boserupian view describes how increasing population pressure is the mechanism that leads to improved technology and increased efficiency in land management and thereby prosperity (Boserup 1965). Evidence can be found to support both positions (Binswanger and Deininger 1997; see also Heath and Binswanger 1996; Sterner and Segnestam 2001; and Tiffen, Mortimore, and Gichuki 1994).[7] A review of more than 70 empirical studies in poor hill and mountain regions was inconclusive about the effects of population growth on land and forest quality (Templeton and Scherr 1999). Other underlying factors must explain why the results are so dramatically different.

A study of Machakos district, Kenya, is a prime example often quoted in support of the Boserupian hypothesis (Tiffen, Mortimore, and Gichuki 1994). Despite a 500% increase in population over 60 years, the land degradation and food insecurity that was predicted for this area in the 1930s has not occurred. Instead, food production has increased manifold due to a threefold increase of livestock holdings per farm and a doubling of the total cropland area. Considerable local investments in water and soil conservation provide the conditions for horticulture and small-scale dairy production. The farmers organize themselves in efficient democratic workgroups (*mweya*) and enjoy secure land rights, infrastructure, and market access on favorable terms, partly because of the proximity between Machakos and Nairobi.

In other areas of Kenya and in the Ethiopian highlands, several of these vital factors are missing; as a result, high population pressure appears to be positively correlated with soil degradation (and increased poverty). In a study of Muranga district, Kenya, between 1960 and 1996, population increase is associated with more soil erosion, not conservation (Ovuka 2000). In a study of some villages in northern Ethiopia, tenure security (proxied by the number of years of land tenure) was positively related to soil conservation measures (Alemu 1999; see also Chapter 5).

Soil conservation practices and increased agricultural production are more easily achieved in communities that collaborate democratically than in those where each farmer works individually. Many studies show the importance of policies, property rights, and good democratic institutions in avoiding the Malthusian spiral of increased population, poverty, and resource degradation (e.g., Ekbom, Knutsson, and Ovuka 2001).

## Risk in Sharecropper Agriculture

Agriculture is inherently risky because weather interacts in a complex way with farmers' efforts. The decisions that farmers make regarding soil conservation, irrigation, fertilizer and pesticide use, and animal husbandry affect expected yields and yield variance in the short run as well as the long run.

The effects of the uncertainty presumably are worse in poor countries, where deviations from an expected harvest that is already low may result in starvation. Asymmetries of information between principals and agents are common. Moral hazard makes it difficult to provide sufficient crop insurance to satisfy the risk-averse farmers (see Chapters 3 and 13). Consequently, crop insurance is rarely found in developing-country agriculture (Hazell, Pomareda, and Valdes 1986; Binswanger and Holden 1998). Farmers may attempt to spread risks through savings, diversification of wealth, and outside employment; however, the institutions that provide these services are commonly inadequate or absent in rural settings. Farmers therefore choose other strategies that help spread risk in the short run, but they may cause externalities or increase risks in the long run[8]:

1. Expand extensively along the agricultural frontier.
2. Use hinterlands, shrublands, and other CPRs for hunting and gathering of plants, wood, and fodder, particularly in times of need.
3. Hold large stocks of cattle as capital.
4. Use pesticides, irrigation, and genetic selection to help minimize variation in harvest. For example, farmers may choose safer (although maybe less high-yielding) varieties of grain or prefer goats to cattle.
5. Organize mutual credit, banking, or insurance schemes.

Different farmers choose different strategies—some of them conflicting. Richer or more powerful farmers may use Strategies 1 and 3, thereby hindering poorer farmers who want to use Strategy 2. Strategies 1 and 2 are discussed in Chapters 30 and 31; the rest of this chapter focuses on Strategies 3–5. All these strategies have implications for ecosystems. They may be acceptable in some cases but cause conflicts of interest in others. In conflicting situations, it may not be enough to regard the resource problem per se; overgrazing cannot easily be solved by a tax on cattle, and to protect or repair a damaged ecosystem, banning or taxing pesticides may not necessarily be the most appropriate policy. To devise instruments, policymakers must understand the two underlying market failures: one related to the environment, and the other related to the unavailability of insurance and other financial services.

## The Economics of Large Herds

Understanding interactions between social and ecological structure is particularly important in managing ecosystems. For instance, in Botswana, the extensive use of cattle in ecosystems to which they are not perfectly adapted has created pressure on the ecosystem (Arntzen 1996). Rangeland degradation, for example, is particularly detrimental to the poor people who live and work in the most degraded areas; lack secure access to resources (and in times of overall scarcity, the high-income groups restrict access to resources); depend heavily on CPRs and on activities such as hunting and gathering, which are on the decline; and lack other alternatives.

Grazing associations have been formed in Lesotho to give exclusive grazing rights to their members. The benefits of these rights are apparently sufficient to persuade people to pay grazing fees, and it may be an important step in the right direction. However, it does not address the fundamental factors that lead to over-stocking, which is at the heart of the problem. In times of drought, a certain per-centage of livestock is at risk of dying. This means a loss of savings to the individ-ual, and the rational, individual response may be to increase savings. The trouble is that if all savings are in cattle, then the pressure on the ecosystems will also rise, increasing the rate of mortality during the next drought. The individual response may thus be self-defeating at an aggregate level. Improved banking or insurance services could be an important mechanism by offering people alternative courses of action to increasing their stock of cattle.

## Crop Insurance as a Substitute for Pesticides

As the tide of chemicals born of the Industrial Age has arisen to engulf our environment, a drastic change has come about in the nature of the most serious public health problems. Only yesterday mankind lived in the fear of the scourges of smallpox, cholera, and plague that once swept nations before them. Now, our major concern is no longer with the disease organ-isms that once were omnipresent: sanitation, better living conditions and new drugs have given us a high degree of control over infectious disease. Today we are concerned with a different kind of hazard that lurks in our environment—a hazard we ourselves have introduced into our world as our modern way of life has evolved. (Carson 1962, 168)

In this quotation, Rachel Carson alludes to the danger of uncontrolled pesticide use. She goes on to give example upon example of humans and birds whose health, reproductive capabilities, and even life have been destroyed as a by-prod-uct of the use of dieldrin and other chemicals, many of which have since been banned, some almost directly as a result of the publication of her book, *Silent Spring*. Still, many thousands of people are poisoned each year by pesticides and other hazardous chemicals. In Brazil alone, more than five thousand cases of pes-ticide poisoning were reported in 1998, and according to the World Health Organization, for every case of poisoning reported, 49 are not (Lins 1996).

One striking example of how pesticides are used where other methods might be preferable is in the fight against desert locusts. The Food and Agriculture Organization (FAO) spends millions of U.S. dollars yearly on spraying pesticides to minimize the threat of locust infestations in Africa. Yet the actual average dam-age to crops is small. In some 50 years of data, a relationship between harvest loss and locusts is not apparent (in fact, the relationship is sometimes inverse because the locusts tend to "attack" in years of good harvest). Expected harvest loss is minimal and, according to some authors, lower than the cost of pesticide spraying (e.g., Heroc and Krall 1995; Joffe 1995). In addition, the widespread use of pesti-cides has been linked to ecosystem effects and other costs.

However, because infestations tend to be locally concentrated, an individual farmer's crops could be almost totally destroyed by desert locusts. In the absence of insurance, income diversification, and banking services, such risks are ulti-mately fatal and thus unacceptable. Hence the urgent need for relevant, effective policy instruments and the reason why governments support the FAO pesticide strategy. However, the convention of pesticide use against desert locusts is being rethought, partly because the locusts are considered a delicacy in the Sahara region but lose their sale value if there is a risk that they have been sprayed. Presumably, farmers in the affected regions would be better served by insurance than by pesticides (Smith and Goodwin 1996). The question is whether insurance companies would be willing to insure farmers given the moral hazard inevitably involved in crop insurance. If donors or governments can help set up the institutions that facilitate the provision of insurance (maybe insurance against desert locust infestations, which would be easier to verify than just crop failure in general), then the need for pesticides would be drastically reduced.

*Financial Institutions To Reduce Risk*

To some degree, insurance and banking are interchangeable options. Savings may act as a buffer that reduces the need for insurance. Even diversification of income or the portfolio of assets results in less variance and thus reduces risk and may act as a substitute for savings or insurance. All the factors that affect risk must be considered in one context. Village money lenders are known to charge exorbitant interest rates, which effectively amount to as much as 100% or even several hundred percent per year (for India, see Pender 1996; for Ethiopia, see Holden, Shiferaw, and Wik 1998, who find that many farmers live with rates of more than 100%). Such rates are typically the result of severe rationing in the credit market in conjunction with absolute poverty (see Box 29-2). The practice is detrimental to progress because many good projects will not get funded at this level of interest.

Why do other banks and bankers not enter the market and thus bring interest down to more reasonable rates? Why do international banks not move in, for instance? The primary reasons are high risk, economies of scale, asymmetric information, asymmetric monitoring, and high costs of information. Local money lenders incur low administrative costs commensurate with the small scale of lending, have better information about which individuals are bad credit risks,

---

## Box 29-2. Investing in Fuel Efficiency and Discount Rates of the Poor

In Cheranástico, a village of 2,400 inhabitants in Michoacán, Mexico, villagers typically spent up to two hours a day collecting firewood for cooking. Researchers from the Universidad Nacional Autónoma de México determined in a field study that fuel needs and collection times could be cut drastically, saving villagers up to 1.5 hours per day (average 50 minutes), if energy-efficient stoves were used. The researchers helped villagers build stoves designed to be completed in a day or two with locally available clay and other materials.

Assuming 25 labor-hours to build each stove, the investment would be extremely "profitable" at a rate of interest of 3% per day (more than 150,000% per year). Most—but not all—villagers built the stoves. Some of the poorest never made the investment because every day was a struggle for survival. This example illustrates starkly how poverty leads to high discount rates.

*Source:* Almeida et al. 1989.

and can monitor their professional effort and performance. Thus, a local money lender can address all the major asymmetries—the moral hazard and the adverse selection problems. Together with freedom from economies of scale and perhaps other barriers to entry (e.g., various indivisibilities, startup costs, and the cost of communication), the local money lender essentially has a monopoly on the lending market, which reinforces the high interest rates and the credit rationing that is so detrimental to development and recovery from bad harvests and deaths as well as the costs associated with weddings, child bearing, illness, and education.

Several cooperative bank movements have been created to overcome such problems. The Grameen Bank in Bangladesh is one of the more successful (see Box 29-3).[9] In the Grameen Bank, entrepreneurs must form groups with separate projects (which helps spread the risks) and apply for loans as a group (which reduces management costs). When loans are approved, typically, only the first two members receive loans. If they manage to meet their repayment schedules, then two more members get loans, and so on until the last member, who is typically the group leader. The structure of lending around groups creates incentives for cooperation within the group and the village as a whole, and for monitoring entrepreneurial activities and repayment behavior reciprocally within the group. This approach also gives group leaders a clear incentive to make the system work (somewhat akin to the practice of giving leaders of irrigation cooperatives a stake in the fields that are last in the line of downstream plots).

Starting banks and insurance companies is not easy. It requires time, patience, and local knowledge and thus is perhaps a task that is best left to the local level. National policymakers (or donors) can facilitate this kind of development through training and education and by creating the necessary institutional preconditions—at least removing the barriers that in many cases make the creation of such institutions difficult. Secure land tenure is essential in this context because it reduces risk directly, creates incentives for productive investments, and can encourage and facilitate banking (because land is a valuable form of collateral). Because poverty and insecurity may lead to many unsustainable and short-sighted practices, the provision of banking and insurance may be important for sustainable development.

## Eco-Taxes in Agroindustry[10]

A common source of pollution in many developing countries is the organic waste from sugar plants, paper mills, palm oil mills, and waste-producing plants that process coffee, tea, and other crops. The waste is biological in origin and, under appropriate conditions, biodegradable. It is not in itself hazardous; however, dumped in large quantities in an inappropriate location (such as a small river or water body), it can effectively kill the existing ecosystem through eutrophication, because its decomposition uses up all the oxygen available in the ecosystem. A measure of oxygen demand such as biological oxygen demand (BOD) is appropriate as a simple measure of this kind of pollution.

Malaysia recovered from the collapse of the rubber market partly by diversifying rather massively into oil palm plantations in the 1960s. The effluent wastes

## Box 29-3. Grameen Bank

In 1976, Professor Muhammad Yunus of the University of Chittagong, Bangladesh, launched an action research project on the design of a credit delivery system to provide banking services to the rural poor. In October 1983, the Grameen Bank Project became an independent bank. The bank is mainly owned by the borrowers (90% of shares), and the remaining 10% is owned by the government.

Grameen Bank (where *grameen* means "rural" or "village" in the local language) has been very successful at providing banking services to clients who normally are thought of as the most difficult: the rural poor. According to the balance sheets, total assets have grown from US$5 million in 1983 to US$400 million in 1999 (more than 30% per year). The bank has provided loans to more than 2 million people, operates in 35,000 villages throughout the country, and has financed more than half a million houses. During the 1990s, the bank was converted into a "movement" with separate funds and organizations that promote irrigation, health, fish farms, and health and education programs. The bank actively seeks out the most deprived in Bangladeshi society—beggars, illiterates, widows—and still succeeds in attaining a loan repayment rate of 99%. It does not look or operate like other banks. There are no carpeted offices with phones and faxes, but village meetings with the borrowers—who are mainly women, because they are considered more responsible than men.

The philosophy of the Grameen Bank is as follows:

> Over the years, representatives of the borrowers have agreed with the bank on certain principles and commitments which they will undertake to help improve their lives and their ability to meet their debts. To Westerners these may seem at best paternalistic; however, the slogans are chanted enthusiastically by the borrowers. They pledge to abide by "the 16 decisions," a set of personal commitments such as "We pledge to send our children to school," and "We pledge not to demand or pay dowry for our daughters' marriage." The most important of these commitments is to join up with four fellow borrowers, none of whom can be a family member, to form a "group." The group dynamic provides a borrower with the self-discipline and courage needed to enter into these uncharted waters. Peer pressure and peer support effectively replace collateral: if one borrower defaults the whole group is penalized. The system also saves the bank the costly business of screening and monitoring borrowers.

*Source:* Grameen Communications 1998.

(roughly 2.5 metric tons per ton of oil) were pumped into the nearest water bodies. By 1977, "42 rivers in Malaysia were so severely polluted that freshwater fish could no longer survive in them" (Aiken and Leigh 1992). Not only did the fish die; freshwater supplies for households, villages, and even municipalities became unusable. Wells had to be dug. Oxygen levels were so low as to encourage anaerobic decomposition, producing hydrogen sulfide, ammonia, and other toxic compounds. The stench was reportedly so bad that entire villages had to relocate, and the mills became a major object of citizens' complaints.

The Malaysian Department of Environment (DOE) could not envision stopping the expansion of the sector that already accounted for around 15% of export earnings in 1975 and was still in rapid expansion, providing both foreign exchange and employment opportunities. However, neither could the agency ignore the enormous effluent problem that this profitable industry presumably should have been able to tackle. The pollution load from the mills in 1975 was equivalent to the raw sewage from 12 million people (the population of the whole country).

The government used a combination of policies to tackle this issue and similar problems in the rubber industry. In 1974, the Environmental Quality Act was passed, which authorized the Malaysian DOE to require firms to be licensed and empowered the agency to set appropriate preconditions for licensing. The Malaysian DOE had the power not only to mandate technology and emissions levels but also to close plants and to differentiate license fees, making it in effect possible to levy fees that were akin to environmental charges. The agency adopted a consensus-building approach by forming an expert committee involving representatives from the industries concerned. This committee and the individual industries searched for possible treatment technologies, and within two years, the agency had seen sufficient progress to feel confident in formulating its conditions.

In July 1978, the Malaysian DOE announced its palm oil regulations. Three policy instruments were used:

- a progressive system of standards that applied to the effluents,
- a varied license fee that corresponded closely to a two-tier effluent charge, and
- subsidies for abatement technology and research into such technology.

The basis for the system was the standards, which applied to several effluent parameters, the most important of which was BOD. The required improvements in effluent were dramatic—reductions of several orders of magnitude resulted within less than a decade (see Table 29-1). The agency did not make the achievement of these standards mandatory the first year in order to give the plants time to make the necessary abatement investments. To stimulate those investments, the agency made the license fee variable, with a small fixed component and a variable component that had two tariff levels, one below the standard and a higher charge above the standard:

$$T = T_0 + T_1 \hat{e} + T_2(e - \hat{e}) \tag{29-1}$$

Table 29-1. *BOD from Effluents and Required Standards for Palm Oil Mills*

| Standard | Year | BOD (ppm) |
|---|---|---|
| None | Before 1977 | 25,000 |
| A | 1978 | 5,000 |
| B | 1979 | 2,000 |
| C | 1980 | 1,000 |
| D | 1981 | 500 |
| E | 1982 | 250 |
| F | 1984 | 100[a] |

*Notes:* BOD = biological oxygen demand (in parts per million [ppm]); None = untreated effluent, typical value.

[a]Even 20 ppm under some conditions.

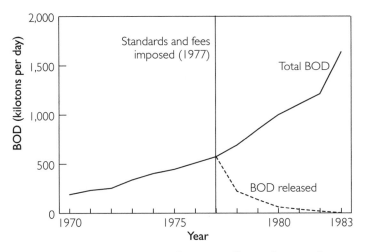

**Figure 29-1.** *Production and Release of BOD Effluent from Malaysian Palm Oil Mills*

*Source:* Adapted from Vincent et al. 1997.

where $e$ is actual effluent; $\hat{e}$ is the allowable effluent according to the standard; $T_0$ is the fixed charge; $T_1$ is the standard fee for allowable pollution; and $T_2$ is the "penalty" charge for "excess" pollution, set at 10 times the charge for allowable pollution ($T_1$).

After the first year, the standards became binding, and the Malaysian DOE demonstrated its statutory powers by closing one plant in late 1979 and a few more between 1981 and 1984. Presumably, this action reduced the role of the fee, although the fee still had to be paid and the agency used its discretion to allow some firms to continue operation despite violating the standards as long as the fees were paid. This discretion was applied to plants that were perceived as making serious efforts in the right direction, and plant closure (refusal to renew licenses) was used as a threat for the more recalcitrant firms.

The third instrument that the Malaysian DOE used was a form of subsidies for research and development (R&D) in abatement technologies. Actual funds were not paid, but under special circumstances (and with the approval of the minister of the environment) the above-mentioned fees could be reduced or waived as a quid pro quo for R&D investments of general interest. The government also established the Palm Oil Research Institute of Malaysia, which increased the capacity to find new abatement technologies.

The total results of all these policies are presented in Figure 29-1, which illustrates how effluents were reduced rapidly despite a continued expansion of the palm oil industry (as shown by the upper curve for total BOD produced). Disaggregating the effects into separate components for the fee, the standards, and the technology subsidies is tricky. In the first year, the fee may have been an important element of the policy, but because plants faced the risk of closure, physical standards presumably became more important. The fees would still act as an additional incentive and as a way of leveling the playing field between the more-compliant (with higher abatement costs) and the (somewhat) less-compliant plants. A survey carried out in 1980 and 1981 showed that the effluents of 90% of the mills were below Standard D (500 parts per million [ppm]), which was applicable starting in July 1981, and the effluents of 40% of the mills were below 100 ppm—a standard that would not come into force until several years later. These extreme reductions could be interpreted as a sign that the fee was still having an incentive effect,

## Supplemental Reading

Andréasson-Gren 1990, 1992
Hanley and Oglethorpe 1999
Howitt and Taylor 1993
Miranowski and Carlson 1993
Pingali and Gerpacio 1998
Randall and Taylor 2000
Rosenzweig 1988a, 1988b

although it could just as well be due to the indivisibilities in technology, suggesting that firms prefer to aim for future targets to avoid more costly investments later.[11]

The success of the policy package has been attributed to its combination of policies. A pure effluent fee probably would not have been sufficient because several plants would have preferred to pay the fee, creating unacceptable local pollution conditions in some areas (Vincent et al. 1997). The timing of the standards and tax subsidies for R&D was the result of industry influence (exercised through the consultative process in the expert panel set up by the Malaysian DOE) and was crucial in securing industry acceptance and support.

Finally, technological progress was an important factor. The nontoxic and biological nature made the problem relatively easy to solve. Several cheap and, in some cases, even profitable abatement technologies have been discovered. The biomass and sludge are excellent raw material for animal feed, fertilizer, biogas, and several other products that have been commercialized successfully. The Malaysian experience should encourage those grappling with similar problems, such as bagasse and other biological wastes from raw material processing and even the water hyacinths of Lake Victoria.

## Notes

1. Relevant topics that are not covered because of space constraints include agricultural water, which is often grossly subsidized and badly mismanaged (see Chapter 26); genetically modified crops and antibiotic resistance, issues that involve subtle externalities (see Brown and Layton 1996 on antibiotics resistance as a depletable resource), and public goods together with difficult ethical and political questions. Trade barriers and trade liberalization are typically vital issues for agriculture in developing countries (see Chapter 12 for a discussion of the underlying principles and Anderson 1998 for an overview).

2. This is one example of the interlinked markets—a linkage among labor, land, and insurance. Other examples are found where markets for credit, inputs, output, savings, insurance, land, and labor are linked in various ways (surveyed in Otsuka, Chuma, and Hayami 1992).

3. For further analysis, see Zilberman and Millock 1997; Zilberman, Templeton, and Khanna 1999; or Wu 1999 on crop insurance and nonpoint-source pollution. See Wu and Babcock 2001 for a focus on spatial heterogeneity and instrument choice.

4. Farming practices may have many other downstream consequences: risk of landslides and flooding, the desiccation of rivers or lakes (e.g., the Aral Sea, emptied largely by massive irri-

gated cotton farming), salinification, desertification, and the destruction of river ecosystems and even coral reefs.

5. Such a combination has been proposed to improve water quality, with a background threat of mandatory controls or taxes if the voluntary approach is unsuccessful in meeting a previously specified water quality goal (Segerson and Miceli 1999). The regulator uses a uniform subsidy rate (i.e., one that does not depend on farm characteristics) to induce participation in a voluntary program to achieve a given water quality goal. This suggestion is somewhat similar to the pollution control system within the industrial estates described in Chapter 25.

6. Thanks to Anders Ekbom and Wilfred Nyangena for interesting comments.

7. The relationship can go either way (Binswanger and Holden 1998). Population increase may speed up technical progress and lead to increased incomes. In many parts of Africa, however, where population increase is coupled with poorly developed institutions, development appears to have been more Malthusian.

8. These strategies are only a small subset of those that can be relevant here. Other important strategies include crop diversification (including intercropping), physical grain storage, and contractual arrangements such as sharecropping (Walker and Jodha 1986). In addition, patterns of marriage and family composition may play a role in evening out income flows and lowering risks (e.g., if they allow extended families to span a large geographical area or several professions). However, these mechanisms appear insufficient to reduce risk in a satisfactory manner (Rosenzweig and Binswanger 1994).

9. India has similar schemes; one is run by the Self-Employed Women's Association (SEWA). Another successful scheme is the Aga Khan Rural Support Program, in Pakistan. The Deutsche Bank runs, as a "proactive" commitment, a microcredit scheme that tries to help set up similar schemes in various countries.

10. This section builds on Vincent et al. 1997, Khalid and Braden 1993, and Khalid and Wan Ali 1992.

11. The charge formula (Equation 29-1) was further modified by a required minimum that was not binding in the first couple of years, when effluent payments were sizeable. However, the minimum payment would apply to a medium-sized firm when the effluents fell below 167 parts per million, and technically, there would be no incentive below that level (Vincent et al. 1997).

# CHAPTER 30

# *Forestry*

G LOBAL FOREST COVER IS ESTIMATED to have decreased by around 40% since preagricultural times (Bryant, Nielson, and Tangley 1997). However, because defining *deforestation* is problematic, one should not rely too heavily on any exact figure. Some authors focus on *forest degradation*, which refers to any decrease in density or disturbance of the forest ecosystem. It is not uncommon for loggers to remove some (particularly valuable) trees, which then opens up an area to ecological and economic changes such as road building, which brings immigration, agriculture, and additional forest clearing. Fewer than 40% of forests globally are relatively undisturbed by human action.

*Tropical deforestation,* the most serious threat as far as biodiversity is concerned, probably exceeds 130,000 square kilometers a year. Tropical forests may be more than 150 million years old, so each hectare of forest has developed an immense and unique biodiversity that may be larger than that of enormous areas that have developed only since the last Ice Age (e.g., the northern hemisphere). Ironically, forests are increasing at the expense of agriculture in many high-income countries, such as the United States and Scandinavia, where high salaries and other factors favor this land use (see Chapter 4). This increase (at least partly) compensates for the loss in carbon due to tropical deforestation but does nothing to limit the loss in biodiversity or other local public good services that forests provide, such as climate stability.

This dichotomy has created a considerable although probably insufficient interest in the donor communities to protect forests, exemplified by the unexpected cooperation between the World Bank and World Wide Fund for Nature in a global alliance for forest conservation and sustainable use. Part of the interest in forest protection is related to carbon and climate change and part to biodiversity and other ecosystem services. This is not to say that all deforestation is inappropriate. It is impossible to define *inappropriate deforestation* because it is an inherently political issue (Kaimowitz and Angelsen 1998). However, deforestation is more likely to be inappropriate under certain circumstances, for example, on lands that

- are unsuitable for agriculture (due to soil quality or slope);
- provide valuable ecosystem services, such as protecting watersheds, or where the downstream effects of deforestation are high;
- contain considerable biodiversity; or
- are inhabited by forest users.

So many economic models explain deforestation that I can do no justice to the wealth of available ideas (see Kaimowitz and Angelsen 1998 for a survey). However, one point that all the models agree on is the importance of access for increasing deforestation; this point is central to the model presented in Chapter 4. The effects of policy instruments or roads will vary depending on the type of forest. An economic analysis focuses on the costs of protecting ownership rights and the intensity of use and typically distinguishes between natural forests (areas where few people live and conventional agriculture has not yet developed) and managed stands (may be densely populated, but the demand for wood has made commercial plantations more profitable than farming).

The economic interests behind public policies are also important. In many temperate forest countries, there has been a pervasive worry about resource depletion. It was partly a tactical concern for the wood processing industry, which was interested in access to cheap raw materials. Lately, however, it has been combined with more general ecological and social concerns. Forests perform several functions, ranging from nutrient recycling to the protection of watersheds, soil quality, and local climate. Clear-cutting in tropical forests can cause massive biodiversity losses, soil erosion, and the destruction of downstream ecosystems, including agriculture, dams, rivers, estuaries, and coral reefs. Such disturbances can have a serious detrimental effect on the local population, most significantly on the poor, who live on marginal lands and depend on ecosystem resources for water, firewood, game, and fishing.

Various instruments are used in forestry management, including subsidies, taxes, and regulation of silvicultural practices. Three additional instruments have received attention in recent years: certification, international carbon offsets, and changes in property rights. Certification or labeling is a developed-country initiative designed to ensure that forest products come from sustainable forests, and carbon offsets are international payments to protect the carbon already sequestered in mature forests to help mitigate climate change. Ownership is important; the government is the largest forest landowner in most countries. Challenges to its effectiveness as a landowner during the past decade have led to greater focus on transferring some rights to (degraded) government forests to local communities and on contractual arrangements for the sale of government timber. The indirect effects on forests from agricultural instruments and roads can be substantial, particularly in developing countries. Certification, carbon rights, and ecotourism potentially offer the opportunity for policymaking in one country to affect forests in another, corresponding to two at least partly global goals: preserving biodiversity and mitigating climate change. In this chapter, I deal with each of these instruments in turn (except ecotourism, which is discussed in Chapter 31).

## Subsidies

The main motivation for domestic forestry subsidies in the well-established forest nations such as Scandinavia and the United States was originally the fear of supply shortage (because the forest was needed for the navies, for heating, or for the timber and paper mill industries). The same fear also led to various other policy instruments, such as regulations for forest management. When subsidies are used, they tend to become permanent through the effects of the strong industry lobbies that benefit from them.

Most industrialized countries have some form of subsidies, often called *forest incentives,* which come in the form of free seedlings, management assistance, or financial aid for management. Financial aid is more common in developed countries that can afford it and where plantation forests are a large source of forest products. Even Chile's well-known free market has been modified with a financial aid program in forestry, and the success of Chile's forest sector leads many foresters in other countries to believe these subsidies are necessary for a successful forestry sector. Free seedlings and management assistance are common to forest policies and rural development programs around the world, regardless of a country's development status.

Despite the popularity of these instruments, response to them is sometimes limited. To understand the problem, refer back to the spatial model in Chapter 4 (particularly Figure 4-4). Incentives diminish management costs for professionally run plantations, but in the first stages of forest development—with large tracts of virgin forest—there are no management costs. Therefore, incentives begin to affect management behavior only during the third stage of forest development, when prices have risen so that landowners begin to manage their forests professionally. In this "mature" stage of forest development, incentives subsidize the cost of managed forests, raise the net forest value function, and extend both the intensive and extensive margins of managed forests. Still, forest incentives do not affect harvest or delivery costs and thus have no effect on extraction from the natural forest.

Developing-country programs that have focused too exclusively on the number of trees planted or on the provision of seedlings have failed because they did not take into account the preferences (discount rates in particular) of local users. If the demand for agricultural products is higher than that for woody biomass, then free seedlings will have little effect on tree planting (and may hamper the domestic production of seedlings). In some settings, however, village plantations have used marginal lands that were of little use for agriculture and where indigenous tree cover would not easily have recovered. For example, some aid-financed village woodlots in the state of Orissa, India, significantly improved the welfare of poor communities (Köhlin 1998). These effects were due primarily to the time saved by collecting fuel close to home but also to the recovery of the natural forest when biomass collection pressure was reduced. A thorough understanding of local conditions is required to plan and site village woodlots properly and thus maximize their benefits. For instance, in general, village woodlots should not be located in the villages that border the natural forest but in those that are more distant, where people would otherwise have the farthest distance to travel.

Other developing-country subsidies that have attracted much attention are "perverse" subsidies that encourage deforestation (e.g., the famous Binswanger 1991 study on subsidies for forest destruction in Brazil mentioned in Chapter 9). There are many criteria to consider when selecting instruments to abolish subsidies on agricultural inputs, for settlement schemes, agricultural extensions, taxes, and credit subsidies for farms on newly cleared land, logging industries, and roads: effectiveness, targetability, costs (direct and indirect), equity issues, and political viability (Kaimowitz, Byron, and Sunderlin 1998). Subsidy reduction meets resistance from powerful beneficiaries but also has the undeniable advantage of not spending but rather saving funds for the public budgets.

## Taxes

Taxation is used mainly on private forestlands, to catch a share of scarcity or land rent (much like a property tax or taxes on other forms of wealth and income) or to avoid or correct for externalities created. The effect and efficiency of taxation depend on exactly how the tax base is defined, the type of forest, and the type of owner. Three common tax bases are land, standing timber, and harvested timber (the taxes for which are known as *severance taxes*). In addition, some communities have introduced preferential tax treatment for ecologically sensitive forestry (see Klemperer 1996 for a summary).

Ad valorem property taxes on forestland are generally expected to be neutral with respect to land use and rotation age (harvest timing). However, these taxes tend to accrue annually, and (at least small) managers obtain forest revenues with infrequent timber harvests. The timing mismatch between tax payments and forest revenues might create an incentive to convert the land to other uses.[1] Severance taxes on in situ resource value are also called *stumpage taxes.* In developing countries with valuable tropical hardwoods, the information asymmetries and monitoring difficulties make the collection of stumpage taxes difficult. Studies have shown that stumpage fees are often so low that the resources are practically given away. In India, Bangladesh, and other Asian countries, large forest industries have paid only a tiny fraction of resource rents, thus in practice benefiting from a form of hidden subsidy (FAO 1993). The Hindustan Paper Corporation in Kerala, India, for instance, paid US$0.50/ton of eucalyptus pulpwood when the Forest Department had direct production costs of almost US$25/ton. Stumpage taxes could conceivably (if poorly designed) encourage highgrading (i.e., extracting only valuable tree species, to the detriment of other trees). Rational loggers leave more and more residual product in the logged forest as they move closer to the margin. It pays to remove only those logs that are worth more than their marginal costs, and because marginal costs are increasing as logging activity advances into the standing forest, they constitute a smaller and smaller share of the total forest. However, it may be possible to use differentiated stumpage taxes to specifically target highgrading (Amacher, Brazee, and Witvliet 2001).

One analysis of export taxes (and bans) indicates that these instruments may be moderately successful in reducing the logging of natural forests, particularly in the short run (Aylward et al. 1994). In the long run, their effects are less certain.

The risk is that they reduce the value of timber and thus discourage serious investments in sustainable management, logging, and processing. This effect might encourage domestic consumption, which might lead to even greater logging (as is thought to have occurred in Indonesia). In principle, importing nations might impose bans or taxes with similar effects. However, this approach requires a fairly large number of countries and raises problems with trade politics.

## Regulations

Forest regulation is common (particularly in temperate country forestry) and often pertains to the regeneration of private lands to ensure rapid reforestation after harvests of the previous stand of mature timber. Regulations initially were imposed in many countries to protect against anticipated timber supply "shortfalls," which usually are not actual shortfalls but an argument used in the price bargaining between forest owners and sawmills or paper industries. However, reforestation regulations also may be motivated by environmental concerns for birds or biodiversity. Recently, some countries (including Sweden) have started to include specifically environmental objectives such as leaving dead trees for birds or leaving protective corridors along streams as a part of overall forestry regulations.

The effect of the regulations depends on the type of forest. For harvests of mature natural forest, requirements may delay harvests by adding a reforestation cost to the harvest cost. The reforestation requirements are an expression of the notion that the discount rate would be too high to warrant the regeneration costs.

It often comes as a surprise to people from tropical countries that forest owners in cold countries, such as Sweden, plant trees despite a rotation period of a hundred years. This practice is partly due to regulations but also is ingrained into the local culture of private forest owners. In the United States, reforestation requirements have had no impact on the long-term standing timber volume on the largely private forestlands of Virginia (Boyd and Hyde 1989). In tropical countries, this form of reforestation regulation is uncommon. Other regulations may typically be included as part of the concession agreements. The "allowable cut," for instance, is a common policy of most forestry agencies around the world. It restricts harvests of mature timber to a fraction of their standing volume (Davis and Johnson 1987; Hyde 1980).

## Forest Concessions and Timber Contracts

One management strategy in countries with large government-owned forests is the granting of forest concessions to private forestry companies. The essential issues for concession policy include securing satisfactory financial proceeds for the treasury, providing incentives for sustainable and efficient harvest methods, and avoiding illegal logging while maintaining the long-term condition of the basic resource. If concessionaires can limit access and competition from illegal loggers, then they should feel sufficient security to manage their concessions sustainably; however, this is not always the case (Kaimowitz, Byron, and Sunderlin

1998). Concessions typically have been allocated noncompetitively, in ways that are nontransparent and possibly corrupt, partly because of the gigantic scale of some of these transactions. In 1968, for instance, Thailand issued 500 timber concessions covering half of the country (FAO 1993).

Associated with an irresponsible handling of concessions is inevitably a short-sighted and rapacious industry. Public forestlands are often either degraded open-access areas or inaccessible natural forest. In pure open-access areas, forest rents normally would already have been extracted, and hardly any would remain. In reality, the extent or ease of access depends on both technical variables (e.g., road quality, which decides logging and transport costs) and socioeconomic ones (e.g., regulations and control effort). As these variables change, forest rents may be created. Major roads such as the Trans-Amazon Highway in Brazil create massive rents. In Indonesia, more than 20% of all concessions operative in 1992 had lost their rights to operate by 1996. The cause for nonrenewal may have been disinterest on the part of the concessionaire, or the concessionaire may have violated acceptable long-term management standards. In either case, it is a sign that rents probably have dissipated.

Most forest ministries charge a fee per unit of log for the right to harvest.[2] The government thereby collects a portion of the rent, and the government typically is tempted to increase this royalty. Doing so might be highly commendable but will decrease the profitability of (legal) logging. Logging will decline, and royalties will increase or decrease depending on the elasticity of the marginal cost function. Unfortunately, this elasticity is seldom estimated. Some developing or formerly planned economies charge almost nothing and thus make no public revenue at all. Lithuania, for example, probably would increase the social profitability of the sector if a small, simple volume-based charge were imposed (Deacon 1996). Royalties also increase the incentive for illegal logging and for highgrading (see the similar discussion as related to fisheries in Chapter 28).

Concern for the condition of the forest at the end of the concession has led some environmentalists to suggest a liability bond on the environmental performance of concessionaires (Ruzicka 1979; Paris and Ruzicka 1989; see also Chapter 10). In the ideal case, the concessionaires would submit a bond at the outset of the harvest agreement. The bond would be returned on demonstration of acceptable performance. If the bond is truly a guarantor of environmental performance, then it must be set high enough to compensate for all costs of returning the forest environment to an acceptable condition in the event of noncompliance. This means higher bonds, for example, for concessions on steep lands and erosive soils and lower bonds for concessions in less erosive regions. Differences in logging effects on other environmental conditions require similar attention, so an assessment of environmental risks and mitigation costs should be conducted at the outset of the concession agreement. However, assessments are rarely conducted.

## Certification

Certification and green labeling might be useful instruments to counteract information asymmetries and give consumers information about the environmental sustainability of various forest practices (see Chapters 10 and 12); its greatest support comes from environmentally oriented consumers. Two of the main certifying organizations are the Pan European Forest Certification Council and the Forest Stewardship Council (FSC), based in Oaxaca, Mexico. The FSC is an international nongovernmental organization, founded in 1993 through the association of environmental groups, timber trade, forestry, indigenous peoples' organizations, community forestry groups, and forest product certification organizations around the world. As of May 2002, the FSC had certified almost 30 million hectares of forest in 56 countries; of this total, 10 million hectares were in Sweden, 4 million hectares in Poland, and 3.5 million hectares in the United States. Among the developing countries, Brazil, Mexico, and Bolivia together had about 2.5 million hectares certified by the FSC, whereas Malaysia and Indonesia together had only 200,000 hectares (FSC 2001).

There is interest in certification in Indonesia, but not through the FSC. Large producers are scrambling to obtain certification and thereby gain European market shares. The Indonesian Ecolabeling Institute (Lembaga Ekolabel Indonesia [LEI]) has developed its own certification program that will comply with the standards of the International Tropical Timber Organization (which "facilitates discussion, consultation, and international cooperation on issues relating to the international trade and utilization of tropical timber and the sustainable management of its resource base" [ITTO 2002]) and will be monitored by an independent body. A protocol was signed in September 2000 between LEI and the FSC to jointly certify wood products under both programs. However, less than 10% of Indonesia's annual harvest comes from plantations; most of Indonesia's timber comes from natural forests. Even if Indonesia's producers obtain certification, monitoring will be a serious problem (to ensure that wood from uncertified lands is not substituted for certified wood). Meanwhile, the small institute is preoccupied with large forestland managers while the smaller private operations (which are likely to be the initial respondents to higher price incentives for sustainable management) have been overlooked.

## Carbon Offsets and Other Forms of International Payment

Carbon offsets, a new kind of policy instrument for forestry, are essentially international subsidies or financial incentives to help mitigate climate change through carbon storage in natural reservoirs (e.g., soils and forests). Deforestation is a significant secondary source of climate gases (mainly carbon dioxide). Reforestation, afforestation, and even increased growth through fertilization are possible "carbon sinks" (i.e., physical means of reducing the carbon content of the atmosphere). In principle, "avoided deforestation" is also a possible sink but is particularly hard to verify. The ideal approach to environmental forest policy is to first analyze a particular forest issue from all angles—local and global externalities, property rights issues, market failure, and monitoring problems—and then formulate policies. A carbon fund or a carbon offset might be a constructive instrument, but it does not necessarily take other issues (such as biodiversity conservation) into account. Global policymaking for climate change issues is so complex that it cannot easily incorporate ancillary benefits such as biodiversity and water quality (see Chapter 24).

The perceived need for some form of early action has led to the development of various funds for carbon payments by the Carbon Fund, the Global Environment Facility (GEF), and even individual countries and firms. Developing countries can apply to receive funds by finding and designating a suitable area that "fits into" the windows of opportunity created by the sponsoring organizations. Although far from ideal, this instrument may still open up interesting possibilities in some cases.

Real forests deliver a multitude of ecosystem benefits, some of which are local or regional (e.g., protection against earth slides, from soil erosion, and of local climate and watersheds). Forest areas that are expanded, especially through reforestation, could increase these benefits. Other global benefits (such as biodiversity or carbon retention) would motivate at least some partial payment from organizations such as GEF (see Chapter 17). The Clean Development Mechanism of the Kyoto Protocol and debt-for-nature swaps are similarly applicable instruments.

The greatest gain in carbon storage and biodiversity potentially would come from protecting mature marginal frontier forests that would have been harvested without the offset payment. Payments to protect the full forest are not necessary because the volume at risk is mainly the forests at the margin. Carbon offsets to accomplish this might possibly be used in combination with other instruments. However, several tricky questions related to additionality will inevitably arise.[3] Developing countries wonder whether the money received is truly additional or just replaces other "aid" they would have received anyway; donor countries wonder whether the forests "saved" simply replace other forests that are cut down or whether the forests would have been left standing anyway. Both parties wonder how to know whether the other is really committing to anything in addition to what would have been done anyway. This is one reason why, at the time of this writing, avoided deforestation of mature forests is not an eligible option for the Clean Development Mechanism, whereas afforestation and reforestation are.

## Clarification of Property Rights

In most countries, the state is the largest single forest landowner and at some time in the past has declared itself the ultimate owner of all unclaimed resources. Much of the national forest was originally land "left over" after private claims were established. On large tracts of prime forest, the government commonly grants forest concessions; other areas of public forests are either degraded or inaccessible, so the management is expensive relative to the production values at stake and appears inefficient. The state may be an unsuccessful forest manager for other reasons. The two most common criticisms are that the state fails to incorporate local demands on the forest and does a poor job of collecting revenues. In addition, government employees sometimes engage in rent-seeking.

The state is effectively an absentee landowner. Many of its problems are related to the vast size of government holdings (see Chapter 5). It does not have the same knowledge and control of the land that a local manager would, even if a ranger is

---

## Box 30-1. Forestry and the Poor: User Rights and Afforestation

The relationship between poverty and deforestation—and the role of gender—are much debated, complex issues. The view that the poor (in their quest for land) are a main cause of deforestation is now widely challenged. Very poor, landless populations are more than proportionately dependent on open-access or common property resources. They suffer most when firewood, medicinal plants, game, nuts, fruit, and other goods become scarce. Women and children are particularly affected because collecting water and firewood is often their task.

In recognition of this, some policymakers have started to involve local communities in forestry programs. In 1978, the Nepali government passed legislation enabling public lands in the middle hills to be handed to the local community organizations, *panchayats*. They were very bureaucratic at first but eventually decentralized all decisionmaking to user groups. In June 1997, 6,000 small informal but autonomous user groups managed 450,000 hectares (Shrestha 1996). In a country that has witnessed drastic deforestation, these user groups appear to be improving the condition of the forests

---

stationed on the land, and often manages the forest for timber regardless of local land use preferences for a multitude of goods and services. Consequently, local users typically trespass to extract whatever they can, treating the forest as an open-access resource even if there is some level of government protection of its property rights.

If forests are already degraded and the government agency is frustrated with its own attempts at management, some or all property rights might be shared with local users. Examples include Panchayat Forestry in Nepal (see Box 30-1), Joint Forest Management in India, community-based forest management in the Philippines, and community forestry in many developing countries. The effect of the transfer of rights is to decrease the cost of protecting land use rights and thus increase forest cover because local management is generally less expensive. Successful examples seem rare but a few from India illustrate the conditions for successful joint community–government forest management (Kant 1996).

If state ownership is inefficient and there is some hesitancy about large private companies, then small-scale private ownership may appear attractive. Many people view such an ownership pattern as hypothetical, but it is common in Europe, particularly in Scandinavia, where a significant share of forests are owned and managed by (small-scale) private owners. When this kind of small private forest owner does not exist, common property resource (CPR) management may be the most adequate solution, particularly for lands that have low and strongly variable productivity (Ostrom 1990). Community control, together with appropriate policy instruments such as user fees (wood taxes), may encourage the sustainable recovery of degraded open-access lands in the Sahel region (Chomitz and Griffiths 2001). The World Bank has abandoned its earlier pro-privatization policy (as expressed in the 1975 Land Reform Policy Paper [Deininger and Binswanger 1999]) and now favors communal tenure systems. The bank recognizes that communal tenure may be more cost-effective than formal titles and that, with imperfect credit markets, titling programs may favor the better off, lead to land concentration, and sometimes cause inequity and inefficiency.

## Notes

1. In this case, the taxes would not be neutral. However, in Sweden, the tax authorities allow forest owners special accounts to even out their flow of income. Ad valorem taxes on land and standing timber may discriminate against forestry because they tax both stock and growth every year until harvest, which could create an incentive for earlier-than-optimal harvesting. A separate problem is that small forest owners (particularly part-time foresters) might fail to react to tax incentives because they lack information.

2. To maximize the rent, it probably would be better to conduct an auction. Some forestry transactions between private landowners and loggers or mills are conducted this way (Hyde and Sedjo 1992).

3. For example, why should grant money be spent on commercial ventures if they are profitable? If they are not profitable, then efficiency will be higher and distortions probably smaller if the grants are used to buy "conservation services" directly rather than to encourage complementary commercial (but nonprofitable) ventures. Numerous complications are related to asymmetrical information, monitoring, and the political economy involved (Simpson and Sedjo 1996).

## Supplemental Reading

Barbier and Burgess 1997
Burgess and Barbier 1999
Hyde and Amacher 2000
Mekonnen 1998
Mendelsohn 1994

# CHAPTER 31

# *Ecosystems*

NATURAL RESOURCES SUCH AS FORESTS and fisheries are complex. Yet ecosystems are much more complicated than suggested by the models discussed in the rest of the chapters in Part Six. Managed ecosystems called "agriculture," "aquaculture," and "silviculture" are purposefully simplified for ease of management and for human enjoyment. The natural ecosystems (on which managed systems are inevitably dependent) are several orders of magnitude more complex than managed systems; many vital functions are derived from them, and thus the future of the human race depends on preserving them (see Chapter 4). Chapter 29 touched on this dependence in the discussion of wetland restoration.

Many anecdotes such as the following illustrate humankind's failure to foresee these complexities in various cases:

> Ernst Mayr tells of a steamer wrecked off Lord Howe Island east of Australia in 1918. Its rats swam ashore. In two years they had so nearly exterminated the native birds that an islander wrote, "This paradise of birds has become a wilderness, and the quietness of death reigns where all was melody." On Tristan da Cunha almost all of the unique land birds that had evolved there in the course of the ages were exterminated by hogs and rats. The native fauna of the island of Tahiti is losing ground against the horde of alien species that man has introduced. The Hawaiian Islands, which have lost their native plants and animals faster than almost any other area in the world, are a classic example of the results of interfering with natural balances. (Carson 1950, 78)

In general, the introduction of exotic species—which is an important element in managed agriculture—introduces numerous risks of this type. Consider the water hyacinth (introduced as an ornamental plant) and the Nile perch, which together have completely transformed the ecology of Lake Victoria, reducing biodiversity considerably. The perch has outcompeted or predated on many of the native species, including a multitude of endemic cyclids, and the hyacinth has spread invasively (see Chapter 29). With modern forms of communication and technology such as genetic engineering, the risk of spreading alien species and disturbances to natural ecosystems may increase drastically.

One of the essential inherent characteristics of many natural ecosystems, particularly the more sensitive ones, is high variability. This variability interacts in a complex way with social variability. On the one hand, people adapt, so culture and social organization reflect the underlying natural conditions. On the other hand, people exert a considerable influence on their landscapes—sometimes improving their productivity and beauty, sometimes degrading them instead.[1]

In this chapter, I illustrate ideas presented elsewhere in this book with examples of ecosystem management. Although negative examples and alarming signals abound in this area, I emphasize selected positive (and, I hope, inspiring) examples of careful and perhaps sustainable natural resource management. Various instruments are particularly suitable for ecosystem protection: forms of zoning; the creation, maintenance, and management of parks; and common property resource (CPR) or community-based range management, particularly wildlife management (e.g., the Communal Areas Management Programme for Indigenous Resources [CAMPFIRE] in Zimbabwe) and coastal zone protection (e.g., coral reefs off of Tanzania and Mexico).

## CPR Management of Wildlife in Zimbabwe[2]

### CAMPFIRE

The crucial importance of appropriate ownership rights and of empowerment of local communities for the resolution of conflicts between farming and wildlife are well illustrated in CAMPFIRE. The program started in 1989 and has been run with some funding by donors but is also generating its own funds—primarily from big trophy hunting—and there is hope that it will find a way to a sustainable and independent existence. This project also illustrates the severe crisis that has resulted from the pursued policies that are creating great political tensions and the declining investments that follow from a lack of confidence in the regime. Under such circumstances, even a well-designed project such as CAMPFIRE that is dependent on hunters who are mainly expatriates is bound to suffer.

In Zimbabwe, as in many other areas, the legacy of colonialism was destructive. All wildlife in Rhodesia was classified as the property of the British crown ("king's game"), and local hunting was banned. Sometimes, even the local inhabitants were removed so as not to disturb the gentlemen's hunting. Local inhabitants often thought of wildlife as the symbol of oppression and "poachers" as heroes, particularly if they ridded the community of animals that were dangerous or competed with livestock for pasture.[3]

In 1975, the Wildlife Department implemented the Parks and Wildlife Act, which gave private landowners "appropriate authority." The purpose of this legislation was to allow and in fact encourage landowners to manage wildlife, including any commercial use they saw fit. The hard-line conservationists (who wanted to ban all commercial exploitation of wild animals) were skeptical but soon found that commercial ranchers started to protect elephants and other animals rather than kill them. Several studies have shown that in some dry areas, wildlife ranching is much more economically profitable than traditional cattle ranching (e.g., Bond 1993; Jansen, Bond, and Child 1992; Skonhoft 1998). The reasons are largely ecological: the local vegetation—shrub and bushes—is not good pasture for cattle, which fall prey to parasites and predators. These problems lead to increased costs for the ranchers and less-than-optimal growth of the cattle. Wildlife is better adapted to the local ecology, with which it has coevolved. Browsing patterns, resistance to disease, and many other characteristics ensure that the wildlife survive better than cattle. They may give less meat, but if that meat fetches a higher price and if money can be made from trophies, hides, hunting, and tourism, then the overall profitability may be higher.

From 1978 to 1986, the Windfall program was implemented, which had only a weak link between wildlife and remuneration and benefited mainly the large farms. In the next phase, starting in 1989, CAMPFIRE extended the devolution of appropriate authority to local communities throughout most of the country. It was a significant step, because small-scale farmers in simple housing bear most of the risk and costs of living with wildlife. Therefore, the resistance to wildlife is strongest in these groups. Even in the 1980s, poaching was still common, and a dozen poachers were wounded, captured, or killed each year by rangers.

For diversified wildlife to thrive, they need not only protection from poaching but more active policies that ensure an appropriate environment. Such policies affect the erection of fences and road construction, pesticide use, and many other aspects of farming and infrastructure planning. Also, human infrastructures need to be built differently to accommodate the inevitable pressure that some types of wildlife represent. For all this to happen, local communities must share in the benefits and must be empowered to make decisions.

To encourage the coexistence of farming with wildlife, CAMPFIRE has built organizations at various levels. This organization parallels the administrative organization but at the same time is its mirror image. CAMPFIRE organizations exist at the local, regional, and national levels. The smallest decisionmaking unit is the village, which typically consists of 100–200 households, even though the CAMPFIRE ward (which typically consists of some 1,000 households, or up to 10 villages) has been more active. On average, 25 wards (10,000–50,000 households) make up a district, and the district council consists of all the traditional chiefs in the area as well as an elected representative from each ward. The Rural District Council (RDC) decides how much money to pass on to the wards or villages. Typically, 20% is retained by the RDC and 80% passed on to the subdistrict units. Of the money retained by the RDC, 15% is earmarked for managing CAMPFIRE in the district and 5% for general council administration and development. The allocation of revenues initially between wards and subsequently between villages is a central administrative issue, and conflicts have arisen over distribution principles between wards, between villages, and between the elected CAMPFIRE representatives and the traditional chiefs.

Nongovernmental organizations (NGOs) have a role to play, mainly in an advisory capacity. CAMPFIRE as an organization has characteristics of a CPR institution, an NGO, and a government organization. One is reminded in this context of Ostrom's seven principles for successful CPR management (Ostrom 1990; see also Chapter 10). Rule 3 recommends democratic and participatory decisionmaking, whereas Rule 7 requires respect for the CPR institutions by higher levels of government and Rule 4 requires local monitoring. These requirements are only partially fulfilled in CAMPFIRE, and additional development would require a greater devolution of appropriate authority. Rule 2 concerns the importance of adapting provision and appropriation rules to the local (ecological and sociocultural) conditions. It is difficult to create a working community when the borders are imposed arbitrarily (Muchapondwa 2001) by government. The CAMPFIRE districts are modeled on the administrative borders within the country, and the borders between districts often have been drawn without regard for ecology, wildlife concentration, or sociocultural structures.

An alternative being considered in Namibia (where there are plans for a somewhat similar wildlife management program) is to let local communities define themselves as districts. This approach may cause difficulties, too, but would have the advantage of creating local ownership and legitimacy.

### Popular Participation in Chikwarakwara

A project such as CAMPFIRE needs "success stories," partly as a pedagogical tool. The following is such a story concerning the build-up of appropriate institutions.

A community in Chikwarakwara, a Zimbabwean village near the border of Kruger National Park, South Africa, has managed democratic decisionmaking—including the distribution of benefits, decisions concerning wildlife, and public goods—in a context where few people can read (Child, Ward, and Tavengwa 1997). In 1989, the district to which Chikwarakwara belongs sold permits to hunt three elephants, some buffalo, and other animals and decided that the revenues should benefit the community where the animals were shot. Because Chikwarakwara had much wildlife, several animals were shot there, and the village earned a large sum of money (60,000 Zimbabwe dollars [Z$]). The whole community gathered for four days beneath a baobab tree, and the money earned from trophy hunting and other sources was piled up in large piles of common notes. The exact sources of the money were carefully accounted for as a way of emphasizing the value of the various wild animals. The next step was to establish a list of village membership. Then, a long, open discussion took place about how to use the money: as cash distributions to village residents, or as funding for common (public good) projects. The almost unanimous decision was to spend half the money on each: thus, each of the 149 households would receive Z$200 in cash, and the equivalent of Z$200 per household would fund a grinding mill and the school. (Characteristically enough, some young men wanted a larger individual cash share).

To legitimize decisions and emphasize the participatory nature of the decision, a final payment ritual was carried out. Each household came forward separately, received its total share (Z$400) in notes, then paid for the public goods that had been agreed upon by splitting its Z$200 contribution into two separate bowls. This unusually transparent procedure was designed to involve every member of the community and avoid corruption and the temptation (and the opportunity) to free ride as well as the hesitancy in decisionmaking and other inefficiencies that accompany even the suspicion of such corruption. The grinding mill was built within three months and has proved to be a catalyst for the village. It is so well managed that it makes a profit (which is unusual for council-run mills). The village has started to strengthen its communal planning in other areas, looking over its water supply, applying for donor funding for a rotating fund to finance the purchase of pumps, limiting cattle numbers, and employing a villager as game warden to protect wildlife.

### Looking Ahead

What happened in Chikwarakwara is probably an ideal seldom attained, and it may be an exception to the rule. CAMPFIRE is still a fledgling and partly donor-supported activity that has yet to prove its long-run sustainability. However, it looks promising and has already achieved some impressive results in a very difficult sociopolitical environment. Before CAMPFIRE, some 50 trophy bulls were shot each year by safari hunters and 200–300 as "problem animals." Now farmers and communities think twice about classifying animals this way because they can make so much money from hunting fees: in the second half of the 1990s, the number of problem animals shot had dropped to around 30 per year, whereas around 90 trophy bulls were shot each year at a value of up to US$10,000 apiece. From 1989 to 1995, CAMPFIRE income grew from US$300,000 to US$1.5 million per year only in the 12 original districts that first started to implement the program, which has since spread to most of the other districts in Zimbabwe. Aside from the financial and ecological benefits, the program is giving the communities valuable experience in marketing, tourism, and negotiation with tour operators. The communities have received increasing shares as their negotiating skills have increased.

Even though CAMPFIRE is a promising program, it faces several threats: conservationists want to save elephants by prohibiting all hunting and by stopping international trade in trophies, hides, and ivory. Since 1999, political turmoil concerning land distribution and elections has surfaced that has scared away hunters, tourists, and other investors, thus presenting a new threat not only to CAMPFIRE but to the Zimbabwean economy as a whole. CAMPFIRE has still shown that the best way to protect wildlife appears to be giving appropriate property rights to local communities, which may involve increased hunting and trade so as to make the animals valuable to those who live with them.

## Protection of Marine Ecosystems[4]

People typically think that nature conservation is a task for national governments and that parks are places you walk or drive in. In this section, I discuss some marine parks—a couple that are privately run and one managed as a CPR—in an attempt to illustrate that many forms of management may be appropriate, depending on the circumstances.

The earliest examples of protected areas were private initiatives, created for recreation exclusively enjoyed by a small group of people.[5] The end result was that some important areas were preserved for posterity. However, this kind of conservation has created considerable animosity that threatens its sustainability because it built on the exclusion of poor people. Public policy is needed to manage the many market failures in ecosystem "provision"; however, the public sector must not necessarily be the manager. In fact, state may be a natural manager (e.g., the 200-mile exclusive zone of coastal states, and the Antarctic appears to be reasonably protected by several nation-states). However, public authorities can be clumsy managers, which is one reason why even traditional public enterprises such as railways or telecommunications are privatized. Many public parks are poorly managed (Braatz 1992), and cooperative or private management should not be excluded. Many developing countries—for example, South Africa, Namibia, Botswana, Costa Rica, and Kenya—have privately protected areas, and in some regions, they are larger than the publicly managed parks (Watkins, Barret, and Paine 1996; see also Langholz et al. 2000).

Many coastal zones in the tropics contain striking biodiversity and perform vital ecosystem functions, such as providing food and protection against erosion for densely populated regions. Developments in biomedicine have created a greater awareness of the potential for new biologically potent compounds from these environments. Yet protected marine areas are a new and perhaps controversial concept. The traditional absence of clearly defined property rights as well as the technical difficulties of implementing any such rights (because of the nonterritorial or migratory nature of many aquatic resources) make the delineation of a park problematic. Management issues such as exclusion, enforcement, and monitoring pose special problems that do not exist for terrestrial parks. Cheap airfares and high incomes in industrialized countries are creating a growing tourism industry that is increasingly sophisticated. The development of safer and cheaper diving gear and underwater photography equipment is creating a greater interest in healthy, untouched reefs (Davis and Tisdell 1996). The concept of marine ecotourism is to create economically viable tourism that provides employment and income to the local community by protecting the natural resources on which the community depends in the long run.

### Mafia and Zanzibar

The experience of protected marine areas off the coast of Tanzania is not very encouraging. In 1975, under the Fisheries Act, eight areas were declared marine reserves. Most of these areas have remained "parks" on paper only, and their condition has deteriorated so much that they are hardly worth protecting any longer. One, Maziwi Island, has even disappeared as a result of erosion, probably caused by damage to reefs (Fay 1992).

Mafia, the park with the most local CPR participation, has been most successful—perhaps because it is not exposed to heavy exploitation (it is farthest from the mainland). Its exceptionally pristine marine ecosystems have attracted considerable attention from the scientific, donor, and conservation communities, leading to the formal protection of Mafia in the Marine Parks and Reserves Act in

1994, a major achievement. Proper representation of all the interested parties—including different groups of fishermen and local cooperative involvement—appears to be key to potential success in Mafia, but progress has been slow, the cooperative approach has its limitations (especially in areas with little population), and much remains to be done (Andersson and Ngazi 1995).

Next to the main island of Zanzibar are two small islands of special interest: Mnemba and Chumbe. Mnemba Island is home to an extremely luxurious exclusive resort (Bill Gates has been there) off the north coast. The island is carefully protected from intruders, which means that the income-generating activities of collecting seashells, octopus, and sea cucumber are no longer available to the local people. The owners presumably take care to protect the local environment for the sake of their wealthy customers, but there is no evidence that they have used any advanced ecological concepts.

Chumbe Island became an area protected by the government of Zanzibar in 1994, and in 2000, it is still the only existing marine park in all of Zanzibar. The project started as a private initiative in 1992 and is the result of one person's ambitions and perseverance. The project holds a 33-year-old lease of 2.5 hectares of cleared land and a management contract of 10 years for the Chumbe reef sanctuary. The site was identified as particularly suitable for its purpose: the marine environment possessed a rare example of pristine coral island ecosystems in an otherwise overfished and overexploited area; the island was uninhabited; and the surrounding waters were not heavily used by fishermen, who might otherwise have felt that they were expelled from their fishing grounds. Thus, the risk for potential conflicts was minimal, and the small size of the island was particularly suitable for a privately managed marine park. The project design and management plan reflect great environmental concern and awareness. And because the venture is privately financed, it must be financially viable in the long run.

The ultimate goals of the project are conservation (of the reef and island) and cultivating an understanding of the ecological importance of the reefs. An important part of the management plan is the construction of an environmental education program for local people. Five former fishermen have been employed and trained as park rangers, and school children regularly visit the island as part of an effort to increase understanding of this project. Although *ecotourism* is defined as an activity that can sustain and finance the other goals and make the project viable and independent in the long run, this result is doubtful, because the costs are driven up by high environmental ambitions. For the same reason, only limited numbers of tourists will be allowed, and thus earning capacity is limited.

The island is 16 hectares, and its carrying capacity for (eco-)tourism is limited. Chumbe did not have any freshwater facilities or electricity, and because coral reefs are extremely sensitive to nutrients, the issue of sanitary and other waste was central to the planning of the venture. The solution to these challenges has been the use of sophisticated eco-technology—state-of-the-art in some countries but definitely advanced by Tanzanian standards. Each of seven bungalows (to house a total of 14 people) is equipped with a composting toilet; rainwater is collected and stored for use in the showers. Graywater is filtered in sand beds and used to irrigate gardens. Each bungalow has solar heating for water and photovoltaic cells that charge batteries for nighttime lighting.

Many other details of the infrastructure on the island (e.g., footpaths, project buildings) reflect the same careful ecological thinking. This effort is appreciated by some people but also, perhaps ironically, has been a source of various problems. The technology is expensive, and the style is not easily recognized by local authorities, who sometimes perceive the technology as being more eccentric than useful. The project has had difficulty being officially recognized as a (luxury) hotel because the usual attributes of luxury hotels (bars, swimming pools, discos, and even a multistory cement dwelling) are all missing. Resulting uncertainty and resistance from authorities causes additional delays and increased project costs. In fact, one option that would have been attractive to the project was to leave the island completely undeveloped and make only day excursions. However, this option was not possible because one requirement of the government lease was the construction of hotel facilities or similar development.

With high costs, another dilemma surfaces concerning pricing: to recover the high costs of this ambitious project with only a limited number of beds, prices would have to be high. Such prices require sophisticated marketing and also may entail some restrictions, for instance, on the acceptability to the customers of daytime tourists and school children. These issues illustrate the difficulty and potential clash of interests between ecological awareness, interaction with the local community, and the interests of those clients who are prepared to pay high rates and demand exclusive treatment.

The experience of Chumbe shows that, in some cases, private protection can be a viable and desirable way to protect a marine ecosystem. One important issue is the site features. Chumbe is a manageable size for a private entrepreneur but would be very small for a public agency. Another feature that made Chumbe ideal is how little the site had been used by the local people. Furthermore, the ecological characteristics of a site determine how successful protection will be in producing ecosystem services. Chumbe appears to have the potential to be an ecologically important retreat and recruitment area for various corals and fish species to replenish the reefs near Zanzibar Island (compare the role of marine sanctuaries in metapopulation models discussed in Chapter 4).

A protected area that produces external benefits or public good services may need some form of compensation—not necessarily financial, but perhaps simplified licensing procedures. A small operator has limited handling capacity for bureaucratic issues. Development projects are most successful when the policy environment allows frictionless and secure operations (Lewis 1988). Without the adequate legal infrastructure, ownership security, and good governmental relations, the transaction costs might become prohibitive. Former socialist countries such as Tanzania still harbor some "antiprivate" attitudes. Among the institutional prerequisites for private protection, the most important would be the security of the ownership. Without it, a private enterprise tends to set up short-term objectives at the expense of long-term sustainability.

There is no single "best" management structure for a national marine park, but many models should be tried and evaluated. Private protection probably would not be the only or even the main method for protection. Important stakeholders should be involved, and CPR schemes might be preferable. However, different parties can learn from each other and collaborate to improve managerial efficiency and to develop ecological engineering and resource-extraction techniques.

In the Tanzanian cases, Mnemba Island has profitability with fairly limited ecological ambitions; Chumbe Island has considerable ecological credibility but is hardly commercially viable; Mafia, which represents a much larger and more complex set of islands, has stakeholder participation but numerous problems concerning viability and sustainability. Initially, some of these projects will be better in technology and others in social conflict resolution. Some will be more ecological, and others may be more profitable—but heterogeneity and collaboration may give great benefits in this area.

### Punta Nizuc, a "Sacrificial Reef" in Cancún, Mexico[6]

Close to the northern tip of the Yucatán Peninsula is the resort of Cancún. The beaches of Cancún draw approximately 2.5 million visitors a year, making it Mexico's fastest growing city and one of the most popular tourist destinations in the Caribbean. With a unique combination of archaeological sites, beautiful beaches, good food, pleasant weather, and proximity to the United States, Cancún provides an opportunity for even further expansion of the tourist trade. However, tourism can "self-destruct" through a form of overexploitation in which the crowding and deterioration of sites lead to a vicious circle of lower expectations, lower pricing of hotels and tours, poor management, and further degradation. It is therefore crucial in terms of not only ecology but also commercial viability to coordinate the tourist industry in such a way as to maintain natural resources.

Restricting the number of tourists is typically the hardest and least popular option, but some other options can minimize the ecological damage done by tourists: zoning, information and education, and technical solutions (e.g., providing buoys for mooring tourist boats). All these options require some way to fund park staff and services, which generally is provided through park fees.

Cancún is a large-scale tourist operation with big hotels lining the beaches. The area contains a diverse and unique combination of coral reefs and marine life as a result of its location: in the Yucatán Channel, at the juncture of the Caribbean Sea with the Gulf of Mexico. It forms part of a coral reef system that extends to Belize and contains a considerable variety of fish species that thrive in the Caribbean Sea. The Cancún area is surrounded by several islands, islets, coral reefs, and other features of potential interest to both ecologists and tourists:

- Isla Mujeres is an island that already has a fully developed although low-key tourist industry with "family" hotels that typically attract European backpackers.
- Nichupté lagoon.
- At the south end of the Nichupté lagoon is Punta Nizuc, just outside of which are readily accessible coral reefs in shallow and protected waters.
- Coral reefs such as the large Arrecife Islache and El Cabezo are relatively difficult to access but attract trained scuba divers. There are also reefs closer to Isla Mujeres.

- Isla Contoy, the most distant of the islands, is uninhabited and has been declared a national park through several laws starting in 1961. It is considered one of the few islands of the Mexican Caribbean whose ecosystems are still relatively natural.

In addition to the park on Isla Contoy, a more recently created park is the National Marine Park of the Western Coast of Isla Mujeres, Punta Cancún, and Punta Nizuc (NMP). These two parks illustrate several important aspects about management, which I summarize in the categories of zoning, damage prevention, and funding.

## Zoning

A growing city that receives millions of visitors each year will not pass unnoticed in the local ecology. Cancún can and should still aspire to have clean beaches and to minimize pollution, but it is impossible to keep the central areas in pristine condition. Although truly pristine conditions are not demanded by the average tourist, the number of people who want to spend at least a small part of their time looking at mangrove forests and coral reefs is increasing. In this situation, which poses a somewhat paradoxical risk for the destruction of resources, zoning allows for advantageous policies. The Isla Contoy park, for example, is used mainly for research and for visits by those with a greater interest in bird watching and so forth. Isla Contoy is farther away from the main Cancún tourist spots, and numerous restrictions automatically reduce the number of tourists.

Snorkeling, scuba diving, and "jungle tours" of mangroves are the main uses of the NMP. (The park also permits some commercial lobster fishing.) Even within the NMP, there is a high degree of specialization: Punta Nizuc is used almost exclusively for snorkeling; most of the scuba diving takes place off Isla Mujeres (only a fraction of the numbers at Punta Nizuc) and Punta Cancún. Because Punta Nizuc receives the greatest flow of visitors, concern has been raised over its deterioration. However, Punta Nizuc has all the characteristics of a perfect "beginners' and tourists' reef" (a shallow area protected from waves, safe and ideal for snorkeling) and therefore is somehow destined to be sacrificed to mass tourism in order to protect other areas and keep them in a more pristine condition. The entire strip is heavily developed with hotels, and it is almost impossible to imagine it being pristine—but it can be kept attractive enough to generate a significant stream of economic benefits in the Cancún tourism industry.

## Damage Management and Monitoring

The vast majority of visitors to NMP come on boat tours, which start inside the lagoon and thus are relatively easy to monitor. They use small two-person motorboats, with one private guide for every five boats, or larger boats that accommodate up to 30 people. Visitors are first taken on a tour of the mangroves inside the lagoon. Next, they are led through a narrow channel that passes under the Nizuc bridge and out into the Punta Nizuc marine park, where they are taken on a snorkeling tour of the coral reefs. This narrow channel is the principal access point to the snorkeling area and makes monitoring the boats fairly easy.

Park management is concerned with controlling the damages caused by visitors to the marine park and to Punta Nizuc in particular. The fear is that excessive damage may eventually lead to the ecological and economic collapse of the park. The main sources of damages are

- physical contact between visitors and the reef, including taking pieces of coral as souvenirs;
- collisions of boats with reef formations;
- hydrocarbon pollution from oil and gasoline from the boat engines;
- pollution due to sunscreen oil from bathers;
- illegal fishing; and
- discharge of sewage into the Nichupté Lagoon, which leads to elevated nutrient levels in the water (the diversity of marine species at Punta Nizuc has declined, and nutrient levels from pollution by organic waste have increased).

To limit physical damage to the reef, the small boats are required to moor at any of 10 fixed-point moorings that have been established to reduce damage from laying anchors on the reef. These fixed moorings are truly crucial for the survival of the corals. They need to be maintained, and enforcement of the no-anchoring rule may be necessary. The large boats dock at a small artificial island that serves the same protective function while offering services (restaurant and sunbathing) to the tourists.

Under the Ministry of the Environment, Natural Resources, and Fishing, park management has undertaken several activities to manage the use of the marine park. In addition to the moorings, the park management requires a minimum of 1 guide per 10 people in the water and requires that each guide take a one- to two-day course in reef ecology and care. All visitors are required to wear life jackets, which limits contact with the reef by preventing people from diving under water. The park also offers biodegradable sunscreens for sale. In addition, the management requires that all boats be licensed by the National Institute of Ecology. However, monitoring is difficult and appears insufficient: of an estimated 1,914 two-person motorboats in the park in 1997, only 638 had permits, even though no fee was then charged for obtaining a permit.

### Park Management and Funding

A combination of zoning and small technological modifications have the potential to protect pristine reefs for posterity, research, ecosystem functions, and advanced ecotourism diving while maintaining the "sacrificial reef" in a sufficiently viable condition to satisfy the average tourist. All that is needed is management—which may turn out to be the most difficult part.

NMP management has long faced a severe shortage of staff and resources to manage tourism activities and monitor ecological conditions. The federal government pays only for a core staff of five people, and federal budget constraints make any additional funding unlikely. An additional part-time staff of about a dozen is paid by the nonprofit Commission for the Protection of the Reefs, established in 1995. This commission initially raised all its revenues from voluntary donations. Boat operators had pledged to contribute 150,000 pesos per month (US$1 ≈ 8.8 pesos), but less than half has been collected. The park's budget was thus based on highly uncertain revenues and barely covered the salaries of its personnel, leaving almost nothing for equipment, fuel, and maintenance costs. An evaluation by John Dixon at the World Bank and Mexican colleagues from the National Marine Park and the ministry in Mexico City suggested the implementation of a park fee.

The commission can legally raise fees, although the constitution prevents a national park from charging a fee of more than 27 pesos (about US$3) per visitor. However, even this modest fee would dramatically change the budget situation. Even US$1 per visitor would increase revenues by more than 300% (under the reasonable assumptions of 90% collection and that the fee would have no effect on demand).[7] The park management has suggested charging such a fee from boat operators for each visitor taken into the park. However, the boat operators have strongly resisted despite the fact that they appreciate the park's services. Their main objection appears to be that they do not want to provide information that would reveal their true level of income and thus lead to higher income taxes.

There is no demand-side problem in the sense that the fee could easily be passed on to the tourists. The reason for a rate increase would be not to reduce demand but to secure funding for a public good (or club good) kind of regulatory agency. The income taxes, however, are a more sensitive issue, particularly from a political viewpoint, because any government would be loathe to accept it as the official problem. One possible solution that has been discussed is to set up a nonprofit organization that would do the monitoring and collect fees to avoid the "prisoner's dilemma" or free-riding risk of depending on voluntary contributions as well as the public insight into the profits of the boat owners. As of late 1999, a voluntary fee (for which the tourists receive plastic bracelets marked "Parque Marino Cancún" to facilitate the monitoring of both tourists and boat operators) was being collected that raised about US$30,000/month. This is 600% of the federal budget of US$50,000!

In principle, earmarking is not desirable as part of policymaking. One could argue that the park charges should go to the central treasury and that the government should pay properly for park rangers. The risks with local fixes include suboptimality of various kinds. Among others, it may be easy to fund parks such as the NMP, which have many visitors, but difficult to finance bioreserves with fewer tourists. However, any pragmatic solution is much better than no solution. If fees can be set so that the park's revenues are roughly sufficient to pay for good management, then a decentralized solution that avoids the transaction costs of dealing with central bureaucracies may have more advantages than disadvantages.

## Shaping Ecosystem Policy

The case studies selected in this chapter concern preservation in one form or another. Understandably, all ecosystems cannot be preserved; some areas will be developed for commercial resource development. This tendency points to the importance of zoning as an overriding policy instrument. The ecosystems and the services they provide (see Chapter 4) must be sufficiently well understood to be able to determine the size of reserves and which other conditions (e.g., buffer zones or connecting corridors) are required to protect ecosystem functions and biodiversity in a satisfactory manner.

## Supplemental Reading

Brown 2000
Dasgupta 1982
Dixon and Sherman 1990

Zoning is thus necessary but in itself insufficient, because protected zones will not necessarily be respected if the appropriate protection and institutions are not provided. To achieve effective protection (and ensure the due process and legitimacy of the decisionmaking), policymakers must consider the interests of the stakeholders (e.g., how would they have used this area in the absence of protection?). It may be done through a CPR management approach, for instance, or through hybrids between CPR and more centralized management, like in CAMPFIRE.

CPR management is best suited to situations where local resource users have evolved cooperative patterns of resource use or at least can be expected to do so. It is not appropriate for protecting the high seas, Antarctica, or outer space. Nor would it make sense in some areas of tourism; where particular skills are required, specialized public or private agents may be best at providing protection. This does not mean that the public should not have the right to share in the benefits. They should, but at the same time, whatever organization operates a park or other area will also need to recover its costs and make a reasonable profit. For private and public bodies alike, financing and cost recovery are vital issues that must be faced.

An ecosystem may contain many resources—water, fish, forests, and wildlife. All the policy instruments that can be used to manage these resources individually may also be considered for marine environments and other ecosystems: taxes, subsidies, permits, appropriate tariffs, regulations, and property rights. However, because of the inherent complexity of ecosystem management, the design of instruments will also be more complicated. Interesting instruments for ecosystem protection include zoning, stakeholder participation, common property rights (and the definition of property rights in general), and payments or subsidies to positive externalities. As mentioned in Chapter 29, many ecologically interesting sites are being protected thanks to payments from municipalities and other downstream organizations. Certification and labeling are other interesting instruments that have been used in fisheries, farming, and forestry to signal some form of management with an ecosystem perspective or care in the commercial handling of natural resources.

## Notes

1. Such failures are an obvious and natural feature of experimentation with cultivation methods and institutional arrangements (Ostrom et al. 1999 is a follow-up to Hardin's famous article). Some failures are a necessary aspect of learning. They become more problematic in the global commons because experiments at that scale leave little room for failure. At the local

level, however, failure may be part of an optimal learning strategy if the number of substitute sites is sufficient.

2. This section has benefited from comments and information from Edwin Muchapondwa.

3. Zimbabwean elephants were estimated at fewer than 4,000 a hundred years ago but now number almost 70,000. They eat 150–300 kilograms of vegetation and drink a few hundred liters of water each day. In some districts, such as Sengwa, about 75% of the tree cover was destroyed by elephants between 1960 and 1975 (Child, Ward, and Tavengwa 1997). Between 1991 and 1996, 368 people were killed in Zimbabwe in connection with protecting their crops or houses from elephants.

4. This section builds on Andersson and Sterner 1998.

5. More recently, multimillionaire (and "deep" ecologist) Douglas Tompkins became the largest private park owner in the world. He owns 21% of the Palena province in southern Chile (323,646 hectares after purchases in December 1999). In 1997, in response to criticism from the Chilean government and the private sector, Tompkins said that he would turn over the land to a local nongovernmental organization, *Fundación Pumalin*, but he has yet to make good on his promise. A striking example of a transaction that is both private and cooperative is the sale of a part of the Easter Island National Park to the islanders, many of whom consider the sale illegal (Goldmann 2002).

6. I am grateful to John Dixon for material and comments on this case study.

7. The park administration estimated more than 550,000 visitors at Punta Nizuc in 1997. Boat operators charge US$38 for a "jungle tour" of the mangroves and US$35 for a direct trip. If the tours are booked through a hotel, the hotel keeps a commission of 10%. Thus, a tax of US$1 would not be a big addition. Estimates of divers' consumer surplus are generally much higher, for instance, US$15 (Davis and Tisdell 1995, 1996) or more (Wright 1994).

# PART SEVEN

# *Conclusion*

# CHAPTER 32

# *Policy Issues and Potential Solutions*

T HE WORLD FACES SERIOUS environmental and resource problems, but new technology offers numerous ways of solving them. Too often, however, potential solutions are not used because the rules that govern modern economies do not provide strong enough incentives. In some cases, the technology is not yet available, and sophisticated socioeconomic structures are needed to create the right incentives for research.

Environmental policymaking is not a simple choice between command-and-control instruments and market-based instruments (MBIs). Fortunately, the range of choices is richer—as in general economic policy, which is not limited to a choice between planning and laissez-faire. To function well, society needs intermediate policies with a lot of fine-tuning. And to meet several goals (e.g., efficiency, sustainability, and fair distribution), a combination of policy instruments usually is required.

An array of policy instruments has been designed specifically for environmental and natural resource issues. In addition, most other policies are highly relevant for resource management, ranging from the definition and enforcement of property rights, to the efficiency of the court system, to macroeconomic variables such as the rate of interest and the exchange rate.

Market-based policy instruments can be designed in many ways, of which Pigovian taxes and tradable permits are just two archetypes. Taxes, charges, deposit–refund systems, and other two-part instruments have been used in northern Europe, the formerly planned economies of Poland and the Baltic Republic of Estonia, and in some developing countries (China, Malaysia, and Colombia). Tradable permit schemes, used for pollution abatement in the United States and in some other countries, are particularly successful in fisheries management. Information provision, labeling, liability, and many other schemes broaden the menu of policies being used. Sometimes policies such as energy taxes and subsidies are "inadvertent" environmental instruments; they were not formulated with primarily environmental goals in mind.

Despite economists' recommendations to have one instrument for each goal, actual policies, such as energy taxes, are shaped out of complex bargaining processes and reflect a multitude of sometimes contradictory goals. They therefore typically lack the purity of textbook environmental policies, but so do policies aimed at furthering democracy, participation, equity, and other goals. Energy taxes still provide a good illustration of the way environmental charges work. In other areas, physical licensing and command-and-control instruments are predominant. This should not upset economists because, according to economic theory, physical regulations are appropriate in many cases. In some cases, economic theory may even lead one to recommend the prohibition of certain processes or substances as the most appropriate instrument.

## Policymaking Criteria

Economists generally assume that the overriding criterion for society is welfare maximization and that this welfare can be measured as a function of individual utilities. However, the utility and welfare functions may be too complicated to be operational, and it is common to have several separate subgoals. The most prominent are cost-effectiveness, efficiency, incentive compatibility, distributional and equity concerns, and administrative feasibility and flexibility. These goals are neither perfectly clear nor completely separable, and the political process is often a struggle in which groups place different emphases on different goals and have different interpretations of them.

For my purposes, *cost-effectiveness* means that the instrument achieves the environmental goals at least cost. *Efficiency* is a more ambitious concept that includes the optimality of the goal (i.e., level of abatement or resource stock) itself. *Incentive compatibility* means that the agents involved (particularly the polluters but also regulators, victims, and others) have an incentive to provide information and undertake abatement as intended. By *distributional and equity concerns* I mean that the distribution of costs should be perceived as "fair." *Administrative feasibility* means that the instrument is practicable, not incurring excessive monetary or informational costs for the operation of the instrument.

These criteria interact because, for example, polluters who think a particular distribution of costs is grossly unfair typically will resist and try to stop implementation. They will not have an incentive to collaborate, and this situation will ultimately lead to inefficiency, particularly if there are asymmetries of information or power. Policy formulation and implementation have many political, cultural, and psychological dimensions. It is therefore important to respect and follow the traditional rules for decisionmaking—sometimes referred to as due process—without naively opening up opportunities for corruptive lobbying. Given the sometimes rapid changes in technology, ecology, and of current understanding of technology and ecology, any instrument used must be flexible enough to allow for adaptation to new circumstances.

The criteria vary in importance depending on the conditions that characterize the issue at hand. In an economy with an even distribution of income and when regulating environmental problems with moderate abatement costs, equity issues are less important. Conversely, when dealing with major issues that affect health (and ultimately life) in countries with large income disparities, distributional concerns may be perceived as equally important as (or even more important than) cost-effectiveness in a narrow sense. When dealing with markets characterized by powerful monopolies or serious information asymmetries, the issues of incentive compatibility may dominate. With complex ecosystem interactions, flexibility is often an overriding concern in determining the design of the appropriate instrument. The policy selection matrix can give some structure to the choice and design of policy instruments. The matrix can be used to compare policy instruments according to different criteria depending on the ecological and economic features of a particular issue. Among the ecological features or conditions that should be taken into account are threshold effects and synergies, for instance. Among the economic features are market conditions, such as the degree of competition, the occurrence of missing markets, and the assignment of ownership rights.

In the rest of this chapter, I present some of the salient issues related to efficiency, risk, information, ecosystem complexities, the provision of public goods, feasibility, market structure, general equilibrium effects, cost distribution, and the politics of national and international policymaking.

## Efficiency

Efficiency is the most classic economic argument, often used in favor of MBIs. In some cases, however, other instruments may be more efficient. Care needs to be taken with respect to the type of efficiency intended (e.g., static or dynamic) as well as other conditions, such as the characteristics of cost and damage curves.

### Efficiency with Heterogeneous Abatement Costs

When pollution is mixed but abatement costs vary strongly, cost savings can be considerable if the firms with the lowest costs do most of the required cleanup. This fact provides a strong argument in favor of MBIs. Taxes and tradable permits are particularly applicable, then deposit–refund schemes and, in principle, subsidies (although they have other disadvantages). These instruments save costs by encouraging "specialization," which is one of the strongest advantages of the market mechanism.

Good examples of specialization are found in the areas of energy economics and global climate change. The marginal cost of decreasing the atmospheric concentrations of nitrogen oxides ($NO_x$) or sulfur oxides ($SO_x$), precursors to acid rain, or the greenhouse gases varies enormously across the transportation, industry, forestry, and agriculture sectors. It also varies within each of these sectors and between countries. Several European countries have a good deal of experience with energy taxes, whereas the United States has experience with permit schemes. Many interesting examples of market-based policy instruments are found in developing or transitional economies.

## Efficiency with Heterogeneous Damage Costs

Sometimes it is the cost of environmental damage, not abatement, that varies strongly (e.g., health costs from vehicle emissions between densely populated and less inhabited areas, or the nuisance of noise between night and day). In these cases, efficiency dictates a corresponding variation in the stringency of abatement efforts. Such a variation is in principle possible with various policy instruments. However, price instruments may be difficult to differentiate (particularly when product taxes are used as proxies for emissions fees), making quantitative rules such as zoning or time-dependent regulations more effective. The environmental (health) damage caused by one mile driven by an average car will be hundreds or thousands of times higher in a city center than in the countryside, yet the price of gasoline can vary only marginally.

Some solutions combine physical requirements (e.g., "clean" cars in some zones) with the use of differentiated road pricing. The banning of polluting vehicles with two-stroke engines is a much-needed zoning policy in some of the world's most polluted cities.

## Intertemporal Efficiency with Technological Change

The cost of abatement technology may change drastically over time. If costs decrease due to technological progress, for instance, then the effect of a tax, permit, or standard will change in different ways. Progress in abatement technology means that achieving abatement becomes cheaper and that the optimal level of abatement increases while the necessary tax level decreases.

With technical progress in abatement, (constant) taxes produce more-than-optimal abatement, whereas (constant) regulation gives less-than-optimal abatement. The size of the respective losses depends on the relative slopes of the damage and abatement cost curves. However, time commonly increases estimates of the damage from emissions through various mechanisms, including increased population, income, and knowledge. In this case, the tax (with its higher abatement level) is preferable. On the other hand, it may be easier to set physical standards, which can be based on observed technological prototypes, whereas the corresponding tax levels must be estimated. The ability of the instrument to give efficient or at least reasonable outcomes in the face of changing conditions (ecological or economic, such as growth or inflation) is a form of flexibility that can be very important.

## Intertemporal Efficiency with Inflation

The effect of inflation on policy is similar to that of technological change. However, one difference is that both abatement and damage costs are changed equally, and thus the optimal abatement or emissions level remains constant while the necessary tax level increases. Consequently, regulation and standards have an advantage for preventing environmental degradation in the face of inflation, whereas monetary instruments such as taxes will be reduced in value. Indexing can be a solution in principle but also entails various problems. In the case of subsidies, which have some undesirable long-run effects, it might actually be an advantage if inflation reduces their real value.

### Incentives for Innovation under Technical Change

Building good incentives for innovation is a crucial feature of policy design when the potential for technical progress is significant. Such incentives are almost completely absent if the policymaker chooses design standards. With emissions or performance standards, they are somewhat greater, but the instruments with the clearest incentives are the MBIs and particularly taxes. This issue is nicely illustrated by the effects of the Norwegian carbon tax, which has led to sequestering and underground storage of carbon in saline aquifers.

## Uncertainty, Risk, and Information Asymmetry

Efficiency is not always enough or even appropriate. In a world of uncertainty and risk, other criteria—such as producing feasible, verifiable, or robust results—may be more important.

### Monitoring Difficulties

Emissions are often difficult to monitor. The consumers of fossil fuels (e.g., gasoline and diesel), for example, are many, small, dispersed, and mobile polluters. In instances where there is a strong one-to-one relationship between the consumption or production of a product and the emissions it produces, then input, output, or product taxes may be good proxy instruments. Taxes on fossil fuels are a good example; they work well to address global climate change, which depends heavily on total (fossil) fuel consumption. However, gasoline and diesel taxes are not very good at internalizing local environmental damage. The emissions from a car and the health impact of those emissions increase with the amount of fuel used, but they vary more significantly with other factors, such as motor and emissions technology. When monitoring becomes sufficiently difficult, the regulator may choose instruments that specifically facilitate monitoring, such as design standards.

### Ex Post Verification

Inappropriate disposal of hazardous wastes is an externality that policymakers must try to reduce. The trouble is that the amounts of effluents and emissions are so small, irregular, and dispersed that they are almost impossible to monitor. It is not the production or use of the product but its disposal that is environmentally damaging, and therefore, product taxes are hardly appropriate. Deposit–refund systems may be the ideal mechanism. They are effectively a tax on disposal but, for the sake of convenience, are levied at the time and point of purchase. The burden of proof concerning appropriate disposal is placed on the buyer, who can avoid paying the "emissions tax" by turning in the item to collect a refund. A similar concept is the "presumptive" tax, whereby consumers are taxed because they are presumed to pollute unless they prove otherwise.

Deposit–refund systems are widely used for glass bottles and aluminum cans. For environmental reasons, they also are used for cars and batteries in some countries. Similar mechanisms could presumably be extended to a wider range of hazardous products. The general principle of the deposit–refund scheme could be used to allow vehicles that can prove they are cleaner to get refunds on annual registration or other taxes, for instance.

### Monitoring Only of Ambient Conditions

A common situation that contains elements of asymmetric information is non-point-source pollution when only aggregate (not individual) emissions can be monitored (e.g., agricultural runoff of nutrients that lead to eutrophication). Information is asymmetric because each party to be regulated (i.e., each farm) has at least some notion of its own emissions, whereas regulators can at best monitor only aggregates, such as the ambient concentrations at some points. In this case, emissions taxes are not feasible.

In principle, regulators might levy some form of an ambient tax based on total ambient levels, but this instrument has considerable problems of practicability. In some cases, input taxes on fertilizers or pesticides might be appropriate, but not all agricultural runoff is caused by purchased inputs. It may result from animal manure or soil erosion, which are difficult to monitor. One possible policy in this case is peer monitoring: neighbors typically know about each others' production methods. If cooperation and reciprocal monitoring can be organized, then this approach may be much more efficient than outside monitoring. It is in fact an essential comparative advantage of common property resource (CPR) regimes. One policy instrument may be to encourage the operation (and perhaps even the formation) of such CPR systems.

Even when no CPR management system is in place, collaboration between neighbors might be elicited; regulators might give a group of polluters a choice between self-regulation and some other policy (e.g., input control or tax). Similar policies may also be applied to the "informal" industrial sector (e.g., the two-tier monitoring in Indian industrial estates and the regulation of Mexican brickmakers).

### Uncertainty When Damage Costs Are Steep

With perfect certainty, there is a symmetry between price-type and quantity-type instruments. A tax of $T$ dollars per ton of effluent will essentially have the same effect as requiring a mandatory emissions limit equal to $\hat{e}$ tons if the marginal cost of abatement is $T$ at the level $\hat{e}$. When there is uncertainty about the abatement costs, the symmetry breaks down; if the marginal damage costs are steep, then quantitative permits should be preferred over taxes or fees (e.g., when there are dramatic threshold values, such as for many banned pesticides, including dieldrin). Policymakers cannot incur the risk of searching for the optimal tax level; they must simply ensure that there is a safe standard. The costs of a small mistake in the quantity used can be terribly high, and the precision of the quantitative instrument in this case is higher.

### Uncertainty When Abatement Costs Are Steep

When the cost of abatement rises steeply and the marginal damage from pollution is relatively flat, an MBI such as a tax is preferable. In this case, the tax has greater precision in avoiding potentially excessive abatement costs. Examples of flat damage curves may be found with stock pollutants (as long as there are no strong threshold effects or nonlinearities). Examples of steep abatement cost curves may be found in the formerly planned economies, where abatement was urgently needed but many companies could not afford new investments. If they were forced (by stringent standards) to switch technologies quickly, it could lead to bankruptcy, unemployment, and significant welfare and political costs.

### Uncertainty and Risk

The value of a firm's formal, legal liability for possible environmental damages is in practice limited by the firm's ability to pay. If the company goes bankrupt, then its liability may essentially be worthless. Other legal instruments (criminal law), mandatory insurance, or liability bonds—deposits of money that can be used in the event of an accident and that are deposited before a company is allowed to start operations—may decrease this risk. Such instruments are infrequently used because they carry a large cost that may be a strong disincentive to business; however, they would be appropriate for those industries where there is a risk of really large releases or accidents, such as the Bhopal tragedy.

### Missing Markets in Insurance and Banking

Insurance usually is not fully supplied by the market. Information asymmetry causes a market failure through the adverse selection of clients (the high-risk people buy the insurance policies) and moral hazard (once people have a policy, they may become reckless), and these factors lead to an undersupply of insurance. Users of natural resources typically operate at high levels of risk and cannot get the insurance they need. Particularly for the poorest farmers in developing countries, who live close to starvation, the risk of crop failure is unacceptable because they lack insurance, and for similar reasons, they have inadequate access to regular banking and savings institutions. They therefore are very risk averse. The result may be unsustainable behavior: people may not dare to invest in new productive methods and may use methods that are damaging to the ecosystem, such as applying excessive amounts of pesticide or keeping large herds of cattle. These practices, although unsustainable, may be individually rational adaptations to missing markets for savings and insurance. In these cases, the best policy will not be to tax or ban pesticides and large herds but to help provide the missing markets by encouraging insurance companies or village banks like Grameen Bank.

## Ecological and Technical Complexities

Ecological and technical complexity foster uncertainty and thus make the simple efficiency criterion hard to apply. The instruments chosen to deal with complex issues typically must consider information, risk, and uncertainty issues.

### Complex and Potentially Hazardous Technology

Designing a sufficiently detailed tax for a complex technology (e.g., many chemicals, large industrial plants) is difficult; a tax must be precise, and it is easiest to levy a tax on a straightforward quantitative measure, such as tons of an input, output, or effluent. Some very sophisticated tax instruments (with different tariffs for each chemical and differentiation with respect to factors such as stack height and sensitivity of the ecosystem) were designed and used, primarily in the former Soviet Union, but the interpretation of taxes in a planned economy is very different from that in a market economy. Passing this kind of tax law in a market economy probably would be difficult. Not only is the level of detail daunting, but regulators explicitly lose control over the quantities emitted. If the tax level is "too low" and a company decides to emit a large amount of some chemical in the middle of a city, then regulators would be in trouble—it would be too late to add extra requirements for abatement. This risk can be thought of in terms of a steep damage curve that favors individual regulation. The relevant criterion is no longer efficiency but rather feasibility and the precautionary principle of avoiding excessive damage risks. For complex technology, individual licensing or liability may be the best options, particularly if the potential damage is serious (steep marginal damages), as with nuclear and some chemical industries.

When the potential damages and risks are not quite as serious, the best available instruments may be various forms of voluntary agreement or information provision, with or without labeling (e.g., the U.S. Toxics Release Inventory, PROPER in Indonesia, or product labeling). As indicated by the comparison of instruments on the solvent trichloroethylene, fees, prohibition, and restrictions on use may be feasible instruments, but each has its advantages and disadvantages.

### Ecological Complexity

Technology is not necessarily the main source of complexity. In many cases, the ecosystem itself is complex, and policies must be guided by the precautionary principle. For biodiversity protection, it would be difficult to construct suitable indicators on which to base taxes. Not only may many features need to be monitored but regulators might want to differentiate between sites as well.

One suitable set of policies is based on the CPR management concept, by which property rights are allocated at the local level and local knowledge of the ecosystem and sustainable harvest methods is harnessed. In fisheries management, individual transferable quotas (ITQs) are a good example of a new policy instrument whereby the issues related to the ecosystem complexity (the total allowable catch) are specifically separated from the issues concerning allocation (of percentage shares) among resource users. In other cases, ecological concerns may favor zoning instruments such as the creation of marine reserves or parks.

## The Provision of Environmental Public Goods

In situations that involve some sort of public good (e.g., a national park), no one polluter can be charged or regulated. Available instruments are public provision and, possibly, private provision that is financed or subsidized by public funds.

## CPRs, Local Public Goods, and Mixed Public Goods

Public goods are not always "pure." Services relating to urban and health infrastructure, such as garbage collection, are impure (or mixed) public goods or, in some cases, club goods or local public goods. In the past, it was often taken for granted that services such as waste collection should be seen as pure public goods. Lately, discussions concerning efficiency as well as incentives and other factors have led to a modified view. In many high-income countries, waste management is experimenting with variable charges to provide an incentive for separation at the source, and many low-income countries are experimenting with innovative solutions such as microenterprises to organize workers in this sector. Frequently there is an inherent conflict between the tariff constructions that would maximize efficiency and those that would help provide necessities for the poor.

Ecosystem resources have many "public" characteristics at the local level. The appropriate policy instruments vary with the circumstances. In some cases, communities have started to pay for ecosystem protection of water supplies or other amenities. In other cases, the reform or clarification of private property rights is an important prerequisite for other policies. For true common pool resources (marginal lands, mangroves, fishing sites, and water sources) that are particularly important for the poorest segment of a society, CPR management may be the most appropriate policy.

It is not easy to "create" new CPRs. Social institutions hinge on reputation and social structure, which create trust and reciprocity; such structures take time to build, and the process may be difficult, as illustrated by CAMPFIRE in Zimbabwe.

## Global Public Goods

At the other end of the scale are several resources or ecosystem attributes that can be described as global public goods, for example, temperature and climate of the atmosphere, and protection against ultraviolet radiation by the stratospheric ozone layer. For global public goods, international cooperation is a sine qua non of any successful policy.

The value of local programs may be to demonstrate that abatement or reduced exploitation is feasible. It is important but cannot replace international negotiations or concerted action. The characteristics of the international treaties interact with the choice of policy instruments at the national level. If, for instance, treaties specify quantity targets for each participating country, then quantity-based instruments such as permits will be favored because they best fulfill quantitative goals.

## No Polluters Accountable

In some cases, polluters are theoretically identifiable, but monitoring is so difficult that they either cannot be identified or are identified too late (e.g., a company that polluted an area decades ago is now out of business). Society must determine whether cleanup warrants the costs of remediation. If so, public provision of abatement or cleanup and subsidies to firms may be the only methods available to achieve the desired goal, although in some cases, firms, entire industries, and even associated insurers, suppliers, bankers, and customers may be involved through extended concepts of liability; joint and several liability has been used in the U.S. Superfund program, for instance.

## Feasibility, Market Structure, and General Equilibrium Effects

The social and economic context within which instruments are to be applied is often a more important determinant than the sector or industry concerned.

### Number of Polluters and the Structure of Markets

The number of polluters and the structure of markets have profound effects on the choice and design of policy instruments. If there is only one polluter (i.e., a monopoly), then a tax will be passed on to consumers and in fact will have perverse incentives because monopolies are characterized by too low an output level—an effect that may be worsened by a tax. Furthermore, if there is only one polluter, decisionmakers would tend to use individual negotiation, licensing, or voluntary agreements instead of going through the whole process of writing a tax law. (Typically, policymakers also strive to introduce competition and break up monopolies which is an argument against adapting policy instruments specifically to them.)

When the number of polluters is intermediate, the analysis of different instruments can become complex, particularly when abatement is carried out in several steps (e.g., in the acquisition of expensive abatement equipment and the operation of such equipment). Permits are one instrument for which the number of participants is crucial. With few players, the permits will be traded in "thin" markets, which may create significant distortions and in some cases very limited trading. Thin markets also may provide an incentive for strategic behavior by the firms.

### Lack of Resources

Countries that are embarking on new or ambitious environmental programs may be constrained by a lack of knowledge and organizational, technical, financial, and human resources. "Sophisticated" instruments might appear to be completely out of their reach, and they might be tempted to conclude that poor agencies should start with command-and-control instruments, leaving supposedly more advanced (market) instruments for later. This approach is unreasonable because all environmental instruments have much common ground; all require systems for monitoring, reporting, verification, and control.

Physical command-and-control instruments are not necessarily easy to administer. They require a system of penalties and enforcement that must be severe enough to act as a deterrent but not so Draconian as to be unenforceable in practice. For this reason, informational, legal, or market-based instruments are sometimes preferred. The sophistication of the instrument may be designed to address an environmental protection agency's lack of resources. The use of product charges (rather than emissions or effluent charges) saves on administrative resources but also reduces allocative efficiency. Earmarking these charges can provide resources for an environmental protection agency.

One should also be wary of the simplistic notion that environmental taxes will always provide a win–win situation with abatement and tax revenues. It is not always the case, even though some countries that are well endowed with natural resources probably would benefit from policies designed to capture a greater share of the rent while providing for more sustainable resource use.

### Only Subgroups of Polluters Targeted

Sometimes only a subset of polluters is targeted as a result of technical or social factors—possibly because of size or nationality, or because the same pollutant can be emitted from a point source in one kind of industry but a nonpoint-source in other industries. The argument might reasonably be made that it is unfair to tax the identified polluters if other (equally important) polluters are not taxed, particularly if the parties are competitors.

In this situation, policymakers may seek instruments or combinations of instruments that provide an incentive for abatement to those firms that have been identified without penalizing them unfairly with respect to other firms outside the program. This goal might be achieved by presumptive taxation or input taxes that affect the nonidentifiable polluters. Otherwise, refunded emissions payments or the free allocation of tradable permits may provide abatement incentives while limiting the cost burden to the group of firms concerned. Such instruments have been used for abatement of $SO_x$ and $NO_x$ in several countries.

### Reducing the Use of Goods that Lead to Pollution

Most analyses of policy instruments are partial analyses, which sometimes are sufficient—for instance, when technical abatement measures or input substitution readily achieve the desired environmental improvement goals. When abatement is technically difficult, environmental goals can be achieved only by reducing the consumption or production of certain goods. The instrument used must have an effect on product output. The general equilibrium repercussions throughout the economy must be taken into account. Simple technology or abatement devices that can reduce carbon emissions sufficiently to slow global warming, for example, are not (yet) available. It will be necessary to reduce the consumption of such goods and services that result in significant emissions of fossil fuel–based carbon (and other greenhouse gases). MBIs such as taxes and traded permits have this output effect and are thus the most appropriate in such cases.

### Recycling Revenue from Environmental Taxation

Emissions can be reduced by a carbon tax or by a system of traded permits. Both systems result in an output effect, but only the tax or auctioned permits lead to the public capture of the associated rent. If this rent is captured, then there is a revenue-recycling effect that may be important in achieving overall efficiency. A tax large enough to substantially lower carbon emissions would generate enormous public funds and thereby considerably change the economy. The revenues raised could be used for public spending or for reducing other taxes, and the way they are used could affect the cost of the policy under discussion through secondary effects on the economy and thereby the politics of policy design.

## Cost Distribution and the Politics of Policymaking

An analysis of policy instruments should not stop at their general allocative properties or overall resource costs. It is crucial for political feasibility to consider also the distribution of costs. Different instruments result in different cost allocations between polluters and among polluters, victims of pollution, and society. These differences often are decisive in implementation.

### Burden of Cost and Issues of Political Feasibility

Technically, congestion and overuse of common goods can be rectified by implementing a tax. However, although the congestion may easily be brought down to an optimal level, the collective welfare of the users may be worse than before the tax. Sometimes users assert historic rights to their level of activity (e.g., fishermen who were "born into" their profession, taking over from generations of fishermen before them), resisting the notion that the government has rights to all the surplus even if it did take fishing down to sustainable levels. In other cases, polluters may be too powerful to tax. In both cases, a tax is politically infeasible, but other policies may be available. Rights can be created—full property rights or at least freely allocated pollution or resource permits. Price-type instruments such as charges also can be used as long as the proceeds are (at least partially) refunded or used in a way that is beneficial to (and determined by) the community concerned.

In the United States, road charges are earmarked for road funds, and in many developing or transitional countries (notably, Poland), environmental fees are earmarked for environmental funds. These funds risk the inefficient use of public money, but in some cases, this risk is the "price" that must be paid to achieve sufficient political acceptability. If modifications in instrument design make the use of an instrument feasible (or feasible at a higher level), then this feasibility may warrant the modification. In a choice between a low but pure tax (that goes to the treasury) and a much higher but refunded emissions payment, the latter may, in many circumstances, be preferable from the perspective of both environmental concern and overall welfare.

### Rent-Seeking and Political Economy

Public policies are not only formed by abstract considerations of optimality but through lobbying and the interplay of various interest groups. Policymakers should anticipate this behavior and be particularly cautious about instruments that tend to promote it. The most obvious example is subsidies, which not only are expensive but also can promote lobbying and even corruption. Even the allocation of permits and the mechanisms for refunding charges may attract considerable lobbying, and these consequences must be taken into account.

These issues are likely to be important if the targeted polluters have any of several characteristics:

- If they are regionally or ethnically different, they may feel discriminated against.
- If they are poor, welfare considerations become particularly prominent.
- If they are rich and powerful, they may have the power to stop or stall implementation.
- If they are a small homogeneous group, they can easily organize.

For each group, the allocation, refunding, and compensation mechanisms may need to be considered separately.

### Fairness and Efficiency for Resource Issues Affecting the Poor

The most urgent class of problems is that in which poverty and environmental degradation occur together. The effects of the two problems commonly reinforce each other, so environmental degradation leads to decreased access to water, fodder, firewood, and other important materials. The desperation and short-sightedness caused by poverty may force poor people into unsustainable practices that worsen resource degradation. All the distinct categories of policy that have been discussed are applicable, depending on the details of the individual case. In rural settings, CPR management may be the most appropriate approach.

Difficulties sometimes arise because efficiency requires market solutions, but some market solutions are socially unacceptable. When distributional and efficiency concerns are equally important and problems are complicated by information asymmetry, then various two-part instruments may provide solutions (e.g., increasing block tariffs or more complicated mechanisms sometimes used for water or electricity, including subsidies for necessary consumption by small consumers). Privatization may be beneficial in introducing market principles but is no panacea. If there are distributional, environmental, or other welfare goals in addition to efficiency, then specific public action will be required because an unregulated market will not reach these goals.

## National and International Policymaking

With technical progress and minimal attention to environmental issues, simple problems are addressed that relate to the most immediate environment. It is not uncommon that technical fixes do not really "solve" problems but rather move them farther away. Transboundary pollution is becoming more important, as is the management of the global commons. In many contexts, policymaking must be considered in terms of a structure at the local, national (federal), and international levels.

### Policymaking in a Small, Open Economy

Trade relations, transboundary pollution, and international treaties are three global layers of restrictions to consider when designing policy instruments. Trade relations are more noticeable in small, open economies that in large, closed economies. In the small, open economy, the price of goods is set by the world market. Any local deviation (e.g., a pollution tax on domestic production) will not change the world market price and therefore will have no effect on consumption. Its only effect will be on profits and thus market shares in the country concerned.

If the pollutant in question is a global one, then imposing an environmental tax risks merely moving production abroad, thus leaving pollution levels constant. In such cases, international negotiation is paramount, but still many other (national) instruments are available, for example, refunded emissions payments, taxes on consumption (rather than production), free permits, voluntary agreements or licensing, labeling, and information provision.

### Mixed Global and Local Public Goods or Environmental Benefits

One important category of projects has benefits at various levels. For instance, local benefits of forest protection may include climate stabilization, maintenance of biodiversity, protection of shorelines, moisture retention, and the halting of soil erosion. At the same time, protecting forests has regional and international benefits (such as carbon retention). Of particular relevance are the cases in which local benefits would not be sufficient to achieve conservation but the total benefits might be. By finding mechanisms to compensate the local economies for global benefits, this category of socially profitable projects may become feasible. Debt-for-nature swaps, the Global Environment Facility, and the Carbon Fund are all (fledgling) examples.

## Conclusion

One factor that commonly hampers policymaking in many countries (particularly the poorer ones) is the weakness and lack of resources (i.e., not only in funding but also staff, training, and laboratories and other facilities) in the environmental protection agencies or the ministries in charge of designing policies. The U.S. Environmental Protection Agency, for example, has some 6,000 staff, whereas the Chinese environmental protection agency has about 200.

Experience has shown the importance of prioritizing and concentrating on a limited number of issues and only the worst polluters. Institutions that are knowledgeable and free from nepotism and corruption must be built. It also is important to build partnerships with the various stakeholders and for the environmental protection agency to be seen not only as a police officer but also as a source of technology and know-how concerning modern sustainable technology. Information and technology dissemination, research support, and extension services are—in addition to the inevitable control function—essential to policymaking.

The setting of fees is a difficult and sensitive issue. In some cases, low fees that are earmarked for abatement, research, and control may be useful. Creating environmental funds in which the polluters have influence may help build political acceptance among regional or sectoral stakeholders who need to be involved rather than alienated. Effective environmental work often builds on functional partnerships. Clarifying rights—with respect to resources as well as to information and to litigation concerning damages—may be an integral part of achieving successful policy implementation.

In this book, I have tried not to overuse the term "economic" instruments. In fact, the term is not clear, because economic theory commonly suggests the superiority of quantitative, informational, or legal instruments. In some cases, the main option may even be to strengthen moral and ethical rules. (Economic incentives or formal rules might possibly "crowd out" moral imperatives so that they reduce altruistic behavior.) Real-life policymaking is seldom a neutral search for the optimal instrument to maximize global welfare. At least as frequent is the battle between lobbying groups striving for survival, personal benefit, power, or environmental goals. Following transparent, democratic, and bureaucratically feasible processes for decisionmaking is crucial for successful policy implementation. The parties affected or interested in legislation must be given the opportunity to influence it for the sake of legitimacy, and because they are the best sources of information, the process must be designed so they have an incentive to reveal at least part of this information. On the other hand, the parties cannot be given too great an influence if it entails the risk of not using effective instruments to attain reasonable goals. To understand the politics of policy design, careful attention must be paid not only to the allocative properties of the instruments but also to the resulting cost distributions between polluters and others in society.

The dominant factors vary strongly from case to case. If abatement costs vary considerably, then efficiency dictates that market mechanisms such as taxes or tradable permits be used. This is the classic argument for the superiority of the market, and if total costs are high, it will be an important factor. If, on the other hand, the damage costs are sufficiently heterogeneous, then more physical, quantitative instruments such as zoning, differentiated regulations, and differentiated licenses may be called for. If important information asymmetries exist, then policy instruments must be designed to be self-revealing, like deposit–refund schemes. If the public good character of the issue is important, then pure research and the dissemination of information will be crucial to successful environmental policy.

In the face of particular technological or ecological complexities, flexibility may be of paramount importance. Policy instruments may be designed to separate decisions concerning the absolute level of resource use from the allocation of rights or the distribution of permits or quotas. In fishing, the separation of these two aspects through ITQs ensures that the catch level can be delegated to expert bodies who can make rapid decisions based on scientific evidence without having to consider or renegotiate all the political complexities surrounding the distribution of costs and rights. This model may be worth considering in other areas, too. In many cases, the power of the polluters is considerably stronger than that of the environmental protection agency. Consequently, instruments such as Pigovian taxes are impossible, whereas tradable permits allow policymakers to fine tune the allocation of rights and cost distributions to make the policy politically acceptable. Similarly, charges that are differentiated and refunded or paid into environmental funds may be used to secure acceptance from important polluters while strengthening the public agencies.

Environmental economics—particularly policy design—is an exciting academic discipline, but it also has a serious purpose. Models are created not (only) to publish in journals but to make a difference in real economies. Sometimes reality requires a great deal of sophistication to match the complexities of technology, ecology, and society. This does not mean that an instrument should be judged only by its complexity; nor should it be judged by how well it fulfills any other single criterion (such as the polluter pays principle). The ultimate test of an instrument is its effect. This effect is determined first by whether the policy gets implemented and second by the level at which the policy is implemented. A perfectly designed Pigovian tax that is set too low may well be much less efficient than another instrument that for some political reason was feasible to set at a higher level.

The resource and environmental problems ahead are significant but probably not insurmountable. It is an important challenge to adapt and develop the general principles discussed here to strive for a more sustainable economy. This ongoing process must be informed by theory as well as experience. The careful evaluation of new policies and the sharing and comparison of experiences must be integral to this process. It will require collaboration not only between countries but also across disciplines and between different segments of society. Researchers have a very important message to communicate and must learn to convey it in a way that is visible and practical to decisionmakers. Researchers must be humble and practical enough to acknowledge that many policy instruments evolve from the everyday process of regulation and negotiation, rather than directly from textbooks.

# References

Aaltonen, J. 1998. *Macroeconomic Instruments for Environmental Management in Developing Countries.* Final Report. Helsinki, Finland: Ministry for Foreign Affairs of Finland, Unit for Sector Policy and Advice Kyo-32.

Adams, J. 1997. Environmental Policy and Competitiveness in a Globalised Economy: Conceptual Issues and a Review of the Empirical Evidence. In *Globalisation and Environment: Preliminary Perspectives.* Paris, France: Organisation for Economic Co-operation and Development.

Adams, M., and Y. Motarjemi. 1999. Foodborne Hazards. In *Basic Food Safety for Health Workers,* chapter 2. WHO/SDE/PHE/FOS/99.1. www.who.int/fsf/BasicFoodSafetyforHealthWorker/2.pdf (accessed May 2002).

Adar, Z., and J.M. Griffin. 1992. Uncertainty and the Choice of Pollution Control Instruments. In *The Economics of the Environment,* edited by Wallace E. Oates. International Library of Critical Writings in Economics, vol. 20. Brookfield, VT: Edward Elgar, 132–142.

ADEME (Agence de l'Environnement et de la Maîtrise de l'Energie). 1998. *Taxe Parafiscale sur la Pollution Atmosphérique.* Rapport d'Activité 1997. Paris, France: ADEME.

Afsah, S., A. Blackman, and D. Ratunanda. 2000. How Do Public Disclosure Pollution Control Programs Work? Evidence from Indonesia. Discussion Paper 00-44. October 2000. Washington, DC: Resources for the Future.

Afsah, S., B. Laplante, and N. Makarim. 1996. Program-Based Pollution Control Management: The Indonesian PROKASIH Program. Policy Research Working Paper 1602. Washington, DC: World Bank.

Afsah, S., B. Laplante, and D. Wheeler. 1999. Controlling Industrial Pollution: A New Paradigm. Policy Research Working Paper no. 1672. Washington, DC: World Bank.

Afsah, S., and J. Vincent. 1997. *Putting Pressure on Polluters: Indonesia's PROPER Program. A Case Study for the HIID 1997 Asia Environmental Economics Policy Seminar.* March. Cambridge, MA: Harvard Institute for International Development. www.worldbank.org/nipr/work_paper/vincent/ (accessed April 2002).

Ahlvik, P., K.E. Egebäck, and R. Westerholm. 1996. *Emissionsfaktorer för Fordon Drivna Med Biodrivmedel.* Stockholm, Sweden: Motortestcenter (MTC) vid AB Svensk Bilprovning.

Ahmad, E., and N.H. Stern. 1984. The Theory of Reform and Indian Indirect Taxes. *Journal of Public Economics* 25: 259–298.

Aiken, R.S., and C.H. Leigh. 1992. *Vanishing Rainforests: The Ecological Transition in Malaysia.* Oxford, U.K.: Clarendon Press.

Akerlof, G. 1970. The Market for Lemons. *Quarterly Journal of Economics* 84: 488–500.

Akimichi, T. 1984. Territorial Regulation in the Small-Scale Fisheries of Itoman, Okinawa. In *Maritime Institutions of the Western Pacific,* edited by K. Ruddle and T. Akimichi. Senri Ethnological Studies no. 17. Osaka, Japan: National Museum of Ethnography, 37–87, 89–120.

Albrecht, J. 1998. Environmental Regulation, Comparative Advantage and the Porter Hypothesis. September. Fondazione Eni Enrico Mattei (FEEM) Working Paper 59/98. Milan, Italy: FEEM. www.feem.it/web/activ/_wp.html (accessed April 2002).

Alemu, T. 1999. Land Tenure and Soil Conservation: Evidence from Ethiopia. Ph.D. dissertation, Department of Economics, University of Gothenburg, Sweden.

All Africa. 2000. A Fitting Reward for Asmal's Water Effort (Editorial). March. allafrica.com/stories/200003240227.html (accessed April 2002).

Almeida, R., J. Cervantes, G. Dutt, L. Garcia, J.F. Garza, R. Joaquin, C. Juarez, M. Martinez, O. Masera, and C. Sheinbaum. 1989. Energy Use Patterns and Social Differences: A Mexican Village Case Study. Manuscript Report 215e. Ottawa, Ontario, Canada: International Development Research Centre.

Amacher, G., R. Brazee, and M. Witvliet. 2001. Royalty Systems, Government Revenues, and Forest Condition: An Application from Malaysia. *Land Economics* 77(2): 300–313.

Amacher, G.S. 1996. Bargaining in Environmental Regulation and the Ideal Regulator. *Journal of Environmental Economics and Management* 30: 233–253.

———. 1998. Instrument Choice When Regulators and Firms Bargain. *Journal of Environmental Economics and Management* 35: 225–241.

———. 2001. Forest Policies and Many Governments. *Forest Science* 48(1): 146–158.

Amacher, G.S., and A. Malik. 1996. Bargaining in Environmental Regulation and the Ideal Regulator. *Journal of Environmental Economics and Management* 30: 233–253.

———. 1998a. Instrument Choice When Regulators and Firms Bargain. *Journal of Environmental Economics and Management* 35: 225–241.

———. 1998b. Taxes versus Standards When Technology Choice Is Endogenous. Paper presented at 8th Annual Conference of the European Association of Environmental and Resource Economists, Tilburg, Netherlands.

Andersen, R. 1976. The Small Island Society and Coastal Resource Management: The Bermudan Experience. In *Marine Policy and the Coastal Community,* edited by D.M. Johnston. London, U.K.: Croom Helm, 255–277.

———. 1979. Public and Private Access Management in Newfoundland Fishing. In *North Atlantic Maritime Cultures,* edited by R. Andersen. The Hague, the Netherlands: Mouton, 299–336.

Anderson, D. 1990. Environmental Policy and the Public Revenue in Developing Countries. Environment Working Paper no. 36. Washington, DC: World Bank.

Anderson, G., and T. Zylicz. 1996. The Role of Environmental Funds in Environmental Policies of Central and Eastern European Countries. Environment Reprint Series, no. 3. April. Cambridge, MA: Harvard Institute for International Development, International Environment Program.

Anderson, K. 1998. Agricultural Trade Reforms, Research Initiatives, and the Environment. In *Agriculture and the Environment,* edited by E. Luntz. Washington, DC: World Bank, 71–83.

Andersson, J. 1995. *Marine Resource Use in the Proposed Mafia Island Marine Park.* Gothenburg, Sweden: University of Gothenburg, Department of Economics, Unit for Environmental Economics.

Andersson, J., and Z. Ngazi. 1995. Marine Resource Use and the Establishment of a Marine Park: Mafia Island Tanzania. *Ambio* 24(7-8): 475–481.

Andersson, J., and T. Sterner. 1998. Private Protection of the Marine Environment: Tanzania, a Case Study. *Ambio* 27(8): 768–771.

Ando, A., V. McConnell, and W. Harrington. 2000. Costs, Emissions, Reductions, and Vehicle Repair: Evidence from Arizona. *Journal of the Air and Waste Management Association* 50: 509–521.

Andréasson-Gren, I.M. 1990. Costs for Reducing Farmers' Use of Nitrogen in Gotland, Sweden. *Ecological Economics* 2: 287–299.

———. 1991. Costs for Nitrogen Source Reduction in a Eutrophied Bay. In *Linking the Natural Environment and the Economy: Essays from the Eco-Eco Group,* edited by C. Folke and T. Kåberger. Dordrecht, the Netherlands: Kluwer Academic.

————. 1992. Profits from Violating Controls on the Use of a Polluting Input. *Environmental and Resource Economics* 2: 459–468.

AQMD (South Coast Air Quality Management District). 1996. CUT-SMOG Smoking Vehicle Program website. ozone.aqmd.gov/smog/cutsmog.html (accessed April 2002).

Arnason, R. 1996. On the Individual Transferable Quota Fisheries Management System in Iceland. *Reviews in Fish Biology and Fisheries* 6(1): 63–90.

Arnason, R., R. Hannesson, and W.E. Schrank. 2000. Costs of Fisheries Management: The Cases of Iceland, Norway, and Newfoundland. *Marine Policy* 24: 233–243.

Arntzen, J. 1995. Economic Instruments for Sustainable Resource Management: The Case of Botswana's Water Resources. *Ambio* 24(6): 335–342.

————. 1996. Ecological Change and Distribution in Southern African Rangelands. *Journal of Income Distribution* 6(2): 305–326.

Arora, S., and T.N. Cason. 1994. A Voluntary Approach to Environmental Regulation: The 33/50 Program. Summer. Washington, DC: Resources for the Future.

Arrhenius, S. 1896. On the Influence of Carbonic Acid in the Air upon the Temperature of the Ground. *The London, Dublin and Edinburgh Philosophical Magazine and Journal of Science* 5th series (April): 237–276.

Arrow, K. 1970. *Essays in the Theory of Risk-Bearing.* Amsterdam, the Netherlands: North-Holland.

Arrow, K., G. Daily, P. Dasgupta, S. Levin, K.-G. Mäler, E. Maskin, D. Starrett, T. Sterner, and T. Tietenberg. 2000. Managing Ecosystem Resources. *Environmental Science and Technology* 34: 1401–1406.

Arrow, K.J. 1951. *Social Choice and Individual Values.* New York, NY: John Wiley & Sons.

Arrow, K.J., and M. Kurz. 1970. *Public Investment, the Rate of Return and Optimal Fiscal Policy.* Baltimore, MD: Johns Hopkins University Press.

Atkinson, A.B., and J.E. Stiglitz. 1972. The Structure of Indirect Taxation and Economic Efficiency. *Journal of Public Economics* 1: 97–119.

————. 1980. *Lectures on Public Economics.* New York, NY: McGraw-Hill.

Aylward, B., J. Bishop, E.B. Barbier, and J.C. Burguess. 1994. *The Economics of the Tropical Timber Trade.* London, U.K.: Earthscan Publishers.

Azar, C., J. Holmberg, and K. Lindgren. 1995. Stability Analysis of Harvesting in a Predator–Prey Model. *Journal of Theoretical Biology* 174: 13–19.

————. 1996. Socio-Ecological Indicators for Sustainability. *Ecological Economics* 18: 89–112.

Azar, C., and S. Schneider. 2001. Are Uncertainties in Climate and Energy Systems a Justification for Stronger Near Term Mitigation Policies? Draft Paper. October. Washington, DC: Pew Center on Climate Change.

Baca, K.A. 1993. Property Rights in Outer Space. *Journal of Air Law and Commerce* 58(4): 1041.

Baines, G. 1982. Coastal Resource Use and Management in Asia and the Pacific. In *Man, Land, and Sea: Coastal Resource Use and Management in Asia and the Pacific,* edited by C. Soysa, C.L. Sien, and W.L. Collier. Bangkok, Thailand: The Agricultural Development Council, 189–198.

Balland, J.-M., and J.P. Plateau. 1996. *Halting Degradation of Natural Resources: Is There a Role for Rural Communities?* Oxford, U.K.: Clarendon Press.

Baltagi, B.H., and J.M. Griffin. 1983. Gasoline Demand in the OECD: An Application of Pooling and Testing Procedures. *European Economic Review* 22(2): 117–137.

Barbier, E.B., and J.C. Burgess. 1997. The Economics of Tropical Forest Land Use. *Land Economics* 73(2): 174–196.

Barrett, S. 1997. Towards a Theory of International Environmental Cooperation. In *New Directions in the Economic Theory of the Environment,* edited by C. Carraro and D. Siniscalco. Cambridge, New York, and Melbourne: Cambridge University Press, 239–80.

————. 2000. Trade and Environment: Local versus Multilateral Reforms. *Environment and Development Economics* 5(4): 349–359.

————. Forthcoming. *Environment and Statecraft: The Strategy of Environmental Treaty-Making.* Oxford, U.K.: Oxford University Press.

Batkin, K.M. 1996. New Zealand's Quota Management System: A Solution to the United States' Federal Fisheries Management Crisis? *Natural Resources Journal* 36(4): 855–880.

Baumol, W.J., and W.E. Oates. 1988. *The Theory of Environmental Policy.* Cambridge, U.K.: Cambridge University Press.

Becker, G. 1983. A Theory of Competition among Pressure Groups for Political Influence. *Quarterly Journal of Economics* 98: 371–400.

Becker, G.S., and K.M. Murphy. 1988. A Theory of Rational Addiction. *Journal of Political Economy* 96(4): 675–700.

Bečvář, J., and Kokine, M. (eds.). 1998. *Role of Economic Instruments in Integrating Environmental Policy with Sectoral Policies.* Proceedings of the workshop organised by the UN/ECE and the Organisation for Economic Co-operation and Development, Czech Republic, Oct. 8–10, 1997. New York, NY: United Nations.

Bemelmans-Videc, M.L., R.C. Rist, and E. Vedung (eds.). 1998. *Carrots, Sticks, and Sermons: Policy Instruments and Their Evaluation.* New Brunswick, NJ: Transaction.

Bennulf, M., A. Biel, N. Fransson, and M. Polk. 1998. Bilismen och Miljön: Attityder och Attitydbildning. KFB Rapport 1998:4. Stockholm, Sweden: KFB (The Swedish Transport and Communications Research Board).

Bergland, H., and P.A. Pedersen. 1997. Catch Regulation and Accident Risk: The Moral Hazard of Fisheries' Management. *Marine Resource Economics* 12(4): 281–292.

Berkes, F. 1977. Fishery Resource Use in a Subarctic Indian Community. *Human Ecology* 5: 289–307.

———. 1985. Fishermen and the Tragedy of the Commons. *Environmental Conservation* 12(3): 199–206.

———. 1986. Local-Level Management and the Commons Problem: A Comparative Study of Turkish Coastal Fisheries. *Marine Policy* 10: 215–229.

Berkes, F., and A. Shaw. 1986. Ecologically Sustainable Development: A Caribbean Fisheries Case Study. *Canadian Journal of Development Studies* 7: 175–196.

Berry, R.A., and W.R. Cline. 1979. *Agrarian Structure and Productivity in Developing Countries.* Geneva, Switzerland: International Labour Organisation.

Beukering, P. 2001. *International Trade, Recycling, and the Environment: An Empirical Analysis.* Amsterdam: Kluwer Academic Publishers.

Beukering, P., M. Sehker, R. Gerlagh, and V. Kumar. 1999. Analysing Urban Solid Waste in Developing Countries: a Perspective on Bangalore, India. CREED Working Paper No. 24. London: CREED/ IIED, International Institute for Environment and Development.

BGS (British Geological Survey). 2001. News Release: Bangladesh Claims against the British Geological Survey. Oct. 3. www.bgs.ac.uk/scripts/news/view_news.cfm?id=116 (accessed March 2002).

Bhattarai, R. 2000. Economics of Municipal Solid Waste Management and Health: A Case of Kathmandu Metropolitan City. Paper presented at the Research Meeting and Training Workshop, Dhulikhel Mountain Resort, Dhulikhel, Nepal, September 21–24, 2000.

Binswanger, H.P. 1991. Brazilian Policies that Encourage Deforestation in the Amazon. *World Development* 19(7): 821–829.

Binswanger, H.P., and K. Deininger. 1997. Explaining Agricultural and Agrarian Policies in Developing Countries. *Journal of Economic Literature* 35: 1958–2005.

Binswanger, H.P., K. Deininger, and G. Feder. 1995. Power, Distortions, Revolt, and Reform in Agricultural Land Relations. In *Handbook of Development Economics,* vol. 3B, edited by J. Behrman and T.N. Srinivasan. Amsterdam, the Netherlands: North-Holland.

Binswanger, H.P., and S.T. Holden. 1998. Small-Farmer Decisionmaking, Market Imperfections, and Natural Resource Management in Developing Countries. In *Agriculture and the Environment,* edited by E. Luntz. Washington, DC: World Bank, 50–71.

Binswanger, H.P., and M.R. Rosenzweig. 1986. Behavioral and Material Determinants of Production Relations in Agriculture. *Journal of Development Studies* 22(3): 503–539.

Bizer, K., and R. Jülich. 1999. Voluntary Agreements: Trick or Treat? *European Environment* 9(2): 59–66.

Bjoerner, T.B., and H.H. Jensen. 2000. Voluntary Agreements, Investment Subsidies or Taxes? An Econometric Evaluation of Their Effect on Energy Use in Danish Industry. Paper presented at the 10th Annual Conference of the European Association of Environmental and Resources Economists (EAERE). Rethymnon, Crete.

Blackman, A., and G.J. Bannister. 1998. Community Pressure and Clean Technology in the Informal Sector: An Econometric Analysis of the Adoption of Propane by Traditional Mexican Brickmakers. *Journal of Environmental Economics and Management* 35: 1–21.

Blackman, A., and W. Harrington. 2000. The Use of Economic Incentives in Developing Countries: Lessons from International Experience with Industrial Air Pollution. *Journal of Environment and Development* 9(1): 5–44.

Bluffstone, R. 1999. Do Pollution Charges Reduce Pollution in Lithuania? In *The Market and the Environment: The Effectiveness of Market-Based Policy Instruments for Environmental Reform*, edited by T. Sterner. Cheltenham, U.K.: Edward Elgar.

Bluffstone, R., and B.A. Larson 1997. *Controlling Pollution in Transition Economies*. Cheltenham, U.K.: Edward Elgar.

Bluffstone, R.A., and T. Panayotou. 2000. Environmental Liability and Privatisation in Central and Eastern Europe. *Environmental and Resource Economics* 17(4): 335–352.

BMF (Bundeministerium der Finanzen). 2002. Gesetzes- und Verordnungstexte zum Mineralöl- und Stromsteuerrecht (in German). www.bundesfinanzministerium.de/Gesetzes-und-Verordnungstexte-.733.htm (accessed March 2002).

Bohm, P. 1981. *Deposit–Refund Systems: Theory and Applications to Environmental, Conservation, and Consumer Policy*. Baltimore, MD: Johns Hopkins University Press for Resources for the Future.

———. 1999. An Emission Quota Trade Experiment among Four Nordic Countries. In *Pollution for Sale: Emissions Trading and Joint Implementation*, edited by S. Sorrell and J. Skea. Cheltenham, U.K.: Edward Elgar.

———. 2000. International Greenhouse Gas Emission Trading—with Special Reference to the Kyoto Protocol. In *Efficiency and Equity of Climate Change*, edited by C. Carraro. Dordrecht, the Netherlands: Kluwer Academic.

Bohm, P., and C.S. Russell. 1985. Comparative Analysis of Alternative Policy Instruments. In *Handbook of Natural Resource and Energy Economics*, vol. 1. Handbooks in Economics Series, no. 6. Amsterdam, the Netherlands: North-Holland, 395–460.

———. 1997. Comparative Analysis of Alternative Policy Instruments. In *The Economics of Environmental Protection: Theory and Demand Revelation*. Cheltenham, U.K.: Edward Elgar, 48–113.

Bokobo Moiche, S. 2000. *Gravámenes e Incentives Fiscales Ambientales*. Madrid, Spain: Ediciones Civitas.

Bonato, D., and A. Schmutzler. 2000. When Do Firms Benefit from Environmental Regulations? A Simple Macroeconomic Approach to the Porter Controversy. *Swiss Journal of Economics and Statistics* 136(4): 513–530.

Bond, I. 1993. The Economics of Wildlife and Land-Use in Zimbabwe: An Examination of Current Knowledge and Issues. Project Paper no. 36. Harare, Zimbabwe: World Wildlife Fund.

Bonilla-Chacin, M., and J.S. Hammer. 1999. Life and Death among the Poorest. Paper presented at the World Bank Economists' Forum. May 3–4, Washington, DC.

Borenstein, S. 1993. Price Incentives for Fuel Switching: Did Price Differences Slow the Phase-Out of Leaded Gasoline? Working Paper Series, no. 93-8. Davis, CA: University of California–Davis, Department of Economics.

Boserup, E. 1965. *The Conditions of Agricultural Growth: The Economics of Agrarian Change under Population Pressure*. London, U.K.: Earthscan Publications.

Bovenberg, A.L., and L.H. Goulder. 1996. Optimal Environmental Taxation in the Presence of Other Taxes: General-Equilibrium Analyses. *American Economic Review* 86(4): 985–1000.

Boyd, J., A.J. Krupnick, and J. Mazurek. 1998. Intel's XL Permit: A Framework for Evaluation. Discussion Paper 98/11. Washington, DC: Resources for the Future.

Boyd, R., and W. Hyde. 1989. *Forestry Sector Intervention: The Impacts of Public Regulation on Social Welfare*. Ames, IA: Iowa State University Press, chapters 2 and 3.

Boyer, M., and J.-J. Laffont. 1996. *Toward a Political Theory of Environmental Policy*. Fondazione Eni Enrico Mattei (FEEM) Note di Lavoro 56/96. Milan, Italy: FEEM.

———. 1997. Environmental Risks and Bank Liability. *European Economic Review* 41(8): 1427–1459.

Braatz, S. 1992. Conserving Biological Diversity, A Strategy for Protected Areas in the Asia-Pacific Region. Technical Paper no. 193. Washington, DC: International Bank for Reconstruction and Development/World Bank.

Brännlund, R., L. Hetemäki, B. Kriström, and E. Romstad. 1996. Command and Control with a Gentle Hand: The Nordic Experience. Research Report 115. Umeå, Sweden: Swedish University of Agricultural Sciences, Department of Forest Economics.

Breen, P.A., and T.H. Kendrick. 1997. A Fisheries Management Success Story: The Gisborne, New Zealand, Fishery for Red Rock Lobsters (Jasus edwardsii). *Marine and Freshwater Research* 48(8): 1103–1110.

Brill, E., E. Hochman, and D. Zilberman. 1997. Allocation and Pricing at the Water District Level. *American Journal of Agricultural Economics* 79(3): 952–963.

Bromley, D.W. 1991. *Environment and Economy: Property Rights and Public Policy.* Oxford, U.K.: Blackwell.

Brown, G. 2000. Renewable Natural Resources Management and Use without Markets. *Journal of Economic Literature* 38: 875–914.

Brown, G., and D. Layton. 1996. Resistance Economics: Social Cost and Evolution of Antibiotics Resistance. *Environment and Development Economics* 3(2): 349–355.

Brown, G., and J. Roughgarden. 1997. A Metapopulation Model with Private Property and a Common Pool. *Ecological Economics* 22: 65–71.

Brunekreef, B. 1986. Childhood Exposure to Environmental Lead. Monitoring and Assessment Research Centre Report no. 34. London, U.K.: University of London, King's College.

Bryant, D., D. Nielson, and L. Tangley. 1997. *The Last Frontier Forests: Ecosystems and Economies on the Edge.* Washington, DC: World Resources Institute.

Buchanan, J. 1965. An Economic Theory of Clubs. *Economica* 32: 1–14.

Buchanan, J., and G. Tullock. 1962. *The Calculus of Consent: Logical Foundations of Constitutional Democracy.* Ann Arbor, MI: University of Michigan Press.

Bundesamt für Güterverkehr. 2002. Bundesamt für Güterverkehr home page (in German). www.bag.bund.de/ (accessed March 2002).

Burgess, J.C., and E.B. Barbier. 1999. Tenure Insecurity and Tropical Deforestation. Working Paper. York, U.K.: University of York.

Burtraw, D. 1998a. Appraisal of the $SO_2$ Cap-and-Trade Market. Paper presented at the University of Illinois at Chicago Workshop on Market-Based Approaches to Environmental Policy. June 19, Federal Reserve Bank of Chicago, Chicago, IL.

———. 1998b. Cost Savings, Market Performance, and Economic Benefits of the U.S. Acid Rain Program. Discussion Paper 98-28-REV. Washington, DC: Resources for the Future.

Burtraw, D., and E. Mansur. 1999. The Environmental Effects of $SO_2$ Trading and Banking. *Environmental Science and Technology* 33(20): 3489–3494.

Button, K.J., and E.T. Verhoef (eds.). 1997. *Road Pricing, Traffic Congestion and the Environment: Issues of Efficiency and Social Feasibility.* Aldershot, Hampshire, U.K.: Edward Elgar.

Byström, O., and D.W. Bromley. 1998. Contracting for Nonpoint-Source Pollution Abatement. *Journal of Agricultural and Resource Economics* 23(1): 39–54.

Calthrop, E., and S. Proost. 1998. Road Transport Externalities: Interaction between Theory and Empirical Research. *Environmental and Resource Economics* 11(3-4): 335–348.

Calthrop, E., S. Proost, and K. Van Dender. 2000. Optimal Urban Road Tolls in the Presence of Distortionary Taxes. Paper presented at the 10th Annual Conference of the European Association of Environmental and Resources Economists (EAERE). Rethymnon, Crete.

Canard, N.F. 1801. *Principes d'Economie Politiques.* Paris, France: Chez F. Boisson.

Carlson, C., D. Burtraw, M. Cropper, and K. Palmer. 2000. $SO_2$ Control by Electric Utilities: What Are the Gains from Trade? *Journal of Political Economy* 108(6): 1292–1326.

Carlsson, F. 1999. *Essays on Externalities and Transport.* Gothenburg, Sweden: University of Gothenburg, Department of Economics.

———. 2000. Environmental Taxation and Strategic Commitment in Duopoly Models. *Environmental and Resource Economics* 15: 24–56.

Carlsson, F., and S. Lundström. 2000. Political and Economic Freedom and the Environment: The Case of $CO_2$ Emissions. Working Paper in Economics no. 29. Gothenburg, Sweden: University of Gothenburg, Department of Economics.

Carraro, C. 1987. Policy Instruments and Coalitions in International Games. *Ricerche Economiche* 41(3-4): 293–314.

Carraro, C., and F. Leveque. 1999. *Voluntary Approaches in Environmental Policy.* Fondazione Eni Enrico Mattei Series on Economics, Energy and Environment, vol. 14. London, U.K.: Kluwer Academic.

Carraro, C., and D. Siniscalco. 1997. *International Environmental Agreements: Incentives and Political Economy.* Nota di Lavoro 96/97. Milan, Italy: Fondazione Eni Enrico Mattei.

Carson, R.L. 1950. *The Sea Around Us.* New York, NY: Mentor Books, Oxford University Press.

———. 1962. *Silent Spring.* Greenwich, CT: Crest Books, Faucett Publications.

Casey, K., C. Dewees, B. Turis, and J. Wilen. 1995. The Effects of Individual Vessel Quotas in the British Columbia Fishery. *Marine Resource Economics* 10: 211–230.

Cason, T. 1995. An Experimental Investigation of the Seller Incentives in the EPA's Emission Trading Auction. *American Economic Review* 85(1, September): 905–922.

CAVA (Concerted Action on Voluntary Agreements). 2002. The European research network on voluntary approaches, funded in the framework of the Environment and Climate programme of DG XII of the European Commission. Coordinated by CERNA. http://www.cerna.ensmp.fr/Progeuropeens/CAVA/Index.html (accessed September 2002).

CDC (Centers for Disease Control). 1991. *Strategic Plan for the Elimination of Childhood Lead Poisoning.* Washington, DC: U.S. Department of Health and Human Services.

Cheung, S.N.S. 1969. *The Theory of Share Tendency.* Chicago, IL: Chicago University Press.

Child, B., S. Ward, and T. Tavengwa. 1997. *Zimbabwe's CAMPFIRE Programme: Natural Resource Management by the People.* World Conservation Union–Regional Office for Southern Africa (IUCN-ROSA) Environmental Issues Series, no. 2. Harare, Zimbabwe: IUCN-ROSA.

Chomitz, K.M. 2000. Personal communication to the author by Kenneth M. Chomitz, Development Research Group, World Bank. November 1.

Chomitz, K.M., and C. Griffiths. 2001. An Economic Analysis and Simulation of Woodfuel Management in the Sahel. *Environmental and Resource Economics* 19(3): 285–304.

Chopra, K. 1991. Participatory Institutions: The Context of Common and Private Property Resources. *Environmental and Resource Economics* 1(4): 353–372.

———. 1998. Environmental Degradation and Population Movements: Hypotheses and Evidence from Rajasthan (India). *Environment and Development Economics* 3(1): 35–57.

Christy, F.T., Jr. 1973. Fisherman Quotas: A Tentative Suggestion for Domestic Management. Occasional Paper no. 19. Honolulu, Hawaii: Law of Sea Institute.

Chua, S., and P. Fredriksson. 1998. The Impact of North American Eco-Labelling on Developing Country Trade. Memo. August. Washington, DC: World Bank, Environment Department.

City of New York. 2002. Department of Environmental Protection, New York City's Water Supply System. http://NYC.gov/html/dep/html/fadplan.html (accessed September 2002).

Clark, C.W. 1990. *Mathematical Bioeconomics: The Optimal Management of Renewable Resources,* 2nd edition. New York, NY: John Wiley & Sons.

Clark, J., and D.H. Cole (eds.). 1998. *Environmental Protection in Transition: Economic, Legal and Socio-Political Perspectives on Poland.* Aldershot, Hampshire, U.K.: Ashgate.

Clarke, F.H., and G.R. Munro. 1987. Coastal States, Distant Water Fishing Nations, and Extended Jurisdiction: A Principal-Agent Analysis. *Natural Resource Modeling* 2: 81–107.

Clarke, T. 2001. Bangladeshis To Sue over Arsenic Poisoning: Up to 75 Million at Risk from Tainted Water. *Nature,* October 11. www.nature.com/nsu/011011/011011-14.html (accessed April 2002).

Cleaver, K., and G. Schreiber. 1994. *Reversing the Spiral: the Population, Agriculture and Environment Nexus in Sub-Saharan Africa.* Washington, DC: World Bank.

Climate Strategies. 2002. Climate Strategies Network home page. www.climate-strategies.org/ (accessed March 2002).

Coase, R. 1960. The Problem of Social Cost. *Journal of Law and Economics* 3: 1–44.

Cohen, F.S. 1954. Dialogue on Private Property. *Rutgers Law Review* IX: 357–387.

Cohen, Y. 1987. A Review of Harvest Theory and Applications of Optimal Control Theory in Fisheries Management. *Canadian Journal of Fish Aquatic Science* 44(2): 75–83.

Cole, J.J., G. Lovett, and S.G. Findlay. 1991. *Comparative Analysis of Ecosystems.* Berlin, Germany: Springer-Verlag.

Collinge, R.A., and W.E. Oates. 1980. Efficiency in Pollution Control in the Short and the Long Runs: A System of Rental Emission Permits. *Canadian Journal of Economics* 15(2): 346–354.

Comisión Federal de Electricidad. 1994. *Tarifas.* Mexico City, Mexico: Comisión Federal de Electricidad.

Conrad, J.M. 1999. *Resource Economics.* Cambridge, U.K.: Cambridge University Press.

Conrad, J.M., and R. Adu-Asamoah. 1986. Single and Multispecies Systems: The Case of Tuna in the Eastern Tropical Atlantic. *Journal of Environmental Economics and Management* 3: 50–68.

Conradie, B., J. Goldin, A. Leiman, B. Standish, and M. Visser. 2001. Competition Policy and Privatisation in the South African Water Industry. Development Policy Research Unit (DPRU) Working Paper 01/48. Rondebosch, South Africa: University of Cape Town.

Convery, F. 2001. Emissions Trading and Environmental Policy in Europe. Paper presented at the pre-summit conference, Knowledge and Learning for a Sustainable Society, Climate and Global Justice Session. June 12–14, University of Gothenburg, Gothenburg, Sweden.

Convery, F., and R. Katz. 2001. Air Emissions Trading in Chile (Santiago Metropolitan Region), Case Study in the Design and Use of Market-Based Instruments for Environmental Policy. Draft 4. Dublin, Ireland: University College.

Copes, P. 1986. A Critical Review of the Individual Quota as a Device in Fisheries Management. *Land Economics* 62(3): 278–291.

Cordell, J. 1978. Carrying Capacity Analysis of Fixed-Territorial Fishing. *Ethnology* 17: 1–24.

Cordell, J.C. 1984. Defending Customary Inshore Sea Rights. In *Maritime Institutions in the Western Pacific,* edited by K. Ruddle and T. Akimichi. Senri Ethnological Studies no. 17. Osaka, Japan: National Museum of Ethnography, 301–326.

CORNARE (Corporación Autonoma Regional del Rionegro-Nare). 2001. Tasas redistributivas: Informe de avance de la meta regional, Noveno Semestre, (Abril–Septiembre 2001). El Santuario, Colombia. http://www.cornare.gov.co/tasar.htm (accessed September 2002).

Coronado, H. 2001. Determinantes del Desempeño y la Inversión Ambiental en la Industria. Master's thesis in environmental economics. Bogotá, Colombia: Universidad de los Andes.

Court of Justice. 1998. *Excise Duty on Electricity.* C-213/96. April 2. Kirschberg, Luxembourg: Court of Justice of the European Community.

CPTEC (Centro de Previsão de Tempo e Estudos Climáticos). 2002. Queimadas no Brasil ("Vegetation Fires in Brazil"). www.cptec.inpe.br/products/queimadas/queimap.html (accessed April 2002).

Crocker, T.D., and H. Wolozin. 1966. The Structuring of Atmospheric Pollution Control Systems. In *The Economics of Air Pollution,* edited by H. Wolozin. New York, NY: W.W. Norton, 61–86.

Dahl, C., and T. Sterner. 1991a. A Survey of Econometric Gasoline Demand Elasticities. *International Journal of Energy Systems* 11(2): 53–76.

———. 1991b. Analysing Gasoline Demand Elasticities: A Survey. *Energy Economics* 13(3): 203–210.

Dahlberg, M., and E. Johansson. 2000. On the Vote Purchasing Behaviour of Incumbent Governments. Working Paper. Uppsala, Sweden: University of Uppsala, Department of Economics.

Dales, J.H. 1968a. Land, Water and Ownership. *Canadian Journal of Economics* 1: 791–804.

———. 1968b. *Pollution, Property and Prices.* Toronto, Canada: University of Toronto Press.

Dasgupta, P. 1982. *The Control of Resources.* Oxford, U.K.: Basil Blackwell.

———. 1993. *An Inquiry into Well-Being and Destitution.* Oxford, U.K.: Oxford University Press.

———. 2000. Population and Resources: An Exploration of Reproductive and Environmental Externalities. *Population and Development Review* 26(4): 643–689.

Dasgupta, P., P. Hammond, and E. Maskin. 1979. The Implementation of Social Choice Rules: Some General Results in Incentive Compatibility. *Review of Economic Studies* 46: 185–216.

———. 1980. On Imperfect Information and Optimal Pollution Control. *Review of Economic Studies* 52: 173–192.

Dasgupta, P., and G.M. Heal. 1979. *Economic Theory and Exhaustible Resources.* Welwyn, U.K.: James Nisbet and Cambridge University Press.

Dasgupta, P., and K.-G. Mäler. 1995. Poverty, Institutions and the Environmental Resource Base. In *Handbook of Development Economics,* vol. 3, edited by J. Behrman and T.N. Srinivasan. Amsterdam, the Netherlands: North-Holland.

———. 2000. Net National Product, Wealth, and Social Well-Being. *Environment and Development Economics* 5(1–2): 69–93.

Dasgupta, P., S. Marglin, and A. Sen. 1972. *Guidelines for Project Evaluation.* New York, NY: United Nations.

Dasgupta, S., H. Hettige, and D. Wheeler. 2000. What Improves Environmental Compliance? Evidence from Mexican Industry. *Journal of Environmental Economics and Management* 39(1): 39–66.

Dasgupta, S., and D. Wheeler. 1997. Citizen Complaints as Environmental Indicators: Evidence from China. World Bank Policy Research Working Paper no. 1704. Washington, DC: World Bank.

Davis, A. 1983. Property Rights and Access Management in the Small-Boat Fishery: A Case-Study from Southwest Nova Scotia. Paper presented at the Learned Societies Conference. Vancouver, B.C., Canada.

Davis, C., and J. Mazurek. 1998. *Pollution Control in the United States: Evaluating the System.* Washington, DC: Resources for the Future.

Davis, D., and C.A. Tisdell. 1995. Recreational Scuba Diving and Carrying Capacity in Marine Protected Areas. *Ocean and Coastal Management* 26(1): 19–40.

———. 1996. Valuing Outdoor Recreational Sites: A New Approach. Discussion Paper. Brisbane, Queensland, Australia: University of Queensland, Department of Economics.

Davis, L., and N. Johnson. 1987. *Forest Management.* New York, NY: McGraw-Hill.

Deacon, R. 1996. Economic Aspects of Forest Policy in Lithuania. Environment Discussion Paper 1996:13. Cambridge, MA: Harvard Institute for International Development.

Deacon, R., and P. Murphy. 1997. The Structure of an Environmental Transaction: The Debt-for-Nature Swap. *Land Economics* 73(1): 1–24.

De Borger, B., I. Mayeres, S. Proost, and S. Wouters. 1996. Optimal Pricing of Urban Transport: A Simulation Exercise for Belgium. *Journal of Transport Economics and Policy* 30: 31–54.

Debreu, G. 1951. The Coefficient of Resource Utilisation. *Econometrica* 19: 273–292.

Deininger, K., and H. Binswanger. 1999. The Evolution of the World Bank's Land Policy: Principles, Experience, and Future Challenges. *World Bank Research Observer* 14(2): 247–276.

Deininger, K., and B. Minten. 1996. Poverty, Policies and Deforestation: The Case of Mexico. Working Paper no. 5. Washington, DC: World Bank, Policy Research Department.

de Mattos, L., and P. Willquist. 1999. Savings Using a GPS Based Logistics System in a Truck Fleet. Master's thesis 99:03. Chalmers University of Technology, Gothenburg, Sweden.

DEPA (Danish Environmental Protection Agency). 1998. UN/ECE Task Force to Phase Out Leaded Petrol in Europe: Main Report. Copenhagen, Denmark: DEPA.

Deutscher Bundestag. 1998. Gesetzentwurf der Fraktionen SPD und Bündnis 90/die Grünen. Drucksache 14/40, Nov. 17. Bonn, Germany: Deutscher Bundestag.

———. 1999. *Gesetz zum Einstieg in die ökologische Steuerreform.* Bonn, Germany: Bürgerliches Gesetzbach, 378.

Diamond, P.A., and J.A. Mirrlees. 1971a. Optimal Taxation and Public Production 1: Production Efficiency. *American Economic Review* 61: 8–27.

———. 1971b. Optimal Taxation and Public Production 2: Tax Rules. *American Economic Review* 61: 261–278.

DieselNet. 2002. Emissions Standards: European Union—Heavy-Duty Diesel Truck and Bus Engines. www.dieselnet.com/standards/eu/hd.html (accessed April 2002).

Dijkstra, B.R. 1999. *The Political Economy of Environmental Policy: A Public Choice Approach to Market Instruments.* Cheltenham, U.K.: Edward Elgar.

Dinda, S., D. Coondoo, and M. Pal. 2000. Air Quality and Economic Growth: An Empirical Study. *Ecological Economics* 34: 409–423.

Dixon, J.A., and C.W. Howe. 1993. Inefficiencies in Water Project Design and Operation in the Third World: An Economic Perspective. *Water Resources Research* 29(7): 1889–1894.

Dixon, J.A., and P.B. Sherman. 1990. *Economics of Protected Areas: A New Look at Benefits and Costs.* Washington, DC: Island Press, East West Center.

Dobkins, B.E. 1959. *The Spanish Element in Texas Water Law.* Austin, TX: University of Texas.

Dosi, C., and T. Tomasi (eds.). 1994. *Nonpoint Source Pollution Regulation: Issues and Analysis.* Dordrecht, the Netherlands: Kluwer Academic.

Durrenberger, E.P., and G. Palsson. 1987. Ownership at Sea: Fishing Territories and Access to Sea Resources. *American Ethnologist* 14: 508–522.

Ebert, U., and O. von dem Hagen. 1998. Pigouvian Taxes under Imperfect Competition if Consumption Depends on Emissions. *Environmental and Resource Economics* 12(4): 507–513.

EEA (European Environment Agency). 2000. *Environmental Taxes: Recent Developments in Tools for Integration.* Copenhagen, Denmark: EEA.

Eggert, H. 1998. Bioeconomic Analysis and Management: The Case of Fisheries. *Environmental and Resource Economics* 11(3-4): 399–411.

Ehrlich, P., and J. Holdren. 1971. Impact of Population Growth. *Science* 171: 1212–1217.

Ekbom, A., P. Knutsson, and M. Ovuka. 2001. Is Sustainable Development Based on Agriculture Attainable in Kenya? A Multi-disciplinary Approach. *Land Degradation and Development* 12(5): 435–447.

Ekins, P. 1999. European Environmental Taxes and Charges: Recent Experience, Issues, and Trends. *Ecological Economics* 31(1): 39–62.

Ellerman, D., and J.-P. Montero. 1998. The Declining Trend in Sulfur Dioxide Emissions: Implications for Allowance Prices. *Journal of Environmental Economics and Management* 36(1): 26–45.

Eltony, M.N. 1996. Demand for Gasoline in the GCC: An Application of Pooling and Testing Procedures. *Energy Economics* 18: 203–209.

*Environment and Development Economics.* 1997. Special Issue: Environmental Kuznets Curve. *Environment and Development Economics* 2(4).

Ercmann, S. 1996. *Pollution Control in the European Community: Guide to the EC Texts and Their Implementation by Member States.* London, U.K.: Kluwer Law International.

Eskeland, G. 1994. A Presumptive Pigovian Tax: Complementing Regulation To Mimic an Emissions Fee. *World Bank Economic Review* 8(3): 373–394.

Eskeland, G., and S. Devarajan. 1996. *Taxing Bads by Taxing Goods: Pollution Control with Presumptive Charges.* Washington, DC: International Bank for Reconstruction and Development/World Bank.

Eskeland, G., and T.N. Feyzioglu. 1997a. Is Demand for Polluting Goods Manageable? An Econometric Study of Car Ownership and Use in Mexico. *Journal of Development Economics* 53: 423–445.

———. 1997b. Rationing Can Backfire: The "Day without a Car" in Mexico City. *World Bank Economic Review* 11(3): 383–408.

Eskeland, G., and E. Jimenez. 1992. Policy Instruments for Pollution Control in Developing Countries. *World Bank Research Observer* 7(2): 145–169.

ETTM (Electronic Toll Collection and Traffic Management). 2002. U.S. Toll Facilities. www.ettm.com/usafac.html (accessed March 2002).

European Commission. 1995. Towards Fair and Efficient Pricing in Transport: Policy Options for Internalising the External Cost of Transport in the European Union. December. Green Paper COM(95)691. europa.eu.int/en/record/green/gp003en.pdf (accessed April 2002).

———. 2002. European Climate Change Programme. europa.eu.int/comm/environment/climat/eccp.htm (accessed September 2002).

Faiz, A., S. Gautam, and E. Burki. 1995. Air Pollution from Motor Vehicles: Issues and Options for Latin American Countries. *Science of the Total Environment* 169: 303–310.

Faiz, A., C.S. Weaver, and M.P. Walsh. 1996. *Air Pollution from Motor Vehicles.* International Bank for Reconstruction and Development/World Bank.

Falkenmark, M. 1999. Forward to the Future: A Conceptual Framework to Water Dependence [Volvo Environment Prize Lecture 1998]. *Ambio* 2: 356–361.

FAO (Food and Agriculture Organization). 1983. *Marine Fisheries and the Law of the Sea: A Decade of Change.* Fisheries Circular no. 853. Rome, Italy: FAO.

———. 1993. Forestry Policies of Selected Countries in Asia and the Pacific. FAO Forestry Paper. Rome, Italy: FAO.

———. 2000. *Small Ponds Make a Big Difference: Integrating Fish with Crop and Livestock Farming.* Rome, Italy: FAO, Farm Management and Production Economics Service.

———. 2002. About Fisheries: Production. http://www.fao.org/fi/prodn.asp (accessed April 2002).

Farrow, S. 1995. The Dual Political Economy of Taxes and Tradable Permits: Applications in Central and Eastern Europe. *Economics Letters* 49(2): 217–220.

———. 1999. The Duality of Taxes and Tradable Permits: A Survey with Applications in Central and Eastern Europe. *Environment and Development Economics* 4(4): 519–535.

Faustman, M. 1849. Berechnung des Werthes, welchem Waldboden sowie noch nicht haubare Holzbestände für die Waldwirtschaft besitzen. *Allgemeine Forst und Jagt-Zeitung* 25: 441–455.

Fay, M.B. 1992. *Maziwi Island of Pangani (Tanzania): History of Its Destruction and Possible Causes.* Regional Report and Studies no. 139. Nairobi, Kenya: United Nations Environment Programme.

Feldstein, M.S. 1976. On the Theory of Tax Reform. *Journal of Public Economics* 6: 77–104.

Fischer, C. 2000a. Climate Change Policy Choices and Technical Innovation. Climate Issue Brief no. 20. June. Washington, DC: Resources for the Future.

———. 2000b. Multinational Taxation and International Emissions Trading. Paper presented at the 10th Annual Conference of the European Association of Environmental and Resources Economists (EAERE). Rethymnon, Crete.

———. 2001. Rebating Environmental Policy Revenues: Output-Based Allocations and Tradable Performance Standards. Working Paper no. 01–22. Washington, DC: Resources for the Future.

Fischer, C., S. Kerr, and M. Toman. 1998. Using Emissions Trading to Regulate U.S. Greenhouse Gas Emissions: An Overview of Policy Design and Implementation Issues. *National Tax Journal* 51(3): 453–464.

Fischer, C., and M. Toman. 1998. Environmentally and Economically Damaging Subsidies: Concepts and Illustrations. Climate Issue Brief no. 14. Washington, DC: Resources for the Future.

Fischer, M.J. 1996. Union Carbide's Bhopal Incident: A Retrospective. *Journal of Risk and Uncertainty* 12: 257–269.

Fisher, A.D. 1981. *Resource and Environmental Economics.* New York, NY: Cambridge University Press.

Flaaten, O. 1988. *The Economics of Multispecies Harvesting: Theory and Application to the Barents Sea Fisheries.* Berlin, Germany: Springer-Verlag.

Flaaten, O., K. Heen, and K.G. Salvanes. 1995. The Invisible Resource Rent in Limited Entry and Quota Managed Fisheries: The Case of Norwegian Purse Seine Fisheries. *Marine Resource Economics* 10(4): 341–356.

Florens, J.-P., and C. Foucher. 1999. Pollution Monitoring: Optimal Design of Inspection: An Economic Analysis of the Use of Satellite Information to Deter Oil Pollution. *Journal of Environmental Economics and Management* 38(1): 81–96.

Folmer, H., P. Mouche, and S. Ragland. 1993. Interconnected Games and International Environmental Problems. *Environmental and Resource Economics* 3(4): 313–335.

Foster, V., and R.W. Hahn. 1995. Designing More Efficient Markets: Lessons from Los Angeles Smog Control. *Journal of Law and Economics* 38(1): 19–48.

Frank, R.H. 1985. The Demand for Unobservable and Other Nonpositional Goods. *American Economic Review* 75: 10–16.

Fredriksson, P.G. 1997. The Political Economy of Pollution Taxes in a Small Open Economy. *Journal of Environmental Economics and Management* 33(1): 44–58.

————. 1998. Environmental Policy Choice: Pollution Abatement Subsidies. *Resource and Energy Economics* 20(1): 51–63.

Freeman, A.M. 1993. *The Measurement of Environmental and Resource Values: Theory and Methods.* Washington, DC: Resources for the Future.

Frey, B.S. 1997. *Not Just for the Money: An Economic Theory of Personal Motivation.* Cheltenham, U.K.: Edward Elgar.

FSC (Forest Stewardship Council). 2001. FSC Receives Environmental Prize. *FSC News.* www.fscoax.org/principal.htm (accessed April 2002).

Fullerton, D., and T.C. Kinnaman. 1994. Curbside Recycling and Unit-Based Pricing. Working Paper 6021. Cambridge, MA: National Bureau of Economic Research.

————. 1995. Garbage, Recycling, and Illicit Burning or Dumping. *Journal of Environmental Economics and Management* 29(1): 78–91.

Fullerton, D., S.P. McDermott, and J.P. Caulkins. 1997. Sulfur Dioxide Compliance of a Regulated Utility. *Journal of Environmental Economics and Management* 34(1): 32–53.

Gajurel, D. 1999. Environment News Service. April 12. Nepal.

Gauvin, J.R., J.M. Ward, and E.E. Burgess. 1994. Description and Evaluation of the Wreckfish (*Polyprion americanus*) Fishery under Individual Transferable Quotas. *Marine Resource Economics* 9(2): 99–118.

Geen, G., and M. Nayar. 1988. Individual Transferable Quotas in the Southern Bluefin Tuna Fishery: An Economic Appraisal. *Marine Resource Economics* 5(4): 365–387.

GEMI (Global Environmental Management Initiative). 1998. Fostering Environmental Prosperity: Multinationals in Developing Countries. www.gemi.org/ (accessed April 2002).

Glasbergen, P. 1999. Tailor-Made Environmental Governance: On the Relevance of the Covenanting Process. *European Environment* 9(2): 49–58.

Gobbi, J.A. 2000. Is Biodiversity-Friendly Coffee Financially Viable? An Analysis of Five Different Coffee Production Systems in Western El Salvador. *Ecological Economics* 33: 267–281.

Goddard, H.C. 1997. Using Tradable Permits To Achieve Sustainability in the World's Large Cities: Policy Design Issues and Efficiency Conditions for Controlling Vehicle Emissions, Congestion, and Urban Decentralization with an Application to Mexico City. *Environmental and Resource Economics* 10(1): 63–99.

Goel, R., and M. Nelson. 1999. The Political Economy of Motor Fuel Taxation. *The Energy Journal* 20(1): 3–59.

Gofman, K.N. 1998. Khozyastvennyi Mekhanizm Prirodopolzovanya: Puti Perestroiki. *Ekonomika i Matematicheskie Metody* 24(3): 389–399.

Goldmann, M. 2002. Hållbara Chile (newsletter from UBV Chile, in Swedish). hem.passagen.se/mgoldmann/ (accessed April 2002).

Golmen, L.G. 1999. $CO_2$ Ocean Sequestration as a Future Method to Reduce Atmospheric $CO_2$. Paper presented at the Minisymposium on Carbon Dioxide Capture and Storage. October 22, Chalmers University of Technology, Gothenburg, Sweden.

Gomez-Ibanez, J.A. 1996. Regulating Private Toll Roads. In *Privatizing Monopolies: Lessons from the Telecommunications and Transport Sectors in Latin America,* edited by Ravi Ramamurti. Baltimore, MD: Johns Hopkins University Press, 317–331.

Gordon, H.S. 1954. The Economic Theory of a Common Property Resource: The Fishery. *Journal of Political Economy* 62: 124–142.

Gornaja, L., E. Kraav, B.A. Larson, and K. Türk 1997. Estonia's Mixed System of Pollution Permits, Standards, and Charges. In *Controlling Pollution in Transition Economies,* edited by R. Bluffstone and B.A. Larson. Cheltenham, U.K.: Edward Elgar.

Gottinger, H.W. 1998. *An Economic Approach to Monitoring Pollution Accidents.* Fondazione Eni Enrico Mattei (FEEM) Note di Lavoro 27/98. Milan, Italy: FEEM.

Goulder, L.H., I.W.H. Parry, R.C. Williams III, and D. Burtraw. 1999. The Cost-Effectiveness of Alternative Instruments for Environmental Protection in a Second-Best Setting. *Journal of Public Economics* 72(3): 329–360.

Grafton, R.Q., and R.A. Devlin. 1996. Paying for Pollution: Permits and Charges. *Scandinavian Journal of Economics* 98(2): 275–288.

Grameen Communications. 1998. Grameen—Banking for the Poor. www.grameen-info.org/ (accessed April 2002).

Green, H.M. 1997. Common Law, Property Rights, and the Environment: A Comparative Analysis of Historical Developments in the United States and England and a Model for the Future. *Cornell International Law Journal* 30(2): 541.

Greene, D.L. 1990. CAFE or Price? An Analysis of Federal Fuel Economy Regulations and Gasoline Price on New Car MPG 197-9. *The Energy Journal* 11: 37–57.

Greenpeace. 1999. Shipbreaking Is Dangerously Polluting—Greenpeace Report Finds. Press Releases. www.greenpeace.org/pressreleases/toxics/1999feb18.html (accessed March 2002).

Grossman, M.G., and E. Helpman. 1994. Protection for Sale. *American Economic Review* 84(4): 833–850.

Grossman, M.G., and A.B. Krueger. 1995. Economic Growth and the Environment. *Quarterly Journal of Economics* 5: 353–377.

Grubb, M. 2001. Review of "The European Union and Global Climate Change: A Review of Five National Programmes," by J. Gummer and R. Moreland. *Climate Policy* 1(3): 426–429.

Grubb, M., J.-C. Hourcade, and S. Oberthür. 2001. Keeping Kyoto: A Study of Approaches to Maintaining the Kyoto Protocol on Climate Change. www.climate-strategies.org/keeping-kyoto.pdf (accessed April 2002).

Guesnerie, R. 1977. On the Direction of Tax Reform. *Journal of Public Economics* 7: 179–202.

Gujarati, D.N. 1988. *Basic Econometrics,* 2nd edition. New York, NY: McGraw-Hill.

Hahn, R., and G. Hester. 1989. Marketable Permits: Lessons for Theory and Practice. *Ecology Law Quarterly* 16(2): 361–406.

Hahn, R., and R. Stavins. 1991. Incentive-Based Environmental Regulation: A New Idea from an Old Idea? *Ecology Law Quarterly* 18(1): 1–42.

Hahn, R.W. 1989. Economic Prescriptions for Environmental Problems: How the Patient Followed the Doctor's Orders. *Journal of Economic Perspectives* 3(2): 95–114.

———.1990. The Political Economy of Environmental Regulation: Towards a Unifying Framework. *Public Choice* 65: 21–47.

Hahn, R.W., and R.G. Noll. 1983. Barriers to Implementing Tradable Air Pollution Permits: Problems of Regulatory Interactions. *Yale Journal on Regulation* 1(1): 63–91.

Hakim, S., P. Seidenstat, and G.W. Bowman (eds.). 1996. *Privatizing Transportation Systems.* Privatizing Government: An Interdisciplinary Series, vol. VIII. Westport, CT: Praeger, 343.

Hammar, H., and Å. Löfgren 2001. The Determinants of Sulfur Emissions from Oil Consumption in Swedish Manufacturing Industry, 1976–1995. *The Energy Journal* 22(2): 107–126.

Hammar, H., Å. Löfgren, and T. Sterner. 2000. Political Economy Obstacles to Fuel Taxation: Using Granger Non-Causality Test to Gauge the Strength of Lobbying and Other Forces. Environmental Economics Unit, Working Paper 2002:8. Gothenburg, Sweden: University of Gothenburg.

Hanemann, W.M. 1995. Improving Environmental Policy: Are Markets the Solution? *Contemporary Economic Policy* 13(1): 74–79.

Hanley, N., and D. Oglethorpe. 1999. Emerging Policies on Externalities from Agriculture: An Analysis for the European Union. *American Journal of Agricultural Economics* 81(5): 1222–1227.

Hanley, N., J.F. Shogren, and B. White. 1997. *Environmental Economics in Theory and in Practice.* London, U.K.: Macmillan.

Hanna, S.S., C. Folke, and K.-G. Mäler (eds.). 1996. *Rights to Nature: Ecological, Economic, Cultural, and Political Principles of Institutions for the Environment.* Washington, DC: Island Press.

Hanna, S.S., and S. Jentoft. 1996. Human Use of the Natural Environment: An Overview of Social and Economic Dimensions. In *Rights to Nature: Ecological, Economic, Cultural, and Political Principles of Institutions for the Environment,* edited by S. Hanna, C. Folke, and K.-G. Mäler. Washington, DC: Island Press, 35–55.

Hannesson, R. 1998. Marine Reserves: What Would They Accomplish? *Marine Resource Economics* 13(3): 159–170.

Harford, J.D. 2000. Initial and Continuing Compliance and the Trade-Off between Monitoring and Control Cost. *Journal of Environmental Economics and Management* 40: 151–163.

Harrington, W. 1988. Enforcement Leverage When Penalties Are Restricted. *Journal of Public Economics* 37(1): 29–53.

————. 1997. Fuel Economy and Motor Vehicle Emissions. *Journal of Environmental Economics and Management* 33: 240–252.

Harrington, W., A. Krupnick, and A. Alberini. 2001. Overcoming Public Aversion to Congestion Pricing. *Transportation Research Part A: Policy and Practice* 35(2): 87–105.

Harrington, W., and V.D. McConnell. 1999. Coase and Car Repair: Who Should Be Responsible for Emissions of Vehicle Use? Discussion Paper 99-22. Washington, DC: Resources for the Future.

Harrington, W., D.R. Morgenstern, and P. Nelson. 2000. On the Accuracy of Regulatory Cost Estimates. *Journal of Policy Analysis and Management* 19(2): 297–322.

Hartman, R.S., D. Wheeler, and M. Singh. 1994. The Cost of Air Pollution Abatement. World Bank Policy Research Working Paper no. 1398. Washington, DC: World Bank.

Hartwick, J.M., and N.D. Olewiler. 1998. *The Economics of Natural Resource Use,* 2nd edition. Reading, MA: Addison-Wesley.

Hau, T. 1992. Congestion Charging Mechanisms for Roads: An Evaluation of Current Practice. Working Paper Series, no. 1071. Washington, DC: World Bank, Infrastructure and Urban Development Department.

Hayes, E.B., M.D. Elvain, H.G. Orbach, A.M. Fernandez, and S. Lyne. 1994. Long-Term Trends in Blood Lead Levels among Children in Chicago: Relationship to Air Lead Levels. *Pediatrics* 93(2).

Hayward, J.E.S. 1983. *Governing Finance: The One and Indivisible Republic,* 2nd edition. London, U.K.: Weidenfeld and Nicolson.

Hazell, P., C. Pomareda, and A. Valdes (eds.). 1986. *Crop Insurance for Agricultural Development: Issues and Experience.* Baltimore, MD: Johns Hopkins University Press.

Heath, J., and H.P. Binswanger. 1996. Natural Resource Degradation Effects of Poverty and Population Growth Are Largely Policy Induced: The Case of Colombia. *Environment and Development Economics* 1(1): 65–83.

Heroc, C.A., and S. Krall. 1995. *Economics of Desert Locust Control.* Germany: Deutsche Gesellschaft für GTZ.

Hettich, W., and S.L. Winer. 1988. Economic and Political Foundations of Tax Structure. *American Economic Review* 78(4): 701–712.

Hettige, M., M. Huq, S. Pargal, and D. Wheeler. 1996. Determinants of Pollution Abatement in Developing Countries: Evidence from South and Southeast Asia. *World Development* 24(12): 1891–1904.

Heyes, A. 1996. Lender Penalty for Environmental Damage and the Equilibrium Cost of Capital. *Economica* 63(250): 311–323.

————. 1998. Making Things Stick: Enforcement and Compliance. *Oxford Review of Economic Policy* 14(4): 50–63.

————. 2001. Honesty in a Regulatory Context: Good Thing or Bad? *European Economic Review* 45(2): 215–232.

Heyes, A., and C. Liston-Heyes. 1999. Corporate Lobbying, Regulatory Conduct and the Porter Hypothesis. *Environmental and Resource Economics* 13(2): 209–218.

Hoel, M. 1991. Global Environmental Problems: The Effects of Unilateral Action Taken by One Country. *Journal of Environmental Economics and Management* 20(1): 55–70.

————.1998. Emission Taxes versus Other Environmental Policies. *Scandinavian Journal of Economics* 100(1): 79–104.

Hoel, M., and K. Schneider. 1997. Incentives to Participate in an International Environmental Agreement. *Environmental and Resource Economics* 9(2): 153–170.

Höglund, L. 2000. *Essays on Environmental Regulation with Applications to Sweden.* Gothenburg, Sweden: University of Gothenburg, Department of Economics.

Holden, S.T., B. Shiferaw, and M. Wik. 1998. Poverty, Market Imperfections, and Time Preferences: Of Relevance for Environmental Policy? *Environment and Development Economics* 3(1): 83–104.

Holman, C., J. Wade, and M. Fergusson. 1993. *Future Emissions from Cars 1990 to 2025: The Importance of the Cold Start Emission Penalty.* Godalming, Surrey, U.K.: WWF-U.K.

Homans, F., and J.E. Wilen. 1997. A Model of Regulated Open Access Resource Use. *Journal of Environmental Economics and Management* 32(1): 1–21.

Hotelling, H. 1931. The Economics of Exhaustible Resources. *Journal of Political Economy* 39: 137–175.

Houston, J.E., and H. Sun. 1999. Cost-Share Incentives and Best Management Practices in a Pilot Water Quality Program. *Journal of Agricultural and Resource Economics* 24(1): 239–252.

Howe, C.W. 1996. Water Resources Planning in a Federation of States: Equity versus Efficiency. *Natural Resources Journal* 36(1): 29–36.

Howitt, R., and C.R. Taylor. 1993. Some Microeconomics of Agricultural Resource Use. In *Agricultural and Environmental Resource Economics*, edited by G.A. Carlson, D. Zilberman, and J.A. Miranowski. Oxford, U.K.: Oxford University Press, 28–68.

Hu, H., A. Aro, M. Payton, S. Korrick, D. Sparrow, S.T. Weiss, and A. Rotnitzky. 1996. The Relationship of Bone and Blood Lead to Hypertension: Normative Aging Study. *Journal of the American Medical Association* 275(15): 1171–1176.

Huber, R.M., J. Ruitenbeek, and R. Seroa da Motta. 1997. Market-Based Instruments for Environmental Policymaking in Latin America and the Caribbean: Lessons from Eleven Countries. Discussion Paper no. 381. Washington, DC: World Bank.

Hughes, G., and K. Lvovsky. 1999. Pricing Air Pollution in the Real World. Paper presented at the World Bank Economists' Forum. May 3–4, Washington, DC.

Huhtala, A., and E. Samakovlis. 1998. *On International Harmonization of Policy Instruments to Promote Paper Recycling in Europe.* Fondazione Eni Enrico Mattei (FEEM) Note di Lavoro 79/98. Milan, Italy: FEEM.

Hurley, T.M., and J.F. Shogren. 1997. Environmental Conflicts and the SLAPP. *Journal of Environmental Economics and Management* 33(3): 253–274.

Hviding, E., and G. Baines. 1994. Community-Based Fisheries Management, Tradition, and the Challenges of Development in Marovo, Solomon Islands. *Development and Change* 25(1): 13–39.

Hyde, W. 1980. *Timber Supply, Land Allocation, and Economic Efficiency.* Baltimore, MD: Johns Hopkins University Press for Resources for the Future.

Hyde, W., and R. Sedjo. 1992. Managing Tropical Forests: Reflections on the Rent Distribution Discussion. *Land Economics* 68(3): 343–350.

Hyde, W.F., and G.S. Amacher (eds.). 2000. *Economics of Forestry and Rural Development: An Empirical Introduction from Asia.* Ann Arbor, MI: The University of Michigan Press.

ICC (International Chamber of Commerce). 1996. *Multiple Criteria-Based Third-Party Environmental Labelling Schemes.* Document no. 210/515, rev. 3. June 20. Paris, France: ICC.

Ingram, V. 1999. From Sparring Partners to Bedfellows: Joint Approaches to Environmental Policy-Making. *European Environment* 9(2): 41–48.

IRS (Internal Revenue Service). 2002. Excise Tax Statistics. www.irs.gov/prod/tax_stats/excise.html (accessed Jan. 25, 2002).

ITTO. 2002. International Tropical Timber Organization website. www.itto.or.jp/index.html (accessed April 2002).

Jaffe, A.B., S.R. Peterson, P.R. Portney, and R.N. Stavins. 1995. Environmental Regulation and Competitiveness in U.S. Manufacturing: What Does the Evidence Tell Us? *Journal of Economic Literature* 33: 132–163.

Jänicke, M. 2000. Ecological Modernization: Innovation and Diffusion of Policy and Technology. Forschungsstelle für Umweltpolitik (FFU) Report 00-08. Berlin, Germany: FFU, Freie Universität Berlin.

Jänicke, M., A. Carius, and H. Jörgens. 1997. *Nationale Umweltplane in ausgewälten Industrieländern.* Berlin, Germany: Springer.

Jänicke, M., and H. Jörgens. 2000. Strategic Environmental Planning and Uncertainty: A Cross-National Comparison of Green Plans in Industrialized Countries. *Policy Studies Journal* 28(3): 612–632.

Jansen, D., L. Bond, and B. Child. 1992. Cattle, Wildlife, Both, or Neither: A Summary of Survey Results for the Commercial Ranches in Zimbabwe. Project Paper no. 30. Harare, Zimbabwe: World Wildlife Fund.

Jenkins, R.R. 1993. *The Economics of Solid Waste Reduction: The Impact of User Fees.* Aldershot, Hampshire, U.K.: Edward Elgar.

Jin, D., and H.L. Kite-Powell. 1995. Environmental Liability. Marine Insurance and Optimal Risk Sharing Strategy for Marine Oil Transportation. *Marine Resource Economics* 4(6): 1–19.

Jodha, N.S. 1988. Poverty Debate in India: A Minority View. *Economic and Political Weekly* 23(45-46-47): 2421–2428.

———. 1992. Common Property Resources: A Missing Dimension of Development Strategies. Discussion Paper no. 169. Washington, DC: World Bank.

———. 1998. Poverty and Environmental Resource Degradation: An Alternative Explanation and Possible Solutions. *Economic and Political Weekly* 5(12): 2384–2390.

Joffe, S.R. 1995. Desert Locust Management: A Time for Change. Discussion Paper no. 284. Washington, DC: World Bank.

Johannes, R.E. 1978. Traditional Marine Conservation Methods in Oceania and Their Demise. *Annual Review of Ecological Systems* 9: 349–364.

Johansson, B., and Mattsson, L.G. (eds.). 1995. *Road Pricing: Theory, Empirical Assessment, and Policy.* Dordrecht, the Netherlands: Kluwer Academic.

Johansson, O. 1997a. Optimal Road-Pricing: Simultaneous Treatment of Time Losses, Increased Fuel Consumption, and Emissions. *Transportation Research Part D: Transport and Environment* 20(2):77–87.

———.1997b. Optimal Pigouvian Taxes with Regard to Altruism. *Land Economics* 73(3): 297–308.

———.1997c. Optimal Road Pricing with Respect to Accidents in a Second-Best Perspective. *International Journal of Transport Economics* 24(October): 343–365.

Johansson, O., and L. Schipper. 1997. Measuring the Long-Run Fuel Demand of Cars: Separate Estimations of Vehicle Stock, Mean Fuel Intensity, and Mean Annual Driving Distance. *Journal of Transport Economics and Policy* 31(3): 277–292.

Johansson, O., and T. Sterner. 1997. What Is the Scope for Environmental Road Pricing? In *Road Pricing, Traffic Congestion and the Environment: Issues of Efficiency and Social Feasibility,* edited by K. J. Button and E.T. Verhoef. Aldershot, Hampshire, U.K.: Edward Elgar.

Johansson, P., and K. Löfgren 1985. *The Economics of Forestry and Natural Resources.* Oxford, U.K.: Basil Blackwell.

Johansson, T.B., H. Kelly, A. Reddy, and R. Williams. 1992. Renewable Fuels and Electricity for a Growing World Economy: Defining and Achieving the Potential. *Energy Studies Review* 4(3): 201–212.

Johansson-Stenman, O., F. Carlsson, and D. Daruvala. 2002. Measuring Hypothetical Grandparents' Preferences for Equality and Status. *Economic Journal* 112(April): 362–383.

Johnson, S.P., and G. Corcelle. 1995. *The Environmental Policy of the European Communities,* 2nd edition. International Environmental Law & Policy Series. London, U.K.: Kluwer Law International.

Joskow, P., and R. Schmalensee. 1998. The Political Economy of Market-Based Environmental Policy: The U.S. Acid Rain Program. *Journal of Law and Economics* 41(1): 37–85.

Joskow, P.L., R. Schmalensee, and E.M. Bailey. 1998. The Market for Sulfur Dioxide Emissions. *American Economic Review* 88(4): 669–685.

Kåberger, T. 1997. A Comment on the Paper by Roger A. Sedjo. *Energy Policy* 25(6): 567–569.

Kågeson, P. 1993. *Getting the Prices Right. A European Scheme for Making Transport Pay Its True Costs.* Transport and Environment Report 93/6. Brussels, Belgium: European Federation for Transport and Environment.

Kågeson, P. and J. Dings. 1999. *Electronic Kilometre Charging for Heavy Goods Vehicles in Europe.* Brussels, Belgium: European Federation for Transport and Environment.

Kågeson, P., and A.-M. Lidmark. 1998. Konsten att Använda 5,4 Miljarder: En Kritisk Granskning av Stödet till de Lokala Investeringsprogramn för Hållbar Utveckling (The Art of Spending 5.4 Billion: A Critical Review of State Support to LIPs). Rapport 9423/98. Stockholm, Sweden: Svenska Naturskyddsföreningen.

Kahn, H., W. Brown, and L. Martel. 1976. *The Next 200 Years: A Scenario for America and the World.* New York, NY: William Morrow.

Kaimowitz, D., and A. Angelsen. 1998. *Economic Models of Tropical Deforestation: A Review.* Bogor, Indonesia: Center for International Forestry Research.

Kaimowitz, D., N. Byron, and W. Sunderlin. 1998. Public Policies to Reduce Inappropriate Tropical Deforestation. In *Agriculture and the Environment,* edited by E. Luntz. Washington, DC: World Bank, 303–323.

Kallaste, T. 1994. Economic Instruments in Estonian Environmental Policy. In *Economic Policies for Sustainable Development,* edited by T. Sterner. Dordrecht, the Netherlands: Kluwer Academic.

Kant, S. 1996. The Economic Welfare of Local Communities and Optimal Resource Regimes for Sustainable Forest Management. Ph.D. dissertation. University of Toronto, Toronto, Ontario, Canada.

Karl, H., and C. Orwat. 1999. Environmental Labelling in Europe: European and National Tasks. *European Environment* 9: 212–220.

Kathuria, V., and G.S. Haripriya. 2000. Industrial Pollution Control: Choosing the Right Option. *Economic and Political Weekly* 35(43-44): 3870–3878.

Kathuria, V., and T. Sterner. 2002. Monitoring and Enforcement: Is Two-Tier Regulation Robust? RFF Discussion Paper 02-17. Washington, DC: Resources for the Future.

Katsonias, G. 1984. The Messolonghi-Etolico Lagoon of Greece: Socio-economic and Ecological Interactions of Cooperatives and Independent Fishermen. In *Management of Coastal Lagoon Fisheries,* edited by J.M. Kapetsky and G. Lasserre. General Fisheries Commission for the Mediterranean (GFCM) Studies and Reviews no. 61. Paris, France: Food and Agriculture Organization, 521–528.

Katsoulacos, Y., and A. Xepapadeas. 1995. Environmental Policy under Oligopoly with Endogenous Market Structure. *Scandinavian Journal of Economics* 97(3): 411–420.

Katz, K., and T. Sterner. 1990. The Value of Clean Air. *Energy Studies Review* 2(1): 39–47.

Kebede, B. 2001. *Land Tenure and Common Pool Resources in Rural Ethiopia: A Study Based on Fifteen Rural Sites.* Gothenburg, Sweden: University of Gothenburg.

Keefer, P., and S. Knack. 1997. Why Don't Poor Countries Catch Up? A Cross-National Test of Institutional Explanation. *Economic Inquiry* 35(3): 590–602.

Keohane, N.O., R.L. Revesz, and R.N. Stavins. 1998. The Choice of Regulatory Instruments in Environmental Policy. *Harvard Environmental Law Review* 22(2): 313–367.

Kern, C., H. Jörgens, and M. Jänicke. 2001. *The Diffusion of Environmental Policy Innovations.* Berlin, Germany: Social Science Research Centre.

Khalid, A.R., and J.B. Braden. 1993. Welfare Effects of Environmental Regulation in an Open Economy: The Case of Malaysian Palm Oil. *Journal of Agricultural Economics* 44(1): 25–37.

Khalid, A.R., and W.M. Wan Ali. 1992. External Effects of Environmental Regulation. *The Environmentalist* 12(4): 277–285.

Kim, R., A. Rotnitzky, D. Sparrow, S.T. Weiss, C. Wager, and H. Hu. 1996. A Longitudinal Study of Low-Level Lead Exposure and Impairment of Renal Function. *Journal of the American Medical Association* 275(15): 1177–1181.

Klaasen, G. 1995. Trade-Offs in Sulphur Emission Trading in Europe. *Environmental and Resource Economics* 5: 191–219.

———. 1996. *Acid Rain and Environmental Degradation.* Cheltenham, U.K.: Edward Elgar.

Klarer, J., J. McNicholas, and E.-M. Knaus (eds.). 1999. *Sourcebook on Economic Instruments for Environmental Policy in Central and Eastern Europe.* Szentendre, Hungary: The Regional Environmental Centre for Central and Eastern Europe.

Klemperer, D. 1996. *Forest Resource Economics and Finance.* New York, NY: McGraw-Hill, chapter 9.

Kneese, A.V., and B.T. Bower. 1968. *Managing Water Quality: Economics, Technology, Institutions.* Baltimore, MD: Johns Hopkins University Press.

Kneese, A.V., and C. Schultze. 1995. Pollution, Prices, and Public Policy. In *Natural Resource Economics: Selected Papers of Allen V. Kneese.* New Horizons in Environmental Economics Series. Aldershot, Hampshire, U.K.: Edward Elgar: 442–468.

Koalitionsvereinbarung. 1998. Aufbruch und Erneuerung: Deutschlands Weg ins 21. Jahrhundert-Koalitionsvereinbarung zwischen der Sozialdemokratischen Partei Deutschlands und Bündnis 90/die Grünen. Oct. 20. Bonn, Germany.

Köhlin, G. 1998. The Value of Social Forestry in Orissa, India. Ph.D. dissertation. Department of Economics, University of Gothenburg, Sweden.

Kolstad, C.D. 2000a. Energy and Depletable Resources: Economics and Policy, 1973–1998. *Journal of Environmental Economics and Management* 39(3): 282–305.

―――. 2000b. *Environmental Economics.* New York, NY: Oxford University Press.

Komanoff, C. 1997. Environmental Consequences of Road Pricing: A Scoping Paper for the Energy Foundation. San Francisco, CA: Energy Foundation.

Köre, T. and S. Widepalm. 1993. Environmental Impact Assessment: A Case Study Concerning a Hydroelectric Power Project in Tanzania. Environmental Economics Unit Working Project Thesis Paper 1993:6. Gothenburg, Sweden: University of Gothenburg, Department of Economics.

Kosmo, M. 1987. *Money To Burn? The High Costs of Energy Subsidies.* Washington, DC: World Resources Institute.

Kraftbörsen. 2001. Svenska Kraftbörsen AB website, Marknadsstatistik. www.kraftborsen.se/ (accessed November 15, 2001).

Krawack, S. 1993. Traffic Management and Emissions. *The Science of the Total Environment* 134: 305–314.

Kriström, B. and P. Riera. 1996. Is the Income Elasticity of Environmental Improvements Less than One? *Environmental and Resource Economics* 7: 45–55.

Kroon, P. 2000. Personal e-mail to the author from Pieter Kroon, project manager, ECN Policy Studies, Energy Research Centre of the Netherlands, Petten, the Netherlands. May 4.

Krupnick, A. 1986. Costs of Alternative Policies for the Control of Nitrogen Dioxide in Baltimore. *Journal of Environmental Economics and Management* 13: 189–197.

Kuik, O.J., M.V. Nadkarni, F.H. Oosterhuis, G.S. Sastry, and A.E. Akkerman. 1997. *Pollution Control in the South and North.* Delhi, India: Sage Publications.

Kuznets, S.S. 1930. Long Swings in the Growth of Population and in Related Economic Variables. *Proceedings of the American Philosophical Society* 102(1): 25–52.

Kwerel, E. 1977. To Tell the Truth: Imperfect Information and Optimal Pollution Control. *Review of Economic Studies* 136: 595–601.

Laffont, J.-J. 1989a. Regulation, Moral Hazard and Insurance of Environmental Risks. *Journal of Public Economics* 58(3): 319–336.

―――. 1989b. A Brief Overview of the Economics of Incomplete Markets. *Economic Record* 65(188): 54–65.

―――. 1994a. The New Economics of Regulation Ten Years After. *Econometrica* 62: 507–537.

―――. 1994b. Regulation of Pollution with Asymmetric Information. In *Nonpoint Source Pollution Regulation: Issues and Analysis,* edited by C. Dosi and T. Tomasi. Dordrecht, the Netherlands: Kluwer Academic.

Land Transport Authority. 2002. Traffic.Smart website. Electronic Road Pricing charts. Singapore. traffic.smart.lta.gov.sg/erprates.htm (accessed March 2001).

Langholz, J.A., J.P. Lassoie, D. Lee, and D. Chapman. 2000. Economic Considerations of Privately Owned Parks. *Ecological Economics* 33: 173–183.

Laurikko, J., L. Erlandsson, and R. Abrahamsson. 1995. Exhaust Emission in Cold Ambient Conditions: Considerations for a European Test Procedure. SAE Technical Paper 950929. Warrendale, PA: Society of Automotive Engineers, Inc.

Lefler, W. 1985. *Petroleum Refining for the Non-Technical Person.* Tulsa, OK: PennWell.

Leksell, I., and L. Löfgren. 1995. Värdering av Lokala Luftföroreningseffekter (Valuation of the Local Effects of Air Pollution; in Swedish, with a 6-page English summary). Swedish Transport and Communication Research Board (KFB) Rapport 1995:5. Stockholm, Sweden: KFB.

Leone, R.A., and T.W. Parkinson. 1990. Conserving Energy: Is There a Better Way? Paper presented at the Association of International Automobile Manufacturers. May, Arlington, VA.

Letson, D., S. Crutchfield, and A. Malik. 1993. Point/Nonpoint Source Trading for Controlling Pollutant Loadings to Coastal Watersheds: A Feasibility Study. In *Theory, Modeling and Experience in the Management of Nonpoint Source Pollution,* edited by C. Russell and J. Shogren. London, U.K.: Kluwer Academic.

Levin, S.A. 1998. Ecosystems and the Biosphere as Complex Adaptive Systems. *Ecosystems* 1(5): 431–436.

———. 1999. *Fragile Dominion: Complexity and the Commons.* Reading, MA: Perseus Books.

Levin, S.A., and S.W. Pacala. 1997. Theories of Simplification and Scaling of Spatially Distributed Processes. In *Spatial Ecology: The Role of Space in Population Dynamics and Interspecific Interactions,* edited by D. Tilman and P. Kareiva. Princeton, NJ: Princeton University Press, 271–296.

Lewis, C. 1993. *Road Pricing: Theory and Practice.* London, U.K.: Thomas Telford.

Lewis, J.P. 1988. *Strengthening the Poor: What Have We Learned?* U.S. Third World Policy Perspectives 10. Washington, DC: Overseas Development Council.

Lewis, T. 2001. Incentives and the Design of Environmental Policy. Paper presented at the European Association of Environmental and Resources Economists (EAERE) 2001 Conference. June 28–30, Southampton, U.K.

Lindeberg, E. 1999. Future Large-Scale Use of Fossil Energy Will Require $CO_2$ Sequestering and Disposal. Paper presented at the Minisymposium on Carbon Dioxide Capture and Storage. October 22, Gothenburg, Sweden, Chalmers University of Technology and University of Gothenburg.

Lindén, O., and A. Granlund. 1998. Building Capacity for Coastal Management: Introduction. *Ambio* 27(8): 589.

Linder, S.H. 1988. Managing Support for Social Research and Development: Research Goals, Risk, and Policy Instruments. *Journal of Policy Analysis and Management* 7: 621–642.

Linder, S.H., and B.G. Peters. 1990. Instruments of Government: Perceptions and Contexts. *Journal of Public Policy* 9: 35–58.

———. 1991. The Logic of Policy Design: Linking Policy Actors and Plausible Instruments. *Knowledge and Policy* 4(1/2): 125–152.

Lins, L. 1996. Agrotóxico Intoxica 300 Mil Brazileiros por Ano. *O Globo,* October 28.

Little, I.M.D., and J.A. Mirrlees. 1969. *Manual of Industrial Project Analysis in Developing Countries.* Paris, France: Organisation for Economic Co-operation and Development.

———. 1974. *Project Appraisal and Planning for Developing Countries.* London, U.K.: Heinemann.

Löfgren, Å., and H. Hammar. 2000. The Phase-Out of Leaded Gasoline in the EU: A Successful Failure? *Transportation Research Part D: Transport and Environment* 5: 419–431.

López, R. 1994. The Environment as a Factor of Production: The Effects of Economic Growth and Trade Liberalisation. *Journal of Environmental Economics and Management* 27(2): 163–185.

———. 2000. The Quality of Growth and the Natural Resources: The Role of the State. Paper presented at the 2nd International Conference on Environment and Development. Sept. 6–8, Beijer Institute, Stockholm, Sweden.

López, R., V. Thomas, and Y. Wang. 1999. Addressing the Education Puzzle: The Distribution of Education and Economic Reforms. Paper presented at the World Bank Economists' Forum. May 3–4, Washington, DC.

Lorenz, K. 1966. *On Aggression.* London, U.K.: Methuen & Co.

Lotka, A.J. 1932. The Growth of Mixed Populations: Two Species Competing for a Common Food Supply. *Journal of the Washington Academy of Sciences* 22: 461–469.

Lovei, M. 1998. Phasing Out Lead from Gasoline: Worldwide Experience and Policy Implications. World Bank Technical Paper no. 397. Pollution Management Series. Washington, DC: International Bank for Reconstruction and Development/World Bank. www.worldbank.org/html/fpd/transport/publicat/b09.pdf (accessed April 2002).

Lovei, M., and B.S. Levy. 1997. Lead Exposure and Health in Central and Eastern Europe: Evidence from Hungary, Poland, and Bulgaria. In *Phasing Out Lead from Gasoline in Central and Eastern Europe: Health Issues, Feasibility, and Policies,* edited by M. Lovei. Washington, DC: World Bank.

Lundqvist, L.J. 2000. *Implementation from Above: The Ecology of Sweden's New Environmental Governance.* Gothenburg, Sweden: University of Gothenburg, Department of Political Science.

Lutz, E., S. Pagiola, and C. Reiche. 1994. *Economic and Institutional Analyses of Soil Conservation Projects in Central America and the Caribbean.* Washington, DC: World Bank.

Lyngfelt, A., and C. Azar (eds.). 1999. *Proceedings of Minisymposium on Carbon Dioxide Capture and Storage.* Oct. 22, 1999. Gothenburg, Sweden: Chalmers University of Technology and University of Gothenburg.

Lvovsky, K. 1996. Effective Pollution Charges: Lessons of Worldwide Experience. Dissemination Notes no. 50. Washington, DC: World Bank, Environment Department.

Maddison, D., D. Pearce, O. Johansson, E. Calthrop, T. Litman, and E. Verhoef. 1996. *The True Cost of Road Transport.* Blueprint Series, no. 5. London, U.K.: Earthscan for the Centre for Social and Economic Research on the Global Environment.

Mäler, K.-G., and A. de Zeeuw. 1998. The Acid Rain Differential Game. *Environmental and Resource Economics* 12(2): 167–184.

Mäler, K.-G., and J. Vincent (eds.). 2001. *The Handbook of Environmental Economics.* Amsterdam, the Netherlands: North-Holland/Elsevier Science.

Malthus, T.R. (1803). *An Essay on the Principle of Population; or, A View of Its Past and Present Effects on Human Happiness; with an Inquiry into Our Prospects Respecting the Future Removal or Mitigation of the Evils Which It Occasions* (1992 printing). Cambridge, U.K.: Cambridge University Press.

Manne, A.S., and L. Schrattenholzer. 1992. International Energy Workshop Projections. In *International Energy Economics,* edited by T. Sterner. London, U.K.: Chapman and Hall, 297–299.

Markandya, A. 1997. Employment and Environmental Protection: The Tradeoffs in an Economy in Transition. Environment Discussion Paper no. 26. Cambridge, MA: Harvard Institute for International Development, International Environment Program.

Markandya, A., and A. Shibli. 1995. Industrial Pollution Control Policies in Asia. Environment Discussion Paper no. 3. Cambridge, MA: Harvard Institute for International Development.

Marshall, A. 1920. *Principles of Economics,* 8th edition. London, U.K.: Macmillan.

Maxwell, J.W., T.P. Lyon, and S.C. Hackett. 2000. Self-Regulation and Social Welfare: The Political Economy of Corporate Environmentalism. *Journal of Law and Economics* 43(2): 583–617.

Mayeres, I. 1993. The Marginal External Cost of Car Use: With An Application to Belgium. *Tijdschrift voor Economie en Management* 38(3): 2–8.

Mayeres, I., and S. Proost. 2001. Marginal Tax Reform, Externalities, and Income Distribution. *Journal of Public Economics* 79(2): 343–363.

Mazurek, J. 1999a.Voluntary Agreements in The United States: An Initial Survey. Concerted Action on Voluntary Approaches (CAVA) Working Paper no. 98/11/1. January. Paris, France: Centre d'Economie Industrielle (CERNA).

———. 1999b. *Making Microchips: Policy, Globalization, and Economic Restructuring in the Semiconductor Industry.* Urban and Industrial Environments Series. Cambridge, MA: The MIT Press.

McGartland, A., and W.E. Oates. 1985. Marketable Permits for the Prevention of Environmental Degradation. *Journal of Environmental Economics and Management* 12: 207–228.

McGuire, A. 1983. The Political Economy of Shrimping in the Gulf of California. *Human Organization* 42: 132–145.

McLeary, W. 1991. The Earmarking of Government Revenue: A Review of Some World Bank Experiences. *World Bank Research Observer* 6: 81–104.

Meade, J.E. 1951. *The Balance of Payments.* The Theory of International Economic Policy, vol. 1. London, U.K.: Oxford University Press.

Meadows, D. 1995. The City of First Priorities. *Whole Earth Review* (Spring).

Meadows, D.H., D.L. Meadows, J. Randers, and W.W. Behrens III. 1972. *The Limits to Growth: A Report for the Club of Rome's Project on the Predicament of Mankind.* London, U.K.: Earth Island Limited.

Mekonnen, A. 1998. Rural Energy and Afforestation: Case Studies from Ethiopia. Ph.D. dissertation. University of Gothenburg, Gothenburg, Sweden.

Mendelsohn, R. 1994. Property Rights and Tropical Deforestation. *Oxford Economic Papers* 46: 750–756.

Mercer, D.E., and J. Soussan. 1992. Fuelwood Problems and Solutions. In *Managing the World's Forests: Looking for Balance between Conservation and Development,* edited by N.P. Sharma. Dubuque, Iowa: Kendall/Hunt, 177–214.

Meyer, M.C., and W.L. Sherman. 1979. *The Course of Mexican History.* New York, NY: Oxford University Press.

Migot-Adholla, S., B. Peter, B. Hazell, B. Blarel, and F. Place. 1993. Indigenous Land Systems in Sub-Saharan Africa: A Constraint on Productivity? In *The Economics of Rural Organisation: Theory, Practice, and Policy,* edited by K. Hoff, A. Braveman, and J.E. Stiglitz. New York, NY: Oxford University Press.

Miller, A.J. 1999. Transferable Development Rights in the Constitutional Landscape: Has Penn Central Failed to Weather the Storm? *Natural Resources Journal* 39(3): 459–516.

Milliman, S.R., and R. Prince. 1989. Firm Incentives to Promote Technological Change in Pollution Control. *Journal of Environmental Economics and Management* 17: 247–265.

Millock, K., and F. Salanié. 1997. *Nonpoint Source Pollution Regulation When Polluters Might Cooperate.* Fondazione Eni Enrico Mattei (FEEM) Working Paper 82.97. Milan, Italy: FEEM.

Millock, K., D. Sunding, and D. Zilberman. 2002. Regulating Pollution with Endogenous Monitoring. *Journal of Environmental Economics and Management* 42(2): 221–241.

Mineta, K. 2002. The Role of Epidemiology in Japan's Pollution Litigation, and the Extent of Its Progress. The Aozora Foundation. www.aozora.or.jp/NewEnglish/eronbun_mi.htm (accessed May 2002).

Miranowski, J.A., and G.A. Carlson. 1993. Agricultural Resource Economics: An Overview. In *Agricultural and Environmental Resource Economics,* edited by G.A. Carlson, D. Zilberman, and J.A. Miranowski. New York, NY: Oxford University Press, chapter 1.

Mirrlees, J.A. 1971. An Exploration in the Theory of Optimum Income Taxation. *Review of Economic Studies* 38: 175–208.

Mkenda, A. 2001. Fishery Resources and Welfare in Rural Zanzibar. Ph.D. dissertation. University of Gothenburg, Gothenburg, Sweden.

Montero, J.P. 2000. Optimal Design of a Phase-In Emissions Trading Program. *Journal of Public Economics* 75(2): 273–291.

Montgomery, H. 1983. Decision Rules and the Search for a Dominance Structure: Towards a Process Model of Decision Making. In *Analyzing and Aiding Decision Processes,* edited by P. Humphreys, O. Svenson, and A. Vari. Amsterdam, the Netherlands: North-Holland.

Montgomery, W.E. 1972. Markets in Licenses and Efficient Pollution Control Programs. *Journal of Economic Theory* 5: 395–418.

Morrison, S.A. 1986. A Survey of Road Pricing. *Transportation Research Part A: Policy and Practice* 20: 87–97.

Mrozek, J.R. 2000. Revenue Neutral Deposit/Refund Systems. *Environmental and Resource Economics* 17: 183–193.

Muchapondwa, E. 2001. Does CAMPFIRE Comply with the Design Principles for Long-Enduring CPR Institutions? Mimeo. Gothenburg, Sweden: University of Gothenburg, Department of Economics.

Mui, V. 1999. Contracting in the Shadow of a Corrupt Court. *Journal of Institutional and Theoretical Economics* 155(2): 249–283.

Mulenga, S. 1999. A Reflection on Electricity Tariffs in Zambia. Environmental Economics Unit (EEU) Working Paper. Gothenburg, Sweden: University of Gothenburg.

Müller, B., A. Michaelowa, and C. Vrolijk. 2001. Rejecting Kyoto: A Study of Proposed Alternatives to the Kyoto Protocol. Climate Strategies website. www.climate-strategies.org/rejectingkyoto2.pdf (accessed April 2002).

Munk, K.J. 1980. Optimal Taxation with Some Non-Taxable Commodities. *Review of Economic Studies* 47: 755–765.

Munro, G.R., and A.D. Scott. 1985. The Economics of Fisheries Management. In *Handbook of Natural Resources and Energy Economics,* edited by A.V. Kneese and J.L. Sweeny. Amsterdam, the Netherlands: North-Holland.

Murty, M.N. 1994. Management of Common Property Resources: Limits to Voluntary Collective Action. *Environmental and Resource Economics* 4: 581–584.

———. 1996. Fiscal Federalism Approach for Controlling Global Environmental Pollution. *Environmental and Resource Economics* 8(4): 449–459.

Murty, M.N., A.J. James, and S. Misra. 1999. *Economics of Water Pollution: The Indian Experience.* Delhi, India: Oxford University Press.

Murty, M.N., and R. Ray. 1989. A Computational Procedure for Calculating Optimal Commodity Taxes with Illustrative Evidence from Indian Budget Data. *Scandinavian Journal of Economics* 91(4): 665–670.

Myles, G.D. 1995. *Public Economics.* Cambridge, U.K.: Cambridge University Press.

Nadaï, A. 1999. Conditions for the Development of a Product Ecolabel. *European Environment* 9: 202–211.

Narayan, D., with R. Patel, K. Schafft, A. Rademacher, and S. Koch-Schulte. 2000. *Can Anyone Hear Us? Voices from 47 Countries.* Voices of the Poor, vol. 1. New York, NY: Oxford University Press for the World Bank.

Naturvårdsverket. 2002. Swedish Environmental Protection Agency home page. www.environ.se/ (accessed March 2002).

Needleman, H.L., C. Gunnoe, A. Leviton, R. Reed, H. Peresie, C. Maher, and P. Barrett. 1979. Deficits in Psychologic and Classroom Performance of Children with Elevated Dentine Lead Levels. *New England Journal of Medicine* 300: 584–695.

NEPA (Beijing National Environmental Protection Agency). 1992. *Pollution Charges in China.* Beijing, China: NEPA.

———. 1994. The Pollution Levy System. Beijing, China: NEPA.

Newbery, D.M. 1990. Economic Principles Relevant to Pricing Roads. Oxford Review of Economic Policy 6(2): 22–39.

Nietschmann, B. 1972. Hunting and Fishing Forms among the Miskito Indians, Eastern Nicaragua. *Human Ecology* 1: 41–67.

———. 1984. Biosphere Reserves and Traditional Societies. In *Conservation, Science and Society.* Paris, France: United Nations Educational, Scientific, and Cultural Organization, 499–507.

Nilsson, M. 1998. Är miljömärkning effektivt som styrmedel? Masters thesis, Department of Economics, University of Gothenburg, Gothenburg, Sweden.

Nivola, P.S., and R.W. Crandall. 1995. *The Extra Mile: Rethinking Energy Policy for Automotive Transportation.* Washington, DC: The Brookings Institute.

Nordic Council of Ministers. 1999. *The Use of Economic Instruments in Nordic Environmental Policy 1997–98.* TemaNord 1999:524. Copenhagen, Denmark: Nordic Council of Ministers.

NRC (National Research Council). 1999. *Sharing the Fish: Toward a National Policy on Individual Fishing Quotas.* Washington, DC: National Academy Press.

Nyborg, K. 2000. Voluntary Agreements and Non-verifiable Emissions. *Environmental and Resource Economics* 17(2): 125–144.

Oakland, W.H. 1972. Congestion, Public Goods, and Welfare. *Journal of Public Economics* 1: 339–357.

Oates, W. 1983. The Regulation of Externalities: Efficient Behavior by Sources and Victims. *Public Finance* 38(3): 362–375.

Oates, W.E. 1972. *Fiscal Federalism.* New York, NY: Harcourt Brace Jovanovich.

———. 1998. Environmental Policy in the European Community: Harmonization or National Standards? *Empirica* 25: 1–13.

Oates, W.E., and R.M. Schwab. 1988. Economic Competition among Jurisdictions: Efficiency-Enhancing or Distortion-Inducing? *Journal of Public Economics* 35: 333–354.

———. 1996. The Theory of Regulatory Federalism: The Case of Environmental Management. In *The Economics of Environmental Regulation,* edited by W.E. Oates. Aldershot, Hampshire, U.K.: Edward Elgar, 319–331.

OECD (Organisation for Economic Co-operation and Development). 1975. An Evolving Context for Environmental Policy. www.oecd.org//env/policy.htm (accessed Jan. 21, 2002).

———. 1989. *Economic Instruments for Environmental Protection.* Paris, France: OECD.

———. 1995. *Environmental Taxes in OECD Countries.* Paris, France: OECD.

———. 1996. *Implementation Strategies for Environmental Taxes.* Paris, France: OECD.

———. 1997. *Eco-labelling: Actual Effects of Selected Programmes.* OECD/GD(97)105. Paris, France: OECD.

Onursal, B., and S.P. Gautam. 1997. Vehicular Air Pollution: Experiences from Seven Latin American Urban Centers. Technical Paper no. 373. Washington, DC: International Bank for Reconstruction and Development/World Bank.

O'Ryan, R. 1996. Cost-Effective Policies to Improve Urban Air Quality in Santiago, Chile. *Journal of Environmental Economics and Management* 31: 302–313.

Östermark, U. 1996. Alkylkate Petrol. Environmental Aspects of Volatile Hydrocarbon Emissions. Ph.D. dissertation, Chalmers University of Technology, Gothenburg, Sweden.

Ostrom, E. 1990. *Governing the Commons: The Evolution of Institutions for Collective Action.* Cambridge, U.K.: Cambridge University Press.

———. 1997. The Comparative Study of Public Economies. Acceptance Paper, The Frank E. Seidman Distinguished Award in Political Economy. Workshop in Political Theory and Policy Analysis. Sept. 26, Rhodes College, Memphis, TN.

———. 1999. Coping with Tragedies of the Commons. *Annual Review of Political Science* 2: 493–535.

Ostrom, E., J. Burger, C.B. Field, R.B. Norgaard, and D. Policansky. 1999. Revisiting the Commons: Local Lessons, Global Challenges. *Science* 284: 278–282.

Ostrom, E., and E. Schlager. 1996. The Formation of Property Rights. In *Rights to Nature: Ecological, Economic, Cultural, and Political Principles of Institutions for the Environment,* edited by S.S. Hanna, C. Folke, and K.-G. Mäler. Washington, DC: Island Press.

Oswald, A.J. 1998. Happiness and Economic Performance. *Economic Journal* 107: 181-831.

Otsuka, K., H. Chuma, and Y. Hayami. 1992. Theories of Share Tenancy: A Critical Survey. *Economic Development and Cultural Change* 37: 31–68.

Ovuka, M. 2000. More People, More Erosion? Land Use, Soil Erosion, and Soil Productivity in Murang'a District, Kenya. *Journal of Land Degradation and Development* 11: 111–124.

Oxman, B.H. 1997. Human Rights and the United Nations Convention on the Law of the Sea. *Columbia Journal of Transnational Law* 35: 399–429.

Pagiola, S. 2000. *Payments for Environmental Services: Environment Matters, Annual Review July 1999–June 2000.* Washington, DC: World Bank.

Palmer, K., W. Oates, and P. Portney. 1995. Tightening Environmental Standards: The Benefit-Cost or the No-Cost Paradigm? *Journal of Economic Perspectives* 9(4): 119–132.

Palmer, K.L., H. Sigman, and M. Walls. 1997. The Cost of Reducing Municipal Solid Waste. *Journal of Environmental Economics and Management* 33(2): 128–150.

Panayotou, T. 1993. *Green Markets: The Economics of Sustainable Development.* San Francisco, CA: ICS Press for the International Centre for Economic Growth.

———. 1995. Effective Financing of Environmentally Sustainable Development in Eastern Europe and Central Asia. Environment Discussion Paper no. 10. Cambridge, MA: Harvard Institute for International Development, International Environment Program.

———. 1998. *Instruments of Change: Motivating and Financing Sustainable Development.* London, U.K.; Earthscan for United Nations Environment Programme.

Panayotou, T., and P.S. Ashton. 1992. *Not by Timber Alone: Economics and Ecology for Sustaining Tropical Forests.* Washington, DC: Island Press.

Pargal, S., and D. Wheeler. 1996. Informal Regulation of Industrial Pollution in Developing Countries: Evidence from Indonesia. *Journal of Political Economy* 104(6): 1314–1327.

Paris, R., and I. Ruzicka. 1989. Barking Up the Wrong Tree: The Role of Rent Appropriation in Sustainable Tropical Forest Management. Occasional Paper no. 1. Manila, Philippines: Asian Development Bank, Environment Office.

Parry, I.W.H., and A. Bento. 2001. Revenue Recycling and the Welfare Effects of Road Pricing. *Scandinavian Journal of Economics* 103(4, December): 645–667.

Parry, I.W.H., and R.C. Robertson. 1999. A Second-Best Evaluation of Eight Policy Instruments to Reduce Carbon Emissions. *Resource and Energy Economics* 21(3-4): 347–373.

Pauly, D. 1997. Points of View: Putting Fisheries Management Back in Places. *Fish Biology and Fisheries* 7: 125–127.

Pearce, D.W. 1999. Economics and Biodiversity Conservation in the Developing World. *Environment and Development Economics* 4(2): 230–233.

Peck, J. 1999. OECS Solid Waste Management Project. Evaluation Memo 990212. Washington, DC: World Bank, Environment Department.

Pender, J.L. 1996. Discount Rates and Credit Market: Theory and Evidence from Rural India. *Journal of Development Economics* 50(2): 257–296.

Perrings, C., and D. Stern. 2000. Modeling Loss of Resilience in Agroecosystems: Rangelands in Botswana. *Environmental and Resource Economics* 16(2): 185–210.

Perrings, C., and B. Walker. 1997. Biodiversity, Resilience, and the Control of Ecological-Economic Systems: The Case of Fire-Driven Rangelands. *Ecological Economics* 22(1): 73–83.

Pesaran, H., and R. Smith. 1995. Alternative Approaches to Estimating Long-Run Energy Demand Elasticities: An Application to Asian Developing Countries. In *Global Warming and Energy Demand,* edited by T. Barker, P. Ekins, and N. Johnstone. London, U.K.: Routledge, 19–46.

Petrakis, E., and A. Xepapadeas. 1996. Environmental Consciousness and Moral Hazard in International Agreements to Protect the Environment. *Journal of Public Economics* 60: 95–110.

Pezzey, J. 1992. The Symmetry between Controlling Pollution by Price and Controlling It by Quantity. *Canadian Journal of Economics* 25: 983–991.

Pierre, J., and B.G. Peters. 2000. *Governance, Politics and the State.* London, U.K.: Macmillan.

Pigou, A.C. 1932. *The Economics of Welfare,* 4th ed. London: Macmillan.

Pingali, P.L., and R.V. Gerpacio. 1998. Towards Reduced Pesticide Use for Cereal Crops in Asia. In *Agriculture and the Environment,* edited by E. Luntz. Washington, DC: World Bank, 254–271.

Pirttilä, J. 1998. Earmarking of Environmental Taxes: Efficient, After All. Discussion Paper no. 4/98. Helsinki, Finland: Bank of Finland.

Pizer, W.A. 1999. The Optimal Choice of Climate Change Policy in the Presence of Uncertainty. *Resource and Energy Economics* 21(3-4): 255–287.

Place, F., and P. Hazell. 1993. Productivity Effects of Indigenous Land Tenure Systems in Sub-Saharan Africa. *American Journal of Agricultural Economics* 75(1): 10–19.

Platteau, J.-P. 1984. The Drive Towards Mechanization of Small-Scale Fisheries in Kerala: A Study of the Transformation Process of Traditional Village Societies. *Development and Change* 15: 65–103.

———. 1999. The Role of the Community in Understanding Poverty Linkages with Natural Resources Management. Working Paper. Namur, Belgium: University of Namur, Centre de Recherche en Economie du Developpement.

———. 2000. Does Africa Need Land Reform? In *Evolving Land Rights, Policy and Tenure in Africa,* edited by C. Toulmin and J. Quan. London, U.K.: International Institute for Environment and Development, Drylands Programme/ Department for International Development.

Porter, M. 1990. *The Competitive Advantage of Nations.* New York, NY: The Free Press.

Porter, M., and C. van der Linde. 1995. Towards a New Conception of the Environment: Competitiveness Relationship. *Journal of Economic Perspectives* 9(4): 97–118.

Portney, P.R. 1990. *Public Policies for Environmental Protection,* 1st edition. Washington, DC: Resources for the Future.

Poterba, J.M. 1991. Is the Gasoline Tax Regressive? In *Tax Policy and the Economy,* Volume 5, edited by D. Bradford. Cambridge, MA: MIT Press, 145–164.

Prefeitura Municipal de Curitiba. 2002. Prefecture of Curitiba, Brazil, website. www.curitiba.pr.gov.br/ (accessed April 26, 2002).

Probst, K.N., and T.C. Beierle. 1999. *The Evolution of Hazardous Waste Programs: Lessons from Eight Countries.* Washington, DC: Resources for the Future, Center for Risk Management.

Prosterman, R.L., and J.M. Riedinger. 1987. *Land Reform and Democratic Development.* Baltimore, MD: Johns Hopkins University Press.

Qu, G. 1991. *Environmental Management in China.* Beijing, China: United Nations Environment Programme and China Environmental Science Press.

Rabin, M. 1998. Psychology and Economics. *Journal of Economic Literature* 36(1): 11–47.

Rabinovitch, J. 1992. Curitiba: Towards Sustainable Urban Development. *Environment and Urbanization* 4(2): 72–73.

Rachlinski, J.J., and F. Jordan. 1998. Remedies and the Psychology of Ownership. *Vanderbilt Law Review* 51: 1541.

Rahmstorf, S. 2000. The Thermohaline Ocean Circulation: A System with Dangerous Thresholds? [Editorial Comment]. *Climatic Change* 46: 247–256.

Ramamurti, R. (ed.). 1996. *Privatising Monopolies: Lessons from the Telecommunications and Transport Sectors in Latin America.* Baltimore, MD: Johns Hopkins University Press, 401.

Randall, A., and M.A. Taylor. 2000. Incentive-Based Solutions to Agricultural Environmental Problems: Recent Developments in Theory and Practice. *Journal of Agricultural and Applied Economics* 32(2): 221–234.

Ravallion, M. 1999. Is More Targeting Consistent with Less Spending? Paper presented at the World Bank Economists' Forum. May 3–4, Washington, DC.

Rees, W.E., and M. Wackernagel. 1994. Ecological Footprints and Appropriated Carrying Capacity: Measuring the Natural Capital Requirements of the Human Economy. In *Investing in Natural Capital: The Ecological Economics Approach to Sustainability,* edited by A.-M. Jansson, M. Hammer, C. Folke, and R. Costanza. Washington, DC: Island Press.

Regeringskansliet. 2001. Local Investment Programmes. miljo.regeringen.se/M-dep_fragor/hallbarutveckling/LIP/inenglish/index.htm (accessed April 2002).

Repetto, R.C., R.C. Dower, R. Jenkins, and J. Geoghegan. 1992. *Green Fees: How a Tax Shift Can Work for the Environment and the Economy.* Washington, DC: World Resources Institute.

Requate, T. 1998. Incentives to Innovate under Emission Taxes and Tradable Permits. *European Journal of Political Economy* 14(1): 139–165.

Requate, T., and W. Unold. 2001. Pollution Control by Options Trading under Imperfect Information. Paper presented at the European Association of Environmental and Resources Economists (EAERE) 2001 conference, Southampton, U.K.

RFF (Resources for the Future). 2002. RFF Environment Library: Climate. www.rff.org/environment/climate.htm (accessed March 2002).

Rhodes, R.A.W. 1996. The New Governance: Governing without Government. *Political Studies* 44: 652–667.

Riksdagsrevisorerna (Auditors of Parliament). 1998/99. Statligt Stöd Till Lokala Investeringsprogram för en Ekologiskt Hållbar Utveckling (State Grants to Local Investment Programs for an Ecologically Sustainable Development, in Swedish). Riksdagens revisorer Rapport 1998/99:8. Stockholm, Sweden: Sveriges Riksdag.

Riksrevisionsverket (National Audit Office). 1999a. *De Lokala Investeringsprogrammen i Praktiken: En Uppföljning av Kommunernas Arbete* (The LIPs in Practice: A Follow-Up of Municipal Activities). Stockholm, Sweden: Riksrevisionsverket.

———. 1999b. *Miljarden Som Försvann* (The Billion That Disappeared). Stockholm, Sweden: Riksrevisionsverket.

Roberts, E., and M. Spence. 1979. Effluent Charges and Licenses Under Uncertainty. *Journal of Public Economics* 5: 193–208.

Rogat, J., and T. Sterner. 1998. The Determinants of Gasoline Demand in Some Latin American Countries. *International Journal of Global Energy Issues* 11(1–4):162–170.

Rogers, G.W. 1979. Alaska's Limited Entry Program: Another View. *Journal of the Fisheries Research Board of Canada* 36: 783–788.

Rosander, P. 1998. Vad har Miljömärkning av Kemtekniska Produkter Inneburit för Reningsverken? Stockholm, Sweden: Svenska Naturskyddsföreningen. www.snf.se/pdf/bmv/rap-bmv-reningsverk.pdf (accessed April 2002).

Rosenzweig, M.R. 1988a. Risk, Implicit Contracts, and the Family in Rural Areas of Low Income Countries. *Economic Journal* 78(2): 245–250.

———. 1988b. Risk, Private Information, and the Family. *American Economic Review* 78(2): 245–250.

Rosenzweig, M.R., and H.P. Binswanger. 1994. Wealth, Weather Risk, and the Profitability of Agricultural Investment. *Economic Journal* 103: 56–78.

Rouwendal, J. 1996. An Economic Analysis of Fuel Use per Kilometre by Private Cars. *Journal of Transport Economics and Policy* 30: 3–14.

Ruddle, K. 1988. Social Principles Underlying Traditional Inshore Fishery Management Systems in the Pacific Basin. *Marine Resource Economics* 5: 351–363.

Ruddle, K., E. Hviding, and R. Johannes. 1992. Marine Resources Management in the Context of Customary Tenure. *Marine Resource Economics* 7: 249–273.

Russell, C.S., W. Harrington, and W.J. Vaughan. 1986. *Enforcing Pollution Control Laws.* Washington, DC: Resources for the Future.

Russell, C.S., and P.T. Powell. 1996. *Choosing Environmental Policy Tools: Theoretical Cautions and Practical Considerations.* Washington, DC: Inter-American Development Bank.

Ruzicka, I. 1979. Rent Appropriation in Indonesian Logging: East Kalimantan. *Bulletin of Indonesian Economic Studies* 15(2): 45–74.

Sadoulet, E., S. Fukui, and A. de Janvry. 1994. Efficient Share Tenancy Contracts under Risk: The Case of Two Rice-Growing Villages in Thailand. *Journal of Development Economics* 45(2): 225–243.

Sairinen, R., and O. Teittinen. 1999. Voluntary Agreements as an Environmental Policy Instrument in Finland. *European Environment* 9(2): 67–74.

Samuelson, P. 1983. Thünen at Two Hundred. *Journal of Economic Literature* 21(4): 1468–1488.

Samuelson, P.A. 1954. The Pure Theory of Public Expenditure. *Review of Economics and Statistics* 36: 387–389.

———. 1955. Diagrammatic Exposition of a Pure Theory of Public Expenditure. *Review of Economics and Statistics* 37: 350–356.

Sanford, R.M., and H.B. Stroud. 2000. Evaluating the Effectiveness of Act 250 in Protecting Vermont Streams. *Journal of Environmental Planning and Management* 43(5): 623–641.

Sankar, U. 1998. Laws and Institutions Relating to Environmental Protection in India. Paper presented at the conference on The Role of Law and Legal Institutions in Asian Economic Development. Nov. 1–4, Erasmus University, Rotterdam, the Netherlands.

Sayeg, P. 1998. Successful Conversion to Unleaded Gasoline in Thailand. Technical Paper no. 410. Washington, DC: International Bank for Reconstruction and Development/World Bank.

Schaefer, M.B. 1954. Some Aspects of the Dynamics of Populations Important to the Management of Commercial Marine Fisheries. Bulletin of the Inter-American Tropical Tuna Commission 1: 25–56.

Scherr, S.J. 2000. A Downward Spiral? Research Evidence on the Relationship between Poverty and Natural Resource Degradation. *Food Policy* 25: 479–498.

Schneider, R.R. 1995. Government and the Economy on the Amazon Frontier. Environment Paper no. 11. Washington, DC: World Bank.

Schug, D.M. 1996. The Revival of Territorial Use Rights in Pacific Island Inshore Fisheries. In *Ocean Yearbook,* no. 12, edited by E.M. Borgese, N. Ginsburg, and J.R. Morgan. Chicago, IL: University of Chicago Press.

Schwartz, J. 1994. Low Level Lead Exposure and Children's IQ: A Meta Analysis and Search for a Threshold. *Environmental Research* 65(1): 42–55.

Scott, A. 1955. The Fishery: The Objectives of Sole Ownership. *Journal of Political Economics* 63: 116–124.

———. 1983. *Natural Resources and the Economics of Conservation.* Ottawa, Ontario, Canada: Carleton University Press.

———. 1993. Obstacles to Fishery Self-Government. *Marine Resource Economics* 8: 187–199.

Segerson, K. 1988. Uncertainty and Incentives for Nonpoint Pollution Control. *Journal of Environmental Economics and Management* 15(1): 87–98.

———. 1993. Liability Transfers: An Economic Assessment of Buyer and Lender Liability. *Journal of Environmental Economics and Management* 25(1): S46–S63.

Segerson, K., and T.J. Miceli. 1999. Voluntary Approaches to Environmental Protection: The Role of Legislative Threats. In *Voluntary Approaches in Environmental Policy.* Fondazione Eni Enrico Mattei Series on Economics, Energy, and Environment, vol. 14. Dordrecht; the Netherlands: Kluwer Academic, 105–120.

Sen, A. 1981. *Poverty and Famines: An Essay on Entitlement and Deprivation.* Oxford, U.K.: Oxford University Press.

———. 1999. *Development as Freedom.* Oxford, U.K.: Oxford University Press.

SEPA (Swedish Environmental Protection Agency). 1997. Environmental Taxes in Sweden: Economic Instruments of Environmental Policy. Report 4745, Stockholm, Sweden: SEPA.

———. 2000. The Swedish Charge on Nitrogen Oxides. Report. Stockholm, Sweden: SEPA.

Seroa da Motta, R., R.M. Huber, and H. Ruitenbeek. 1999. Market Based Instruments for Environmental Policymaking in Latin America and the Caribbean: Lessons from Eleven Countries. *Environment and Development Economics* 4(2): 177–201.

Shah, J.J., and T. Nagpal (eds.). 1997. Urban Air Quality Management Strategy in Asia: Metro Manila Report. Technical Paper no. 380. Washington, DC: International Bank for Reconstruction and Development/World Bank.

Shah, J.J., T. Nagpal, and C.J. Brandon (eds.). 1997. *Urban Air Quality Management Strategy in Asia: Guide Book.* Washington, DC: International Bank for Reconstruction and Development/World Bank.

Shell. 2001. People, Planet, and Profits. The Shell Report 06-04-2001. www.shell.com (accessed July 7, 2001).

Shibli, A., and A. Markandya. 1995. Industrial Pollution Control Policies in Asia: How Successful Are the Strategies? *Asian Journal of Environmental Management* 3(2): 87–117.

Shiferaw, B., and S.T. Holden. 1999. Soil Erosion and Smallholders' Conservation Decisions in the Highlands of Ethiopia. *World Development* 27(4): 739–752.

Shogren, J.F., and T.D. Crocker. 1999. Risk and Its Consequences. *Journal of Environmental Economics and Management* 37(1): 44–51.

Shortle, J.S., R.D. Horan, and D.G. Abler. 1998. Research Issues in Nonpoint Pollution Control. *Environmental and Resource Economics* 11(3-4): 571–585.

Shrestha, K.B. 1996. Nepal Madhyasthata Samuha: Community Forestry in Nepal—An Overview of Conflict. Discussion Paper Series, no. MNR 96/2. Katmandu, Nepal: International Centre for Integrated Mountain Development.

Sida (Swedish International Development Cooperation Agency). 1998. Policy on Urban Transport Development. Draft. December. Stockholm, Sweden: Sida.

Simpson, R. 1995. Optimal Pollution Taxation in a Cournot Duopoly. *Environmental and Resource Economics* 6: 359–369.

Simpson, R.D., and R.A. Sedjo. 1996. Paying for the Conservation of Endangered Ecosystems: A Comparison of Direct and Indirect Approaches. *Environment and Development Economics* 1: 241–257.

Sinclair-Desgagné, B., and H.L. Gabel. 1997. Environmental Auditing in Management Systems and Public Policy. *Journal of Environmental Economics and Management* 33(3): 331–347.

SIWI (Stockholm International Water Institute). 2000. SIWI website. www.siwi.org/menu/menu.html (accessed April 2002).

Skonhoft, A. 1998. Resource Utilization, Property Rights, and Welfare: Wildlife and the Local People. *Ecological Economics* 26: 67–80.

Slunge, D., and T. Sterner. 2001. Implementation of Policy Instruments for Chlorinated Solvents. *European Environment* 11(5): 281–296.

Small, K.A. 1992. Urban Transportation Economics. Reading, U.K.: Harwood Academic.

Small, K.A., and C. Kazimi. 1995. On the Costs of Air Pollution from Motor Vehicles. *Journal of Transport Economics and Policy* 29: 7–32.

Smith, J.M. 1982. *Evolution and the Theory of Games.* Cambridge, U.K.: Cambridge University Press.

Smith, R., and Y. Tsur. 1997. Asymmetric Information and the Pricing of Natural Resources: The Case of Unmetered Water. *Land Economics* 73(3): 392–403.

Smith, R.B.W., and T.D. Tomasi. 1995. Transaction Costs and Agricultural Nonpoint-Source Water Pollution Control Policies. *Journal of Agricultural and Resource Economics* 20(2): 277–290.

Smith, V.H., and B.K. Goodwin. 1996. Crop Insurance, Moral Hazard, and Agricultural Chemical Use. *American Journal of Agricultural Economics* 78(2): 428–438.

Smith, V.K., and R. Walsh. 2000. Do Painless Environmental Policies Exist? *Journal of Risk and Uncertainty* 21(1): 73–94.

Smith, V.L., and J.M. Waker. 1993. Monetary Rewards and Decision Cost in Experimental Economics. *Economic Inquiry* 31(2): 245–261.

Solyom, P., and L.-G. Lindfors. 1998. Evaluation of Nordic Eco-Labeling for Eight Groups of Products Containing Surface Active Agents. Report to the Swedish Environmental Research Institute (IVL) no. B1307. Stockholm, Sweden: IVL.

SOU (Statens Offentliga Utredningar). 1991. Konkurrensneutral Beskattning. SOU 1991:90. Stockholm, Sweden: Allmänna Förlaget.

———. 1997. Skatter, Miljö och Sysselsättning. SOU 1997:11. Stockholm, Sweden: Allmänna Förlaget.

Spulber, D.F. 1985. Optimal Environmental Regulation under Asymmetric Information. *Journal of Public Economics* 35: 163–181.

Stavins, R.N. 1996. Correlated Uncertainty and Policy Instrument Choice. *Journal of Environmental Economics and Management* 30(2): 218–232.

———. 1998. What Can We Learn from the Grand Policy Experiment? Lessons from SO2 Allowance Trading. *Journal of Economic Perspectives* 12(3): 69–88.

———. 2001. Experience with Market-Based Environmental Policy Instruments. In *The Handbook of Environmental Economics,* edited by K.-G. Mäler and J. Vincent. Amsterdam, the Netherlands: North-Holland/Elsevier Science.

Stavins, R.N., and B. Whitehead. 1997. *Thinking Ecologically: The Next Generation of Environmental Policy.* New Haven, CT: Yale University Press.

Stavins, R.N., and T. Zylicz. 1995. Environmental Policy in a Transition Economy: Designing Tradable Permits for Poland. Environment Discussion Paper no. 9. Cambridge, MA: Harvard Institute for International Development, International Environment Program.

Sterner, T. 1989. Oil Products in Latin America: The Politics of Energy Pricing. *The Energy Journal* 10(2): 25–45.

———. 1991. Gasoline Demand in the OECD: Choice of Model and Data Set in Pooled Estimations. *OPEC Review* XV(2): 91–101.

———. 1993. Policy Instruments for a Sustainable Economy. In *Economic Instruments for Environmental Management in Developing Countries,* Proceedings of a Workshop Held at OECD Headquarters in Paris on 8 October 1992. Paris, France: Organisation for Economic Cooperation and Development, chapter 5.

——— (ed.). 1994. *Economic Policies for Sustainable Development.* Dordrecht, the Netherlands: Kluwer Academic.

———. 1996. Environmental Tax Reform, Theory, Industrialized Country Experience, and Relevance in LDCs. In *New Directions in Development Economics? Growth, Environmental Concerns, and Government in the 1990s,* edited by M. Lundahl and B.J. Ndulu. London, U.K.: Routledge.

———. 1999. *The Market and the Environment: The Effectiveness of Market-Based Policy Instruments for Environmental Reform.* Cheltenham, U.K.: Edward Elgar.

Sterner, T., and H. Bartelings. 1999. Household Waste Management in a Swedish Municipality: Determinants of Waste Disposal, Recycling and Composting. *Environmental and Resource Economics* 13(4): 473–491.

Sterner, T., and M. Belhaj. 1989. Les Prix de l'Energie en Afrique. *Revue de l'Energie* 415(November): 3–11.

Sterner, T., and C. Dahl. 1992. Modelling the Demand for Highway Transport Fuels. In International Energy Economics, edited by T. Sterner. London, U.K.: Chapman and Hall

Sterner, T., C. Dahl, and M. Franzén. 1992. Gasoline Tax Policy, Carbon Emissions, and the Environment. *Journal of Transport and Economic Policy* 26: 109–120.

Sterner, T., and M. Franzén. 1995. Long-Run Demand Elasticities for Gasoline. In *Global Warming and Energy Elasticities,* edited by T. Barker, N. Johnstone, and P. Ekins. London, U.K.: Routledge.

Sterner, T., and L. Höglund. 2000. Output-Based Refunding of Emission Payments: Theory, Distribution of Costs, and International Experience. Discussion Paper no. 00-29. Washington, DC: Resources for the Future.

Sterner, T., and M. Segnestam. 2001. *Miljö Och Fattigdom.* Stockholm, Sweden: Sida.

Stevenson, G.G. 1991. *Common Property Economics: A General Theory and Land Use Applications.* Cambridge, U.K.: Cambridge University Press.

Stiglitz, J.E. 1974. Incentives and Risk-Sharing in Sharecropping. *Review of Economic Studies* 41(2): 219–255.

————. 1999. Whither Reform? Ten Years of the Transition [keynote address]. In *Annual World Bank Conference on Development Economics 1999,* edited by B. Pleskovic and J.E. Stiglitz. Washington, DC: World Bank.

Stone, C.D. 1995. What To Do about Biodiversity: Property Rights, Public Goods, and the Earth's Biological Riches. *Southern California Law Review* 68: 577–620.

Strand, J. 1994. Environmental Accidents under Moral Hazard and Limited Firm Liability. *Environmental and Resource Economics* 4(5): 495–509.

————. 1996. On First-Best Environmental Taxation under Moral Hazard. In *Law and Economics of the Environment,* edited by E. Eide and R. van den Bergh. Oslo: Juridisk, 231-243.

Svenska Naturskyddsföreningen. 1999. *Hushållskemikalier i Förändring.* Stockholm, Sweden: Svenska Naturskyddsföreningen.

Tarhoaca, C., and C. Zinnes. 1996. *Revenue Considerations for Establishing an Environmental Fund: The Case of Romania.* Environment Reprint Series, no. 2. April. Cambridge, MA: Harvard Institute for International Development, International Environment Program.

Teclaff, L.A. 1985. *Water in Historical Perspective.* Buffalo, NY: W.S. Hein.

Temple, F. 1999. Comments by Fred Temple, World Bank country director, quoted by Reuters. March 8.

Templeton, S., and S. Scherr. 1997. *Population Pressure and Micro-Economy of Land Management in Hills and Mountains of Developing Countries.* Washington, DC: International Food Policy Research Institute.

————. 1999. Effects of Demographic and Related Microeconomic Change on Land Quality in Hills and Mountains in Developing Countries. *World Development* 27(6): 903–918.

Tenbrunsel, D.M., and A.E. Messick. 1999. Sanctioning Systems, Decision Frames, and Cooperation. *Administrative Science Quarterly* 44: 684–707.

Thomas, A. 1995. Regulating Pollution under Asymmetric Information: The Case of Industrial Wastewater Treatment Activity. *Journal of Environmental Economics and Management* 28(3): 357–373.

Thomas, V., and Y. Wang. 1998. Missing Lessons of East Asia: Openness, Education, and the Environment. Paper prepared for the World Bank Annual Conference on Development in Latin America and the Caribbean, Trade: Towards Open Regionalism. June 29–July 1, 1997, Montevideo, Uruguay.

Thomas, V.M. 1995. The Elimination of Lead in Gasoline. *Annual Review of Energy and Environment* 20: 301–324.

Thomson, D. 1980. Conflict within the Fishing Industry. *ICLARM Newsletter* 3: 3–4.

Thorpe, S.G. 1997. Fuel Economy Standards, New Vehicle Sales, and Average Fuel Efficiency. *Journal of Regulatory Economics* 11: 311–326.

Tiebout, C.M. 1956. A Pure Theory of Local Expenditures. *Journal of Political Economy* 64: 416–424.

Tietenberg, T.H. 1985. *Emissions Trading, an Exercise in Reforming Pollution Policy.* Washington, DC: Resources for the Future.

————. 1990. Economic Instruments for Environmental Regulation. *Oxford Review of Economic Policy* 6(1): 17–33.

————. 1992. *Environmental and Natural Resource Economics.* New York, NY: HarperCollins.

————. 1994. *Economics and Environmental Policy.* Aldershot, Hampshire, U.K.: Edward Elgar.

————. 1995. Transferable Discharge Permits and Global Warming. In *The Handbook of Environmental Economics,* edited by D.W. Bromley. Oxford, U.K.: Basil Blackwell, 317–352.

————. 1998. Disclosure Strategies for Pollution Control. In *The Market and the Environment,* edited by T. Sterner. Cheltenham, U.K.: Edward Elgar.

Tiffen, M., M. Mortimore, and F. Gichuki. 1994. *More People, Less Erosion: Environmental Recovery in Kenya.* New York, NY: John Wiley & Sons.

Tlaiye, L., and D. Biller. 1994. Successful Environmental Institutions: Lessons from Colombia and Curitiba, Brazil. LATEN Dissemination Note 12. Washington, DC: World Bank, Latin and the Caribbean Environment Unit.

Toman, M., R. Morgenstern, and J. Anderson. 1999. The Economics of "When" Flexibility in the Design of Greenhouse Gas Abatement Policies. *Annual Review of Energy and Environment* 24: 431–460.

Toman, M.A. (ed.). 2001. *Climate Change Economics and Policy: An RFF Anthology.* Washington, DC: Resources for the Future.

Toulmin, C., and J. Quan (eds.). 2000. *Evolving Land Rights, Policy, and Tenure in Africa.* London, U.K.: International Institute for Environment and Development, Drylands Programme/Department for International Development.

Transport for London. 2002. Transport for London website. www.transportforlondon.gov.uk (accessed March 2002).

Transportation Research Board. 1994. *Curbing Gridlock: Peak Period Fees To Relieve Traffic Congestion,* Volume 2. Transportation Research Board Special Report 242. Washington, DC: National Research Council, Transportation Research Board.

Tuck, L., and K. Lindert. 1996. From Universal Food Subsidies to a Self Targeted Program. Discussion Paper no. 351. Washington, DC: World Bank.

Tullock, G. 1965. *The Politics of Bureaucracy.* Washington, DC: Public Affairs Press.

———. 1981. Why So Much Stability? *Public Choice* 37(2): 189–202.

Turvey, R. 1963. On Divergences between Social Cost and Private Cost. *Economica* 30: 309–313.

Tversky, A., and D. Kahneman. 1981. The Framing of Decisions and the Psychology of Choice. *Science* 211: 453–458.

Ulph, A. 1996. Environmental Policy Instruments and Imperfectly Competitive International Trade. *Environmental and Resource Economics* 7(4): 333–355.

UNDP/World Bank ESMAP. 1996. Elimination of Lead in Gasoline in Latin America and the Caribbean. Report no. 194/97EN. December. Washington, DC: United Nations Development Programme/World Bank Energy Sector Management Assistance Program.

———. 1998. Harmonization of Fuels Specifications in Latin America and the Caribbean. Report no. 203/98EN. June. Washington, DC: United Nations Development Programme/World Bank Energy Sector Management Assistance Program.

UNEP (United Nations Environment Programme). 1999. Basel Convention on the Control of Transboundary Movements of Hazardous Wastes and Their Disposal. www.basel.int/meetings/sbc/liab10-2.htm (accessed April 2002).

UNFCCC (United Nations Framework Convention on Climate Change). 2002. United Nations Framework Convention on Climate Change home page. www.unfccc.de/ (accessed March 2002).

United Nations. 1948. Universal Declaration of Human Rights. www.un.org/Overview/rights.html (accessed April 2002).

U.S. EPA (Environmental Protection Agency). 1998. *Partners for the Environment: Collective Statement of Success.* Washington, DC: U.S. EPA, Office of Reinvention.

———. 2000a. EPA Sets Enforcement Records in 1999. Press release. Jan. 19. yosemite1.epa.gov/opa/admpress.nsf/016bcfb1deb9fecd85256aca005d74df/4fe231640d8982c48525686b005a3b8c?OpenDocument (accessed September 2002).

———. 2000b. Toxics Release Inventory (TRI) Program, 1998 Public Data Release Report. www.epa.gov/triinter/tridata/tri98/pdr/index.htm (accessed March 2002).

———. 2001. *The United States Experience with Economic Incentives for Protecting the Environment.* EPA-240-R-01-001. January. Washington, DC: U.S. EPA.

———. 2002a. Emissions Exchange, Cantor Fitzgerald EBS, and Fieldston Publications. www.epa.gov/docs/acidrain/ats/pricetbl.html (accessed March 2002).

———. 2002b. Pollution Prevention Directory. www.epa.gov/ChemLibPPD/ppdir.txt (accessed Jan. 24, 2002).

———. 2002c. Monthly Average Price of Sulfur Dioxide Allowances. www.epa.gov/airmarkets/trading/so2market/prices.html (accessed April 2002).

———. 2002d. Superfund home page. Construction Completions at National Priorities List (NPL) Sites—by Number. www.epa.gov/superfund/sites/query/queryhtm/nplccl1.htm (accessed March 2002).

Verhoef, E. 1994. External Effects and Social Costs of Road Transport. *Transportation Research Part A: Policy and Practice* 28: 273–287.

Verhoef, E., P. Nijkamp, and P. Rietveld. 1995. Second-Best Regulation of Road Transport Externalities. *Journal of Transport Economics and Policy* 29: 147–167.

Vickrey, W.S. 1963. Pricing in Urban and Suburban Transport. *American Economic Review Papers and Proceedings* 43: 452–465.

———. 1969. Congestion Theory and Transport Investment. *American Economic Review* 59(2): 251–260.

Victoria Transport Institute. 2001. Online Transportation Demand Management Encyclopedia. www.vtpi.org/tdm/tdm35.htm (accessed March 2002).

Vincent, J.R., R.M. Ali, and associates. 1997. *Environment and Development in a Resource-Rich Economy: Malaysia Under the New Economic Policy.* Cambridge, MA: Harvard Institute for International Development.

Vincent, J.R., and S. Farrow. 1997. A Survey of Pollution Charge Systems and Key Issues in Policy Design. In *Controlling Pollution in Transition Economies,* edited by R. Bluffstone and B.A. Larson. Cheltenham, U.K.: Edward Elgar.

Viton, P.A. 1995. Private Roads. *Journal of Urban Economics* 37(3): 260–289.

Vitousek, P.M., P.R. Ehrlich, A.H. Erlich, and P. Matson. 1986. Human Appropriation of the Products of Photosynthesis. *BioScience* 36: 368–373.

Vitousek, P.M., H.A. Mooney, J. Lubchenco, and J.M. Melillo. 1997. Human Domination of Earth's Ecosystems. *Science* 277: 494–499.

Walker, T.S., and N.S. Jodha. 1986. How Small Farm Households Adapt to Risk. In *Crop Insurance for Agricultural Development: Issues and Experience,* edited by P. Hazell, C. Pomareda, and A. Valdes. Baltimore, MD: Johns Hopkins University Press.

Wallart, N. 1999. *The Political Economy of Environmental Taxes.* Cheltenham, U.K.: Edward Elgar.

Walsh, M. 1994. Environmental Regulation of Transport: US, OECD and Global Trends in Legislation and Policy. In *Economic Policies for Sustainable Development,* edited by T. Sterner. Dordrecht, the Netherlands: Kluwer Academic.

Walsh, M.P. (ed.). 2000. *Car Lines.* Issue 3. walshcarlines.com (accessed April 2002).

Walters, A.A. 1961. The Theory and Measurements of Private and Social Cost of Highway Congestion. *Econometrica* 29: 676–699.

Wang, H., and D. Wheeler. 1996. Pricing Industrial Pollution in China. Policy Research Working Paper 1644. Washington, DC: World Bank.

———. 1999. Pricing Industrial Pollution and Economic Development in China: An Econometric Analysis of the Pollution Levy System. Paper presented at the World Bank Economists' Forum. May 3–4, Washington, DC.

Wang, S.D. 1995. The Surf Clam ITQ Management: An Evaluation. *Marine Resource Economics* 10: 93–98.

Watabe, A. 1992. On Economic Incentives for Reducing Hazardous Waste Generation. *Journal of Environmental Economics and Management* 23(2): 154–160.

Watson, R., and D. Pauly. 2001. Systematic Distortions in World Fisheries Catch Trends. *Nature* 414: 534–536.

Watkins, C.W., M. Barret, and J.R. Paine. 1996. Private Protected Areas: A Preliminary Study of Private Initiatives To Conserve Biodiversity in Selected African Countries. December. Cambridge, U.K.: World Conservation Monitoring Centre.

Watson, D.J. 1982. Subsistence Fish Exploitation and Implications for Management in the Baram River System, Sarawak, Malaysia. *Fisheries Research* 1: 299–310.

Weiss, Y., and C. Fershtman. 1998. Social Status and Economic Performance: A Survey. *European Economic Review* 42: 801–820.

Weitzman, M.L. 1974. Prices vs. Quantities. *Review of Economic Studies* 41: 225–234.

———. 2002. Landing Fees vs. Harvest Quotas with Uncertain Fish Stocks. *Journal of Environmental Economics and Management* 43(2): 325–338.

We the Peoples. 1995. Chipko Movement, India. We the Peoples: 50 Communities project website. www.iisd.org/50comm/commdb/desc/d07.htm (accessed April 2002).

White, T., R. Anderson, and C. Ford. 1995. The Emergence and Evolution of Collective Action: Lessons from Watershed Management in Haiti. *World Development* 23(10): 1683–1698.

White, M.J., and D. Wittman. 1982. Pollution Taxes and Optimal Spatial Location. *Economica* 49(195): 297–311.

WHO (World Health Organization). 1987. *Air Quality Guidelines for Europe*. Copenhagen, Denmark: WHO.

———. 2001. An Anthology on Women, Health, and Environment: Water. www.who.int/ environmental_information/Women/9411wat.pdf (accessed May 2002).

Wiel, S.C. 1914. Theories of Water Law. *Harvard Law Review* XXVII: 530–540

———. 1918. Origin and Comparative Development of the Law of Watercourses in the Common Law and in the Civil Law. *California Law Review* VI: 254–255

Wiener, J.B. 1999. Global Environmental Regulation: Instrument Choice In Legal Context. *Yale Law Journal* 108: 677–690.

Wilen, J.E. 2000. Renewable Resource Economists and Policy: What Differences Have We Made? *Journal of Environmental Economics and Management* 39: 306–327.

Wilske, Å. 1999. Miljöutvärdering av Bra Miljöval för Husållskemikalier. A report for the Svenska Naturskyddsföreningen. Gothenburg, Sweden: Svenska Naturskyddsföreningen.

Wolf, M.A. (ed.). 2001. *Powell on Real Property*. Newark, NJ: Mathew Bender.

Wood, R., J. Handley, and S. Kidd. 1999. Sustainable Development and Institutional Design: The Example of the Mersey Basin Campaign. *Journal of Environmental Planning and Management* 42(3): 341–354.

World Bank. 1991. *World Development Report 1990: Poverty*. New York, NY: Oxford University Press for the World Bank.

———. 1995. Staff Appraisal Report for a Solid Waste Management Project, Latin America and the Caribbean Region Division. April. Washington, DC: World Bank.

———. 1996. *Sustainable Transport: Priorities for Policy Reform*. Development in Practice Series. Washington, DC: World Bank.

———. 1997. *Five Years after Rio: Innovations in Environmental Policy*. Environmentally Sustainable Development Studies and Monograph Series, no. 18. Washington, DC: World Bank.

———. 1998a. National Commitment Building Program to Phase Out Lead from Gasoline in Azerbaijan, Kazakhstan, and Uzbekistan: Summary Report. Washington, DC: World Bank.

———. 1998b. *Pollution Prevention and Abatement Handbook*. Washington, DC: World Bank.

———. 2000. *Greening Industry: New Roles for Communities, Markets, and Governments*. World Bank Policy Research Report. New York, NY: Oxford University Press.

———. 2002. Using Market-Based Instruments in the Developing World: The Case of Pollution Charges in Colombia. www.worldbank.org/NIPR/lacsem/columpres (accessed March 2002).

World Commission on Environment and Development. 1987. *Our Common Future*. Oxford, U.K.: Oxford University Press.

Worldwatch Institute. 1993. *State of the World 1993: A Worldwatch Institute Report on Progress Toward a Sustainable Society*. New York, NY: W.W. Norton.

——— (L.R. Brown, J. Abramovitz, C. Bright, C. Flavin, G. Gardner, H. Kane, A. Platt, S. Postel, D. Roodman, A. Sachs, and L. Starke). 1996. *State of the World 1996: A Worldwatch Institute Report on Progress Toward a Sustainable Society*. New York, NY: W.W. Norton.

Wright, M.G. 1994. An Economic Analysis of Coral Reef Protection in Negril, Jamaica. Master's thesis. Williams College, Williamstown, MA.

Wright, P.A. 1996. Regulation of Petroleum Product Pricing in Africa: A Proposed System Based on Studies of Four Sub-Saharan Countries. IEN Occasional Paper no. 7. August. Washington, DC: World Bank, Industry and Energy Department.

Wu, J.J. 1999. Crop Insurance, Acreage Decisions, and Nonpoint-Source Pollution. *American Journal of Agricultural Economics* 81(2): 305–320.

Wu, J.J., and B.A. Babcock. 2001. Spatial Heterogeneity and the Choice of Instruments to Control Nonpoint Pollution. *Environmental and Resource Economics* 18(2): 173–192.

Xepapadeas, A. 1991. Environmental Policy under Imperfect Information: Incentives and Moral Hazard. *Journal of Environmental Economics and Management* 20: 113–126.

———. 1997. *Advanced Principles in Environmental Policy*. Aldershot, Hampshire, U.K.: Edward Elgar.

Xepapadeas, A., and A. de Zeeuw. 1999. Environmental Policy and Competitiveness: The Porter Hypothesis and the Composition of Capital. *Journal of Environmental Economics and Management* 37(2): 165–182.

Xu, K., R.J. Windle, C.M. Grimm, and T.M. Corsi. 1994. Re-evaluating Returns to Scale in Transportation. *Journal of Transport Economics and Policy* 28(3): 275–286.

Yohe, G.W., and P. MacAvoy. 1987. A Tax Cum Subsidy Regulatory Alternative for Controlling Pollution. *Economic Letters* 25: 177–182.

Zamparutti, A. 1999. *Environment in the Transition to a Market Economy: Progress in Central and Eastern Europe and the New Independent States.* Paris, France: OECD.

Zhang, Z. 1998. *The Economics of Energy Policy in China: Implications for Global Climate Change.* Cheltenham, U.K.: Edward Elgar.

Zhao, J. 2000. Trade and Environmental Distortions: Coordinated Intervention. *Environmental and Development Economics* 5(4): 361–375.

Zilberman, D., and K. Millock. 1997. Pesticide Use and Regulation: Making Economic Sense Out of an Externality and Regulation Nightmare. *Journal of Agricultural and Resource Economics* 22(2): 321–332.

Zilberman, D., S. Templeton, and M. Khanna. 1999. Agriculture and the Environment: An Economic Perspective with Implications for Nutrition. *Food Policy* 24(2-3): 211–229.

Zinnes, C.F. 1997. The Road to Creating an Integrated Pollution Charge and Permitting System in Romania. In *Controlling Pollution in Transition Economies,* edited by R. Bluffstone and B.A. Larson. Cheltenham, U.K.: Edward Elgar.

Zylicz, T. 1994. Environmental Policy Reform in Poland. In *Economic Policies for Sustainable Development,* edited by T. Sterner. Dordrecht, the Netherlands: Kluwer Academic.

# Index

In page references, n. indicates an endnote, and italicized page numbers indicate figures, tables, or boxes.

# About the Author

THOMAS STERNER is professor of environmental economics at the University of Gothenburg, Sweden, where he directs the Environmental Economics Unit (EEU). The EEU specializes in the economics of the environment and natural resource management in both high-income and developing countries. Through extensive collaboration with organizations such as the Swedish International Development Cooperation Agency, the World Bank, the Beijer Institute of the Royal Swedish Academy, Resources for the Future (RFF), the International Institute for Environment and Development, Centro Agronómico Tropical de Investigación y Enseñanza (Tropical Agriculture Research and Higher Learning Center), and the Economy and Environment Program for Southeast Asia, the EEU runs a number of research and training programs for capacity building in environmental economics that particularly focus on developing countries.

Sterner is chairman of the board of the Centre for Environmental and Sustainability in Gothenburg and a university fellow at RFF. He served on the board of the European Association of Environmental and Resource Economists from 1997 to 1999. During 1998 and 1999, Sterner was Gilbert White fellow at RFF and a consultant to the World Bank. His previous books include *The Market and the Environment: The Effectiveness of Market-Based Policy Instruments for Environmental Reform* and *Economic Policies for Sustainable Development*.